NORTH CAROLINA
STATE BOARD OF COMMUNITY COLLEGES
LIBRARIES
ASHEVILLE-BUNCOMBE TECHNICAL COLLEGE

DISCARDED

JUN 26 2025

SOURCE BOOK ON INNOVATIVE WELDING PROCESSES

Other Source Books in This Series:

Electron Beam and Laser Welding
Brazing and Brazing Technology
Gear Design, Technology and Performance
Copper and Copper Alloys
Materials for Elevated-Temperature Applications
Maraging Steels
Powder Metallurgy
Wear Control Technology
Selection and Fabrication of Aluminum Alloys
Industrial Alloy and Engineering Data
Materials Selection (Volumes I and II)
Nitriding
Ductile Iron
Forming of Steel Sheet
Stainless Steels
Heat Treating (Volumes I and II)
Cold Forming
Failure Analysis

Metals Science Source Book Series:

The Metallurgical Evolution of Stainless Steels
Hydrogen Damage

SOURCE BOOK ON INNOVATIVE WELDING PROCESSES

A comprehensive collection of outstanding articles from the periodical and reference literature

Compiled by
Consulting Editor

MELVIN M. SCHWARTZ

Manager, Manufacturing Technology
Manufacturing Planning/Engineering
Sikorsky Aircraft
Division of United Technologies

American Society for Metals
Metals Park, Ohio 44073

Copyright © 1981
by the
AMERICAN SOCIETY FOR METALS
All rights reserved

No part of this book may be reproduced, stored in a retrieval system, or transmitted, in any form or by any means, electronic, mechanical, photocopying, recording, or otherwise, without the prior written permission of the publisher.

Nothing contained in this book is to be construed as a grant of any right of manufacture, sale, or use in connection with any method, process, apparatus, product, or composition, whether or not covered by letters patent or registered trademark, nor as a defense against liability for the infringement of letters patent or registered trademark.

Library of Congress Cataloging in Publication Data

Main entry under title:

Source book on innovative welding processes.

    Includes bibliographical references and index.
    1. Welding—Technological innovations—Addresses, essays, lectures.   I. Schwartz, Mel M.
TS227.2.S67   671.5′2   81-3535
ISBN 0-87170-105-7   AACR2

PRINTED IN THE UNITED STATES OF AMERICA

Contributors to This Source Book

A. S. BAHRANI
The Queen's University of Belfast

S. K. BANERJEE
National Metallurgical Laboratory
Jamshedpur, India

SCHOLER BANGS
Welding Design & Fabrication

P. M. BARTLE
British Welding Institute

B. J. BASTIAN
Induction Process Equipment Co.

J. T. BERRY
Georgia Institute of Technology

C. S. BEUYUKIAN
North American Rockwell

R. BROSILOW
Welding Design & Fabrication

STEVE H. CARPENTER
University of Denver

B. CROSSLAND
The Queen's University of Belfast

PAUL DENT
Grumman Aerospace Corp.

G. S. DOBLE
TRW Inc.

D. S. DUVALL
Pratt & Whitney Aircraft Div.
United Technologies Corp.

T. M. EBERHART
Beatle Plastics

HILMER F. EBLING
A. O. Smith Corp.

GERALD S. ELLSWORTH
Advanced Technologies, Inc.

W. G. FASSNACHT
South Bend Div.
The Bendix Corp.

ELTON S. GAMBLE
Olin Metals Research Center
Olin Corp.

ROBERT J. GASSER
Sciaky Bros., Inc.

TATSUYA HASHIMOTO

T. H. HAZLETT
University of California

CLINTON R. HEIPLE
Rocky Flats Div.
Dow Chemical U.S.A.

JOHN C. JENKINS
Nelson Stud Welding Co.

L. E. JENSEN
E. F. Industries, Inc.

IVAR W. JOHNSON (ret.)
General Electric Co.

L.J. KORB
North American Rockwell

C. D. LOYD
Caterpillar Tractor Co.

D. J. McMULLAN
The Queen's University of Belfast

K. E. MEINERS
Battelle Columbus Laboratories

D. R. MILNER
University of Birmingham

J. D. MOTE
E. F. Industries, Inc.

E. D. NICHOLAS
The Welding Institute

M. G. NICHOLAS
University of Birmingham

E. D. OPPENHEIMER
Consulting Engineer
AMF Thermatool Corp.

H. B. OSBORN, JR.
TOCCO Div.
The Ohio Crankshaft Co.

W. A. OWCZARSKI
Pratt & Whitney Aircraft Div.
United Technologies Corp.

R. A. QUEENEY
Pennsylvania State University

G. W. REEDER
American Welding &
  Manufacturing Co.

ROBERT R. REINKING
McDonnell Douglas-Tulsa
McDonnell Douglas Corp.

J. ROWE
North American Rockwell

WALLACE C. RUDD
AMF Thermatool Inc.

NOTE: Affiliations given were applicable at date of contribution.

J. SEARLE
Friction Welding Co., Ltd.

D. A. SEIFERT
Battelle Columbus Laboratories

GEORGE W. STACHER
Rockwell International Corp.

KINJI TANUMA

I. J. TOTH
TRW Inc.

C. A. TUDBURY
AMF Thermatool Inc.

R. F. TYLECOTE
University of
   Newcastle-upon-Tyne

H. N. UDALL
AMF Thermatool Inc.

L. R. VAIDYANATH
University of Birmingham

EDWARD D. WEISERT
Rockwell International Corp.

A. L. WILLIAMS
Wean Industries, Inc.

W. F. WILLIAMS
Buick Motors Div.
General Motors Corp.

ROBERT H. WITTMAN
University of Denver

E. J. WYNNE
Tube Investments Research
   Laboratories

J. A. YOBLIN
E. F. Industries, Inc.

A. C. YOUNG
General Electric Co.

CONTENTS

SECTION I: FRICTION WELDING

Friction Welding: An Introduction to the Process (*E. D. Nicholas*) 3
Fundamentals of Friction Welding (*T. H. Hazlitt*) 11
Friction Welding (*ASM Committee on Flash, Friction and Stud Welding*) 37
The Mechanics of Friction Welding Dissimilar Metals (*D. J. McMullan and A. S. Bahrani*) 49
Inertia-Welding of P/M Parts (*T. M. Eberhart and R. A. Queeney*) 55
Orbital-Motion Technique for Friction Welding Non-circular Components (*J. Searle*) 60
Advances in Friction Weld Monitoring and Control (*Gerald S. Ellsworth*) 64
Friction Welding of Maraging Steel for Small Missile-Systems Applications (*D. A. Seifert and K. E. Meiners*) 76
Friction-Welded Heat Pipes Stabilize Arctic Soils (*Robert R. Reinking*) 92

SECTION II: EXPLOSIVE WELDING

Fundamentals of Explosive Welding (*B. Crossland and A. S. Bahrani*) 99
Review of the Present State-of-the-Art in Explosive Welding (*B. Crossland*) 116
Explosive Welding and Cladding — Overview of the Process and Selected Applications (*J. A. Yoblin, J. D. Mote and L. E. Jensen*) 129
Mechanical Properties of Explosively-Cladded Plates (*S. K. Banerjee and B. Crossland*) 145
Recent Developments in the Theory and Application of Explosion Welding (*Steve H. Carpenter and Robert H. Wittman*) 151

SECTION III: HIGH-FREQUENCY WELDING

A High Speed Welding System for the Production of Custom Designed HSLA Structural Sections (*H. N. Udall, J. T. Berry and E. D. Oppenheimer*) 171
High-Frequency Welding of Pipe and Tubing (*H. B. Osborn, Jr.*) 199
HF Contact or Induction Tube Welding: Which Is Better? (*C. A. Tudbury*) 233
High-Frequency "Bar-Butt" Welding (*Wallace C. Rudd*) 236
How To Make a Wheel Rim — in 4 Seconds (*R. Brosilow*) 240
Melt Welding — a New High Frequency Process (*W. C. Rudd and H. N. Udall*) 243

SECTION IV: DIFFUSION WELDING

Introduction to Diffusion Bonding (*P. M. Bartle*) 255
Diffusion Bonding: No Longer a Mysterious Process (*Scholer Bangs*) 259
Pressure Welding by Rolling (*L. R. Vaidyanath, M. G. Nicholas and D. R. Milner*) 263
Effect of Heat Treatment on Cold Pressure Welds (*R. F. Tylecote and E. J. Wynne*) 279
Diffusion Welding of Molybdenum (*Tatsuya Hashimoto and Kinji Tanuma*) 289
Diffusion Bonded Columbium Panels for the Shuttle Heat Shield (*L. J. Korb, C. S. Beuyukian and J. Rowe*) 298
Advanced Diffusion-Welding Processes (*W. A. Owczarski and D. S. Duvall*) 309
Roll Diffusion Bonding of Boron Aluminum Composites (*G. S. Doble and I. J. Toth*) 338
Fabricating Titanium Parts With SPF/DB [Superplastic Forming/Diffusion Bonding] Process (*Edward D. Weisert and George W. Stacher*) 353
Mechanical Properties of Diffusion-Bonded Beryllium Ingot Sheet (*Clinton R. Heiple*) 358

INDEX 365

PREFACE

Innovative welding processes have been developed to meet joining requirements when, for either economic or technical reasons, or both, standard welding procedures will not suffice. Differences among the many welding processes are based principally on how each closes the distance that separates atoms. Traditionally, this has been done by a casting operation, with filler metal poured into the joint. Fusion welding still is used extensively today, but in addition to the conventional electric arc process, other techniques such as the use of high-energy beams, plasma arcs, ultrasonic waves, and frictional heat are also being employed. Solid-state diffusion welding (and brazing) has also come on the scene.

It does no commercial good to weld two pieces of metal if the process of coalescence detracts from the usefulness of the components. To a large extent, the engineering function of welding metallurgy activity is concerned not with the invention of joining mechanisms for metals but rather with techniques for minimizing the complex and costly side effects of the joining process. The existence of such a great number of distinctly different welding processes attests to the complexity of the problem.

To oversimplify, anyone can stick two pieces of metal together. However, some make a commercial success of this activity, while others either price themselves out of the marketplace or offer an inferior, nonfunctional, and noncompetitive product.

SOURCE BOOK ON INNOVATIVE WELDING PROCESSES describes a variety of these methods in order to help the reader choose the one that will be most cost effective for his particular application. Each of these processes is reviewed briefly in the following summary.

Friction Welding. Friction welding is a well-established commercial process used extensively in many industries throughout the world. At current estimates, more than 2000 friction welding machines have been installed in factories and laboratories and produce more than 400 million joints each year. Although originally introduced as a mass-production tool for the automobile industry, friction welding is now widely used for batch production as well.

After initial promotion in eastern Europe in the late 1950's, the friction welding process was introduced into the United States in 1960. The mid-1960's brought a new version of the friction welding machine, using a flywheel attached to a rotating spindle to store as kinetic energy all the energy required before the welding process actually begins. Because the power required for heat generation at the interface comes from the "inertia" of the rotating mass, this process is called inertia welding or flywheel friction welding, as opposed to conventional or continuous-drive friction welding. The nine articles in Section I of this book provide information on both the old and the new friction welding processes.

Explosive Welding. Although not in any usual sense a mature welding technique, explosive welding is important because of its suitability for joining large areas and its potential economic advantages.

Controlled application of the enormous power generated by detonating explosives gives rise to phenomena not intuitively expected by those accustomed to more traditional velocities and pressures. With the use of explosive welding, metal surfaces can be thoroughly cleaned of contaminants such as oxide and nitride films, oil, and adsorbed gases and then be immediately pressed together at pressures of several million pounds per square inch. Furthermore, the relatively small amount of energy involved gives rise to minimum melt accumulation at the interface. These are ideal conditions for effective welding.

Although explosive welding has been under development for less than a decade, it is now in general use by the technological community. The five articles in Section II of this book include basic and state-of-the-art information about this important process.

High-Frequency Welding. High-frequency welding can be divided into two categories:
1. High-frequency resistance welding, in which the current is introduced into the work by direct electrical contact
2. High-frequency induction welding (sometimes called induction resistance welding), in which the current is induced in the work by an inductor coil, without electrical contact.

High-frequency resistance welding is an important but not widely known process that can provide substantial economic advantages to its users because of its high welding speed. Many thousands of tons of steel pipe are produced by this process every year. The six articles in Section III of this book present both general and specialized discussion of high-frequency welding.

Diffusion Welding. Diffusion welding has been used primarily in the atomic energy and aerospace industries. To meet the stringent requirements of these industries, it was necessary not only to use new materials but, equally important, to devise methods of fabricating them into useful engineering components. Diffusion welding is one such technique developed to keep pace with the requirements of advancing technology.

There are many names for the various applications of diffusion welding, but the essential details are the same for all. The ten articles in Section IV of this book provide information on key variables, materials, and applications for this very significant process.

The American Society for Metals extends most grateful acknowledgment to the many authors whose work is presented in this book and to their publishers.

MELVIN M. SCHWARTZ
Consulting Editor

Cover illustrations are from articles that begin on pages 3, 116, 171 and 259 in this book.

SOURCE BOOK ON INNOVATIVE WELDING PROCESSES

SECTION I:
Friction Welding

Friction Welding: An Introduction to the Process	3
Fundamentals of Friction Welding	11
Friction Welding	37
The Mechanics of Friction Welding Dissimilar Metals	49
Inertia-Welding of P/M Parts	55
Orbital-Motion Technique for Friction Welding Non-circular Components	60
Advances in Friction Weld Monitoring and Control	64
Friction Welding of Maraging Steel for Small Missile-Systems Applications	76
Friction-Welded Heat Pipes Stabilize Arctic Soils	92

Friction welding: an introduction to the process

E.D. Nicholas

INTRODUCTION

There are some forty different welding processes which can be broadly divided into two groups: fusion welding, which relies on the formation of a molten bridge, and deformation, relying on the flow of metal.

Pressure welding is one of the oldest forms of joining. It is known that joints can be made in very ductile materials at room temperature using high pressure and appreciable deformation. In theory, all ductile metals could be cold pressure welded but the forces needed would be prohibitive. Therefore materials were heated to improve ductility, allowing pressure welds to be made with lower forces. Several heating methods are used commercially such as flame heating, induction, resistance, and, more recently, friction.

Apart from being a very efficient heat source friction provides an effective cleaning action, removing the surface contamination which prevents metals from bonding together.

The thermal energy developed from the interaction of two surfaces rubbed together under pressure can be harnessed to provide conditions suitable for welding. The technique is autogenous because no additional filler material is used, and welding takes place in the solid phase because no macroscopic melting is observed.

Throughout the literature on friction welding continual reference is made to early knowledge of the process, but industrial application was absent until the mid '50s. At that time Soviet work predominated in the field, leading to the reported[1] introduction in 1961 of over thirty friction welding machines into Soviet industry. Machines were also being installed in Czechoslovakia and China. Today the process is finding worldwide acclaim as industry in many countries recognises its special advantages: efficient energy utilisation, excellent and reproducible weld strength, high production rates, and versatility in successfully joining a wide variety of similar and dissimilar metal combinations.

In terms of industrial exploitation the number of machines estimated to have been installed in the following countries is of interest:

Country	Number of machines
Japan	>1000
USSR	>1000
United Kingdom	200
Western Europe	>300
Eastern Europe	>300
United States	>200

An example of the growing acceptance of friction welding is shown by the growth rate of equipment commissioned in Japan, Fig.1, since the start of the involvement of that country in the early '60s. In the United Kingdom several well-known manufacturers sell welding machines throughout the world,[2] and research programmes are in progress at The Welding Institute to investigate the effects of welding parameters on the metallurgical and mechanical (static and dynamic) properties of various similar and dissimilar material combinations; the evaluation of in-process quality monitoring and control procedures; welding in unusual environments,

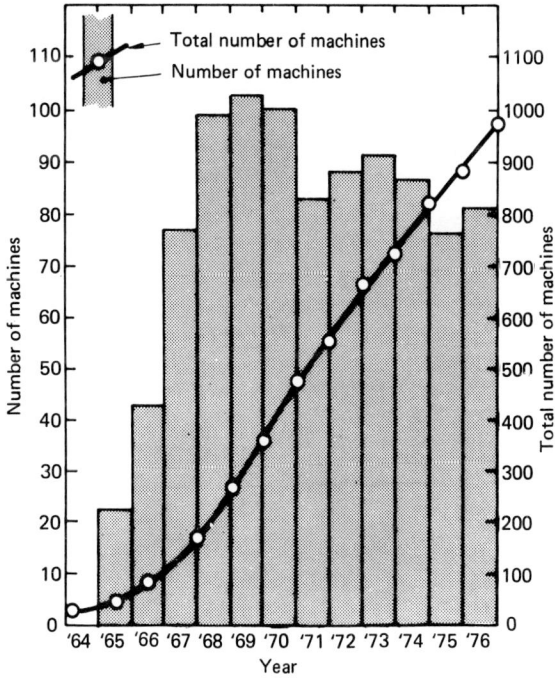

1 Annual production of friction welding machines in Japan

Source: The Welding Institute

e.g. under water, and the development of a new hybrid system* to be known as radial friction welding. The latter is a particularly exciting project since, if successful, it will lead to the manufacture of more portable friction welding machines which are especially suitable for welding pipes and other components on site.

In connection with site applications it is interesting that in the German Democratic Republic a special machine[3] has been commissioned to site weld plastic pipelines up to 225mm diameter. This equipment utilises the technique of rotating a nonconsumable insert disc between the pipe ends to generate heat. After a predetermined heating time the disc is removed and the pipes forced together under an increased axial pressure.

THE PROCESS

The simplest mechanical arrangement for continuous drive friction welding, Fig.2, involves two cylindrical bars held in axial alignment: one of the bars is rotated while the other is advanced into contact under a pre-selected axial pressure. Rotation continues for a specific duration sufficient for the frictional heat to achieve an interfacial temperature at which the metal in the joint zone is in a plastic state, both at the interface and for some depth behind. Having achieved this condition the rotating bar is stopped while the pressure is either maintained or increased to consolidate the joint. Other arrangements which may be employed, Fig.3, include rotating a centre piece between two stationary end pieces, a technique suitable to join wire coils and long

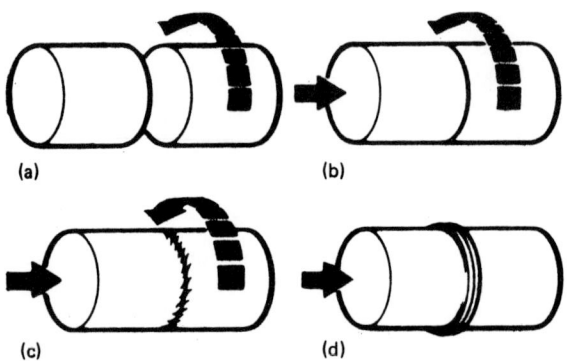

2 Simplest method of friction welding: (a) rotary member spun, (b) advanced into contact under load, (c) contact pressure maintained, and (d) rotation stopped (switching off motor, or flywheel coming to rest), pressure maintained to produce weld

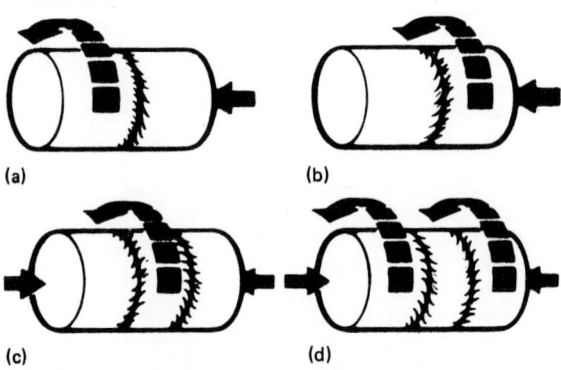

3 Alternative mechanical arrangements: (a) one member rotated, other advanced axially, (b) one member stationary, other rotated and advanced, (c) intermediary member rotated, end pirces advanced, and (d) internal member stationary, ends rotated and advanced

*Patent obtained.

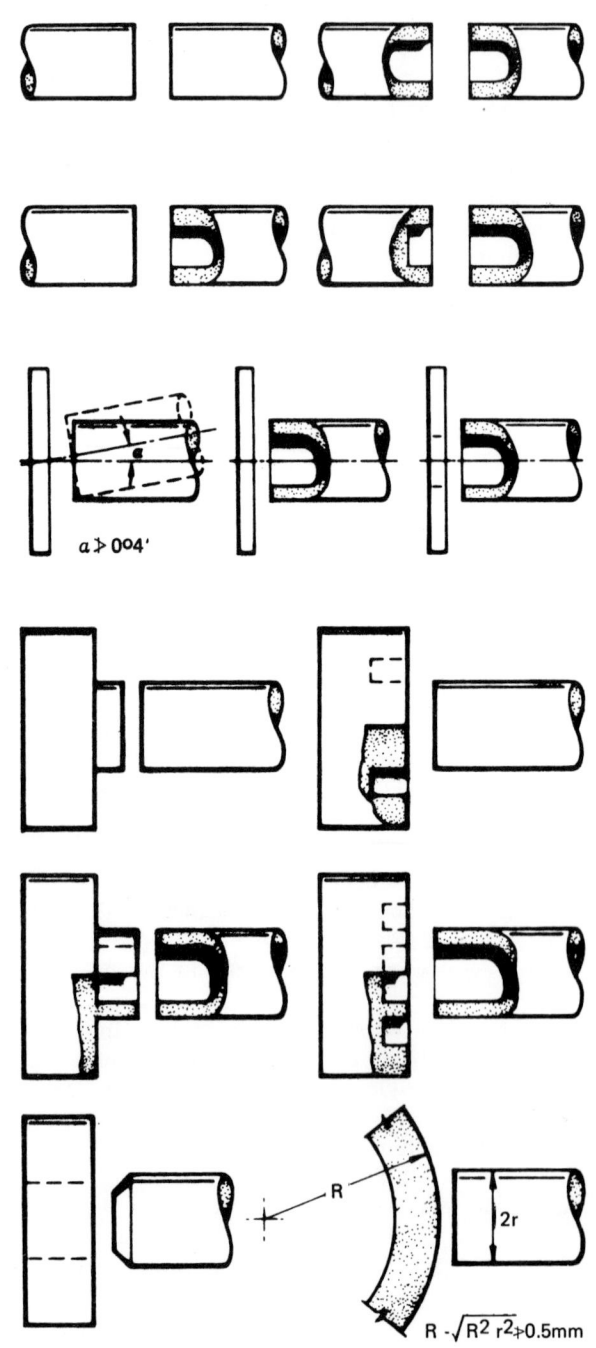

4 Part designs for friction welding

part lengths, and rotating two end pieces against a stationary centre piece, as used for the fabrication of prop shafts and rear axle casings. Friction welding is sufficiently versatile to be used to join a wide range of part designs, examples of which are shown in Fig.4.

Table 1 presents a classification of friction welding processes which quite clearly shows the two process variants under the energy classification The most widely used system in Europe, the USSR, and Japan is continuous drive welding, but the stored energy method (often known as inertia or flywheel welding[4]) is mainly followed in the USA.

With the inertia process it is usual for the round component to be placed in the chuck of a flywheel system of the correct inertia and accelerated to a predetermined peripheral velocity. All the energy to make the weld is stored in the flywheel system, and the correct initial sliding velocity at the weld interface is

Table 1   Classification of friction welding processes

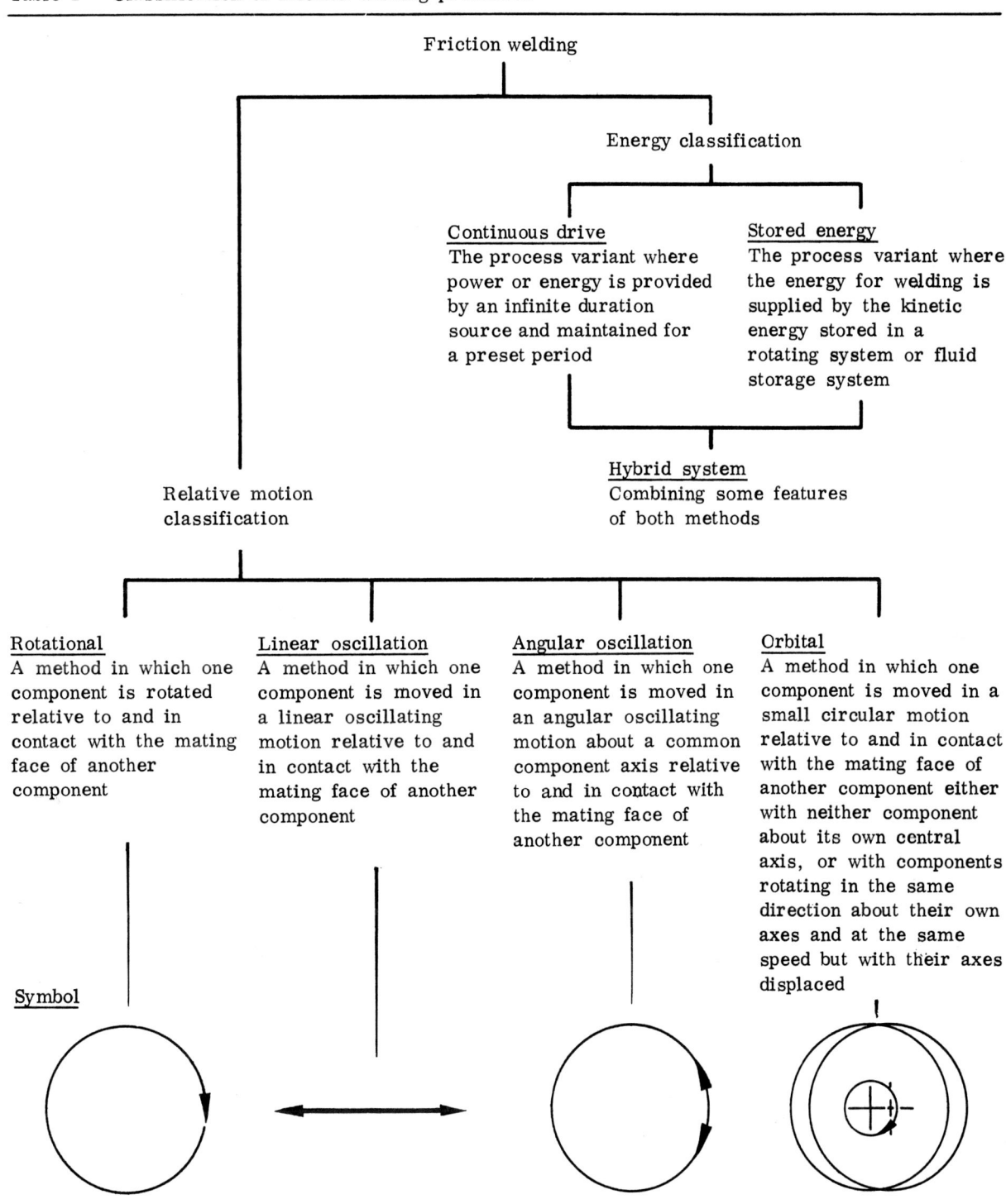

Source: The Welding Institute

also achieved. The flywheel system is uncoupled from the accelerating power source and a thrust force is applied rapidly against a second fixed member. The braking action at the interface decelerates the flywheel system to rest, converting the kinetic energy to heat and torsional forging action at the weld interface. The technique is controlled by three parameters: flywheel system inertia, angular velocity, and thrust force.

The main feature of the inertia method is the use of the flywheel since it:

(a) supplies all the energy for heating at whatever rate is demanded by the weld
(b) supplies the torsional forging force near the end of the weld
(c) supplies process control
(d) permits the speed to decline from the beginning to the end of the weld

The flywheel's power capability is limited only by the rate at which it can be decelerated. This is strongly influenced by the weld interface which produces the retarding torque.

For inertia welding, minimum surface speeds of 1.4m/sec for tool steels and about 1.8m/sec for low carbon steels are suggested. For every material there is a minimum rotary speed under which poor welds occur. Today 5 and 7.6m/sec are not considered as upper limits of initial speed for steel. The minimum speed for welding is dependent upon the physical properties of the material to be welded. Of less influence are the yield strength, the parts shape, and the pressure used. Tubes, for example, normally require higher initial speeds than bars. Added pressure will flatten the heat pattern and thus permit higher initial welding speed.

A.D. Little* has proposed the use of:

$$(k \rho c)^{1/2} \frac{T_{mp}}{10^4}$$

as 'Q' or a rough measure of power required, where k is thermal conductivity, ρ density, c specific heat, and T_{mp} the melting point of the material to be welded. The 10^4 figure is used simply to reduce the size of the number. This power requirement, Q, is that power which will produce the temperature gradient required to localise plastic deformation to a shallow zone at the weld interface constituting an adiabatic shear condition. The power requirement, which in this situation is a product of torque and velocity, must be fulfilled via torque values below that which cause excessive depth of deformation.

---
*Arthur D. Little Inc., Cambridge, Mass, under contract for Caterpillar Tractor Co.

Studies have shown that, when torques create shear stresses beyond a reasonably consistent fraction of the room temperature yield strength of the material, the desired adiabatic conditions do not occur. An empirical equation has been developed whereby the lower critical speed is a function of the Q power requirement and the yield strength:

Calculated Q values are shown in Table 2 with the lower critical speed for different metals and show good correlation. Lead has the lowest calculated Q value at 50 and tungsten the highest at 1750. Lead can be welded at very low speeds, but tungsten requires the highest initial speed of any material yet welded using this technique.

Although each variant has it advocates they will both provide welded parts of very high integrity and at production rates which are far better than alternative joining methods such as flash and resistance butt welding. Cost savings can also be made because of lower power consumption, reduced machining, and greater conservation of material.

The Russians Voinov and Boldyrev, aware of the differences in welding times between continuous drive and stored energy systems, are investigating another variant known as pulsed friction welding.[5] They claim that the use of energy generated from the centrifugal forces acting on a rotating eccentric mass increases the thermal power available for welding, which will thus reduce the welding period. In the UK Foister has proposed a heat-under-power welding process[6] which is essentially another hybrid system containing some features of both main process variants. Essentially this method makes use of the infinite power that is available from continuous drive for the heating cycle while, during the deceleration period, the rotating component is coupled to a rotating flywheel of fixed stored energy to provide additional deceleration inertia.

Table 2 Q values and minimum surface interface velocity for welding various materials

Material	10^4 Q	Lower critical speed, m/sec
Lead	50	0.25
Stainless steel	260	1.00
Aluminium	380	1.25
Tool steel	430	1.4
Low carbon steel	470	1.75
Nickel	650	3.25
Titanium	800	3.75
Copper	950	9.00
Molybdenum	1250	10.00
Tungsten	1750	12.5

Another interesting development has been reported from Japan[7] in which one of the components is driven at a constant speed and the initially nonrotating part, which is attached to a flywheel, is accelerated up to the predetermined speed of rotation. The aim is to provide rotation to both parts so that the upset collar which forms can be removed during the welding sequence. By using this method increased productivity and more effective use of the welding forces are claimed. Also, increased tool life can be achieved because the upset is removed while it is hot and relatively ductile.

From Table 1 it can be seen that four relative motion categories are available for friction welding. The most widely used is that of conventional rotation as it is the simplest system to engineer and the most versatile to apply. The major limitations of this system are that only parts of essentially circular geometry are suitable and precise angular orientation between parts cannot easily be achieved. Angular oscillation and orbital motion systems[8] have been successfully developed to overcome these limitations. Angular motion has been incorporated in a production machine for welding plastic thermostat housings, and orbital friction welding equipment is now available to join non-circular parts up to 2000mm² cross-section.

Interesting features of orbital motion are the development of a uniform surface velocity (which can be of advantage when welding plastics and materials prone to reduced strength if subjected to overheating) and the capability of the process to perform many welds in one welding sequence. This latter characteristic will enable production rates per machine to be substantially increased.

In the context of correct angular alignment of parts after welding Japanese researchers have developed a continuous drive machine[9] which is capable of synchronising the parts to an angular accuracy of ±0.1°. A schematic arrangement of such a machine is presented in Fig.5. Alignment is achieved by engagement of the two 'correction' cams at the end of the friction stage just prior to forging.

Metallurgical and mechanical properties of the welds are adequate and this technique can be used to fabricate connecting rods, propeller shafts, and any assemblies requiring angular orientation.

WELDING PARAMETERS

For all friction welding systems relative velocity, axial pressure applied to the parts, and heating duration are the three principal controlling variables which influence the metallurgical and mechanical properties of friction welded joints.

Work at The Welding Institute with the continuous drive system has demonstrated that sound welds can be produced in a wide range of materials and workpiece sizes at peripheral velocities in the range 1.0-3m/sec. For example, 2mm diameter mild steel rod requires a speed of about 76 000rev/min, but for 76mm diameter bar the speed is reduced to 320-450rev/min. At lower velocities higher torques are developed which can give rise to workholding problems and nonuniform deformation. However, for certain dissimilar metal combination such as aluminium alloy (½%Si-½%Mg) to mild steel, low surface velocities (⅓-⅔m/sec) are advantageous to minimise the formation of brittle intermetallic compounds at the rubbing surfaces, thus ensuring good weld strength with acceptable ductility. Care must be taken in the selection of axial pressure and duration of heating when welding at high velocities to avoid overheating the weld region. However, for some alloy steels prone to hardening these conditions can be used to advantage as they will assist in lowering the rate of cooling throughout the weld region, thus reducing the risk of cracking.

Axial pressure applied to the parts is important as it controls the temperature gradient in the weld zone and thus governs the drive power (torque) demanded and the metal displacement characteristics. Its value is dependent on the materials being joined and their configuration, e.g. bar to bar, tube to tube, or bar/tube to plate.

The pressure level must be selected to maintain the two surfaces in intimate contact to prevent atmospheric contamination. With many materials and combinations it is often an advantage to apply a higher pressure on termination of heating to forge the joint. Typically, for continuous drive friction welding of mild steel, friction (heating) pressures of 31-60N/mm² and forging pressures of 77-150N/mm² are used. For stainless steel and nickel-based alloys characterised by their

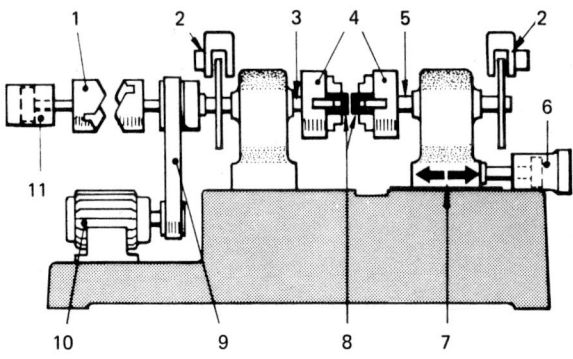

5 Synchronised friction welding machine. 1 – correction cam; 2 – brake; 3 – driving spindle; 4 – chuck; 5 – stationary spindle; 6 – oil hydraulic cylinder; 7 – slide; 8 – material; 9 – belt; 10 – motor; 11 – oil hydraulic cylinder for correction and upset

high hot strength greater forging pressures are generally required.

The weld heating cycle during rotation is usually determined experimentally to give optimum thermal conditions. Any unnecessary extension of this period results in lower productivity and greater material consumption. However, failure to reach the optimum temperature conditions can give rise to irregular heating with the occurrence of oxide inclusions and unbonded regions at the interface of the completed weld. In continuous drive friction welding the duration of rotation (heating) is controlled by time or preselected axial metal displacement, whereas with the stored energy system this duration is controlled by flywheel inertia.

The values of the process variables are dependent on the type of weld geometry, materials, and sizes being joined. Once the working limits are established, welds of the highest integrity are produced with a high level of reliability. Variables such as torque (power), temperature distribution, and displacement rate during heating and forging, which are influenced by the welding conditions, have been widely reported in the literature.[10] A concise appraisal of the literature on this subject has been compiled by Wang.[11] In contrast, researchers at The Queen's University of Belfast, and in Japan, have studied the mechanics of the weld interface for both continuous drive and stored energy systems respectively. The former workers established that a 'plasticised' annular area is generated during the heating phase. Its radial location and width depend on the speed of rotation and the applied axial pressure. The Japanese investigators used insert pieces of nickel at the centre of their specimens to establish the interfacial metal movement characteristics for both process variants. They found that with the stored energy process greater radial movement took place when compared with that observed for the continuous drive process.

MATERIALS

One of the inherent features of friction welding is the efficient utilisation of the thermal energy developed. By careful control of the main welding variables the weld region does not suffer from excessive heat buildup. This is particularly important when welding mild, low alloy, and stainless steels since overheated microstructures are avoided. Moreover the process is readily amenable to join widely dissimilar metal combinations such as aluminium to stainless steel, copper to steel, aluminium to titanium. For guidance the weldability chart, Fig. 6, refers to materials that have been joined at The Welding Institute to yield high integrity welds. Further information regarding materials' weldability is given by Wang for both continuous drive and stored energy processes.

Exclusion of the atmosphere during welding allows the formation of sound high strength bonds in the more likely reactive metals such as zirconium, titanium, and their alloys. Welding oxygen-free copper can be undertaken while still maintaining its excellent thermal and electrical properties since oxygen contamination does take place. However, a note of caution should be sounded when attempts are made to weld materials with high nonmetallic inclusion contents, such as free-machining steels, because the redistribution of the inclusions will provide transverse planes of weakness at the weld. This will result in a reduction of ultimate tensile strength and loss of both tensile and notch ductility. Sulphur-bearing steels have been found to be more susceptible to these problems than leaded steels.

Sound welds characterised by good mechanical properties can be obtained in the hardenable low and high alloy steels without the incidence of cracking. For these particular metals postweld heat treatment is generally required to reduce hardness and restore ductility.

Most of the methods of motion used under continuous drive conditions have been successfully applied to join various types of thermoplastic. In particular, angular and orbital motion have been very successful so that equipment based on the former is now used in production. Experience at The Welding Institute has shown that the following plastics are readily weldable by friction: monocast nylon, glass fibre reinforced nylon, polycarbonate, and polyvinylchloride. Many other plastics have been reported as successfully joined by other researchers.

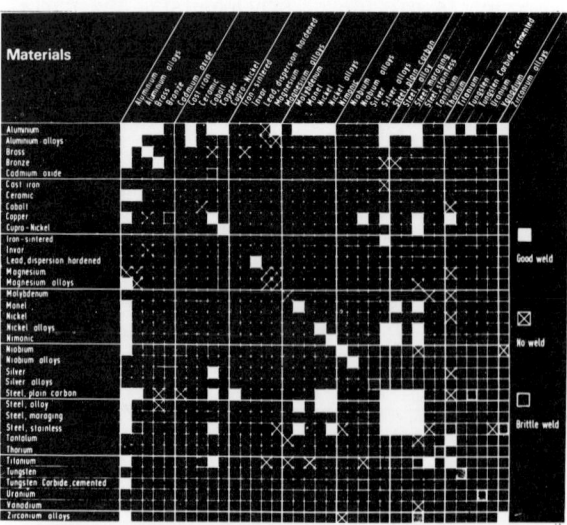

6 Relative weldability of some of the materials investigated at The Welding Institute. Range of combinations which can be successfully joined has not been fully explored

In general the results of static and dynamic mechanical tests that have been reported confirm the integrity of the friction weld.[12] Tests have demonstrated that parent metal properties can be achieved and the joints in dissimilar metals are stronger than the weaker of the two parent materials. Welds are leak-tight to levels which satisfy the helium mass spectrometer leak test. Fatigue results have shown that properties approaching that of the parent metal can be achieved from friction welds made by both continuous drive and stored energy systems.[13]

MACHINE SIZE RANGE

It has been established that, as the cross-sectional area of the parts to be welded increases, greater welding force and transmission drive power are required. Thus no one size of friction welding machine could be considered for the wide range of sizes (for example 1 to 150mm diameter solid bar) that might require to be welded. To demonstrate the range of machines that has been developed, consider those capable of welding at the extremes of the sizes. At the smallest end an industrial prototype microfriction welding machine was developed at The Welding Institute, Fig. 7, which will weld within the size range of 0.75-2mm diameter, utilising welding forces from 15-850N and rotation speeds of 40-100 x 10^3 rev/min. Figure 8 illustrates various components which have been welded with this machine, and describes the materials used.

In complete contrast, a very large machine capable of welding up to 150mm diameter solid mild steel bar or pipes up to 450mm OD x 10mm wall thickness is illustrated in Fig. 9. This machine develops 1800kN axial thrust and provides, over a speed range of 100-500mm/min, a peak power equivalent of 740kW at the chuck.

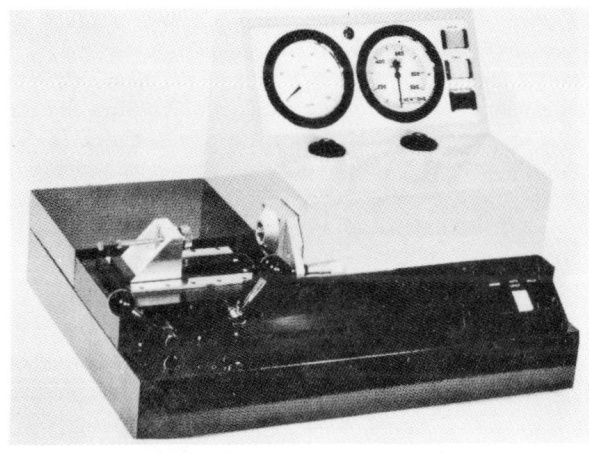

7 Industrial prototype microfriction welding machine

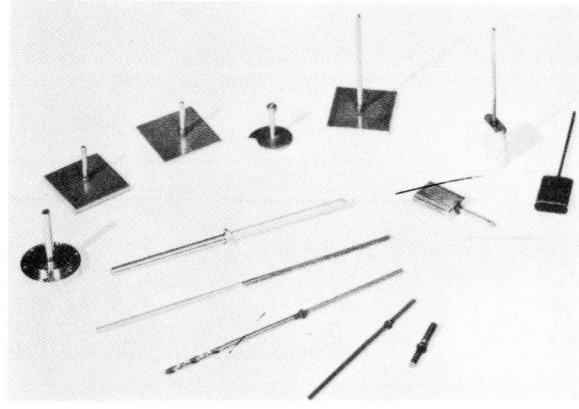

8 Typical examples of microfriction welds

9 1800kN axial thrust friction welding machine. 1 – control console; 2 – fixed headstock; 3 – moving carriage: 4 – rigid tailstock; 5 – fabricated bed

REFERENCES

1 VILL, V.I. 'Friction welding of metals'. Translated from the Russian and published by the American Welding Society, 1962.

2 'Friction welding', special feature. Metal Constr., 2 (5), 1970, 185-8.

3 NICKEL, S. and SCHULD, N. 'Friction welding of plastic pipes' (in German). Schweisstechnik (Berlin), 25 (4), 1975, 164-6.

4 KIWALLE, J. 'Designing for inertia welding'. Machine Design, 40 (26), 7 November 1968, 161-6.

5 VOINOV, V.P. and BOLDYREV, R.N. 'The role of the mechanical drive in the friction welding process'. Welding Production, 21 (1), 1974, 58-61.

6 FOISTER, P.B. 'Heat under power (HUP) friction welding'. Procs 3rd Int'l Conference 'Advances in Welding Processes', Harrogate, 7-9 May 1974, 243-8.

7 HASUI, A. et al. 'Some experiments with a new-designed friction welding machine'. Bull. JSME, 12 (51), 1969, 656-64.

Source: The Welding Institute

8   SEARLE, J.G. 'Friction welding non-circular components using orbital motion'. Welding and Metal Fab., 39 (8), 1971, 294-7.

9   'Friction welding machine in Japan (1972)'. IIW Doc.III-469-73, presented at IIW Annual Assembly, Düsseldorf, September 1973.

10 ELLIS, C.R.G. 'Continuous drive friction welding of mild steel'. Welding J., 51 (4), 1972, 183s-97s.

11 WANG, K.K. 'Friction welding'. Welding Research Council Bull. (204), April 1975.

12 JENNINGS, P. 'Some properties of dissimilar metal joints made by friction welding'. Procs 2nd Int'l Conference 'Advances in Welding Processes', Harrogate, 14-16 April 1970, 147-52.

13 OBERLE, T.C. 'Inertia welding improves bond toughness fatigue strength'. Automotive Eng'g, 80 (12), 1972, 39-41.

FUNDAMENTALS OF FRICTION WELDING

T. H. Hazlett

The only comprehensive summary of friction welding developments was published by Vill in 1959. Since that time many studies have been made to more clearly define and understand the role of the various process parameters. The present paper is an attempt to integrate subsequent contributions into the general knowledge of the process and to define some of the areas that require additional investigation. It is shown that work published during the past five years adds considerably to an understanding of the various process parameters and of the bonding mechanism. However, the initial heating is explainable in a very general manner only; theory is incapable of defining the specific role of the welding parameters in a quantitative manner.

The author is Professor of Mechanical Engineering, University of California, Berkeley. Presented at the National ASM Metals/Materials Congress, Chicago, Illinois, November 1966.

Friction welding is the creation of an intermetallic bond between two metals utilizing friction as the heat source. This definition assumes movement between the contacting surfaces and the simultaneous application of a force normal to those surfaces. No specific mechanism of bonding is implied by the term 'friction welding', but rather, the mechanics by which the bond is generated, as shown in Figure 1. The possible mechanisms involved, which necessarily concern the role of the properties of the materials being joined and of the welding parameters, are the subject of the following presentation.

A. I. Chudikov is credited by Vill[1] for first proposing the practical application of frictional heat as an energy source for welding in 1956. Subsequently a research team under Vill's leadership made a comprehensive study of the process, which work resulted in the most complete compilation published on the subject to date[1]. Numerous individual papers were also published in Russia about the same time. These were primarily concerned with energy utilization and distribution[2-5] during welding.

The process was first publicized in the United States by Tesmen[6,7] in 1959, but the first published word of work being carried out in this country was in late 1960[8]. Thereafter a number of papers appeared in the technical literature describing various features of the process[9-14]. Today it is recognized and accepted for production, but as is frequently the case, application has far surpassed a clear understanding of the fundamentals involved.

MODERN FRICTION THEORY

Glaeser[15] has pointed out that in spite of the fact that friction plays such an important role in modern metal processing, progress in analyzing this phenomenon has been difficult to achieve. This is primarily because of the lack of a good basic theory that embraces the severe boundary conditions involved. Although Amoton's Law, which states that frictional resistance is proportional to the load and independent of the contact area, is frequently obeyed, it is well recognized that it is a special case and a much more general theory is required. This lack of universality of Amoton's Law is supported by the fact that the measured coefficient of friction decreases with load in the case of steel on ice and steel on Teflon. Further, Thomsen et al[16] have shown that the effective coefficient

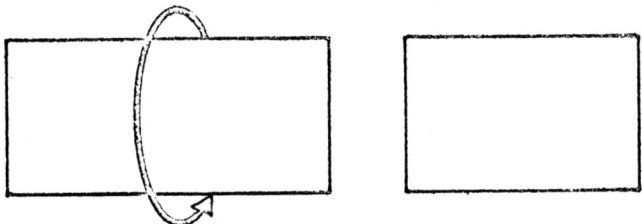
Rotating member brought up to desired speed.

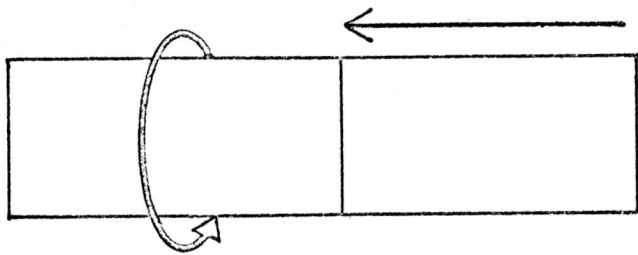
Non-rotating member advanced to meet the rotating member and pressure applied (weld pressure).

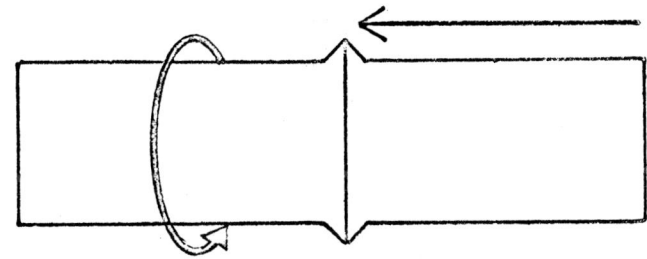
HEATING PHASE: Pressure and rotation maintained for a specified period of time or specified upset.

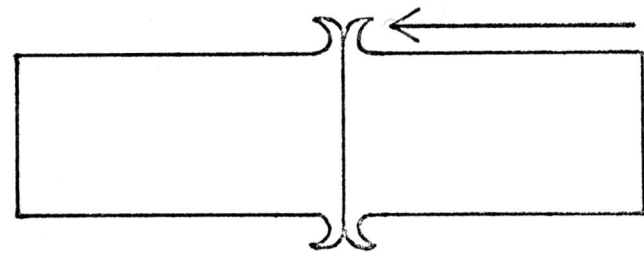
FORGING PHASE: Rotation stopped and pressure either maintained or increased.

FIG. I   PRESSURE WELDING PROCESS

Source: ASM Technical Report No. C6-30.1, 1966

of friction is also reduced in machining when the normal stress exceeds a critical value, above which shearing is induced in a layer of material below the surface. Hollander et al[17] have also found a decrease in the effective coefficient of friction when welding stainless steel to low-alloy steel.

Kragel'skiy and Vinogradova[18] proposed a very general equation for determining the coefficient of friction which involves four coefficients (A,B,C,D) and the sliding velocity (v).

$$u = (A + Bv)e^{-Cv} + D \qquad \text{(Equation 1)}$$

Based upon assigned values of the coefficients in this equation, the change of the friction coefficient with sliding velocity is shown in Figure 2 for differing degrees of normal pressure.

Modern friction theory is based on the complex geometry of real surfaces and the interaction of these surfaces. The importance of material properties, surface films and sliding velocity is acknowledged, but their introduction into the theory in a quantitative manner has not been satisfactorily achieved. Thus, although Kragel'skiy and Vinogradova indicate that at least nine factors have an influence on the coefficient of friction, and Vill has stated that five of these are directly applicable to friction welding, theory to date does not permit the inclusion of even this number in an explicit manner to the welding process parameters.

Bowden and Tabor[19] in 1954 recognized two significant mechanisms contributing to friction: (1) shearing of asperity junctions and (2) ploughing of hard asperities through a softer surface. Without going into detailed derivation, they found that for the real three-dimensional case of shearing a single asperity junction, the coefficient of friction between dry solids may be expressed as:

$$u = \frac{S_i}{p} = \frac{1}{(aK^{-2} - a)^{1/2}} \qquad \text{(Equation 2)}$$

where S_i = interfacial shear strength

p = normal pressure.

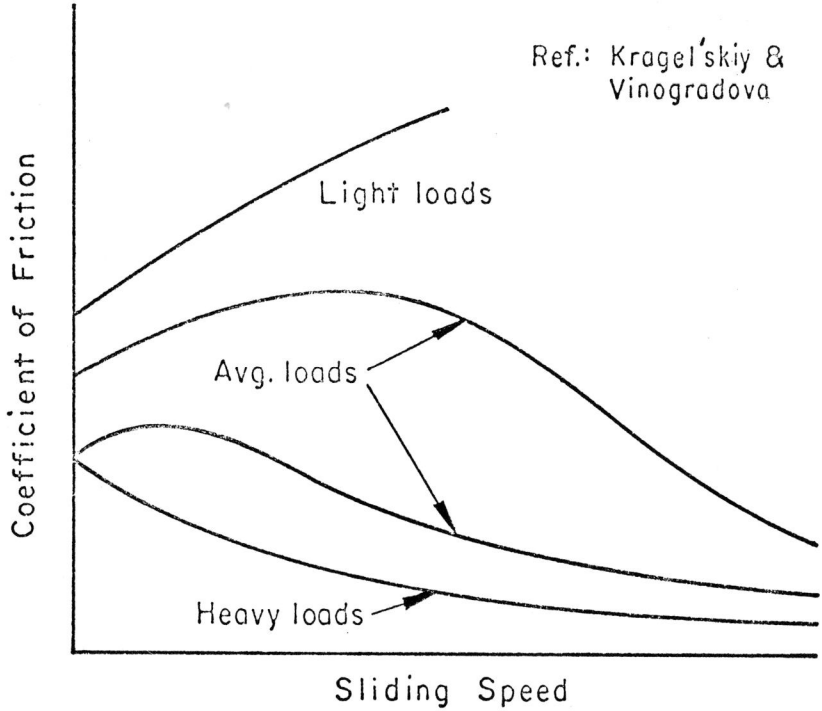

FIG. 2 EFFECT OF SPEED AND NORMAL LOADING ON FRICTION

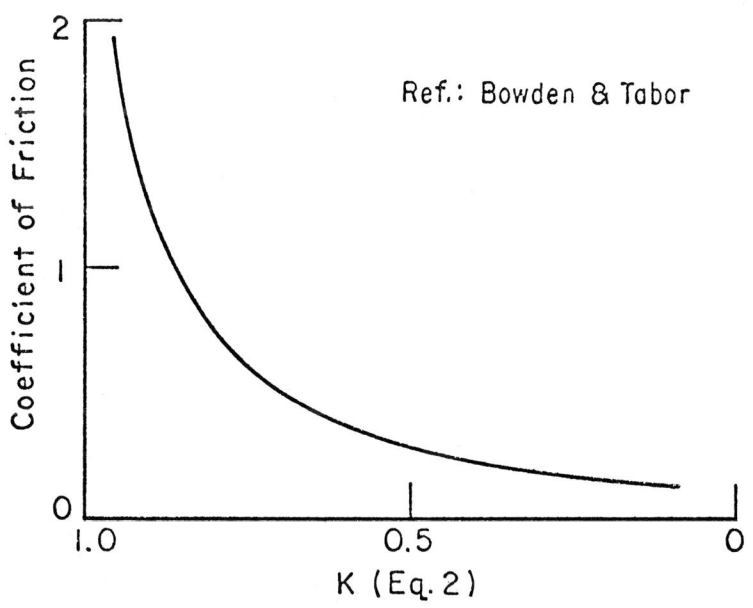

FIG. 3 COEFFICIENT OF FRICTION AS FUNCTION OF SURFACE CLEANLINESS

a = a constant whose value depends upon the softer metal in the junction and has an assigned value of approx. 9.

K = proportionality factor between the interfacial shear strength (S_i) and the critical shear strength and is thus a function of surface cleanliness in the metallurgical sense.

The sensitivity of the friction coefficient to the ratio (K) is shown in Figure 3. From this it may be seen that when S_i = critical shear stress the coefficient of friction (u) is very large. This case assumes complete welding between the mating asperities.

The above relationships have been derived for a single asperity-junction. However, the mating of real surfaces involves many such junctions and an increase of the true contact area as the normal load is increased. This may be visualized in an idealized case as shown in Figure 4, in which a true flat contacts a rough surface. Profiles of the two surfaces and the contact areas resulting are shown schematically under three loading conditions: initial, heavy and very heavy. Generally, the coefficient of friction will follow Amonton's Law between stage one and stage two, but the area of contact changes rapidly thereafter as the load is increased. As a consequence, this linear relationship between frictional resistance and normal load is no longer maintained.

Cocks[20] has also observed the formation of wedges of sheared material at moving contact interfaces when soft metals are in contact with each other. In this case, the friction force is that required to shear the solid junction at the wedge root.

When joining dissimilar materials by friction welding, one member may be considerably harder than the other and the "ploughing" mechanism could conceivably occur. In this case an asperity of the hard material will dig a trough in the soft mating member and displace a volume of the softer material. The total friction force will then consist of the sum of the asperity shearing force described above and the ploughing force, the latter of which is dependent upon the shear strength of the metal being deformed, the contact area involved and the volume displaced. It is reasonable to conclude that this ploughing mechanism is a minor consideration in friction welding except when joining materials

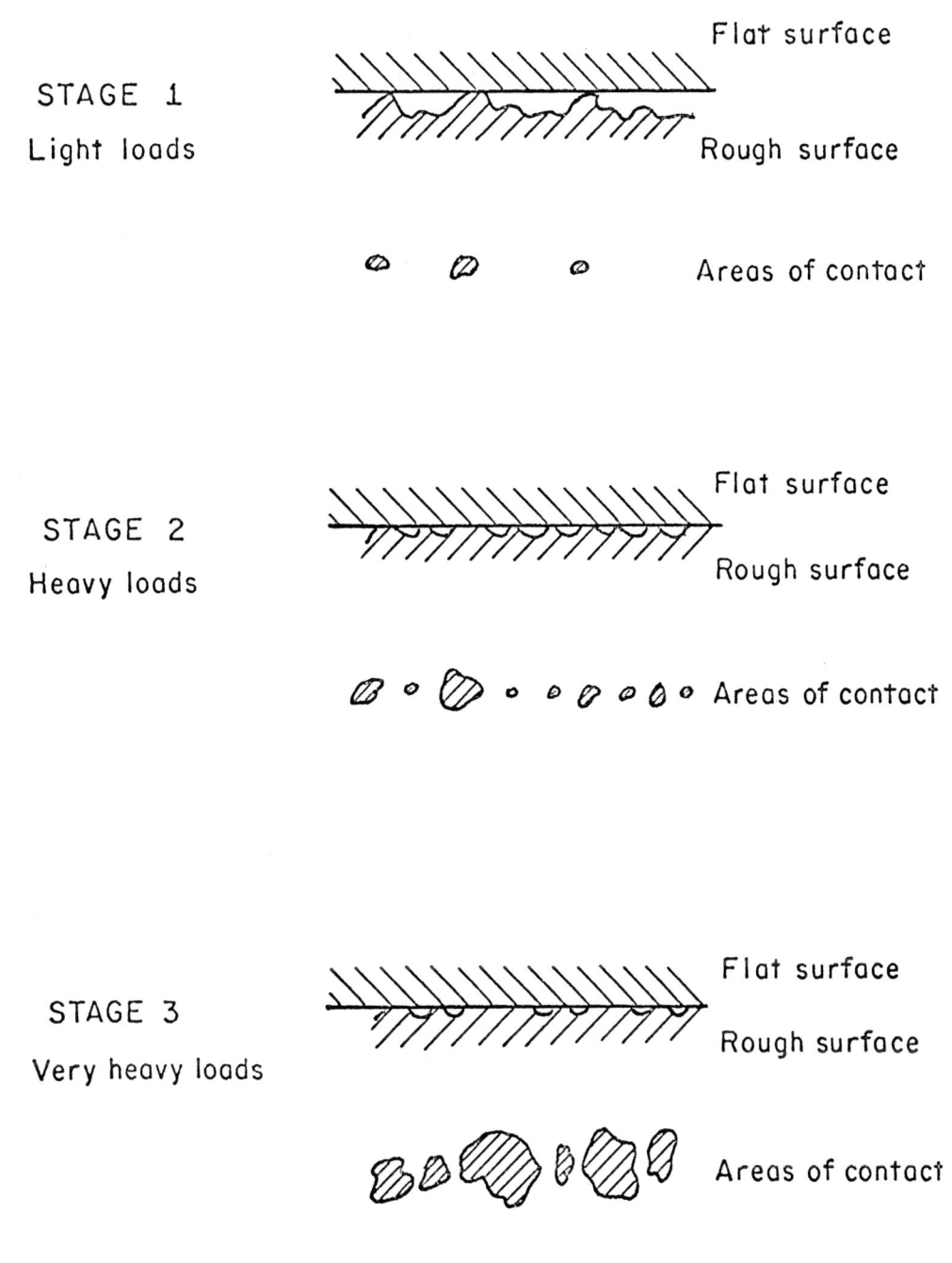

FIG. 4 CONTACT CONDITIONS FOR FLAT SURFACE LOADED AGAINST A ROUGH SURFACE.

Source: ASM Technical Report No. C6-30.1, 1966

having widely differing mechanical properties or if one of the members being joined forms a hard, temperature resistant oxide.

Alison et al[21] have shown that the abrasion-friction coefficient, which is essentially the ploughing contribution, decreases with increased hardness for polycrystalline metals having a cubic lattice structure, whereas those with hexagonal structures are independent of hardness. Insufficient data are available at this time to determine the influence of crystal structure on friction welding, but in view of this work it cannot be ignored.

Goddard and Wilman[22] have attempted to incorporate material properties into both the shearing and ploughing contribution to frictional resistance. They have represented the coefficient of friction as

$$u = K_1 S L^{(2-n)/n} + K_2 P_m L^{(3-n)/n} \quad \text{(Equation 3)}$$

(Total) = (Shearing) + (Ploughing)

where:

n = Meyer's index; an exponent in the Meyer hardness expression which is a measure of the state of hardness and usually lies between 2 (fully hardened) and 2.5 (dead soft).

S = shear strength of interface

P_m = dynamic shear strength of softer material

L = normal load

K_1 & K_2 = constants

This relationship suggests that when $n = 2$ (hardened materials) the shearing term is essentially independent of load, but when soft material is involved ($n = 2.5$) the shearing term decreases with load. Results of this are in agreement with the effect of applied load as expressed by Kragel'skiy and Vinogradova (Figure 2) and found by Thomsen[16] and by Hollander et al[17] (Figure 5) if a constant velocity is assumed. Further, Hollander et al found the effective coefficient of friction decreases as sliding velocity increases (Figure 6), as predicted by Kragel'skiy and Vinogradova.

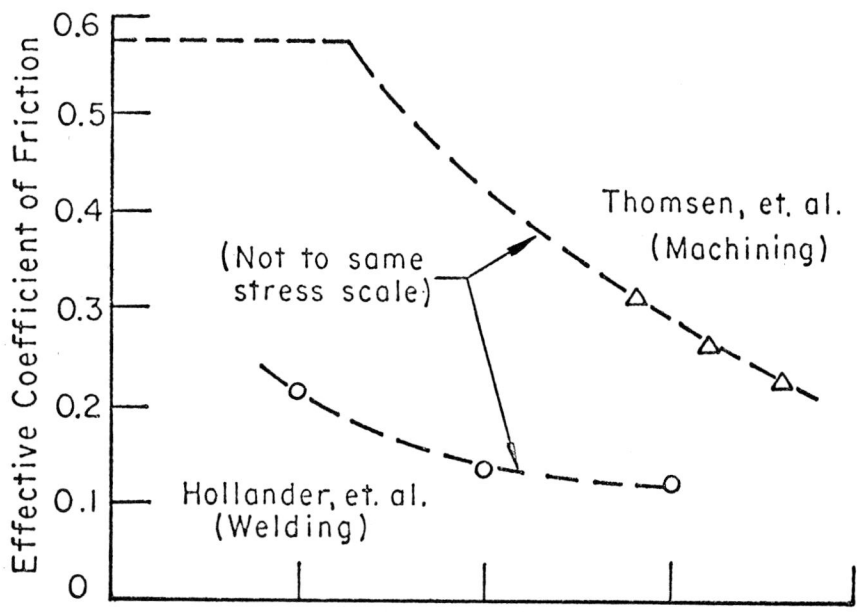

FIG. 5 EFFECT OF NORMAL STRESS ON COEFFICIENT OF FRICTION

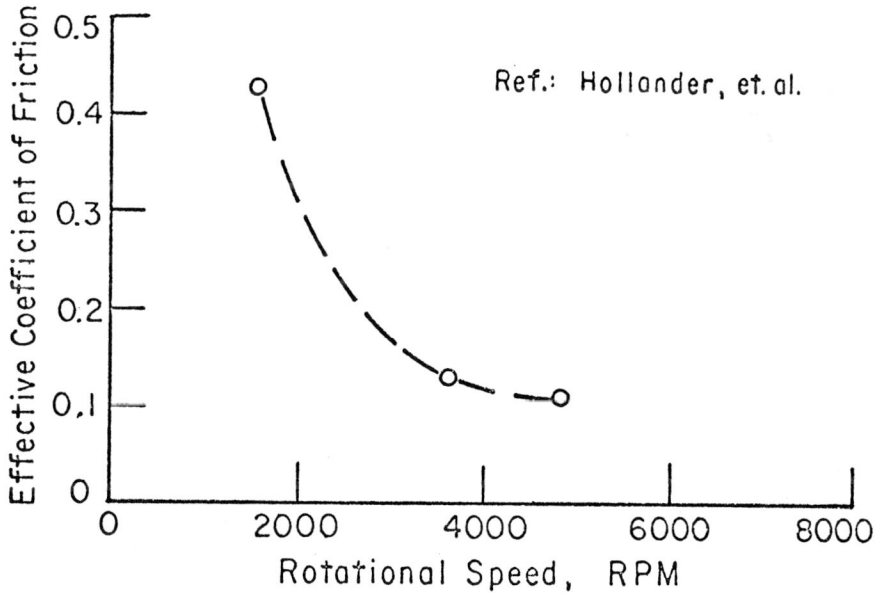

FIG. 6 EFFECT OF VELOCITY ON THE COEFFICIENT OF FRICTION

Source: ASM Technical Report No. C6-30.1, 1966

Thus, it may be concluded that none of the existing theories or derived expressions is sufficiently explicit to introduce all of the five factors in a quantitative manner that, according to Vill, are applicable to friction welding. These are:

1. The relative speed of motion of the friction surfaces

2. The temperature of the friction surfaces

3. The nature of the material and the presence of surface films

4. The magnitude of the normal pressure force

5. The rigidity and elasticity of the friction surfaces

The validity of the first item (dependence on speed) is substantiated by Hollander et al (Figure 6) but has generally been ignored except by Kragel'skiy and Vinogradova and by the (P_m) term of Equation 3 may be thus construed. The four remaining factors could all conceivably be included in Equation 3. However, values of (n), (S) and P_m) must be determined over a temperature range varying from room temperature to that approaching the melting point of the materials involved, and the reality of non-equilibrium heating must be faced.

The recognition of sub-layer flow as proposed by Shaw and by Thomsen[16] is helpful after the first stages of the welding process have passed, but a much more complex phenomenon takes place prior to that time. The important role of material properties on frictional force is illustrated in Figure 7 in which the welding parameters (speed and pressure), specimen diameter and surface condition (roughness and cleanliness) were held constant. The difference in the measured torque, which in this case is directly proportional to the nominal coefficient of friction, is startling when the initial portions of the curves are compared. In all of these tests the rotating member was initially rotated at 3,500 rpm which did not change when the normal pressure was applied. Modern theory is incapable of explaining such experimental results in a quantitative manner, much less making reasonably accurate predictions.

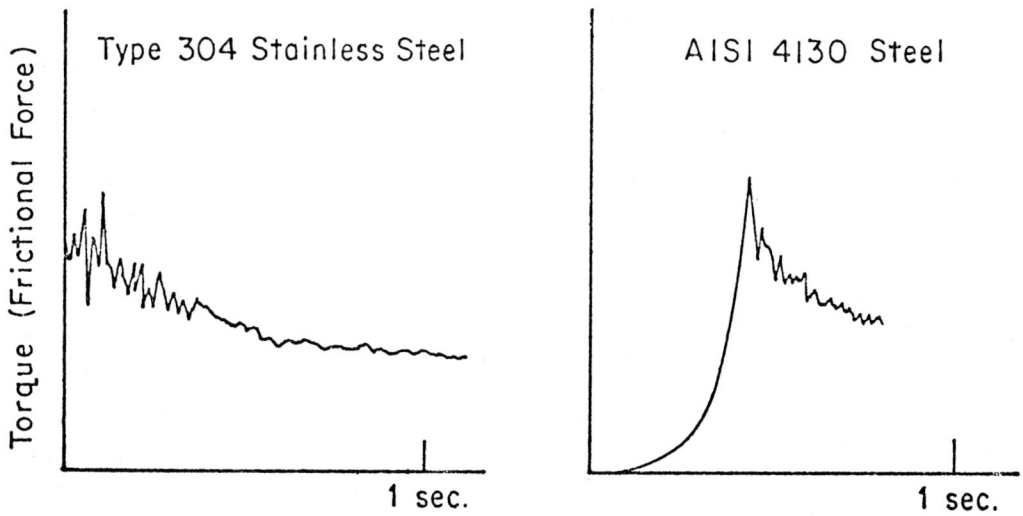
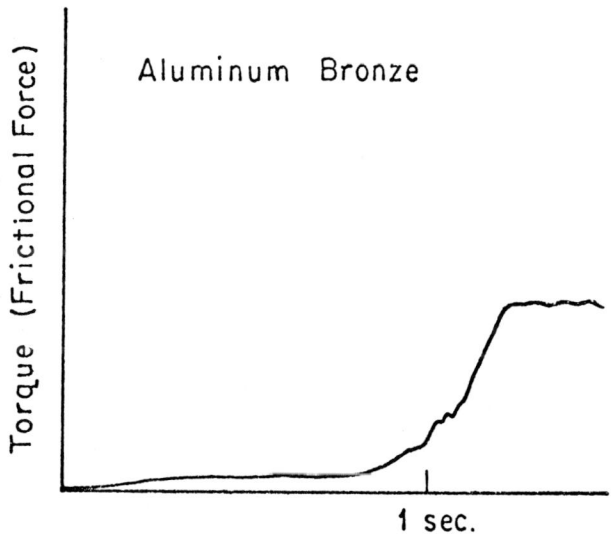

FIG. 7 FRICTION CHARACTERISTICS FOR THREE COMMERCIAL ALLOYS UNDER CONDITIONS OF CONTINUOUS HEATING FROM ROOM TEMPERATURE AND CONSTANT NORMAL LOAD.

Source: ASM Technical Report No. C6-30.1, 1966

TEMPERATURE GENERATION AND DISTRIBUTION

Rukalin et al[5] have calculated the axial temperature distribution in a weld made between two solid low-carbon steel bars. They also measured the radial and axial temperature by imbedding thermocouples at various locations. A significant temperature gradient was measured in the radial direction (Figure 8) and the axial temperature calculations were in fair agreement with measured values. The latter is rather surprising in view of the drastic simplications assumed.

Cheng[23] later made a more sophisticated analysis based upon variable thermal properties of the members being welded and computed by the finite difference method. He assumed the welding of one inch diameter SAE 4140 tubing and the formation of a melted zone at the interface. His calculated and experimental values, also measured by imbedded thermocouples, are shown in Figure 9 in which the dimension (X) is the distance from the welding interface. It may be seen that the agreement between calculated and measured values is quite good considering the complexities involved, especially for the first half of the welding time. He attributes the later deviation to the relative displacement between the thermocouples and the welding interface. Subsequent calculations which considered the much more complex problem of welding AISI 4140 low alloy steel to an austenitic stainless steel are not in as good agreement with experiment[24].

However, these calculations assume that the interface is molten while the United States Welding Delegation to the Soviet Union stated that "Dr. Vill is firm in his belief that molten metal is not formed during the welding process."[25] This 'belief' is substantiated by the results reported by Weiss and Hazlett[14] who found that when welding dissimilar metals the average interface temperature never reaches the melting point of either material being joined nor of eutectic alloys that may be expected. They further found that the maximum average interface temperature over the outer 25% of the solid bar radius was less than ten percent greater than the average of the solid bar.

Consideration of all of the experimental temperature data available leads to the conclusion that the radial temperature distribution measured by Rykalin et al is reasonable when extrapolated to the interface; i.e. the temperature of the interface at the center of a solid bar does not reach the maximum interface temperature, or even

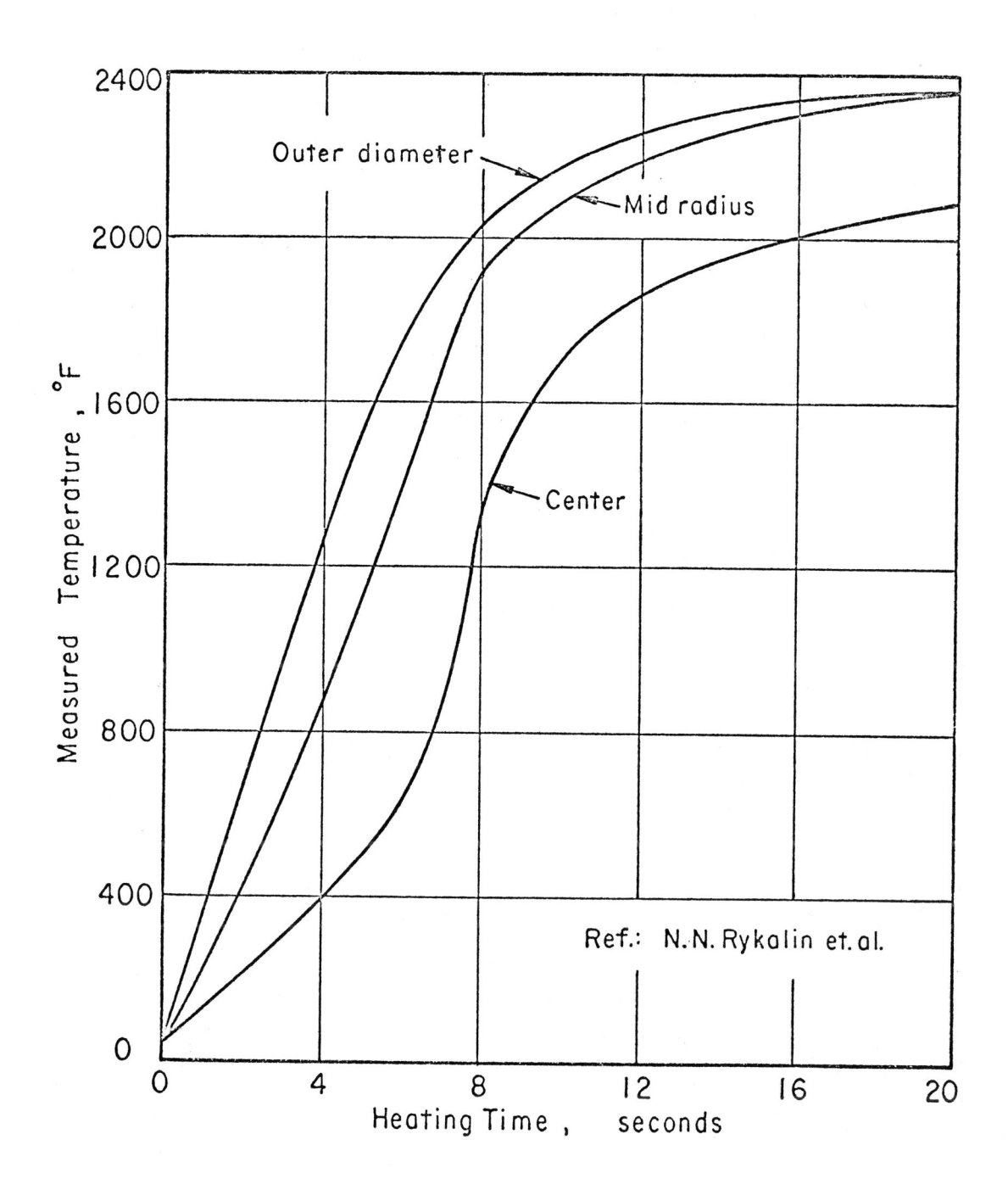

FIG. 8 RADIAL TEMPERATURE DISTRIBUTION IN 0.8 IN. DIA. FRICTION WELDED LOW CARBON STEEL – 0.041 IN. FROM INTERFACE

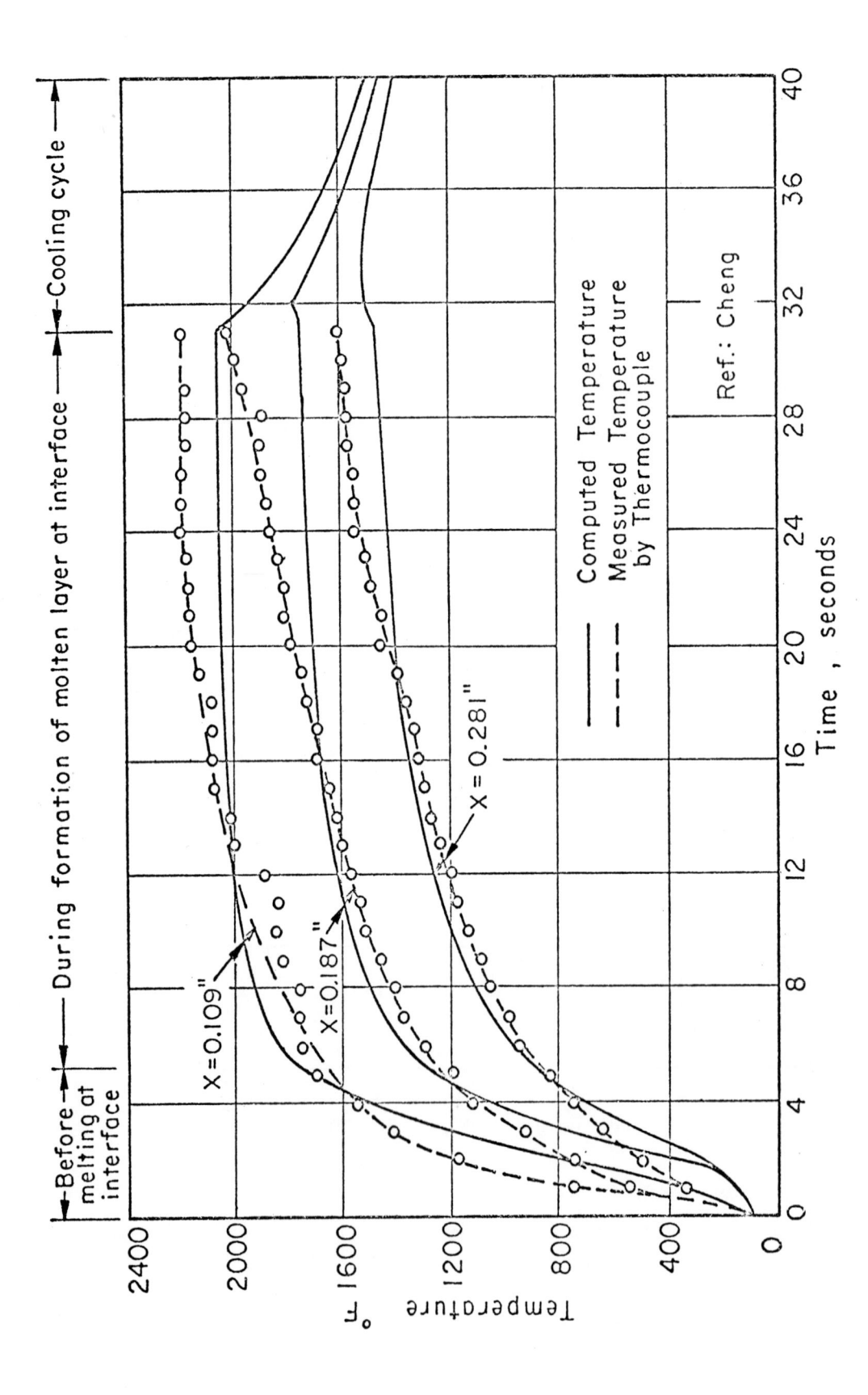

FIG. 9 TEMPERATURE-TIME HISTORY OF THE FRICTION WELDING PROCESS

that of the measured average, by a significant amount during rather long welding times (20 sec.). It must be presumed that this temperature difference increases, for the same time, as the diameter of the work piece increases and the gradient will vary with the conductivity and diffusivity of the materials being joined. This could be a major contributor to the condition frequently encountered in which the outer portion of a solid bar is metallurgically bonded but the center is not welded.

The same investigators also found that the maximum average temperature of the interface was essentially independent of applied force but the time required to reach that temperature decreased as the welding pressure increased. This is shown in Figure 10, which was measured during the welding of iron to nickel.

These maximum temperatures are essentially the same as that achieved when an even greater welding force was applied (18,000 psi) and the interface was permitted to reach an equilibrium temperature; a good weld both metallurgically and mechanically was achieved across the entire interface. The stable temperature maximum, which is below the melting point, may be explained by the introduction of relatively cool metal forced into the interface by the axial pressure.

EFFECT OF WELDING SPEED AND PRESSURE

Although Vill[1] has stated unequivocally that slow welding speeds and high pressures are the most efficient, he bases this conclusion on the results of Zackson and Voznesenskii[3] and on the heat affected zone pattern of welds made between solid bars, both results of which were obtained on plain low-carbon steel. It is recognized that the optimum values of both speed and pressure will vary with the diameter of the parts being welded since they are a function of effective radius and area, respectively. However, there is no evidence to support Vill's view on speed and pressure for all materials. The various publications of Hollander and Cheng do not substantiate this contention when their data are analyzed. Further, Hazlett[12] speculated that although he was unsuccessful in welding high conductivity copper to itself at a speed of 3500 rpm, it could probably be achieved if higher speeds were available. Cheng later reported to the writer that indeed he had succeeded in making such welds at 5000 rpm. On the other hand, the limited information available on "Interia Welding" indicates that slow speeds can be very effective in the formation of a sound weld. However, no systematic studies of this parameter have been reported in the recent literature. It must, therefore, be concluded that this is an area that requires more extensive study.

THE ROLE OF MATERIAL PROPERTIES

Many examples have been cited in the literature of friction welding a wide variety of metals to themselves and to each other[8,26] and many more detailed studies have been made of specific combinations.[1-5,9,10,13] Vill has pointed out that numerous complex interactions occur during the friction welding process. The friction forces generated during the initial portion of the welding cycle are at the present time best represented by Equation 3, but three terms of this expression depend upon the properties of the materials being joined which are in turn dependent upon the instantaneous temperature which is continuously changing during this portion of the welding cycle. The importance of temperature is shown by the change of flow strength with temperature as illustrated for a number of metals in Figure 11. Further, the strain rate dependence of flow stress for metals at elevated temperatures has been recognized and reported for more than two decades. Thus the practical application of this, or any other existing equation, to the initial portion of the welding cycle is currently impractical.

That portion of the cycle during which the temperature remains essentially constant and the torque shows a relatively small decrease may be best explained at this time on the basis of sublayer flow as proposed by Shaw and Thomsen[16]. The existence of this layer is clearly evident in Figure 12, which is a cross-section of the outer portion of a weld between aluminum and titanium. The presence of several regions having differing metallurgical structures may be seen in the aluminum: a) a highly worked area adjacent to the interface which shows marked banding, b) the also severely worked region in the curled upset which extends quite far back from the interface, and finally c) the structure of the parent material in the lower right-hand corner. The decrease of torque which occurs after the maximum has been reached may be due to an increase of the sublayer thickness due to heat flow, and is thus a dependent function of the thermal diffusivity.

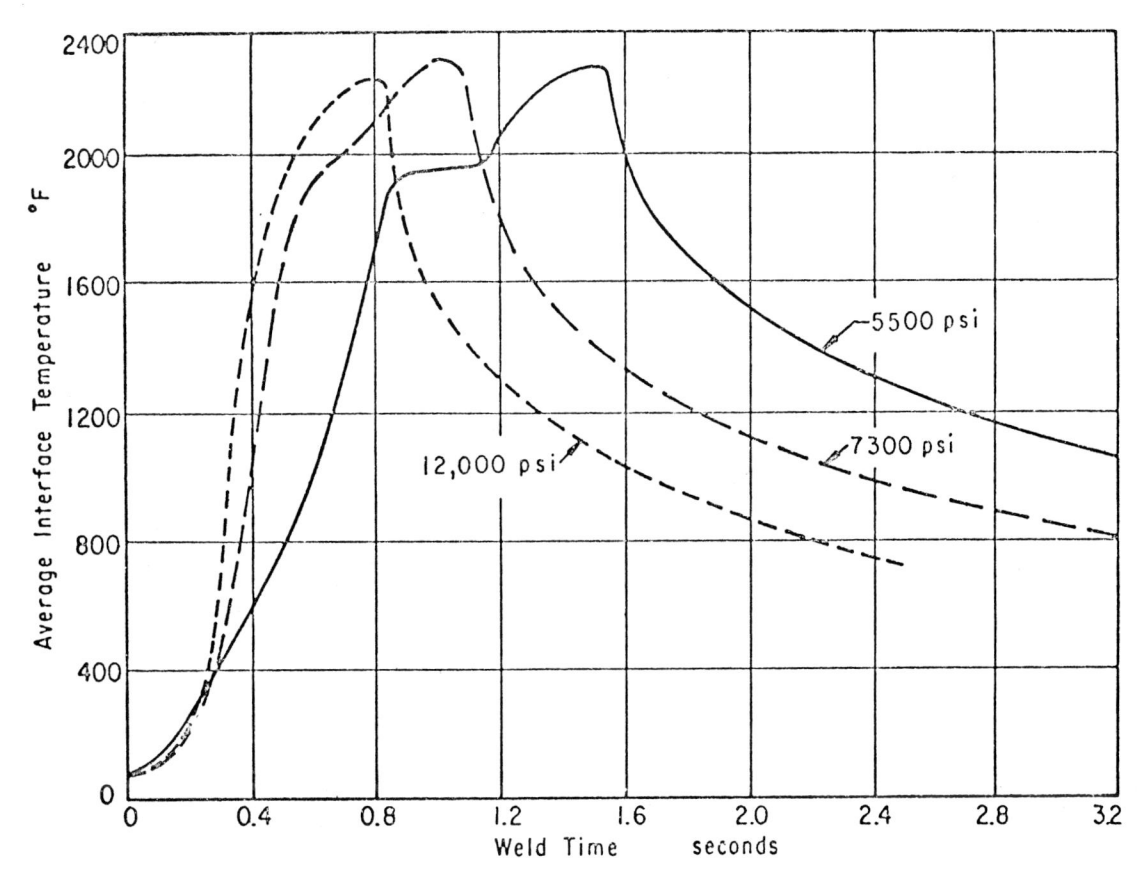

FIG. 10 EFFECT OF VARING WELDING PRESSURE ON INTERFACE TEMPERATURE

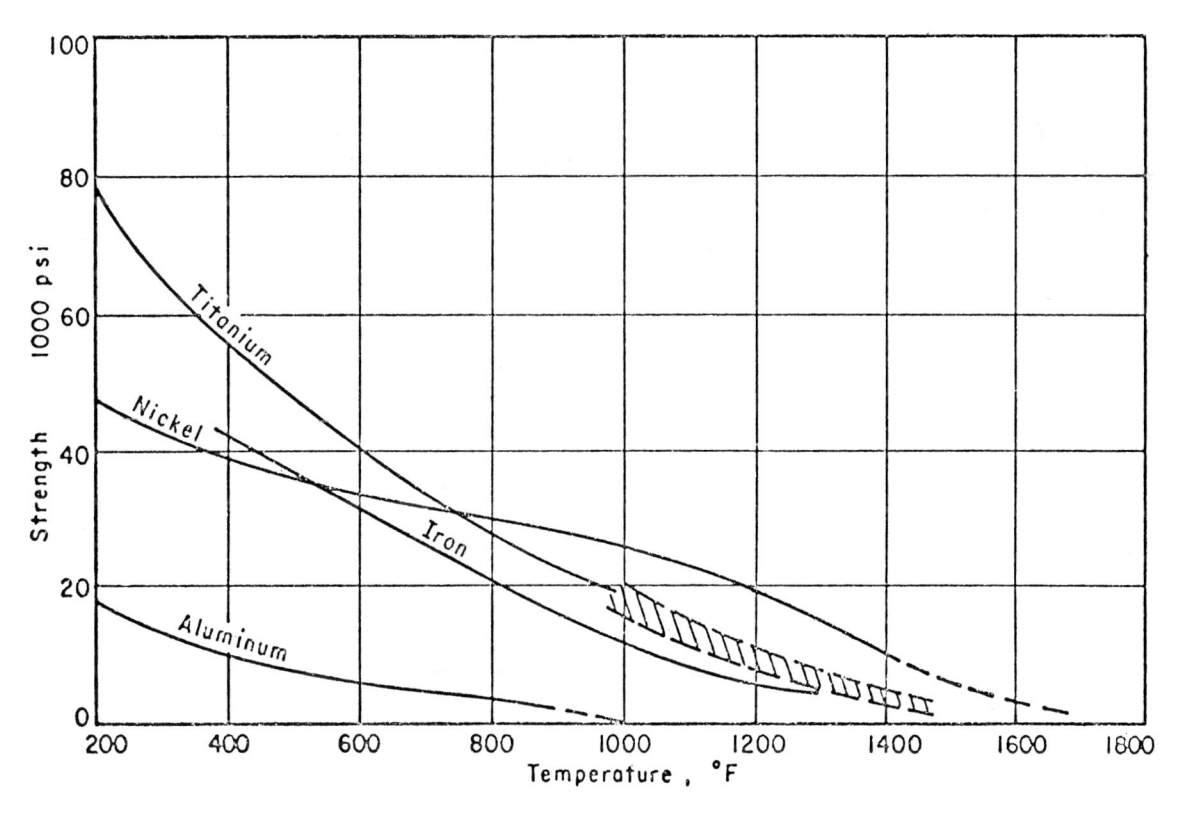

FIG. 11 FLOW STRESS vs TEMPERATURE

Source: ASM Technical Report No. C6-30.1, 1966

Titanium — Aluminum

FIG. 12 SUB-LAYER FLOW IN FRICTION WELD

Parent Material
(100 X)

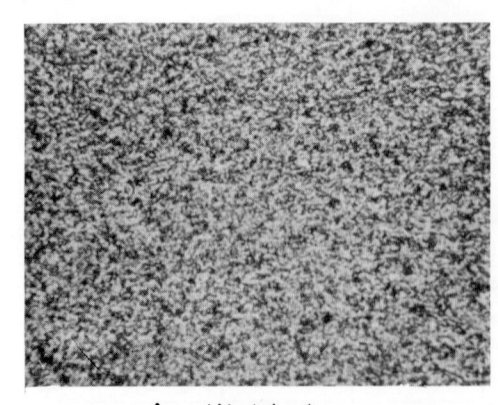

As Welded
(500X)

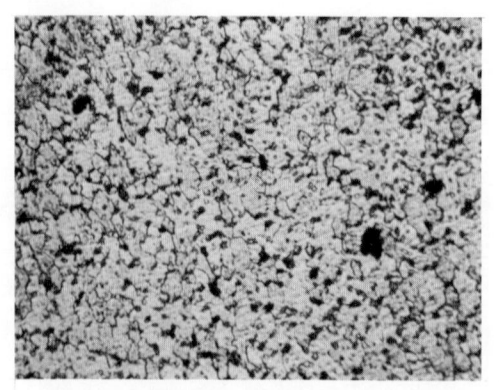

Welded and Recrystallized
1 hr @ 600°F   (500X)

FIG. 13   FRICTION WELDED 2024-T4 ALUMINUM

An example of the severe cold work at the interface of a sound friction weld is shown in Figure 13, in which the micro-structure of the parent material, the 'as welded' interface and the recrystallized interface is shown for 2024-T4 aluminum alloy. This high degree of cold work is undoubtedly the reason for the very high joint strengths found by Hazlett and Gupta[13] for this alloy in the 'as welded' condition.

Although mutual solubility between the members being welded is not a factor when joining similar materials, Weiss and Hazlett have shown that the ease of making a mechanically sound weld is increased as the mutual solubility of dissimilar materials is increased. As a consequence, they have attributed such bonds to high temperature solid state diffusion. Winter and Nielsen[27], in their study of ultra-sonic welding, also concluded that mutual solubility is a major factor in the bonding process.

Such a diffusion layer may be seen in Figure 14, which is the interface between friction welded nickel to titanium. The micrograph indicates that the nickel has diffused into the titanium to a limited degree. A study of the Ni-Ti phase diagram shows that a eutectic exists at 28.5% Ni, the melting point of which is about the maximum interface temperature reached during friction welding. Thus, this narrow band may be of a composition near the eutectic.

NATURE OF THE FRICTION WELDED BOND

It has been generally presumed that the mechanism of all pressure welding is the production of two nascent surfaces in intimate contact with each other due to the applied normal force. It has further been assumed that the clean nascent surfaces were created by severe deformation of the original contact areas which extruded surface contamination away from the area of structural importance. In the case of friction welding this is aided by the rotational shearing action, while the high temperatures at the interface may also assist in placing some contaminants in solution. The latter may, for example, be a contributing factor in the welding of steel where iron oxide is readily dissolved in the parent material. To these mechanisms, diffusion as shown in Figure 14 and to a slightly less definitive extent in Figures 15 through 18 must now be added.

Nickel　　　　Titanium

FIG. 14 WELD INTERFACE (X400)

Aluminum　　　　Iron

FIG. 15 WELD INTERFACE (X400)

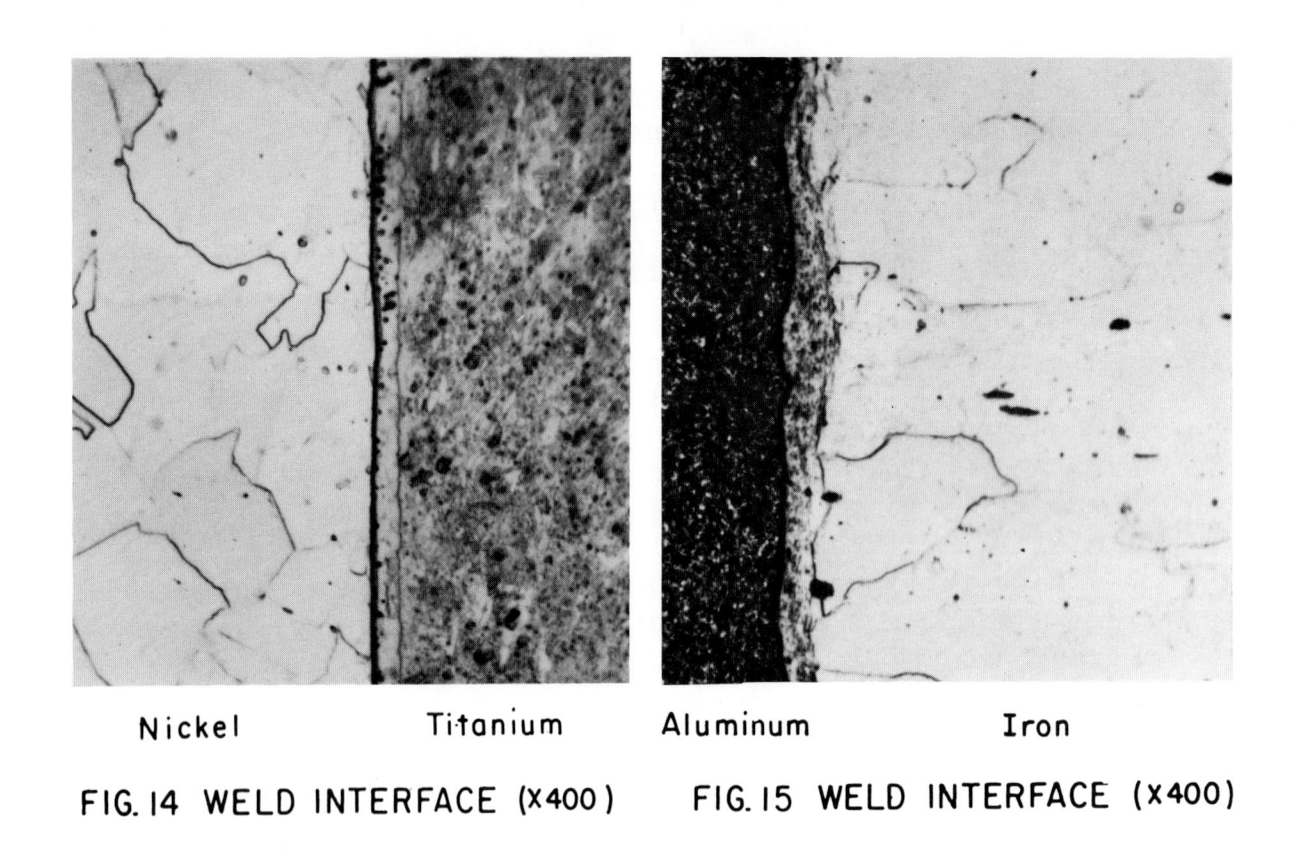

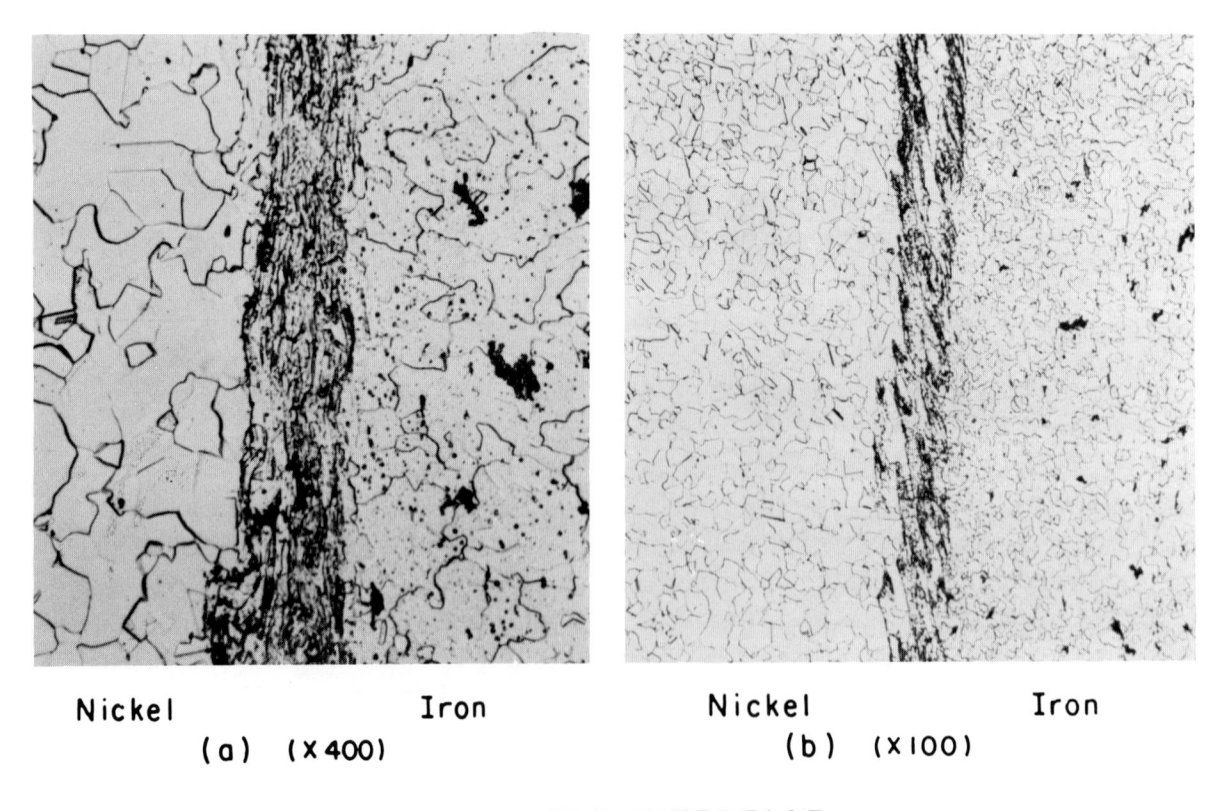

Nickel　　　　Iron
(a) (X400)

Nickel　　　　Iron
(b) (X100)

FIG. 16 WELD INTERFACE

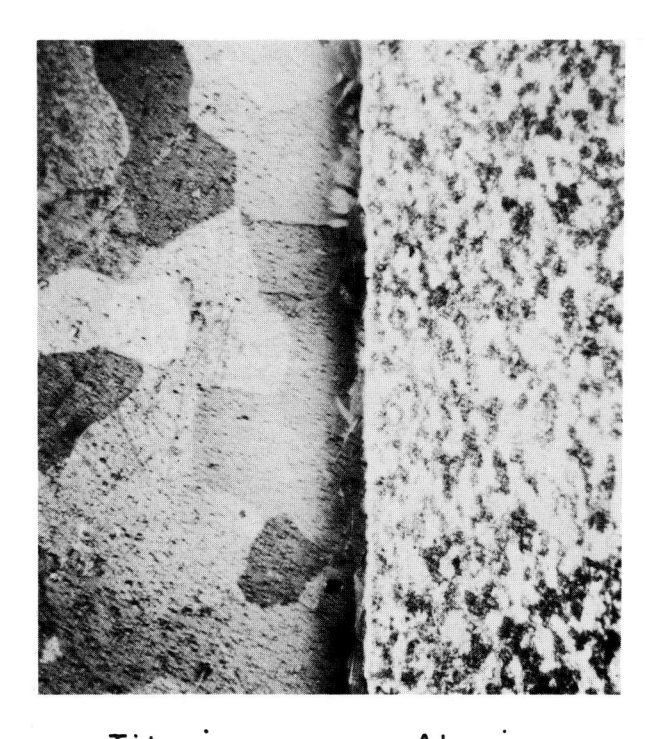

Titanium   Aluminum

FIG. 17 WELD INTERFACE (X400)

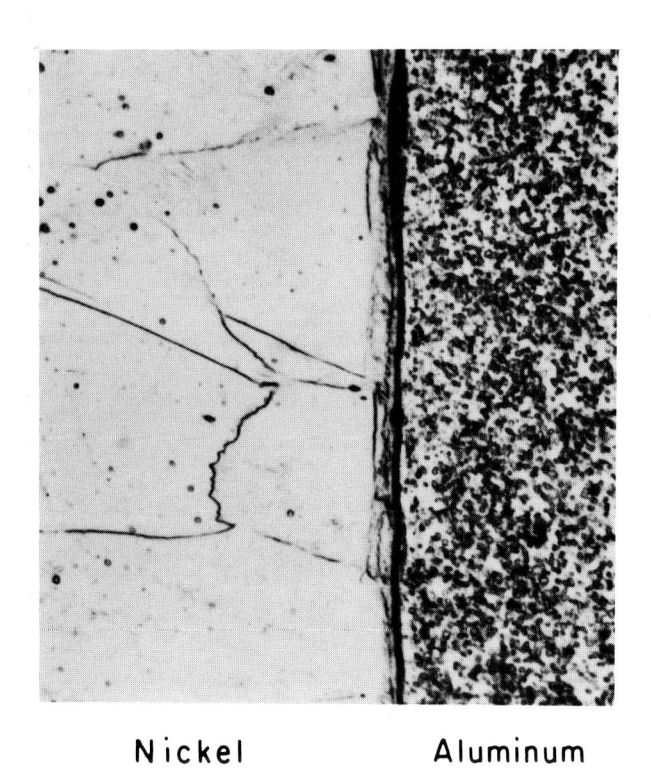

Nickel   Aluminum

FIG. 18 WELD INTERFACE (X400)

Source: ASM Technical Report No. C6-30.1, 1966

The relative importance of these mechanisms is not clear and probably varies with the materials being joined, but metallographic evidence of diffusion at dissimilar metal weld interfaces has been found for all metal combinations studied.

Iron and nickel, which have a high degree of mutual solubility, appear at higher magnifications to exhibit a severely cold worked interface as shown in Figure 16a. However, micrographs at lower magnification (Figure 16b) indicate migration of the iron or the nickel, resulting in what may be a second phase at a considerable distance from the original interface. Although the titanium-aluminum weld interface shown in Figure 17 is very sharp, close examination reveals that there is clear evidence of diffusion. The presence of this mechanism is revealed by "rays" of what appear to be a different phase extending into the titanium matrix along the interface. It may be presumed, in view of the relatively high mutual solubility, that the aluminum has diffused into the titanium where the orientation of the titanium grains is most favorable. The sharp interface is probably due to the very large difference in the absolute melting points of the metals being joined, which precludes the presence of a well defined diffusion zone as seen in Figures 14 and 15. Figure 18 shows a friction welded nickel-aluminum interface, and although the mutual solubility of these metals is relatively low, a diffusion zone is clearly evident. Thus, there is very strong evidence that diffusion accompanies successful friction welding of dissimilar metals.

Considering the measured interface temperatures for dissimilar metals[14] and the extrapolated temperatures for steel,[5,23] combined with the results of metallographic studies, it must be concluded that diffusion is a primary mechanism contributing to friction welding. This is undoubtedly aided by the forging pressure. However, forge pressure probably serves at least a dual role. It increases the strength of the weld zone by imparting cold work as shown in Figure 13. In addition, many studies of solid state diffusion bonded junctions have shown that pressures higher than that required for complete area contact have a pronounced influence on bond quality; higher pressures produce higher quality bonds. It may, therefore, be reasonably concluded that high forging pressures also aid diffusion in the case of friction welding.

CONCLUSIONS

1. The mechanism of bonding is primarily high temperature diffusion. This is aided by the cleansing action created by the simultaneous relative rotation and applied normal pressure, plus the probable additional driving force provided by the forge pressure.

2. Sublayer flow as proposed by Shaw and Thomsen is undoubtedly a controlling mechanism of the frictional force between the members being welded during the 'steady state' stage.

3. The temperature distribution at the weld interface is not uniform but increases with distance from the center to the outer radius.

4. Analytical calculations of linear temperature distribution have been refined to a degree that acceptable engineering predictions can be made when the material properties are known. The accuracy of such calculations decreases as materials of differing properties are joined.

5. Maximum temperatures achieved at the weld interface are essentially independent of welding pressure and are below the melting point of the materials being welded or of anticipated eutectic compositions.

6. Modern friction theory fails to account for the frictional forces that have been measured during the first stage of welding except in a very general manner. Thus it is impossible to analytically predict satisfactory welding parameters at this time.

7. The specific role of welding speed and pressure on materials of differing properties is not clear and additional work is required in this area. However, this may be intimately related to the first stage of welding referred to in (6) above and is therefore only amenable to empirical treatment at this time.

Source: ASM Technical Report No. C6-30.1, 1966

REFERENCES

1. Vill, V. I., "Friction Welding of Metals," Translated by AWS, New York (February 1962)

2. Vill, V. I., "Energy Distribution in the Friction Welding of Steel Bars," *Welding Prod.*, (October 1959)

3. Zakson, R. I., and Voznesenskii, V. D., "Power and Heat Parameters of Friction Welding," *Ibid.*, (October 1959)

4. Gel'dman, A. S. and Sander, M. P., "Power and Heating in the Friction Welding of Thick-Walled Pipe," *Ibid.*, (October 1959)

5. Rykalin, N. N., Pugin, A. I. and Vasil'eva, V. A., "The Heating and Cooling of Rods Butt Welded by the Friction Process," *Ibid.*, (October 1959)

6. Tesmen, A. B., "Welding and Metal Forming in Russia," *Metal Progress*, (January 1959)

7. Tesmen, A. B., "Friction Welding: A Lesson in Economics from Russia," *Ibid.*, (August 1960)

8. Irving, R. R., "How Friction Welding Joins Bar Stock and Tubing," *The Iron Age*, (December 29, 1960)

9. Hazlett, T. H., "Properties of Friction Welded Plain Carbon and Low Alloy Steels," *Welding Journal*, Research Supplement, Vol. 41, No. 2, (1962)

10. Hollander, M. B., "Developments in Friction Welding," *Metals Eng. Quart.*, (May 1962)

11. Hollander, M. B. and Cheng, C. J., "Friction Welding Low Alloy Steel to Austenitic Stainless Steel," ASTME Engineering Conf., Cleveland, Ohio, (May 1962)

12. Hazlett, T. H., "Properties of Friction Welds Between Dissimilar Metals," *Welding Journal*, Research Supplement, (October 1962)

13. Hazlett, T. H. and Gupta, K. K., "Friction Welding of High Strength Structural Aluminum Alloys," *Welding Journal*, Research Supplement, (November 1963)

14. Weiss, H. D. and Hazlett, T. H., "The Role of Material Properties and Interface Temperatures in Friction Welding Dissimilar Metals," ASME Metals Engineering Conference, Cleveland, Ohio, (April 1966); ASME Pub. No 66-Met-8

15. Glaeser, W. A., "Friction and Wear in Metal-Deformation Processes," DMIC Report No 226, (July 7, 1966)

16. Thomsen, E. G., Yang, C. T., and Kobayashi, S., *Plastic Deformation In Metal Processing*, Macmillan (1965)

17. Hollander, M. B., Cheng, C. J., and Wyman, J. C., "Friction Welding Parameter Analysis," *Welding Journal*, Research Supplement, (November 1963)

18. Kragel'skiy, I. V. and Vinogradova, I. E., "Friction Coefficients," Mashgiz, (1955)

19. Bowden, F. P. and Tabor, D., *The Friction and Lubrication of Solids*, Part I, Oxford University Press, London, (1954)

20. Cocks, M., "Formation of Wedges Between Sliding Surfaces," *Wear*, 8, (1965)

21. Alison, P. J., Stroud, M. F., and Wilman, H., "Abrasion of Metals and Binary Alloys," Third Convention on Lubrication and Wear, Inst. Mech. Engrs., (London), (May 1965)

22. Goddard, J. and Wilman, H., "A Theory of Friction and Wear During The Abrasion of Metals," *Wear*, 5, (1962)

23. Cheng, C. J., "Transition Temperature Distribution During Friction Welding of Two Similar Materials in Tubular Form," *Welding Journal*, Research Supplement, (December 1962)

24. Cheng, C. J., "Transition Temperature Distribution During Friction Welding of Two Dissimilar Materials in Tubular Form," *Ibid.*, (May 1963)

25. *Report of United States Welding Delegation on Visit to Soviet Union, July 1962*, American Welding Society, New York

26. Hollander, M. B., Cheng, C. J., and Quimby, J. A., "Friction Welding - A 'Natural' for Production," Soc. of Auto. Engrs., Nat'l Automobile Week Meeting, (1964)

27. Winter, J. and Nielsen, J. P., "Preliminary Study on the Mechanics of Ultrasonic Welding," New York University, College of Engineering, Division of Research, (November 1959)

Friction Welding

*By the ASM Committee on Flash, Friction and Stud Welding**

FRICTION WELDING is a process in which the heat for welding is produced by direct conversion of mechanical energy to thermal energy at the interface of the workpieces without the application of electrical energy, or heat from other sources, to the workpieces. Friction welds are made by holding a nonrotating workpiece in contact with a rotating workpiece under constant or gradually increasing pressure until the interface reaches welding temperature, and then stopping rotation to complete the weld. The frictional heat developed at the interface rapidly raises the temperature of the workpieces, over a very short axial distance, to values approaching, but below, the melting range; welding occurs under the influence of a pressure that is applied while the heated zone is in the plastic temperature range.

Friction welding is classified as a solid-state welding process, in which joining occurs at a temperature below the melting point of the work metal. If incipient melting does occur, there is no evidence in the finished weld, because the metal is worked during the welding stage.

A section through a friction weld joining two dissimilar steels is shown in Fig. 1. The steel with the greater forgeability has the greater amount of weld upset. When similar metals are friction welded, the amount of weld upset is about the same on both sides of the bond line.

Applications. Friction welding has been used in high production of hollow precombustion chambers for diesel engines, in welding trunnions to mounting blocks for air and hydraulic cylinders, in welding connectors to piston rods, and in fabricating track-roller hubs and ball-shaft linkages.

In the automotive industry, friction welding is used in fabricating drive shafts, axles, steering shafts and bimetal valves, and for joining hubs to gears. Another application is welding bar stock to small forgings or to plate to produce parts that would otherwise be forged. Blanks for cutting tools are made by welding low-carbon or low-alloy steel shanks to tool steel bodies.

Jet-engine parts are made by welding components made of a heat-resisting alloy to components made of a hardenable or wear-resistant alloy.

Welding Methods. There are three methods of joining workpieces by friction welding: (a) conventional friction welding, (b) inertia welding, and (c) flywheel friction welding.

In conventional friction welding, mechanical energy is converted to heat energy by rotating one workpiece while pressing it against a nonrotating workpiece. After a specific period of time, rotation is suddenly stopped and the pressure is increased and held for another specified period of time, producing a weld.

In inertia welding, the workpiece component that is to be rotated is held in a collet-chuck – flywheel assembly. The assembly is then accelerated to a predetermined speed, at which time the flywheel is disconnected from the power supply and the workpieces are brought into contact under a constant force. Flywheel energy is rapidly converted to heat at the interface, and welding occurs as rotation ceases.

Flywheel friction welding incorporates features of both the conventional and the inertia processes. Flywheels are connected to the drive motor and to the spindle, and are coupled through an integral clutch. The drive-motor – flywheel system rotates continuously and is coupled to the flywheel-spindle system to bring the rotating workpiece to the proper speed. The motor flywheel is disengaged from the spindle flywheel after the desired energy has been extracted. The spindle flywheel, having a low moment of inertia, comes quickly to rest without braking, to complete the weld.

Process Capabilities

Many ferrous and nonferrous alloys can be friction welded. Friction welding also can be used to join metals of widely differing thermal and mechanical properties. Often combinations that can be friction welded cannot be joined by other welding processes because of the formation of brittle phases that would make such joints unserviceable. The submelting temperatures and short weld times of friction welding allow many combinations of work metals to be joined.

End preparation of workpieces, other than that necessary to ensure reasonably good alignment and to produce the required length tolerance for a specific set of welding conditions, is not critical. Frictional wear removes irregularities from the joint surfaces and leaves clean, smooth surfaces heated to welding temperature. In some applications where weld integrity is important, a small projection at the center of one of the weld members is used to ensure proper heating and forging action, and to eliminate center defects. This projection is especially helpful in welding large-diameter bars.

Automatic loading and unloading of the welding machines permit high production rates. For instance, bimetal valves are produced two-at-a-time at a rate of 1200 per hour.

Other advantages of friction welding include the following:

1. Flux, filler metal or protective atmospheres are not needed.
2. Electric-power and total-energy requirements are a fraction of those needed for other welding processes.
3. The operation is relatively clean, there is little spatter, and no arcs, fumes or scale are developed.
4. The heat-affected zone is very narrow and has a grain size that frequently is smaller than that in the base metal.

Limitations of friction welding are:

1. One workpiece must be round (or nearly round) at the interface and must have a size and shape that can be clamped and rotated. (Hexagon-shape bars have been friction welded to billets.)
2. Workpieces must be able to withstand the torque and axial pressure imposed during heating and forging. (See "Sections Welded", on the next page.)
3. Workholding devices must be strong enough to withstand heavy shock and torque loads.
4. The process is restricted to flat and angular butt welds that are concentric with the axis of rotation (see Fig. 13).
5. Conventional friction welding machines require expensive modifications to be able to weld workpieces that must have final angular alignment.

Economy in Operation and Material. Friction welding can be done at high production rates, and therefore is economical in operation. In applications where friction welding has replaced other joining processes, the production rate has been substantially increased. For instance, in Example 461, production rate was increased from 3 to 30 welds per hour when friction welding replaced pressure gas welding.

Savings in material also can be realized by the use of friction welding. In Example 454, ¾ in. less metal was needed in friction welding than in flash welding. Also, as described in Example 460, considerable metal was saved by friction welding two pieces of 4140 steel together to make a square-head bolt instead of machining the bolt from a single piece.

Substantial savings in material and machining time can be realized by using friction welding to join stub shafts to large-diameter rotor bodies or other rotating members, or to join components of valve stems — as in the applications described in the two-part example that follows.

Example 451. Changes From Machining Complete From Bar to Friction Welding of Machined Components, for Lower Metal and Machining Costs

Example 451a — Rotor Blank (Fig. 2). A blank for an air-motor rotor (Fig. 2) was made from 8620 steel by joining stub shafts to the ends of the main body by conventional friction welding. In finish machining of the welded blank, the diameter of each of the three pieces was reduced by only 0.060 in., which illustrates the concentricity to which parts can be friction welded. The use of

Fig. 1. Cross section through a friction weld joining two dissimilar steels

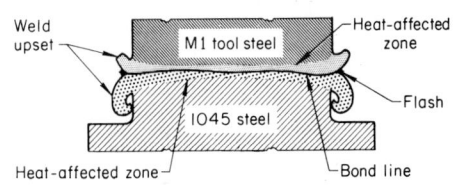

The greater amount of weld upset occurred in the 1045 steel because of its greater forgeability. When workpieces made of the same or closely similar metal are welded, approximately equal amounts of weld upset are formed on both sides of the bond line.

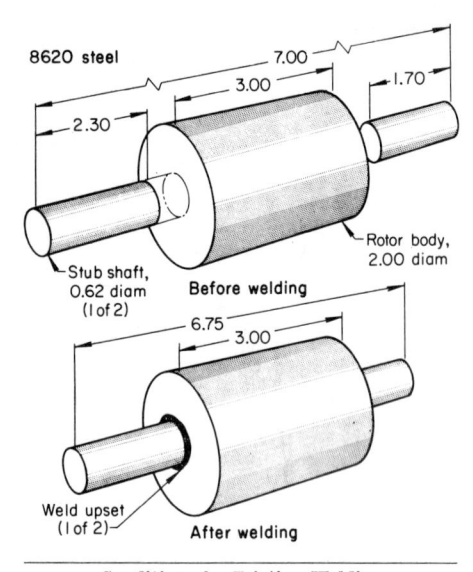

Conditions for Friction Welding

Spindle speed	2400 rpm
Axial force:	
Heating	2480 lb
Welding	4400 lb
Heat-and-weld time	45 sec per part
Weld area	0.31 sq in. per weld
Metal lost(a)	0.125 in. per weld
Production rate	65 parts per hour

(a) Total axial shortening of the workpieces during welding.

Fig. 2. Rotor blank made by friction welding three pieces instead of by machining from solid bar, to reduce work-metal and machining costs (Example 451a)

friction welding resulted in a 20% material saving and a 39% reduction in machining costs compared with the original method, which was machining from a solid bar.

Time for making the two welds, one at a time, on each rotor blank was 45 sec. The welding machine was manually loaded and unloaded and was equipped with manually actuated chucks for gripping the workpieces. Conditions for friction welding are given in the table that accompanies Fig. 2.

Example 451b — Valve Stem (Fig. 3). The use of friction welding reduced material and machining costs in the manufacturing of the bronze valve stem shown in Fig. 3. Originally, the valve stem was machined complete from a ⅝-in.-diam bar. By friction welding two pieces together and producing a weld upset large enough to provide material from which the flange could be machined to the required final dimensions (see Fig. 3), machining time was reduced 95% and material loss was reduced 98%. The friction weld was made in 7 sec, using a spindle speed of 2200 rpm, a heating pressure of 14,000 psi (about 2100 lb) and a welding pressure of 65,000 psi (about 9750 lb). Total axial shortening of the workpieces during upsetting was 0.150 in. Production rate was 180 valve stems per hour.

Weld Strength. For most metals, the strength of a friction welded joint is about the same as that of the base metal. The metal at the weld interface is hot worked, which refines the grain structure. During the final portion of the weld cycle, upsetting and extrusion of flash ensures removal of oxidized metal that may have been produced during heating. This flash usually appears in the valley at the intersection of the two weld upsets (see Fig. 1).

The relatively large unheated areas adjacent to the joint extract heat quickly from the small mass of the heat-affected zone, thus keeping the zone small in the welded part.

In the following example, the mechanical properties of joints made by friction welding were superior to those of joints made by flash welding.

Example 452. Comparison of Properties of Joints Made by Flash Welding and by Friction Welding (Table 1)

A 1026 steel shaft 0.750 in. in diameter was joined to the stub shaft of a worm gear made of 5120 steel by conventional friction welding and by flash welding. Tension tests, rotating-beam fatigue tests and reversed-torsion fatigue tests were made on the weldments. The results of these tests are given in Table 1.

Examination of the microstructures of the welded joints showed a much narrower heat-affected zone and a finer grain structure in the friction welded joint than in the flash welded joint. The friction welded part had a higher tensile strength than the flash welded part.

For flash welding, the shaft had a sheared end and the stub shaft on the worm gear had a specially tapered section that reduced the flashing contact diameter from 0.750 in. to 0.375 in. This tapered section was flashed off during welding. For friction welding, the end of the shaft was used as-sheared but the end of the stub shaft on the worm gear was machined flat. Thus, the need for tapering the end was eliminated by changing from flash welding to friction welding.

In both methods, the worm gear was carburized and hardened before welding, but the stub shaft was flash copper plated,

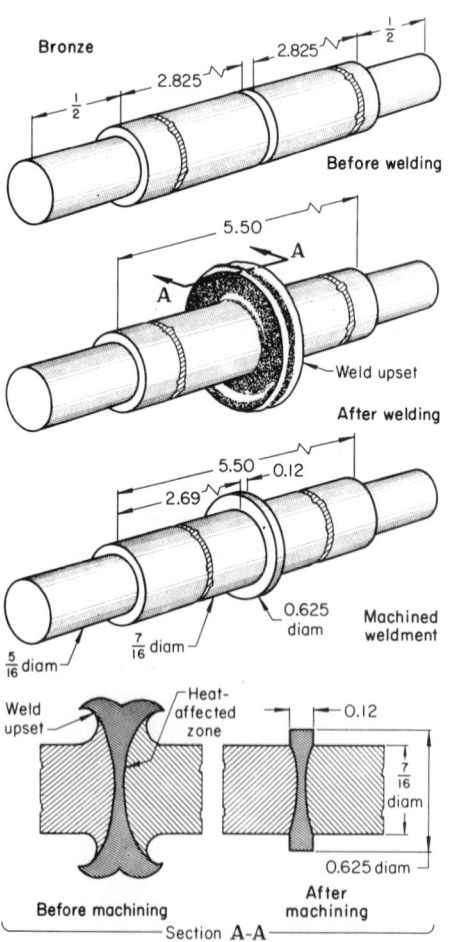

Fig. 3. Friction welded bronze valve stem (Example 451b)

Work-metal and machining costs were reduced substantially when friction welding replaced machining from bar stock. In the improved method, the center flange was machined from the upset produced by friction welding.

Table 1. Comparison of Properties of Flash and Friction Welded Joints Made Between 1026 and 5120 Steel Bars (Example 452)

Test or property	Flash welding	Friction welding
Tensile strength	70,000 psi	75,000 psi
Bending fatigue time, minutes	32	180
Torsional fatigue time, cycles	1,000,000(a)	4,000,000(b)
Metal lost, in.(c)	0.75	0.25

(a) Cycles to failure. (b) Cycles without failure. (c) Axial shortening of bars during welding.

or covered with a tight-fitting cup, to prevent it from being carburized. Thus, the weld areas and the heat-affected zones produced by both welding methods had essentially the same hardness. No heat treatment was needed after welding, because both steels had low carbon contents (although one was a low-alloy steel); the as-welded hardness was acceptable in both the shallow heat-affected zone made by friction welding and in the wider zone made by flash welding.

Line-voltage variations frequently caused a severe drop in the flow of current to the flash welding machine, which resulted in a cold weld. Line-voltage regulators were needed to stabilize the current and shut off the power when the voltage dropped below a predetermined level. In friction welding, line-voltage fluctuations had little effect on the amount of frictional energy produced and voltage regulators were not needed.

The scrap rate for flash welding ranged from 3 to 6%, which was attributed to equipment maintenance, line-voltage variations, and defective components. In friction welding, scrap rate was less than 1%.

Because there is no spatter of weld flash during friction welding, the time required to clean the machines was reduced from 18 hours to less than 5 hours per week. The increased productivity that resulted from less downtime and higher production rate made it possible to use two friction welding instead of five flash welding machines.

Sections Welded. In friction welding, the joint face of at least one of the workpieces must be essentially round. The rotating workpiece should be somewhat concentric in shape because it revolves at a relatively high speed. Workpieces that are not round, such as hexagon-shape workpieces, have been friction welded successfully, but the resulting weld upset is rough, asymmetrical, and difficult to remove without damaging the welded assembly. For a few special applications, welding machines have been modified so that the spindle stops at the same place each time, thus making it possible for workpieces to be oriented to each other.

Solid bars of 1018 steel from ¼ to 4 in. in diameter can be friction welded in the available welding machines. Welding of larger diameters, although feasible, is limited by machine cost.

Wire and tubing of like and unlike metals 1 to 2.5 mm in diameter have been friction welded in special machines to plates 0.2 to 2 mm thick. Wires of unlike metals 1.5 to 2 mm in diameter have been joined.

Tubular sections can be much larger in diameter than the rated capacity of the welding machine for solid bars, and the maximum weldable tube diameter depends primarily on wall thickness. For example, a machine capable of welding a 4-in.-diam 1018 steel bar can weld a 1018 steel tube 30 in. in diameter with a 3/16-in.-thick wall. The maximum diameter decreases to about 7½ in. when the wall thickness is 1 in.

The size of section that can be friction welded depends somewhat on the distance the plastic metal must travel to be extruded from the weld interface. Metal in solid bars must travel outward from the center of the bar; metal in tubes can travel both inward and outward from the center of the wall.

Metals Welded

Friction welding can be used to join almost any metal that can be forged and that is not a good dry-bearing metal. The alloying elements that provide dry lubrication (or do not seize under normal operating conditions when without grease or other lubricants) prevent the interfaces from being heated to welding temperature by friction. Metals that contain free-machining additives are likely to be hot short and are generally unsatisfactory for welding.

Many similar and dissimilar metal combinations can be friction welded, and in most combinations a sound metallurgical bond is formed. In some combinations, the bond is not as strong as the base metal, and postweld heat treatment may be needed to develop full weld-zone strength in alloy steels and hardenable stainless steels.

Carbon and alloy steels are relatively easy to friction weld. Low-carbon and medium-carbon steels can be welded under a wide range of welding conditions. High-carbon and alloy steels are easily joined but the welding conditions must be controlled within narrower ranges than are permissible for welding low-carbon steels, and the axial pressure must be increased to compensate for lower forgeability.

High speed tool steel can be welded to carbon and alloy steel shanks for making drills, reamers and other cutting tools. Steel balls made of 52100 steel, which is normally difficult to weld, are welded to one or both ends of carbon steel rods to make linkage rods. Frequently the rods are made of 1045 steel and one end is induction hardened before the 52100 steel ball is welded to the opposite end. The weldments are tempered after welding.

Free-machining steels, except those having a high sulfur and low manganese content, can be welded, but the free-machining elements result in undesirable directional properties in the weld zone. Friction welds in free-machining steels have fatigue strength less than 80% that of the base metal and should not be used in applications where high stresses are involved and high fatigue strength is required. Friction welding of free-machining steels should be limited to those with 0.08 to 0.13% sulfur, lead or tellurium. For example, 1141 steel is welded satisfactorily, but 1144 is not.

Heat treated steels can be friction welded with only localized changes in hardness because the heating is confined to a very narrow zone. Also, the rapid quenching restores hardness to the weld zone. For instance, as shown in Fig. 4, when 8630 steel bars were hardened to Rockwell C 35 or C 50 and then were friction welded, the minimum hardnesses (Rockwell C 33 for the Rockwell C 50 steel; Rockwell C 31 for the Rockwell C 35 steel) occurred at

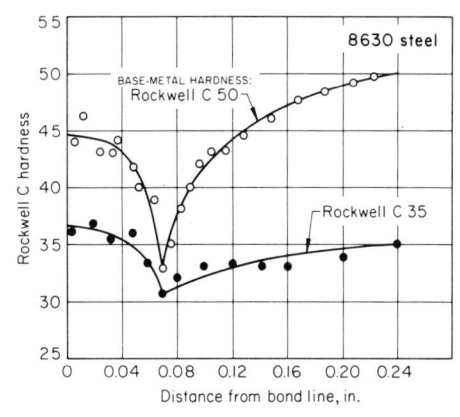

Bars for upper curve postweld tempered at 400 F; those for lower curve, at 1000 F.

Fig. 4. Relation of metal hardness in the heat-affected zone to distance from the bond line after friction welding 8630 steel bars having hardnesses of Rockwell C 35 and C 50

about 0.065 in. from the bond line; the heat-affected zone extended about 0.24 in. into the base metal. In welding hardenable steel, the weld upset usually reaches a high hardness because of rapid quenching. Therefore, the weld upset usually must be removed by grinding, or by machining after annealing. In some applications, the upset is removed by hot shearing before it has cooled from the welding temperature to the M_s temperature.

Sintered metals are further compacted during friction welding and have a wrought structure in the weld zone. The joint usually is stronger than the base metal.

The results of friction welding steel forgings and steel castings are about the same as those of friction welding steel bars of similar composition.

Stainless steels are comparatively easy to friction weld, and good weld properties can be obtained under a wide range of welding conditions. The heat treatable stainless steels are sensitive to heat and pressure and, for good weld properties, require postweld heat treatment in the heat-affected zone. An important application of friction welding of stainless steel is the production of bimetal shafts that are exposed to a corrosive atmosphere or to wear in service. To provide the correct type of resistance where needed, to reduce work-metal cost, or to increase machinability, an alloy that will withstand a corrosive atmosphere can be joined to a metal that is less expensive and easier to machine, as in the example that follows.

Example 453. Comparison of Costs for Making a Pump Shaft by Machining From One Piece of Stainless Steel and by Machining an Inertia Weldment (Fig. 5)

The cost of making the pump shaft shown in Fig. 5 from one piece of stainless steel (0.05 C, 0.75 Mn, 1.0 Si, 20 Cr, 29 Ni, 2.2 Mo, 3.2 Cu, rem Fe) was compared with the cost of making the shaft by inertia welding a bar made of the same stainless steel to a bar made of 1018 steel. In making the bimetal shaft, most of the machining was done on the 1018 steel, which was less expensive and easier to machine than the stainless steel.

When the shaft was made from the two metals instead of from stainless steel only, material cost was reduced by 45% and machining cost was reduced by 10%. Metallographic examination of the inertia welded joint showed good fusion, and when the shafts were field tested in pumps running at 3500 rpm, no failures were reported.

Before welding, the ends of both bars were ground to remove any mill scale. After welding, weld upset was removed as the shaft was machined to size. In order to produce good welds, leaded low-carbon steel was not used. The conditions for inertia welding and the production rates are given in the table that accompanies Fig. 5.

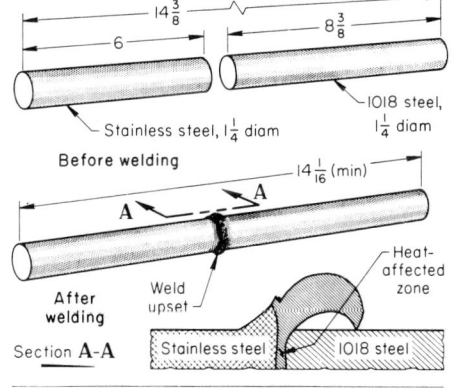

Cast iron in any form — gray, ductile or malleable — has not been friction welded satisfactorily in production. (Joining of ductile iron to steel in laboratories has been reported.) Free graphite gathers at the interface and acts as a lubricant, which limits friction heating. Also, these materials are not forgeable, which is a general requirement for friction welding.

Nonferrous Metals and Alloys. Aluminum alloys are friction welded to similar and dissimilar aluminum alloys, copper alloys to similar and dissimilar copper alloys, and aluminum alloys to copper alloys. Most applications of friction welding these metals are in joining aluminum and copper alloys to steel, although problems are presented by high thermal conductivity, large differences in forging temperatures, and the formation of brittle intermetallic compounds.

Joints between aluminum alloy 6061 and copper have a tensile strength near that of the copper. Joints between aluminum alloy 1100 and stainless steel have a strength near that of the aluminum alloy. Friction welding of other aluminum alloys may develop a joint strength of only 60 to 70% that of the weaker base metal. Even though these

Conditions for Inertia Welding

```
Flywheel moment of inertia ............50 lb-ft²
Spindle speed ........................3150 rpm
Weld energy(a) ...................84,000 ft-lb
Axial force ..........................60,000 lb
Heat-and-weld time ................2 to 4 sec
Weld area ...........................1.23 sq in.
Metal lost, max(b) ....................5/16 in.
Production rate:
  Manual ...................120 welds per hour
  Automatic ...............360 welds per hour
```

(a) Calculated from flywheel size (moment of inertia) and spindle speed. (b) Total axial shortening of the workpieces during welding.

Fig. 5. Pump shaft that was made by machining an inertia weldment of stainless steel and 1018 steel, instead of by machining from stainless steel only, to reduce costs (Example 453)

Source: *Metals Handbook*, 8th Ed., Vol. 6, ASM, 1971

joints are relatively weak, they are useful for pressure sealing and for joining assemblies that require good electrical and thermal conductivity, rather than high strength.

Titanium, titanium alloys, zirconium alloys and magnesium alloys can be friction welded to themselves.

Most nickel-base and cobalt-base alloys, including the heat-resisting alloys, are easily welded to themselves and to alloy steels. The nickel-base alloy GMR-235 can be welded to 1040 steel, Inconel 718 to Inconel 713C, and Inconel 713C to 8630 steel in producing jet-engine parts that require high-strength bonds.

The refractory metals — tungsten, molybdenum, columbium and tantalum — can be welded to themselves. Friction welds between molybdenum rods are ductile enough to withstand substantial reduction by wire drawing.

Conventional Friction Welding

Conventional friction welding requires a machine resembling an engine lathe equipped with an efficient spindle-braking system, a means of applying and controlling axial pressure and a weld-cycle timer and control. The equipment is simple in principle, but the machines are complex when big enough to weld large workpieces.

Principle of Operation. The workpiece to be rotated is clamped in the spindle chuck, and the spindle is brought to a predetermined speed. The nonrotating component is clamped in a chuck or fixture mounted to a hydraulically actuated tailstock slide. To heat the workpieces to welding temperature, the tailstock slide is advanced to bring the workpieces in contact under a constant or gradually increasing axial pressure. When the workpieces are at, or slightly above, the welding temperature, the spindle brake is applied, which suddenly stops the spindle rotation. Simultaneously, the tailstock pressure is increased to complete the weld.

The spindle speed, axial pressure and length of time the pressure is applied for a given weldment depend on (a) the cross-sectional area of the workpieces to be welded; (b) the melting point and thermal conductivity of the work metal; and (c) the metallurgical changes that occur during the heating cycle, particularly when dissimilar metals are being welded.

Process variables to be controlled are rotational speed, initial (heating) pressure, length of time that heating pressure is applied, and welding pressure. The time needed to stop the spindle can affect the temperature to which the workpieces are heated, timing of application of the welding pressure, and weld properties.

Rotational speed, or peripheral velocity, is the least sensitive process variable and can be varied over a wide range if heating time and pressure are properly adjusted. However, heating time must be limited to prevent excessive depth of heating. The peripheral velocity recommended for welding most low-carbon, medium-carbon, and high-carbon steels is 400 to 1400 sfm.

Heating pressures used for welding low-carbon and low-alloy steels are from 3000 to 20,000 psi. Welding pressures for these steels are from 5000 to 25,000 psi. Usually, the welding pressure is higher than the heating pressure, but sometimes they are nearly the same.

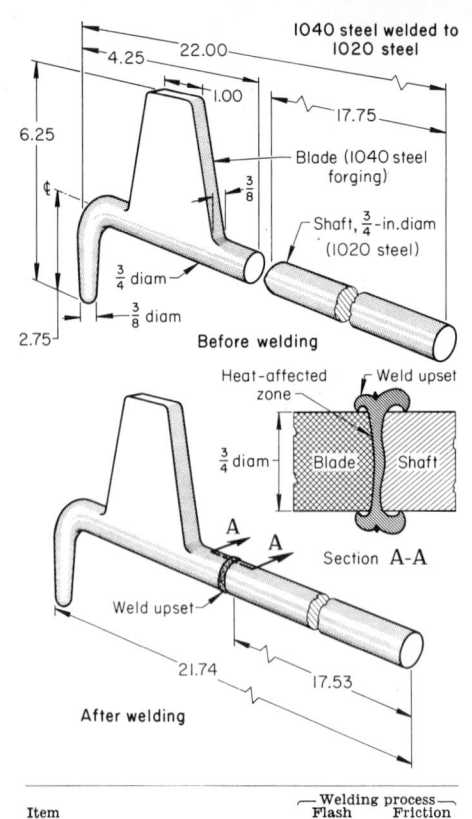

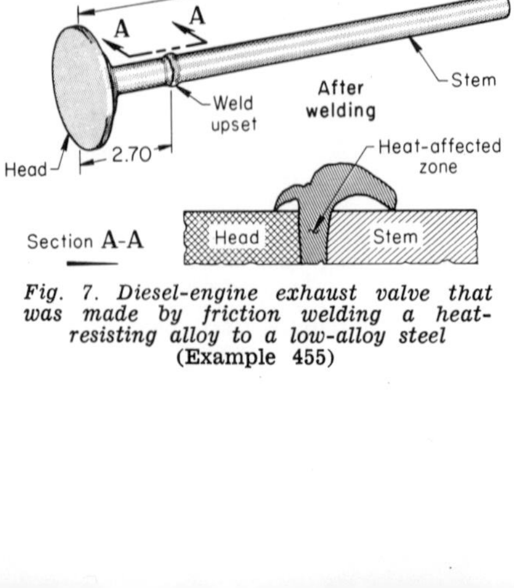

Fig. 6. Shutdown-control shaft that was made by joining a steel forging to a steel shaft at lower cost by friction welding than by flash welding (Example 454)

Item	Welding process	
	Flash	Friction
Cost Comparison		
Production, parts per hour	240	300
Steel savings(a)	...	$5500
Scrap loss(b)	6 to 7%	½ to 1%
Forging scrap loss	3.5%	1.5%
Fixture maintenance(c)	$1830	$1050

(a) For each 100,000 units. (b) From misalignment, burns and other causes. (c) Cost per year.

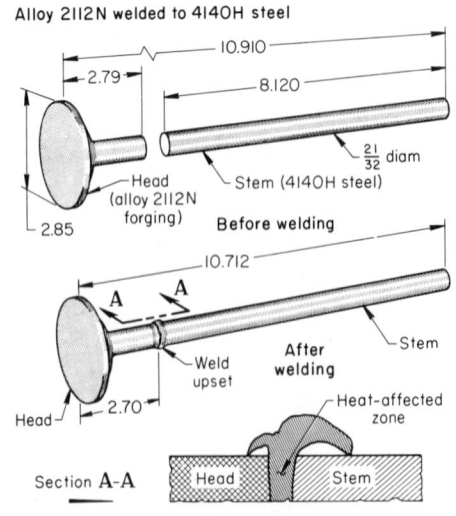

Fig. 7. Diesel-engine exhaust valve that was made by friction welding a heat-resisting alloy to a low-alloy steel (Example 455)

For medium-carbon and high-carbon steels, heating pressures are from 10,000 to 30,000 psi, and welding pressures are from 15,000 to 60,000 psi. Lower heating pressures are sometimes used for large workpieces so that the power requirements do not exceed the capacity of the welding machine.

Heating time varies with the heating pressure, the carbon and alloy content of the steel, and the diameter of the workpiece. Usually, heating time is determined by trial.

The spindle should be stopped rapidly to keep the weld from twisting or tearing. For a workpiece less than ½ in. in diameter, stopping time should be within 1½ sec; a 3-in.-diam bar should be stopped within 3 sec.

Examples of Practice. Only minor alterations in design are needed to adapt to friction welding a workpiece that previously was butt welded by other processes. Generally, in friction welding less metal is lost during heating and upsetting than in flash welding, and it is not necessary, as it is in flash welding, to machine the interface so that heating will start at the center of the workpiece section. Where allowances are made in the size of a forging for differences in metal lost and for welding in the as-forged condition, forging costs can be reduced, as in the following example.

Example 454. Comparison of Friction Welding and Flash Welding of Shutdown-Control Shafts (Fig. 6)

The shutdown-control shaft for a textile loom was made by butt welding a ¾-in.-diam 1020 steel shaft to a forged 1040 steel blade, as shown in Fig. 6. In service, these shafts received severe impact loading and thus required excellent weld quality. Satisfactory joints were made by both flash welding and conventional friction welding, but scrap loss was lower with friction welding than with flash welding because of better accuracy of alignment.

For friction welding, the stub shaft on the forging was ¾ in. shorter than that needed for flash welding because no metal was lost during flashing and the forging was not machined before welding. The over-all length of the control shaft was not held to close tolerances, and therefore, the stub shaft on the blade was friction welded in the as-forged condition. The stub shaft was machined before flash welding so that flashing would start at the center. The shorter forging used for friction welding was easier to produce and required less material. The 1020 steel shaft was used as-sheared for both processes.

The only machining required on the friction welded shaft was removal of the weld upset by grinding. This finished surface served as a bearing journal for the blade end of the shaft.

A comparison of costs for flash and friction welding of the shafts is given in the table that accompanies Fig. 6.

Workpieces made of dissimilar metals are frequently used in order to minimize costs while still providing, where needed, a work metal that meets the necessary service requirements. Many corrosion-resisting and heat-resisting alloys are welded to less expensive alloys to reduce material and machining costs, or to provide a wear-resistant surface. This was done in Example 453 (in which a high-alloy stainless steel was friction welded to 1018 steel), and in the exhaust-valve application described in the example that follows.

Example 455. Production of Diesel-Engine Exhaust Valves by Friction Welding Dissimilar Alloys (Fig. 7)

A large exhaust valve for a diesel engine, shown in Fig. 7, was made by conventional friction welding a head made of a heat-resisting alloy to a low-alloy steel stem. This permitted using the more expensive (heat-resisting) alloy only where required.

The head was forged from alloy 2112N (a 21Cr-12Ni austenitic iron-base alloy used for exhaust valves), and the stem was made of 4140H steel 21/32 in. in diameter. The initial weld area was 0.34 sq in. Overall length of the part after joining by friction welding was 10.712 in.

A 25-hp friction welding machine was used, although the peak requirement was only 5 hp.

In operation, the valve head was clamped in an air-actuated fixture mounted in the tailstock, and the stem in an air-actuated chuck attached to the machine spindle.

The machine cycle was started, and the spindle was rapidly accelerated to 2700 rpm as the tailstock advanced to bring the workpieces into contact. The pressure of contact initiated a signal in the control circuit that began cycle timing and started a gradual increase in axial pressure by a slope-control unit.

The pressure, applied through the tailstock by two hydraulic cylinders, was smoothly increased to 11,000 psi (about 3740 lb) in 4 sec. The pressure level was maintained for an additional 4 sec, during which time the workpieces were heated to welding temperature.

At the end of the 4-sec heating period, the braking system quickly stopped spindle rotation, and a welding pressure of 32,000 psi (about 10,900 lb) was applied. The resulting weld had a good metallurgical bond that was free of oxides, cast metal structure, porosity, inclusions and other defects. Total axial shortening of the workpieces during welding was 0.195 to 0.200 in.

The weld upset was removed in an automatic lathe prior to heat treating of the valve. After heat treatment, the stem was ground to finished diameter.

The production rate, with an unskilled operator, was 190 to 200 pieces per hour. The total input energy per weld was 6900 watt-sec with a peak power draw of 5 hp, including all losses. The cost per weld, including removal of the weld upset, was less than $0.05. The welding machine cost $20,000, and the semiautomatic loading and unloading equipment cost $3000 (1967).

Inertia Welding

Inertia welding makes use of the kinetic energy of a freely rotating flywheel for all of the heating required to produce a weld.

Welding Machine. The machine is constructed with a horizontal bed and overhead tie bars to contain the axial-pressure and torque reactions and to ensure accurate spindle-to-bed alignment. The spindle is driven by a hydrostatic motor through a change-gear transmission. A hydraulically actuated tailstock retracts on adjustable ways for loading and unloading. A self-centering vise can be attached to the tailstock for clamping cylindrical parts, and fixtures are used for holding asymmetrical parts for which the vise is not suitable. The spindle has a means for mounting a collet chuck, and a draw bar is used for opening and closing the collet.

Flywheel size (moment of inertia of the flywheel or spindle) is adjusted by adding or removing flywheel disks. The spindle speed and axial pressure are adjusted by dials on the control panel.

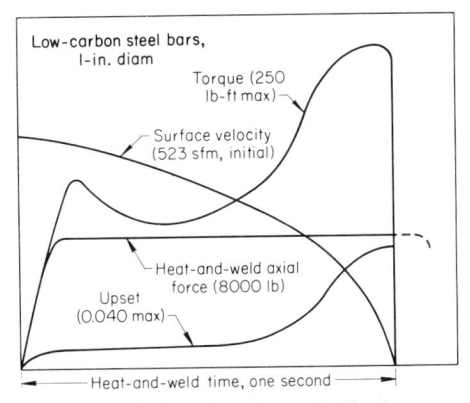

Fig. 8: Variation of surface velocity, torque, axial force and weld upset in a one-second heat-and-weld time in inertia welding

Principle of Operation. In inertia welding, as in conventional friction welding, one workpiece is clamped in a nonrotating vise or fixture and the other workpiece is clamped in a chuck mounted to a rotating spindle.

The drive motor accelerates the rotation of the flywheel-spindle assembly to a predetermined speed, and then the rotating drive power is shut off. The surfaces to be welded are brought together and the kinetic energy of the freely rotating flywheel is rapidly converted to heat at the weld interface as axial pressure is applied. Once the axial pressure, flywheel moment of inertia, and spindle speed have been established for a given workpiece, uniform welds are produced repetitively.

Two characteristics of inertia welding — continuously decreasing surface velocity of the workpiece and continuously changing torque at the weld interface — are illustrated in Fig. 8. Surface velocity begins at some initial value and decreases along an essentially parabolic curve to zero, at which time the weld is completed. Heating and welding time is usually 0.2 to 4 sec. Torque has a peak value of short duration early in the cycle, gradually decreases, and then increases until the velocity has decreased to the value at which welding begins, at which time torque rises sharply. This high torque is accompanied by forging in the weld zone, and is responsible for much of the upsetting. The high-torque phase, present only in inertia welding, refines the grain structure and expels any oxides at the weld interface. The gradually decreasing and increasing part of the torque curve is essential to the formation of good welds. The second (low-torque) phase generally will not develop if initial velocity is too low. Figure 8 is typical for inertia welding of 1-in.-diam low-carbon steel bars.

Other differences between inertia welding and conventional friction welding are input power at the weld interface and heating time. The power needed for the weld itself is not of concern in inertia welding, because whatever power is required can always be supplied by deceleration of the flywheel at the required rate. In conventional friction welding, power is limited by the size of the drive motor.

The high power used in inertia welding is a result of a relatively high, rapidly applied axial pressure. Power demands in conventional friction welding are controlled and limited to motor capacity by applying the axial pressure slowly; usually 2 to 4 seconds elapse before the full pressure is applied. The lower heating rates of conventional friction welding require more energy because much of the heat is conducted away from the weld interface. By rapid application of small amounts of energy, inertia welding produces narrower heat-affected zones than those produced in conventional friction welding.

In inertia welding, intense hot working of the weld zone in conjunction with rapid cooling immediately after hot working results in a very small grain size in the as-welded condition. Subsequent heat treatment will restore the grains to their normal size.

Process Variables. Three variables control the characteristics of an inertia weld: initial peripheral velocity of the rotating workpiece, axial pressure, and flywheel size (moment of inertia).

Table 2. Conditions for Inertia Welding 1-In.-Diam Bars in Combinations of Similar and Dissimilar Metals

Work metal	Spindle speed, rpm	Axial force, lb	Flywheel size, lb-ft²(a)	Weld energy, ft-lb	Metal lost, in.(b)	Total time, sec(c)
Metals Welded to Themselves						
1018 steel	4600	12,000	6.7	24,000	0.10	2.0
1045 steel	4600	14,000	7.8	28,000	0.10	2.0
4140 steel	4600	15,000	8.3	30,000	0.10	2.0
Inconel 718	1500	50,000	130.0	50,000	0.15	3.0
Maraging steel	3000	20,000	20.0	30,000	0.10	2.5
Type 410 stainless steel	3000	18,000	20.0	30,000	0.10	2.5
Type 302 stainless steel	3500	18,000	14.0	30,000	0.10	2.5
Copper, commercially pure	8000	5,000	1.0	10,000	0.15	0.5
Copper alloy 260 (cartridge brass, 70%)	7000	5,000	1.2	10,000	0.15	0.7
Titanium alloy Ti-6Al-4V	6000	8,000	1.7	16,000	0.10	2.0
Aluminum alloy 1100	5700	6,000	2.7	15,000	0.15	1.0
Aluminum alloy 6061	5700	7,000	3.0	17,000	0.15	1.0
Dissimilar-Metal Combinations						
Copper to 1018 steel	8000	5,000	1.4	15,000	0.15	1.0
M2 tool steel to 1045 steel	3000	40,000	27.0	40,000	0.10	3.0
Nickel alloy 718 to 1045 steel	1500	40,000	130.0	50,000	0.15	2.5
Type 302 stainless to 1020 steel	3000	18,000	20.0	30,000	0.10	2.5
Sintered high-carbon steel to 1018	4600	12,000	8.3	30,000	0.10	2.5
Aluminum 6061 to type 302 stainless	5500	5000 & 15,000(d)	3.9	20,000	0.20	3.0
Copper to aluminum alloy 1100	2000	7,500	11.0	7,500	0.20	1.0

(a) Moment of inertia of the flywheel. (b) Total axial shortening of workpieces during welding. (c) Includes heat time and weld time. (d) The 5000-lb force is applied during the heating stage of the weld; force is increased to 15,000 lb near the end of the weld.

Source: *Metals Handbook*, 8th Ed., Vol. 6, ASM, 1971

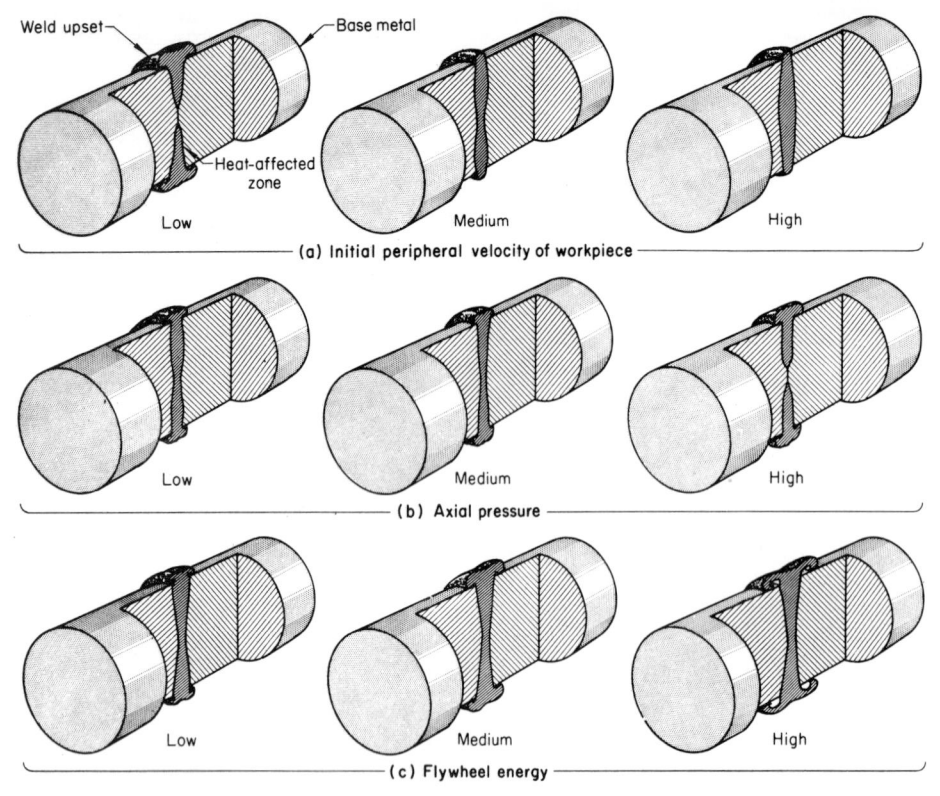

Fig. 9. Effect of low, medium and high levels of the three welding variables on depth and uniformity of heating, and on size and shape of the weld upset, in inertia welding of steel

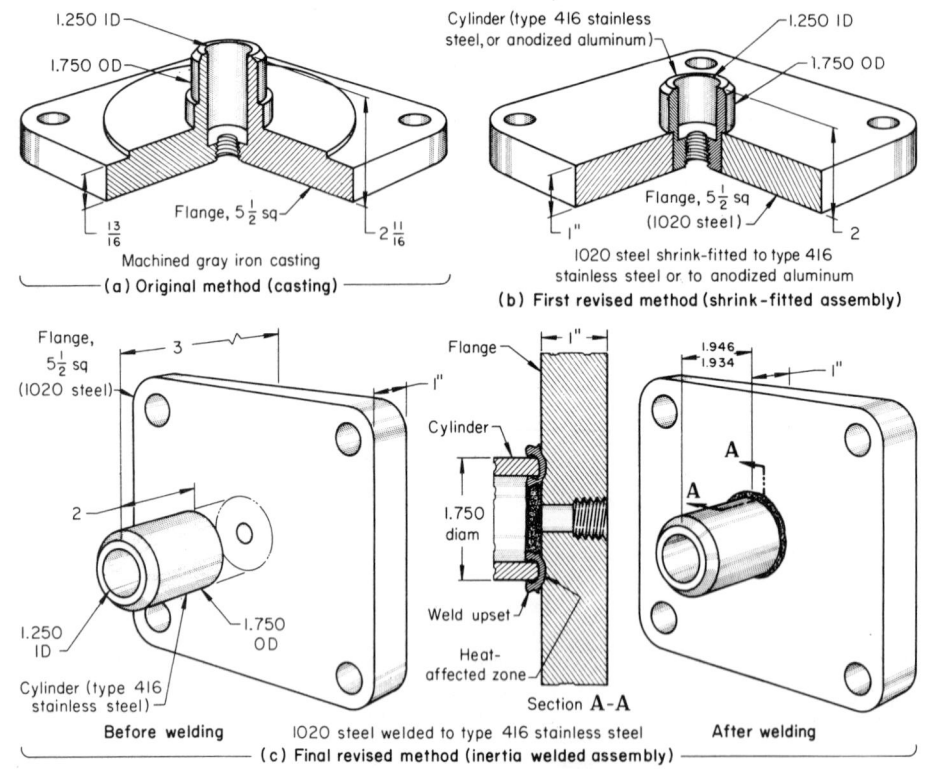

Conditions for Inertia Welding

Machine capacity:
  Part diameter .... 0.43 to 1.50 in.
  Spindle speed .... 8000 rpm max
  Axial force .... 45,000 lb max
  Flywheel moment of inertia ..... 20 lb-ft²

Spindle speed .... 3900 rpm
Weld energy(a) .... 51,800 ft-lb
Axial force .... 35,000 lb
Weld area .... 1.18 sq in.
Metal lost(b) .... 0.060 ± 0.006 in.

(a) Calculated from flywheel size (moment of inertia) and spindle speed. (b) Total axial shortening of the workpieces during welding.

Fig. 10. Suction-valve cover that was produced by three different processes, using the designs shown. Inertia welding produced the highest-quality part at lowest cost. (Example 456)

For each weld, the minimum energy input required is provided by using the proper combination of flywheel size and spindle speed. Additional energy may be needed if the surfaces to be welded are rough or out of square with the axis of rotation. Very high energy input causes excessive loss of metal, but generally does not affect the strength or quality of the weld.

Table 2 gives representative conditions for inertia welding of 1-in.-diam bars of various metals and alloys, in similar and dissimilar combinations.

Peripheral Velocity of Workpiece. For each combination of work metals, there is a range of peripheral velocity that produces the best weld properties. For welding steel to steel, the recommended initial peripheral velocity of the workpiece ranges from 500 to 1500 sfm; however, welds can be made at velocities as low as 275 sfm. As illustrated in Fig. 9(a), low velocities (less than 300 sfm) can reduce center heating and produce rough, uneven weld upset. At medium velocities (300 to 900 sfm), the heating pattern in steel has an hourglass shape at the lower value and gradually flattens as the upper velocity is approached. The heating pattern is essentially flat and uniformly thick across the workpiece at velocities of 900 to 1200 sfm. At high initial velocities (above 1200 sfm), the weld becomes rounded and is thicker at the center than at the periphery.

Spindle speeds in revolutions per minute for inertia welding 1-in.-diam bars of various metals and alloys are given in Table 2. (For speed conversions — sfm to rpm for many diameters — see the inside back cover of Volume 3 of this Handbook.) The relationship of total time and weld upset to bar diameter is approximately linear.

Axial Pressure. The effect of varying the axial pressure is similar but opposite to the effect of varying the velocity. As Fig. 9(b) shows, welds made at low axial pressure resemble welds made at medium velocity, in regard to formation of weld upset and heat-affected zones. Use of excessive pressure produces a weld that is poor at the center and has a large amount of weld upset, similar to a weld made at a low velocity.

The axial force (in pounds) for welding 1-in.-diam bars is given in Table 2. Axial pressure (in psi) varies as a function of the square root of workpiece diameter. For instance, a 2-in.-diam bar uses 1.414 times the axial pressure needed for a 1-in.-diam bar.

Effect of Flywheel Energy. The flywheel moment of inertia is selected to produce the desired amount of kinetic energy and the desired amount of forging. Forging results from the characteristic increase in torque (see Fig. 8) that occurs at the weld interface as the flywheel slows and comes to rest. This increased torque, in combination with the axial pressure, produces forging as depicted by the upset curve in Fig. 8. Because forging begins at some critical velocity (about 200 sfm for low-carbon steel), the amount of forging depends on the amount of energy remaining in the flywheel, which is a linear function of the flywheel moment of inertia. Large, low-speed flywheels produce greater forging than small, high-speed flywheels even though they contain the same amount of kinetic energy. Although low, medium and high amounts of flywheel energy produce similar heating patterns, the amount of energy greatly affects the size and shape of the weld upset, as shown in Fig. 9(c).

Examples of Practice. Inertia welding is used in the manufacture of bimetal exhaust valves for internal-combustion engines, bimetal shafts for pumps, and cluster and ring gears for automotive and aircraft applications. The process has been incorporated into the redesign of many parts in order to reduce costs or improve service life, as in the following two examples. Other applications of inertia welding are described in the section on Friction Welding vs Other Processes, beginning on page 516.

In the example that follows, to produce corrosion-resistant parts at low cost, inertia welding was selected over casting or shrink-fit assembly.

Example 456. Use of Inertia Welding Instead of Casting or Shrink-Fit Assembly (Fig. 10)

A high-production part, consisting essentially of a 1¼-in.-ID cylinder with a bolting flange, served as a suction-valve cover. The manufacturing process and design of the part evolved through three successive stages as shown in Fig. 10. The first stage was a one-piece casting (Fig. 10a); the second, a shrink-fitted assembly (Fig. 10b); and the third, the inertia welded assembly shown in Fig. 10(c). The slight differences among the three designs are the results of changes in the design of the mating piece. The inertia welded component was used as-welded, without removing weld upset from either the inner or the outer surface of the cylinder.

Originally, the part was machined from a gray iron casting as shown in Fig. 10(a). Some difficulty was experienced in obtaining the desired finish on the inner surface of the cylinder, but the most serious problem was rusting of the inside of the cylinder. Attempts were made to improve corrosion resistance by burnishing the cylinder wall, and then plating with cadmium, zinc or tin, or phosphate coating. Although phosphate coatings were best on the cast iron surface, none of these coatings proved satisfactory in service.

The second method consisted of using for the flange a low-carbon steel plate that was bored to accept a shrink-fitted cylinder (Fig. 10b). Cylinders were made of anodized aluminum alloy or of stainless steel. When the anodized aluminum alloy was used, the cost per part was slightly higher than that of the gray iron casting. Cylinders made of type 416 stainless steel gave excellent results, but cost considerably more than the castings.

In the final design, shown in Fig. 10(c), inertia welding was used to join a type 416 stainless steel tube to a 1020 steel plate. Valve covers made by this method cost 15% less than the original cast iron covers and the corrosion and surface-finish problems were eliminated.

For inertia welding, flanges were gas cut from 1020 steel plate, and the cylinders were saw cut from type 416 stainless steel tubing. Both components were finish machined and the joint surfaces were cleaned carefully before welding.

Conditions for friction welding are given in the table with Fig. 10. With these settings, an upset of 0.060 ± 0.006 in. was a good indication of an acceptable weld. Part specifications required that a 100-psi pressure test be applied to 5% of the weldments, and that 1% of the weldments be examined for weld configuration by sectioning. The test samples were selected on a random basis. Under these conditions, the rejection rate was less than 0.3%, and most rejections were caused by defects in the base metal rather than in the weld.

The manufacture of a long, slender shaft with a spline on one end is described in the example that follows. Use of inertia welding permitted joining a ground and polished steel shaft to another steel part without damaging the polished surface. Changing from forging to inertia welding reduced the over-all cost and eliminated nine manufacturing operations.

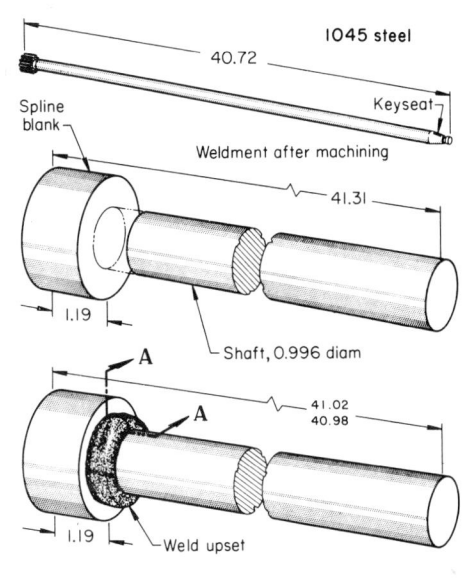

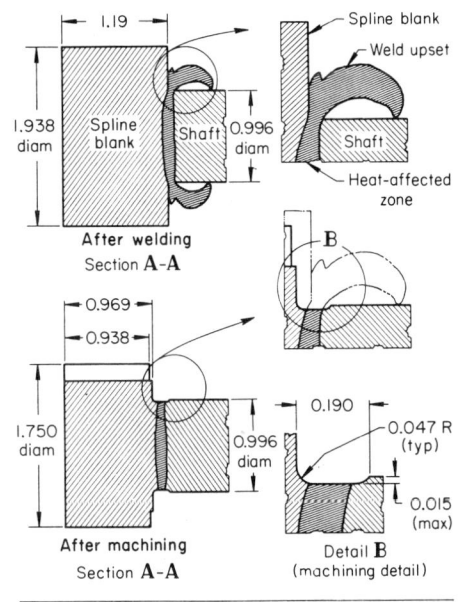

Conditions for Inertia Welding(a)

Flywheel moment of inertia .... 150 lb-ft²
Spindle speed .... 1850 rpm
Weld energy(b) .... 87,500 ft-lb
Axial force .... 21,000 lb
Weld area .... 0.78 sq in.
Metal lost(c) .... 0.31 ± 0.02 in.
Production rate .... 70 welds per hour

(a) In this application, lower-than-normal spindle speeds and larger-than-normal flywheels were used for convenience in changeover from one workpiece to another. (b) Calculated from flywheel size (moment of inertia) and spindle speed. (c) Total axial shortening of the workpieces during welding.

Fig. 11. Drive shaft for which change from forging to inertia welding reduced costs by $12.17 (Example 457)

Example 457. Change From Forging to Inertia Welding To Reduce the Cost of Producing a Long, Slender Shaft (Fig. 11)

The long, slender power-control drive shaft shown in Fig. 11 originally was made from a 1045 steel forging, but was severely warped during forging and heat treating. Nineteen manufacturing operations were needed to complete the forged shaft, including hardening and tempering of the forging to Rockwell C 23 to 30, cleaning by shot blasting, three straightening operations and ten machining operations.

The number of manufacturing operations was reduced to ten when forging was replaced by inertia welding. A spline blank of 1045 steel 1.938 in. in diameter by 1.19 in. long was inertia welded to a 1045 steel shaft 40.12 in. long (see Fig. 11). The shaft had a hardness of Rockwell C 27 and, before welding, was ground and polished to 0.996 in. diameter. The jaws for holding the shaft were designed so as not to damage the ground and polished surface.

Hardening, tempering, shot blasting, straightening and three grinding operations were eliminated, and the cost of making each shaft was reduced by $12.17, when forging was replaced by inertia welding. Three operations — hardening a bearing surface, deburring, and washing — were included in both manufacturing sequences.

Machining of the spline end of the weldment included removing the weld upset and relieving the area to a minimum diameter of 0.966 in. and to a width of 0.19 in., which eliminated a possible stress raiser. A taper 1.375 in. long, a seat for a Woodruff key, and ⅝-11 threads were machined on the other end of the shaft.

Conditions for inertia welding are given in the table with Fig. 11. These conditions were selected to produce a weld upset of 0.31 ± 0.02 in., which was greater than usual, so that the upset could be machined in the as-welded condition. Failure of the shaft during fatigue testing occurred at the keyseat, not in the weld.

Flywheel Friction Welding

Flywheel friction welding is done with a machine in which mechanical energy is stored in, and released by, a flywheel in amounts predetermined and gaged by flywheel speed. The amount of energy released by the flywheel is determined by its speed when axial pressure is first applied, and by the speed at which the clutch disengages the spindle from the motor. (For additional description of the mechanics of flywheel friction welding, see "Welding Methods" on page 507.)

In operation, one workpiece is clamped in a collet chuck mounted to the spindle and the clutch is engaged, which causes the workpiece to be rotated at a predetermined speed. The mating workpiece is clamped in the tailstock and then brought in contact with the rotating workpiece, and pressure is applied to heat the workpieces. At a predetermined time or spindle speed, the clutch is disengaged and a welding pressure is applied to stop rotation and complete the weld. After the flywheel is disengaged, the heating pressure can be continued or the welding pressure can be applied immediately. Thus both kinetic and direct mechanical energy can be used to heat the workpieces to welding temperature, although the machine is designed with the intent that all heating be derived from the kinetic energy of the freely rotating flywheel, and none from direct mechanical energy.

Source: *Metals Handbook*, 8th Ed., Vol. 6, ASM, 1971

Fig. 12. Original and improved methods of welding trunnions to a mounting block to make a trunnion mount for air and hydraulic cylinders. Cost per mount was reduced by 71% when flywheel friction welding replaced gas metal-arc welding. (Example 458)

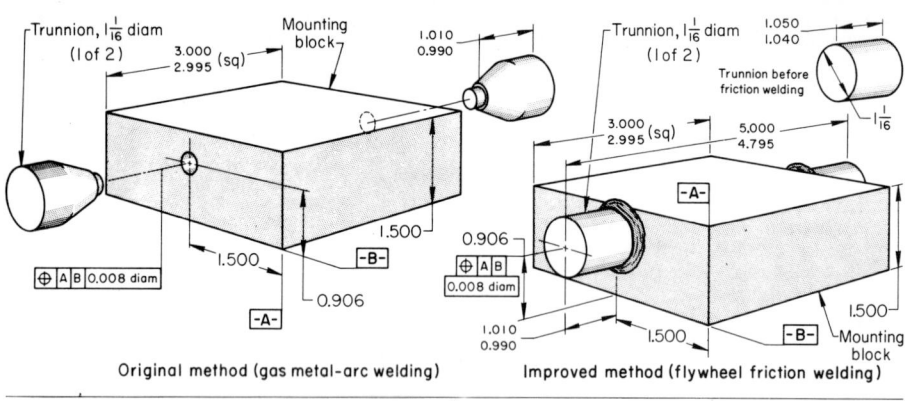

Conditions for Flywheel Friction Welding	
Flywheel moment of inertia	40 lb-ft²
Spindle speed	1800 rpm
Weld energy(a)	21,800 ft-lb
Axial force	18,000 lb
Time per weld	1 sec
Time, floor to floor	30 sec
Metal lost (b)	0.045 ± 0.005 in.

(a) Calculated from flywheel size (moment of inertia) and spindle speed. (b) Total axial shortening of the workpieces during welding.

Operation	Gas metal-arc welding	Flywheel friction welding
Comparison of Costs per Mount		
Machine two trunnions	$0.90	$0.06
Drill two holes in block	0.12	...
Assemble trunnions to block	0.10	...
Weld	0.85	0.50
Total cost per mount	$1.97	$0.56
Saving per mount		$1.41

The variables in flywheel friction welding are rotational speed, flywheel size (moment of inertia), cutout speed and axial pressure. The effects of each variable (except cutout speed) on the weld are the same as in inertia welding. The continuous rotation of the drive-motor–flywheel assembly makes the weld energy almost immediately available as soon as the workpieces are loaded into the collets or fixtures.

In the example that follows, gas metal-arc welding was replaced by flywheel friction welding for joining trunnions to a mounting block. In both of the methods used to join the three components, the location of each trunnion was held to a close tolerance, to minimize the machining necessary to make the trunnions concentric.

Example 458. Joining of Trunnions to Mounting Blocks by Flywheel Friction Welding at Lower Cost Than by Gas Metal-Arc Welding (Fig. 12)

A trunnion mount (see Fig. 12) for air and hydraulic cylinders originally was made by gas metal-arc welding two trunnions to a mounting block. When the joining process was changed to flywheel friction welding, the total cost of making the mount was reduced by 71%.

In the original method, the trunnions were machined from bar stock and then gas metal-arc welded to a mounting block. Each trunnion had a locating boss and a weld chamfer. The block was drilled to accept the trunnions, which were pressed in, and then welded. After welding, the trunnions were machined concentric. Before welding, the blocks were broached square and parallel within 0.005 in. To minimize the machining necessary to make them concentric, the trunnions were held within 0.008 in. of true position. As shown in the table with Fig. 12, the cost of machining and arc welding the trunnions was $1.97 per block.

When the process was changed to flywheel friction welding, the trunnions were saw cut to 1.050 +0.000, −0.010 in. in length and the broached blocks were not drilled. The cost of cutting two trunnions to length and friction welding them to a block was $0.56.

The trunnions were held by a collet chuck mounted to the spindle of the flywheel friction welding machine and the mounting block was located and held in a fixture mounted to the tailstock. The length of each trunnion was held to 1.000 +0.000, −0.010 in., which eliminated the need for facing the ends after welding. Time for each weld was 1 second; floor-to-floor time for each piece was 30 seconds. Additional conditions for flywheel friction welding, and a comparison of costs for the two processes, are given in the table that accompanies Fig. 12.

Joint Design

The mechanics of friction welding restrict its use to flat and angular butt welds that are perpendicular to and concentric with the axis of rotation. Flat joints are the most common and can be classified as: (a) bar to bar, (b) bar to tube, (c) tube to tube, (d) bar to plate, and (e) tube to plate, as shown in Fig. 13. These classifications refer to the joint itself, and not to the shape of the parts. The joint used in making the rotor bodies described in Example 451 is classified as a bar-to-plate joint because a 0.62-in.-diam rod was joined to a 2-in.-diam rod. Friction welding the cluster gears in Example 463 illustrates a tube-to-plate joint on the OD of the gear and a tube-to-tube joint on the ID.

Joint surface conditions, such as surface finish, squareness and cleanness, are not critical for friction welding because the original abutting surfaces are rubbed off and extruded in the process. As-forged, sheared, gas-cut, abrasive-cut or sawed surfaces are acceptable, but extra heat is needed to remove irregularities and allow uniform heating to occur. Also, if the face of a workpiece is not perpendicular to the axis of rotation, forces are produced that can affect the concentricity of the components after welding.

Projections left by cutoff tools present no problem, and in some applications may even help to heat the center of the bar. However, center-drill holes must be avoided; when the upset metal compresses the air entrapped in a center-drill hole during upsetting, a weld defect usually occurs.

Heavy mill scale, thick chromium plating or a thin carburized or nitrided case acts as a bearing surface and cannot be extruded from the weld area. When deeply carburized workpieces are to be welded, the joint surface, and the adjacent surfaces, must either be machined before welding (or before hardening, if it precedes welding) to remove the carburized metal, or be copper plated or otherwise covered during the carburizing operation to prevent them from being carburized. In Example 463, the weld area was machined before hardening to remove the carburized metal. In Example 451, the end to be welded was copper plated before carburizing.

Tubular Welds. Tube-to-plate welds are not as strong as tube-to-tube welds, because rounding of the end of the tube during heating and upsetting reduces weld effectiveness. Fatigue life of a tube-to-plate weldment increases with removal of the sharp notch at the tube base before the part is put to use. On some metals, when heating is too slow, or when excessive weld energy is applied, the upset metal flows up and around the outside of the tube and forms, in effect, a tube within a tube.

On tubular welds, the upset is extruded equally toward the bore and to-

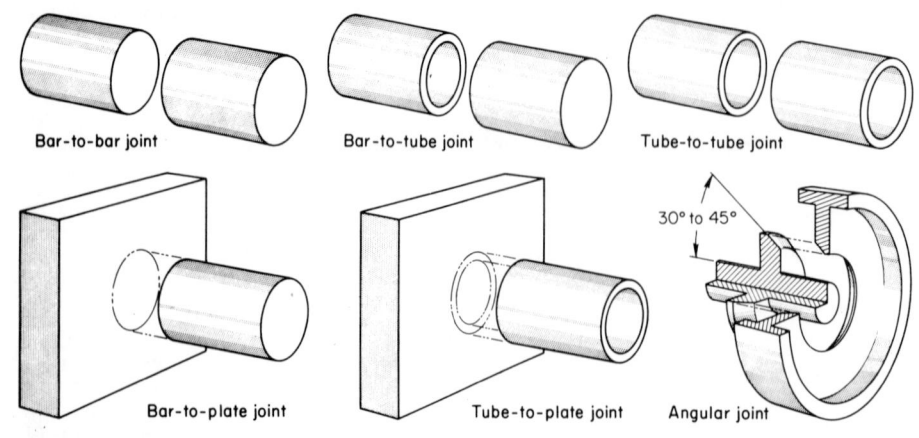

Fig. 13. Types of joints commonly made by friction welding. Bars and tubes, and joint surfaces of angular joints, must be concentric with the axis of rotation; bar, tube and plate surfaces to be welded must be perpendicular to the axis of rotation.

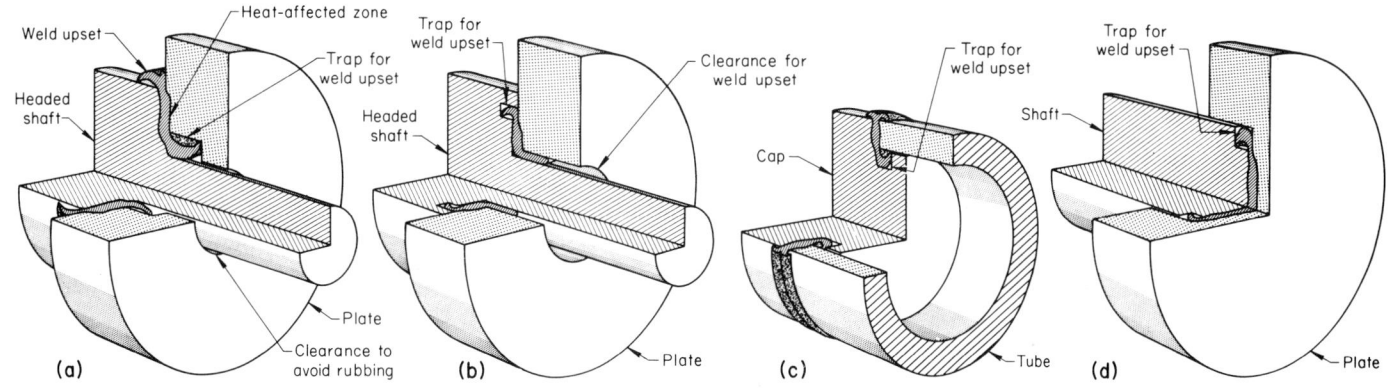

Fig. 14. Recesses used as traps for weld upset in friction welded joints. (See discussion in text, under "Design for Flow of Weld Upset".)

ward the periphery. When weld upset cannot be permitted to remain in the bore and cannot be reached for removal after welding, a trap must be incorporated into the joint design. (See the subsection on Design for Flow of Weld Upset, this page.)

Angular joints (see Fig. 13) are used in welding the inside diameter of one component to the outside diameter of another — for instance, welding the rim of a jet-engine fan to the flange of the hub section or welding a flange to a shaft at some position between the two ends.

The joint can be made where a chamfered shoulder can be mated with an equally chamfered bore in the flange. The interfaces of angular joints must have equal included angles, and the tapered bore in the outer component must have sufficient strength to withstand the axial pressure required to make the weld.

Angular joints are usually designed with faces 30° to 45° from the centerline (an included angle of 60° to 90°) to prevent one part from being pushed through the hole in the other part. Some nickel-base alloys have been welded using smaller angles.

Design for Heat Balance. When similar metals, or dissimilar metals having about the same forging temperature and thermal conductivity, are friction welded, there are no restrictions on the relationship of the cross-sectional area of a bar or tube to the size of plate to which it is welded. Heating rates as high as 100,000 F per second make the weld cycle very short and keep heat losses to the cold metal in the plate adjacent to the weld very low. Thus, bar-to-plate and tube-to-plate welds are feasible for joining similar and some dissimilar metal combinations.

In welding dissimilar metals with widely different forging temperatures and thermal conductivities, adjustment of size or area adjacent to the weld interface may be necessary in one of the workpieces. Area differentials must be determined experimentally because each metal reacts differently. For instance, in welding a nickel-base alloy shaft 1 in. in diameter to an alloy steel shaft, the alloy steel shaft should be 1/16 to 1/8 in. in diameter larger than the nickel-base alloy. If the components are tubes, the inside diameter of the alloy steel tube should be 1/16 to 1/8 in. less than that of the nickel alloy tube. Thus the alloy steel tube would have a smaller inside diameter and a greater wall thickness and outside diameter than the nickel alloy tube. In making a cutting-tool blank, a tool steel shank can be readily welded to a tool steel body using a bar-to-plate joint, but if the shank is changed to alloy steel, the tool steel body must be modified to produce a bar-to-bar joint.

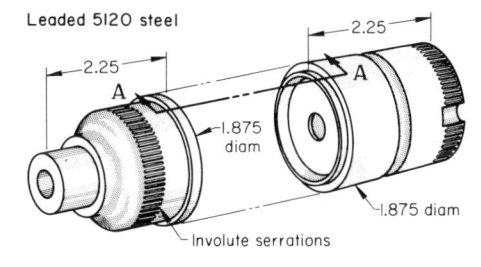

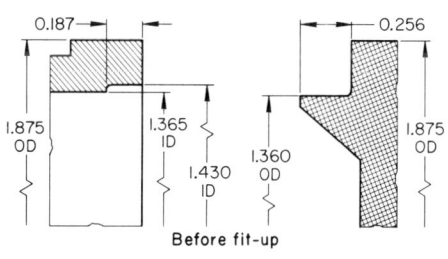

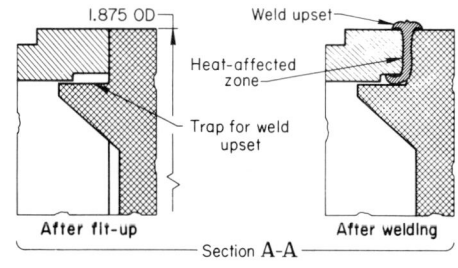

Conditions for Inertia Welding

Flywheel moment of inertia ....10.75 lb-ft²
Spindle speed ....2800 rpm
Weld energy(a) ....14,300 ft-lb
Axial force ....17,000 lb
Heat-and-weld time ....0.7 sec
Machine cycle time ....4 sec
Weld area ....1.36 sq in.
Metal lost(b) ....0.020 ± 0.003 in.
Production rate ....280 parts per hour

(a) Calculated from flywheel size (moment of inertia) and spindle speed. (b) Total axial shortening of the workpieces during welding.

Joint was designed with a trap that prevented the formation of weld upset on the internal surface of the chamber.

Fig. 15. Components of a diesel-engine precombustion chamber that were joined by inertia welding (Example 459)

When a tube is welded to a thin plate with a hole of the same diameter as the inside of the tube, the hole in the plate frequently is made smaller than the ID of the tube to avoid excessive heating of the plate around the hole. This difficulty does not occur with thick plates because the metal around the hole is not heated through.

Design for Flow of Weld Upset. In applications where the presence of weld upset on one or more work-metal surfaces is undesirable, and where the upset cannot be removed after welding, traps can be incorporated into the joint design, to provide clearance for the flow of weld metal from the interface. When a plate is welded to a shaft extending through a hole in the plate, or a boss on one workpiece extends into a hole in the other workpiece, enough clearance must be provided to prevent rubbing of the shaft or boss against adjacent surfaces. Rubbing of adjacent surfaces parallel to the axis of rotation uses flywheel energy unpredictably and diminishes reproducibility of the weld.

Typical designs of traps for weld upset are shown in Fig. 14. An assembly of a headed shaft and a plate is shown in Fig. 14(a) and (b). In Fig. 14(a), weld upset could not be tolerated on the surface of the plate opposite the joint and could not be removed after welding. A counterbore in the hole in the plate, extending inward from the joint surface of the plate, served as a trap for weld upset, which was permitted to flow out around the head of the shaft. In Fig. 14(b), weld upset was not permitted on the joint surface of the plate, and the trap was formed in the head of the shaft. A surface on the head touched the plate as rotation ceased, and provided a seal.

An end-cap-to-tube weld is shown in Fig. 14(c). A tight internal corner joint free of weld metal was obtained by making the trap in the boss on the end cap. In Fig. 14(d), the step on the joint end of the shaft provided the trap for weld upset, leaving the outside corner formed by the intersection of the plate and the shaft free from upset metal.

Parts that require difficult-to-machine internal recesses or chambers can be made by joining two specially designed and machined parts. Removal of weld upset from the inside surfaces of such parts is difficult; therefore, the joint must be designed with a trap so that weld upset cannot form on the

Source: *Metals Handbook*, 8th Ed., Vol. 6, ASM, 1971

internal surface. A trap was used effectively in the example that follows.

Example 459. Inertia Welding of a Precombustion Chamber Using a Joint Designed To Prevent Weld Upset From Forming on an Internal Surface (Fig. 15)

Precombustion chambers for diesel engines (see Fig. 15) originally were made by furnace brazing two sections together. The brazed joint was subjected to precombustion temperatures as well as a combustion pressure of about 1700 psi. Precise machining of the joint surfaces and careful inspection of the parts were needed.

When furnace brazing was replaced by inertia welding, close joint tolerances and meticulous inspection procedures were no longer needed, and product quality was improved. The mating surfaces were designed to contain the internal weld upset so that it could not flow into the chamber.

The two components were made of leaded 5120 steel in a multiple-spindle bar machine. The outside diameter of the chamber at the weld interface was 1.875 in., and the width of the weld interface was about 0.22 in. A small enclosed internal cavity, shown in Fig. 15 (section A-A, after fit-up), contained the internal weld upset and prevented it from flowing into the chamber. Involute serrations were rolled onto each part during the machining operation to aid the hydraulically operated welding-machine collet chucks in gripping the workpieces, and mating serrations were machined in the collet pads of the chucks. Dependable holding techniques were needed to prevent undesirable malfunctions, to reduce wear on the gripping surface of the chuck, and to minimize upkeep cost.

In operation, the two sections were clamped in their respective collet chucks, and the tailstock was moved forward until the surfaces to be welded were 0.045 in. apart. The inertia flywheel and spindle were accelerated to 2800 rpm (in about 1.2 sec), the energy source was disconnected from the spindle and, immediately, the two sections were brought into contact under a force of 17,000 lb, which remained constant until rotation stopped. Heat-and-weld time was 0.7 sec. Strength of the welded joint was equal to that of the base metal.

The spindle chuck unclamped automatically and the tailstock retracted to the starting position. Then the tailstock chuck unclamped automatically and the workpiece was removed manually and was placed in an automatic lathe where the weld upset was machined off while the next piece was being welded.

Machine cycle time, exclusive of handling time, was 4 sec. Metal lost was 0.020 ± 0.003 in. Total cost of operation, including overhead, was $0.035 per weld.

Additional welding conditions are given in the table with Fig. 15.

Design for Machine Size. The amount of energy needed for friction welding depends partly on the distance metal must be extruded from the innermost point of the interface of the workpieces. Thus, more weld energy is required, per square inch of weld area, for joining two 1-in.-diam bars than for joining a 1-in.-diam bar to a 1-in.-diam tube. The joints are of nearly equal strength, because the metal at the center of the bar is near the neutral axis and contributes very little to torsional or bending strength.

Some joints can be designed with a center relief in one or both workpieces to eliminate the nonworking but hard-to-weld metal at the center. This also may permit the part to be welded on a smaller-capacity machine, or the length of the welding cycle to be reduced. The size of the center relief must be large enough so that compression of the entrapped air will not cause a weld defect, and to provide adequate space for the weld upset.

Similarly, a bar-to-plate joint can be redesigned as a tube-to-plate or a bar-to-bar joint if desirable to bring it within the capacity of a particular machine. A bar-to-bar joint requires less energy than a bar-to-plate joint.

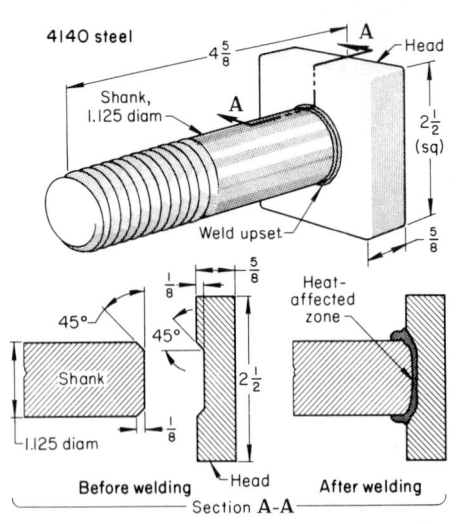

Fig. 16. Square-head bolt that was made by inertia welding to reduce costs from original method of machining from solid bar stock (Example 460)

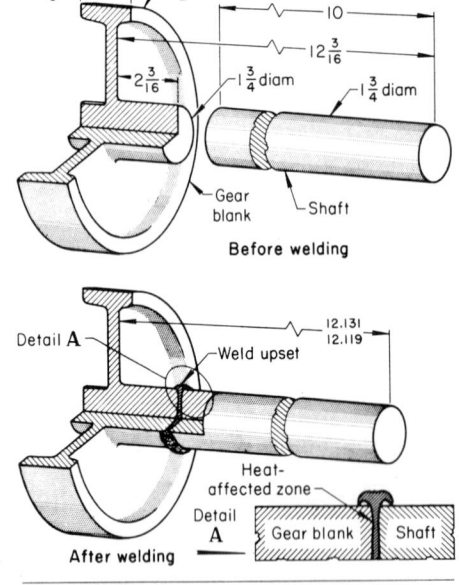

Conditions for Inertia Welding

Machine capacity:
Part diameter .... 0.875 to 2.5 in.
Spindle speed .... 4800 rpm max
Axial force .... 160,000 lb max
Flywheel moment of inertia .... 170 lb-ft²
Spindle speed .... 2500 rpm
Weld energy(a) .... 180,900 ft-lb
Axial force .... 45,000 lb
Floor-to-floor time .... 2 min (approx)
Weld area .... 2.4 sq in.
Metal lost(b) .... 0.062 ± 0.006 in.

(a) Calculated from flywheel size (moment of inertia) and spindle speed. (b) Total axial shortening of the workpieces during welding.

Fig. 17. Gear-blank-and-shaft assembly that was joined by inertia welding. Floor-to-floor time was reduced by 90% compared with original method of pressure gas welding. (Example 461)

Control of Weld Quality

Inspection of friction welds usually consists of visual examination of the weld upset and measurement of the over-all length of the assembly to determine the variation in axial shortening of the members during welding.

Variation in axial shortening, or metal lost, is a good indication of weld quality. Because rotational speed and heating and welding pressures are readily controlled within close limits, the amount of metal lost usually is within ±5% of the nominal value. Therefore, when the lengths of the workpieces are held to close tolerances, the expected variation in the over-all length of the welded assembly can be calculated using the tolerances of the workpieces and weld upset. The nominal amount of metal lost can be determined while making tests to establish welding conditions.

Tension, bend, impact or fatigue tests frequently are used to check weld quality. Sectioning for microscopic examination and for hardness tests is also performed on test samples or on randomly selected production parts.

Proof testing by bending just enough to cause slight yielding in the weld zone has been used for testing friction welded pump shafts. Defective welds are discovered but no harm is done to satisfactory welds. Magnetic-particle inspection can be done on steel parts after the weld upset has been removed.

Friction Welding vs Other Processes

Friction welding can replace or supplement other manufacturing processes to reduce machining and material costs, to permit the use of work metals that meet the application requirements, and to provide reliable joints.

In Example 451, machining and material costs were reduced by friction welding stub shafts to a rotor body instead of machining from solid bar stock. The quality of friction welded parts can be equal to that of parts machined complete from bar stock, as illustrated by the following example.

Example 460. Change From Machining to Inertia Welding That Reduced the Cost of Making Square-Head Bolts (Fig. 16)

The square-head bolt shown in Fig. 16 originally was machined from 2½-in.-sq 4140 steel bar and heat treated to the required mechanical properties. Cost of each bolt was $6.03.

When machining from solid bar stock was replaced by inertia welding a shank to a bolt head, there were no changes in the over-all quality, serviceability and appearance of the bolt, and cost was reduced to $2.707, for a saving of $3.323 per bolt. The welded bolts, when tested in accordance with SAE standard J429, met the requirements for Grade 7 threaded fasteners. Annual production was 1000 bolts.

In the improved method, the two workpieces were made of 4140 steel in an automatic bar machine. The head was faced, recessed and cut off from a 2½-in.-sq bar. The shank was turned, chamfered, threaded and cut to length from a 1¼-in.-diam bar. The mating surfaces of the shank and head had matching contours, as shown in Fig. 16, to assist in alignment. After inertia welding, the bolt was normal-

end was machined to 0.875 in. in diameter, which produced a weld area of 0.60 sq in. The parts were degreased before being inertia welded.

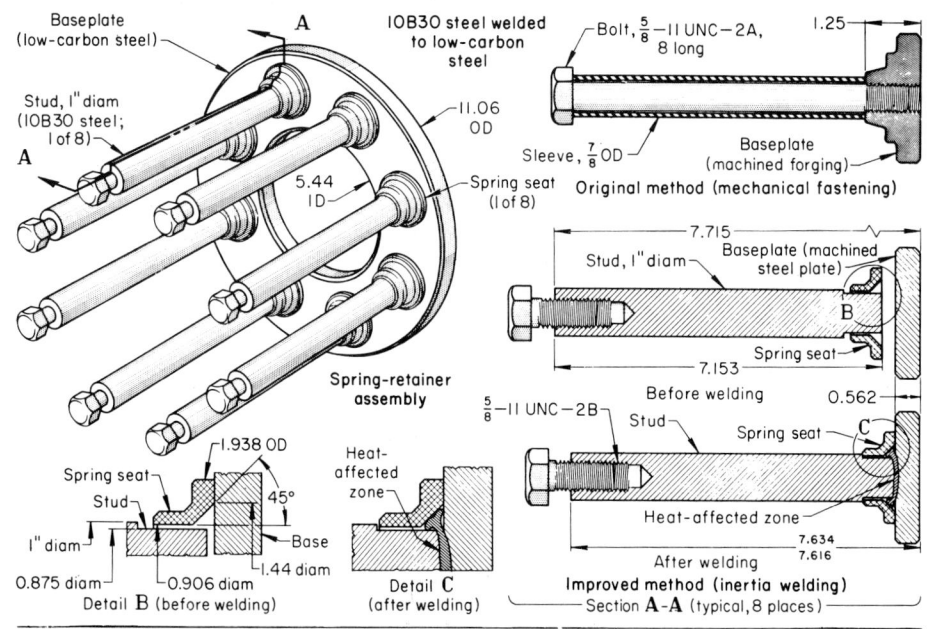

The baseplate was located by its inside diameter in an indexing fixture that was manually indexed 45° between successive welds. A manually loaded, hydraulically actuated collet chuck was mounted on the spindle for gripping the 1-in.-diam studs.

In operation, the baseplate was secured in the indexing fixture, a stud was clamped in the collet chuck, and a spring seat was placed over the 0.875-in.-diam end of the stud. Rotation of the spindle did not begin until the tailstock had advanced to a position that prevented the spring seat from falling off the end of the stud. The weld upset was not removed, because it was covered by the spring seat. Heat-and-weld time was 1.5 sec per weld, and production rate was 15 assemblies per hour. Additional welding conditions are given with Fig. 18.

Weld quality was checked by visually inspecting the weld upset, and the distance from the bottom of the baseplate to the top of the studs was measured for compliance with a specification of 7.625 ± 0.009 in.

In the example that follows, the method of making a flanged transmission gear was changed from forging to

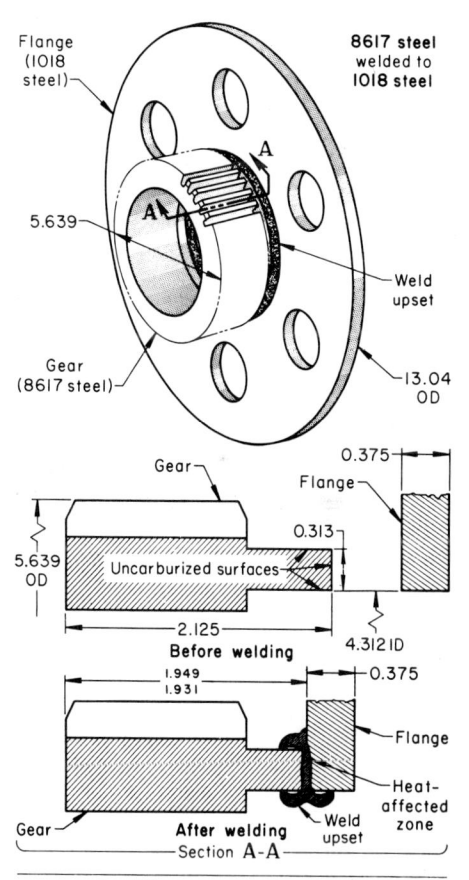

Fig. 18. Spring-retainer assembly made by inertia welding eight studs to a baseplate at lower cost than by tapping a forged plate and inserting bolts (Example 462)

Conditions for Inertia Welding

Flywheel moment of inertia	8.5 lb-ft²
Spindle speed	3900 rpm
Weld energy(a)	22,000 ft-lb
Axial force	11,000 lb
Heat-and-weld time	1.5 sec per weld
Weld area	0.60 sq in.
Metal lost(b)	0.090 ± 0.009 in.
Production rate	15 parts per hour

(a) Calculated from flywheel size (moment of inertia) and spindle speed. (b) Total axial shortening of the workpieces during welding.

ized, the weld upset was trimmed to a ⅛-in.-radius fillet, and then the bolt was quenched and tempered. About ¹⁄₁₆ in. of metal was lost during upsetting.

Sectioning and macroetching the weld zone showed a sound weld. Microscopic examination after heat treatment disclosed an area of transition of finer grain size between the base metal and weld area.

In the example that follows, production time was reduced 90% and rejection rate was reduced slightly when pressure gas welding was replaced by inertia welding for joining a forged gear blank to a shaft.

Example 461. Change From Pressure Gas Welding to Inertia Welding That Reduced Production Time by 90% (Fig. 17)

The 94B17 steel gear-blank-and-shaft assembly shown in Fig. 17 originally was joined by pressure gas welding. The two parts were machined flat at the interface without use of a cutting fluid, and then were chucked in a modified lathe equipped with a hydraulically actuated tailstock and an adjustable oxyacetylene ring burner on the bed. The parts were brought together, heated to slightly below the melting range and held under pressure until the weld was completed. Weld quality was good, and rejection rate was approximately 0.5%. However, 20 min was required for loading, heating, welding and unloading each part.

To increase the production rate, the joining method was changed to inertia welding. No changes in part design or joint preparation were required.

The gear blank was placed in a chuck that was attached to the spindle, and the shaft was chucked in the nonrotating tailstock. The spindle was accelerated to 2500 rpm, the drive mechanism was disengaged, and the workpieces were brought together with an axial force of 45,000 lb. The weld was completed when the flywheel energy was exhausted and the flywheel came to rest.

Although some latitude existed in the selection of machine settings, excellent results were obtained by using the conditions given in the table with Fig. 17. A relatively high axial force was needed to obtain the desired weld upset because of the strength of the 94B17 steel at elevated temperature. Total axial shortening of the workpieces (metal lost) was 0.062 ± 0.006 in.

The change from pressure gas welding to inertia welding reduced floor-to-floor time per assembly to approximately 2 min, and reduced the rejection rate to 0.2%. Weld quality was checked by ultrasonic and magnetic-particle inspection after removal of the weld upset.

Redesign of the workpieces in the following example allowed inertia welding to replace forging and machining. An assembly of equivalent quality was produced at lower cost.

Example 462. Redesign of a Spring-Retainer Assembly for Production by Inertia Welding To Reduce Costs (Fig. 18)

A spring-retainer assembly (see Fig. 18) for a heavy-duty tractor steering clutch originally was made from a forged, machined and tapped baseplate into which hexagon-head bolts were inserted. Spring seats were coined in the base plate prior to machining. The height to which the bolts extended above the baseplate was maintained by a sleeve made of ⅞-in.-OD tubing. When the product design was changed so that studs could be inertia welded to a steel plate, the cost of each assembly was reduced by $0.71.

In the improved method, the baseplate was gas-cut from a ⅞-in.-thick low-carbon steel plate. The inside and outside diameters were turned to the dimensions shown in Fig. 18, and were then ground flat to a thickness of 0.562 in. The spring seats were made in an automatic bar machine from 1¹⁵⁄₁₆-in.-diam 1213 steel bar stock. The studs were machined from 1-in.-diam 10B30 cold finished steel that had been heat treated to a hardness of Rockwell C 23 to 30. One end of the stud was tapped with a ⅝-11 UNC-2B thread, and the other

Conditions for Inertia Welding

Flywheel moment of inertia	668 lb-ft²
Spindle speed	1450 rpm
Weld energy(a)	239,000 ft-lb
Axial force	105,000 lb
Heat-and-weld time	2 sec
Weld area	4.55 sq in.
Metal lost(b)	0.185 ± 0.009 in.
Production rate	63 parts per hour

(a) Calculated from flywheel size (moment of inertia) and spindle speed. (b) Total axial shortening of the workpieces during welding.

Fig. 19. Flanged transmission gear that was made by inertia welding instead of by forging, to reduce costs and avoid distortion of the flange during heat treatment (Example 463)

Source: *Metals Handbook*, 8th Ed., Vol. 6, ASM, 1971

inertia welding to allow selective hardening of the gear section and so avoid flange distortion. The inertia weldment weighed and cost less than the forging.

Example 463. Change From Forging to Inertia Welding To Avoid Workpiece Distortion (Fig. 19)

A flanged transmission gear originally was machined from a one-piece 8617 steel forging. Hardening of the gear section was required, and distortion of the large-diameter flange necessitated press-die quenching. Although not required functionally, a 0.64-in.-thick by about 1-in.-wide rim was needed to minimize distortion of the 0.335-in.-thick flange during heat treatment. To locate the forging during die quenching, annular surfaces were machined on the flange opposite the gear and on the rim to an axial dimension of 0.100 ± 0.001 in. Because of the cost of alloy steel forgings and the problems from distortion during heat treatment, the gear was redesigned for inertia welding, as shown in Fig. 19.

The flange section was gas-cut from scale-free 1018 steel plate 0.375 in. thick, machined to a 13.04-in. OD and a 4.312-in. ID, and degreased. The gear section was hobbed from 8617 steel tubing and carburized. Before hardening, the carburized case was machined off at the end and sides of the joint area, so that the weld would be made on the low-carbon 8617 core, not on the carburized surface. The area of the weld surface was about 4.55 sq in. The gear section was hardened before welding, and thus heat treatment of the entire part and the resulting distortion of the flange were avoided.

The flange section was held in a nonrotating fixture mounted to the tailstock of the inertia welding machine and the gear was held in a rotating fixture mounted on the spindle. Locating and driving were done on the pitch diameter of the gear. The machine was loaded and unloaded manually. After the gear was welded to the flange, the weld upset was machined from the inside and outside surfaces, and spline teeth were hobbed around the periphery of the flange.

Welding conditions (see table with Fig. 19) and machine performance were checked for each production lot. The weld upset on each part was visually inspected, and random measurements were made to determine whether the length of the gear conformed to the specified 1.940 ± 0.009 in.

Inertia welding and machining produced a part with quality equivalent to that of the machined forging, but at a cost reduction of $4.25 per piece.

Cost of Inertia Welding vs Other Methods of Fabrication

The cost advantages of inertia welding in comparison with other methods of fabrication are shown in the following projections of costs for the production of three different parts.

Drill-to-Shank Welding. Costs for flash welding and inertia welding of 4140 steel shanks to M10 high speed tool steel drill bodies, indicating the savings provided by inertia welding, are given in Table 3.

Assuming the quantity of 0.547-in.-diam drills (drill A, Table 3) to be about three times that of drills B and C, the average saving per drill would be $0.136 — which, at 300,000 drills per year, would give annual gross savings of $40,800. The inertia welding machine and semiautomatic tooling would cost about $60,000.

Shaft-and-Pinion: One-Piece Forging vs Weldment. Table 4 compares costs for two sizes of each of two designs

Table 3. Cost Comparison for Joining M10 High Speed Tool Steel Drill Bodies to 4140 Steel Drill Shanks by Flash Welding and by Inertia Welding

Item	Drill A Flash welding	Drill A Inertia welding	Drill B Flash welding	Drill B Inertia welding	Drill C Flash welding	Drill C Inertia welding
Drill Dimensions, In.						
Diameter	0.547		0.832		1.062	
Over-all length	8.25		9.75		11.00	
Body length	4.25		5.75		6.25	
Welding Conditions						
Metal lost (total), in.	0.400	0.120	0.510	0.175	0.620	0.200
Metal lost (high speed tool steel), in.	0.240	0.030	0.305	0.045	0.370	0.060
Production rate, welds per hour	165	225	138	190	84	160
Scrap rate, %	5	0.5	5	0.5	5	0.5
Costs per 100 Pieces(a)						
Body stock	$38.88	$37.06	$121.10	$115.90	$216.08	$205.96
Shank stock	4.58	4.49	10.93	10.74	21.48	21.00
Welding labor	9.34	6.65	10.72	7.90	17.69	9.38
Cutoff labor	5.00	5.00	6.00	6.00	7.00	7.00
Scrap loss	2.89	0.26	7.44	0.70	13.11	1.21
Total cost	$60.69	$53.46	$156.19	$141.24	$275.36	$244.55
Savings	...	$7.23	...	$14.95	...	$30.81

(a) Values used in calculations: M10 high speed tool steel, $1.30 per pound; 4140 steel, $0.17 per pound; labor plus burden, $12 per hour; labor efficiency, 80%.

of pinion-and-shaft one-piece upset forgings made of 8630 steel and equivalent parts made by inertia welding an 8630 steel pancake or upset forging to a 1035 steel tube or shaft. The two designs, identified as designs A and B, are illustrated in Table 4 as the weldments after cutting of gear teeth.

The cost savings shown for the inertia welded pinion-and-shaft assemblies accrue from the use of less expensive forgings, the lower cost of 1035 steel than of 8630 steel for the shafts, and elimination of a drilling operation for the hollow shaft on design A. The inertia welding machine and tooling would cost about $100,000.

Cluster Gear. Following preliminary machining, a heavy gear-and-pinion cluster, upset forged in one piece from 8822 carburizing steel, presented problems in quench hardening after carburizing to provide a case hardness of Rockwell C 60 and a core hardness of Rockwell C 35. A rather severe quench was needed to develop the required core hardness in the heavier pinion section of the forging, and this resulted in excessive distortion and occasional cracking at the intersection of the web and gear. Although increasing the web thickness might have solved the problem, it would have increased the as-forged weight of the one-piece cluster from 130 lb to 152 lb.

To solve the problem and reduce manufacturing costs, investigation was made (a) on a change to an inertia welded assembly of a forged gear blank and a forged pinion blank, and (b) on the selection of a carburizing steel with slightly higher hardenability for the heavier pinion forging.

The cost of the inertia welding machine, equipped for manual loading and with semiautomatic tooling, was about $225,000. With a machine cycle time of 4 min, production rate was projected at 13.8 assemblies per hour.

Following is a comparison of costs for the two methods of fabrication:

Cost factor	Cost
Original forging, 130 lb at $0.25 per pound	$32.50
Improved forging, 122 lb at $0.23 per pound	28.06
Material saving per piece	$ 4.44
Inertia welding labor and burden, 0.075 hr at $11 per hour	0.82
Cost saving per piece	$ 3.62
Cost saving per hour (13.8 assemblies per hour × $3.62)	$50.00

Table 4. Costs for Making Two Sizes of Two Designs of Pinions as One-Piece Upset Forgings and as Inertia Weldments of Pancake or Upset Forgings and Shafts (Illustrated Below)

Cost factor	Pinion design A 7 in. long; 7500/yr Upset forging	Pinion design A 7 in. long; 7500/yr Inertia welding	Pinion design A 5½ in. long; 10,000/yr Upset forging	Pinion design A 5½ in. long; 10,000/yr Inertia welding	Pinion design B 11 in. long; 6000/yr Upset forging	Pinion design B 11 in. long; 6000/yr Inertia welding	Pinion design B 15 in. long; 2000/yr Upset forging	Pinion design B 15 in. long; 2000/yr Inertia welding
Weight of forging, lb	14.5	12.0	5.6	3.6	11.3	6.9	29.1	19.3
Comparison of Costs(a)								
Cost of forging	$3.19	$2.28	$1.22	$0.68	$2.49	$1.52	$6.40	$4.25
Cost of drilling shaft	0.92	...	0.92	...	...	...	...	...
Cost of shaft stock	...	0.36	...	0.29	...	0.31	...	0.43
Welding labor cost	...	0.09	...	0.09	...	0.09	...	0.09
Total cost per piece	$4.11	$2.73	$2.14	$1.06	$2.49	$1.92	$6.40	$4.77
Savings per piece	...	$1.38	...	$1.08	...	$0.57	...	$1.63
Annual savings, total	...	$10,350	...	$10,800	...	$3420	...	$3260

(a) Based on unit costs as follows: upset forgings, $0.22 per pound; pancake forgings, $0.19 per pound; bar stock, $0.07 per pound; 1.5-in.-diam tubing, $0.86 per foot; and labor and burden, $11 per hour.

The Mechanics of Friction Welding Dissimilar Metals
by   D. J. McMullan and A. S. Bahrani
Mechanical Engineering Department, The Queen's University of Belfast (U.K.)

1.0   Introduction

Friction welding is now well established as a mass production welding process which can be used for the joining of similar as well as dissimilar metals. However although the process is widely used and the welding parameters for a wide variety of combinations of metals are firmly established, there are many aspects of the process still inadequately understood. In recent years many investigations have been carried out with the object of understanding the mechanics of the process (1-5) and some progress has been made. In this paper some of the results of experiments on the friction welding of similar and dissimilar metals are reported and a theory for the mechanics of friction welding dissimilar metals is presented.

2.0   Frictional behaviour of metals in friction welding:

Recently Duffin and Bahrani (3,4) studied the frictional behaviour of mild steel under the conditions encountered in friction welding and based on the results of their experiments they put forward the following theory: When the rotating specimen makes contact with the stationary specimen sliding takes place between their unlubricated surfaces. Since the axial force is large adhesion junctions are formed at the points of real contact and seizure develops. At some of the junctions the adhesion between the surfaces is stronger than the metal on either side and shearing takes place at a short distance from the interface so that fragments of metal are transferred from one surface to the other and vice versa. The rubbing surfaces become rough and as the high spots on one surface ride over the high spots on the other they tend to force the specimens apart and thus increase the axial force. The adhesion and seizure between the rubbing surfaces increases the frictional force and consequently the resisting torque. They also raise the temperature of the material at the interface.

The size of the metal fragments which are transferred and re-transferred from one rubbing surface to the other depends on the temperature of the material, the rubbing speed and the axial pressure. Increasing the temperature of the material and increasing the axial pressure have the effect of increasing the size of the transferred fragments while increasing the rubbing speed produces a reduction in the size of the transferred fragments. Thus the zone of contact and mutual transfer of fragments will not extend over the whole area of the rubbing surfaces, but will be confined to a certain annular area where the size of the transferred fragments is large. The size and location of the annular area of active rubbing depends on the speed of rotation and on the axial pressure. Increasing the speed tends to move the annular area towards the centre of the specimen thus decreasing the resisting torque. On the other hand increasing the axial pressure does not move the annular area radially but causes it to spread and increase in width thus again producing an increase in the resisting torque.

As the rubbing continues the temperature rises and eventually reaches a value at which the transferred fragments and the material in the annular rubbing region become soft and under the applied high shear stresses it starts to flow in a similar manner to a liquid of high viscosity. This material is called plasticised material.

The plasticised material from both sides of the interface merge with each other. Thus the condition at the interface changes from two solid surfaces rubbing against each other to that of two solids separated by a viscous fluid which is being churned between them. When this state is established, the resisting torque begins to decrease and the equilibrium phase of the welding cycle starts. This also marks the start of the axial shortening of the specimens being welded and the formation of the weld collar.

During the equilibrium phase both the resisting torque and rate of axial shortening remain substantially constant. The annular region of rubbing transforms into an annular plasticised layer and most of the shear resistance and generation of heat is produced in the plasticised layer. When the speed of rotation decreases during the deceleration phase, the annular plasticised layer moves radially outwards thus causing an increase in torque and producing the terminal peak torque.

It is confidently believed that the hot and severely worked metal in the plasticised layer forms the joining medium between the two parts being welded. Thus, if the welding conditions are such that a continuous plasticised layer is formed at the interface then sound welds are produced.

Experiments carried out by the present authors on the friction welding of tubular and solid specimens of aluminium alloy E9IE (0.5% Si, 0.5% Mg in the solution and precipitation treated condition) have shown that the frictional behaviour of this material under friction welding conditions is similar to that of mild steel. Fig.1 shows two macro - sections of welds carried out at the same axial pressure of 36MN/m^2 but two speeds of rotation of 1000 rev/min and 4000 rev/min. The shape and radial movement of the plasticised zone will be noted.

3.0 Experimental work on the friction welding of dissimilar metals:

Experiments were carried out on the friction welding of aluminium alloy E91E to copper, aluminium alloy E91E to mild steel and copper to mild steel. Solid specimens having a diameter of 19mm were used. The speed of rotation was varied in turn within the range 500 - 4000 rev/min and the axial pressure within the range 22 - 103.5 MN/m^2. The forging pressure used was the same as the pressure during the welding cycle. The amount of axial shortening (up-set) was kept constant at 4.5mm for all the tests. The experiments were carried out using a fully instrumented research friction welding machine and during every test the variation of speed, torque, axial force, and axial shortening with time were measured and recorded. Fig.-2- shows some of the results obtained from the experiments on the friction welding of aluminium alloy to mild steel while Fig -3- shows some of the results obtained from the tests on the welding of aluminium alloy to copper. It will be noted that for both combinations of materials, increasing the axial pressure produces an increase in the initial peak torque, the equilibrium torque and the rate of up-set. It will also be noted that both the initial peak torque and the equilibrium torque increased with decrease in the speed of rotation.

At each of the welding conditions used, specimens were produced for metallographic examination. Fig. -4- shows macro sections of welds between aluminium alloy E91E and mild steel and Fig. -5- of welds between the aluminium alloy and copper. It was observed that a plasticised annular zone was formed near the weld interface and in the specimen having the lower melting point, in this particular case in the aluminium alloy. There are also indications that a stagnant layer of aluminium alloy is formed at the face of the mild steel or the copper specimen. Increasing the speed of rotation had the effect of moving the plasticised layer radially towards the centre. Increasing the axial pressure produced the effect of spreading the plasticised layer and decreasing the thickness of the stagnant layer at the interface. It was also observed that when intermediate compounds (intermetallics) are formed at the interface, they tend to form opposite the annular plasticised layer.

4.0 The mechanics of friction welding dissimilar metals:

Most of the observed results in the friction welding of dissimilar metals can be explained by the following proposed theory for the mechanics of the process.

When the two dissimilar specimens start rubbing against each other adhesion junctions are formed at the points of real contact. As the relative movement proceeds these junctions are sheared either at the interface or at a short distance from the interface in the softer specimen. Thus in the early part of the welding cycle fragments of the softer specimen are transferred to the surface of the harder specimen and within a very short period of time the surface of the harder specimen is completely covered with a coating of the softer metal (6,7). This sequence of events happens during the welding of the aluminium alloy to mild steel and to copper.

The coating of soft metal on the harder specimen forms a stagnant layer and consequently the plane of rubbing moves axially into the softer specimen. Thus the rubbing becomes between two similar materials and eventually a plasticised annular zone is developed on the rubbing plane. The plasticised material is made up completely of the softer metal. The location and thickness of the plasticised zone will depend on the rubbing speed and axial pressure (see Figs -4- and -5-). Most of the heat is generated by the churning of metal in the plasticised layer.

The harder specimen does not participate in the formation of the plasticised layer or the up-set collar. However, it does exert some influence on the size of the plasticised material by restraining the stagnant adhering layer. It also acts as a heat sink and thus influences the rate of up-setting (see Figs. -2- and -3-).

From the above, it is clear that there is an important difference in the mechanics of joint formation between similar metals and dissimilar metals. In the welding of similar metals the plasticised layer is at the interface of the parts being welded and forms the joining material between them. In the welding of dissimilar metals the joint is made in the early part of the welding cycle by the transfer of fragments from the soft metal to the harder metal and this is followed by a weld between layers of similar material.

Thus in the welding of dissimilar metals it is the early part of the welding cycle which is most important in forming the joint between the two different metals. The latter part of the cycle determines the quality of the joint between the two surfaces of the same metal. However, if the incorrect welding parameters are used and the weld time is prolonged then there is the danger of the formation of intermediate compounds at the dissimilar metal interface. These tend to form opposite the annular plasticised layer where the temperature is highest.

5.0 Acknowledgements:

The research work reported here was supported by a generous grant from the Science Research Council (U.K.) The authors also wish to acknowledge the help and encouragement of Professor B. Crossland and the help of Mr. R. Garret in preparing the specimens.

6.0   References:

1. Eichhorn, F. and Schafer, R., Schweissen and Schneiden
        May 1969, Vol. 21, No. 5 pp. 189 - 198
2. Hasui, A., Fukushima, S. and Kinugawa, J., Trans. of the Nat.
        Research Inst. for Metals (Japan) Vol. 10 No.4 (1968).
3. Duffin, F. D. and Bahrani, A. S., Wear, Vol. 26 (1973) pp 53 - 74.
4. Duffin, F. D. and Bahrani, A. S., Proc. Third International Conf.
        on Advances in Welding Processes, U.K. 1974 pp. 228 - 242.
5. Rao, M., and Hazlett, T. H., Welding Journal, Vol. 49, No.4 (1970)
        pp. 181s - 188s.
6. Rabinowicz, E. "Friction and wear of materials" Wiley 1965.
7. Bowden, F. P. and Tabor, D. "The friction and lubrication of solids"
        Oxford University Press 1964.

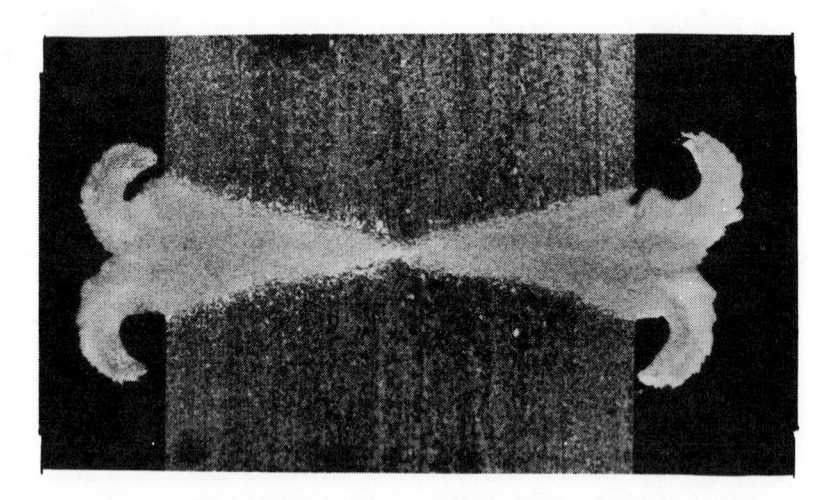

(a) Speed 1000 rev/min Pressure 36 MN/m^2

(b) Speed 4000 rev/min Pressure 36 MN/m^2

Fig.1. ALUMINIUM ALLOY E91E WELDED TO ITSELF

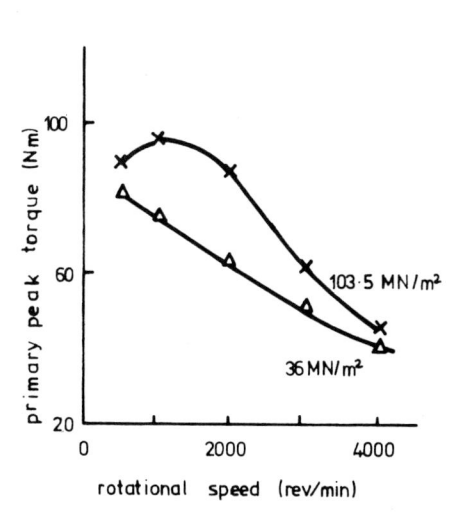

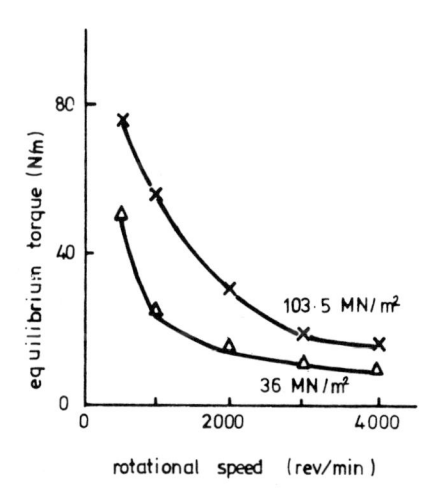

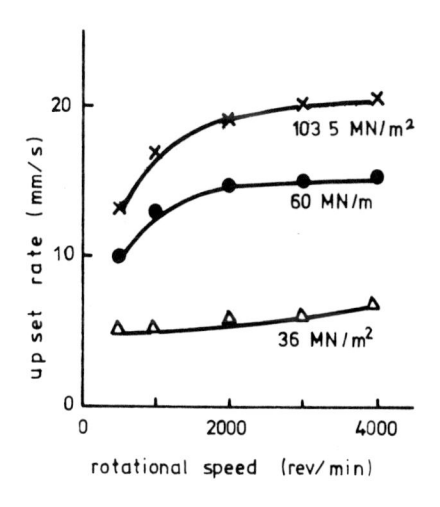

Fig.2 RESULTS FOR ALUMINIUM ALLOY E91E TO MILD STEEL EN3B

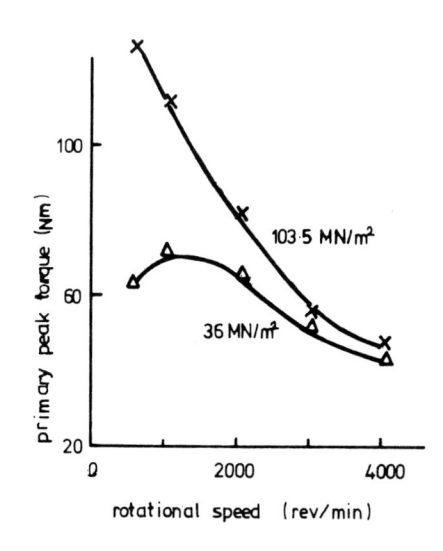

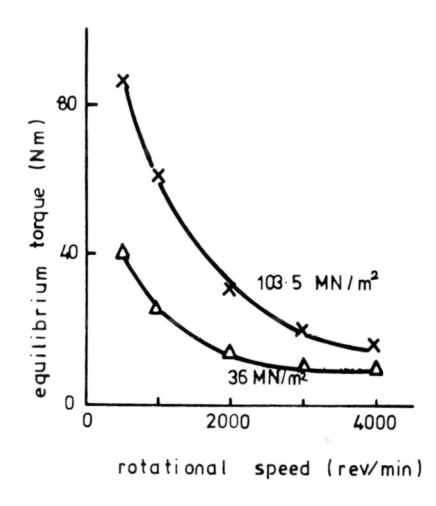

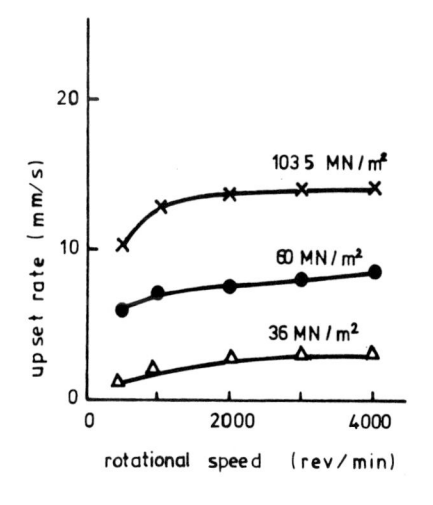

Fig.3 RESULTS FOR ALUMINIUM ALLOY E91E TO COPPER

Source: *Advanced Welding Technology*, The Japan Welding Society, 1975

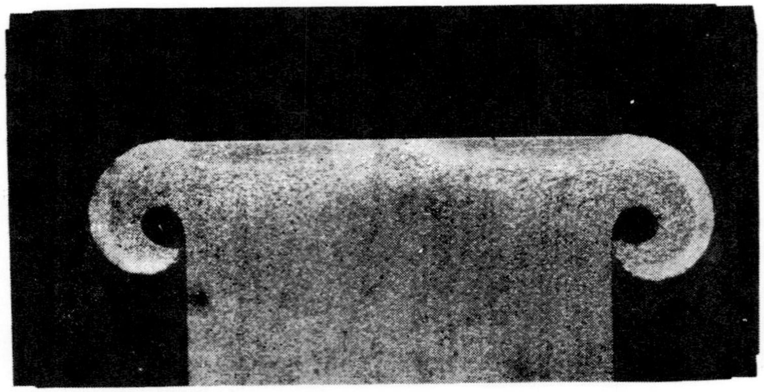

(a) Speed 500 rev/min Pressure 60 MN/m^2 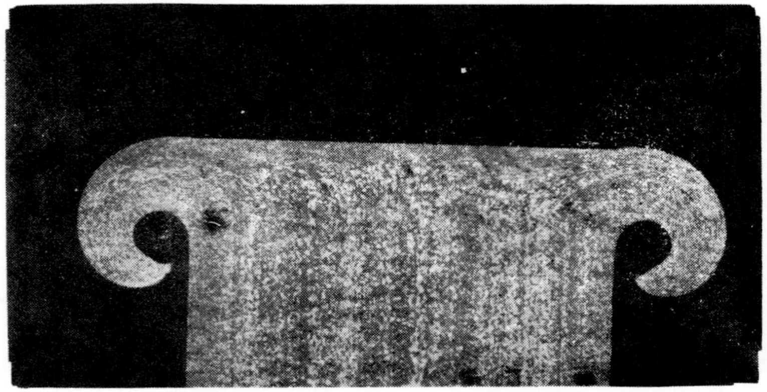   (b) Speed 4000 rev/min Pressure 60 MN/m^2

Fig.4 MACROSECTIONS OF ALUMINIUM ALLOY E91E TO MILD STEEL

(a) Speed 500 rev/min Pressure 103.5 MN/m^2 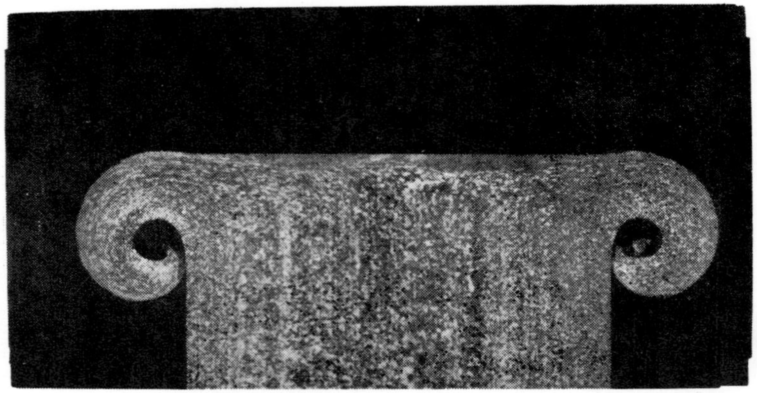   (b) Speed 4000 rev/min Pressure 103.5 MN/m^2

Fig.5 MACROSECTIONS OF ALUMINIUM ALLOY E91E TO COPPER

Inertia-Welding of P/M Parts+

T. M. EBERHART* and R. A. QUEENEY**

Abstract

Additional design flexibility can be realized in the utilization of sintered powdered metals if complex geometries are fabricated from simple component shapes joined by welding. Inertial welding techniques were employed in this study to demonstrate the feasbility of joining sintered metals. The mechanical integrity of the weld regions was measured and related to the main micro-feature of the system, the porosity. The welded samples exhibited improved mechanical properties in the weld region and a gradation in response from the sintered to welded portions that did not lead to any degradation in response.

Introduction

That structural shapes can be produced from compacted and sintered metal powders to finish, or near-finish, dimensional tolerances, is well-recognized as a positive feature of their utilization. Over-all production costs can often be lowered due to reduced metal cutting finishing operations. However, structure geometry is limited by the necessity to remove the green pressed part from the die, and shape complexity is constrained.

Further structural shape generality can be achieved, however, by the use of conventional joining techniques commonly applied to wrought density metallics. For example, the brazing of sintered compacts has been recently described (1). In addition, weldments of sintered structural shapes joined to wrought materials have been successfully produced using a variety of welding techniques (2). To date, however, no reports have appeared to discussing the feasibility of joining two sintered bodies by welding. The present study examines this possibility and the mechanical stability of the weldments produced. Inertia welding was selected for the joining process to avoid the added complexity of weld characterization that would be encountered if filler metal is introduced, as it would be in many other welding schemes.

Experimental Procedure

The alloy powder used in this study was Grade 300M EMP steel supplied by A. O. Smith Corporation. Composition analysis indicated 99.5% Fe, 0.01% C, 0.2%Mn, 0.01%P, 0.029%S, and 0.12%O_2, by weight. Sufficient carbon, in the form of 99.999% pure graphite, was blended with the alloy powder to yield alloy compositions after sintering equivalent to AISI 1020 and 1040 steels (2). Billets 2.54 cm in diameter by 7.62 cm long were pressed at 345 and 690 MN/m^2 in a double-acting ram die. All billets were sintered in dried hydrogen for one hour at 1100 C and slowly cooled in flowing hydrogen.

The sintered billets were machined to 1.77 cm diameter and welded in an AMF vertical inertial welder. The specially modified welder

---
*Project Engineer, Beatle Plastics, Cleveland, Ohio.
**Associate Professor of Engineering Science and Mechanics, The Pennsylvania State University, University Park, Pennsylvania.

+ Inertia welding, a type of friction welding, utilites the complete inertial energy stored in a rotating fly wheel to generate frictional heat.

had a spindle speed of 2850 rpm and an inertial capacity of 2.4Nm^2. Axial loads were variable, and were adjusted between 8451 N and 8896 N to produce satisfacory welds. Fig.1 shows the weld flash appearance of the sintered material and, for comparison, that of a wrought density steel.

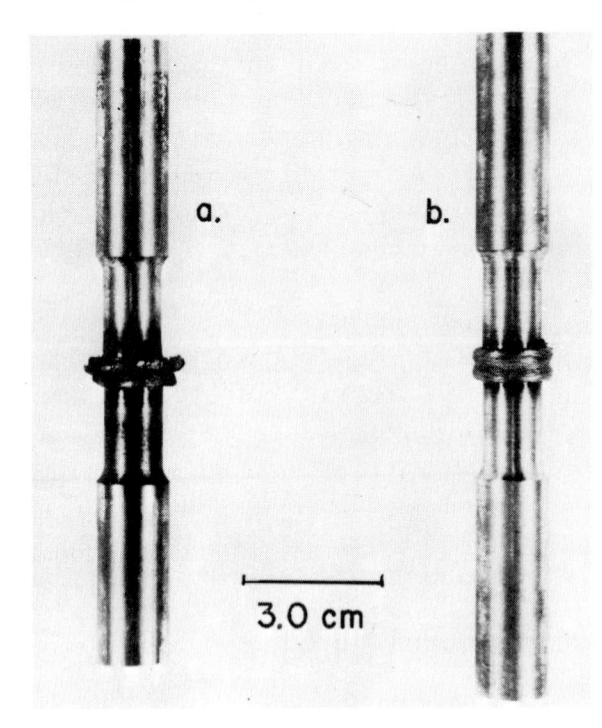

FIGURE 1 Inertial Welded Samples of: a) Sintered AISI 1020 steel; b) Wrought AISI 1020 steel.

Weldment strength was characterized by Vickers hardness measurements on the sintered billet, weld region, and weld-billet interface. These measurements were performed on axially ground and lapped flats with a 100 gram indenter load. Circumferential vee-grooves were machined into the weld zones and into samples of the sintered billets to determine the fracture resistance levels of the base material before and after welding. All samples were machined to 0.89 cm diameter and grooved 0.076 cm deep for fracture testing. Groove root radii were kept to less than 1.27 x 10^{-3} cm. Fracturing was accomplished in 3-point bend testing with a crosshead speed of 0.05 cm/minute.

Finally, porosity was characterized by light optical metallographic examination of lapped plane sections. Pore diameters were individually measured from photomicrographs and the area-fraction of pores was computed.

Results and Discussion

Fig. 2 shows a composite photomicrograph qualitatively indicating the change in porosity character seen moving from the sintered billet to the completely melted and re-solidified weld zone. The fraction of plane section area occupied by porosity, normalized to unity relative to the fraction in the undisturbed sintered billet region, is shown in Fig. 3. As expected, the porosity diminished and nearly vanishes as the weld zones, subjected to melting temperatures and high shear strains, are approached. The AISI 1020 composition, compacted at 345 MN/m^2, developed an increase in porosity as the weld zone is approached. This is thought to be due to pore extension in the plastic shear strain field and the appearance of this phenomena was felt to be due to the dynamics of the welding process. These dynamics (particularly the fixed total energy input through inertial transfer) were not amenable to control in the experimental welding apparatus used.

Vickers hardness numbers scanned through the sintered base material, transition region, and weld zone are shown in Fig. 4, for the AISI 1020 steel composition, and are compared there to the same measure in a welded wrought sample of the same composition. The hardness, an indirect approximate measure of yield strength, deteriorates slightly as one approaches the weld zone, presumably due to the high shear strains accumulated in this region of elevated temperature, but where

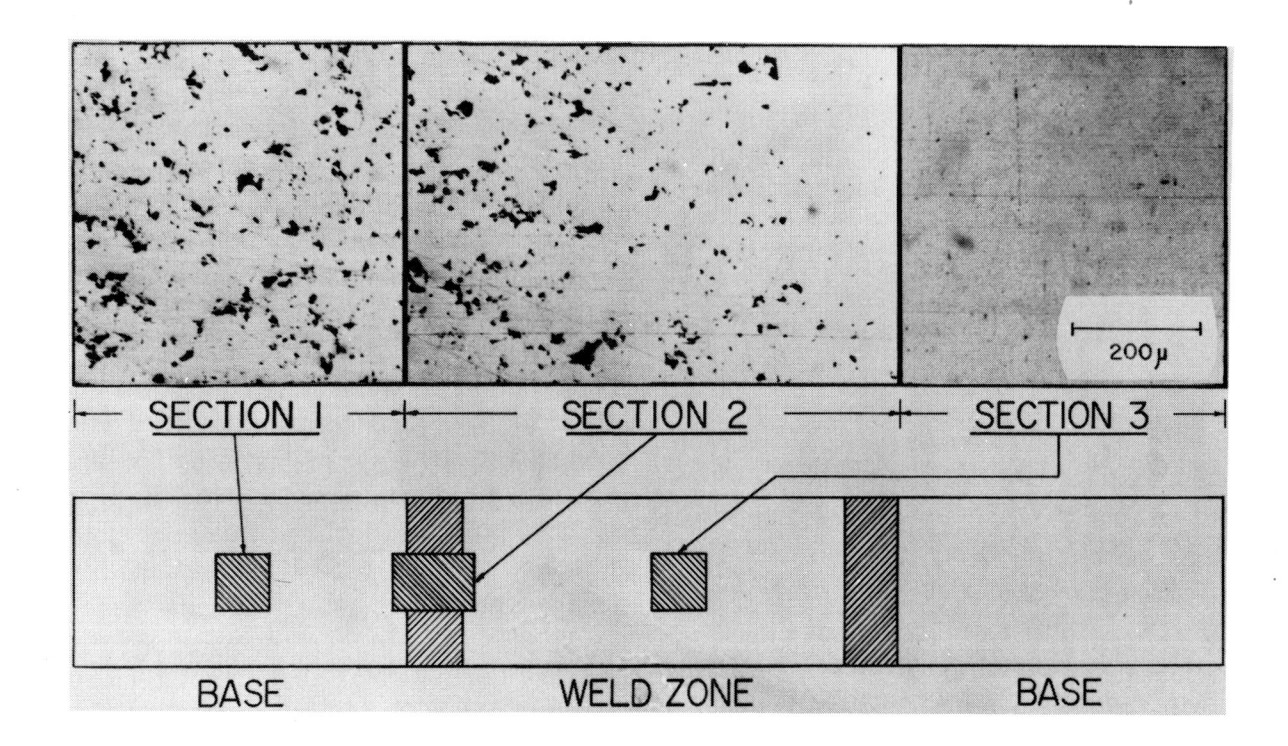

FIGURE 2 Composite Photomicrograph showing porosity character of a Sintered AISI 1020 Steel Weldment: Section 1, Sintered Base Alloy; Section 2, Interface between Sintered Base Alloy and Weld Zone; Section 3, Weld Zone.

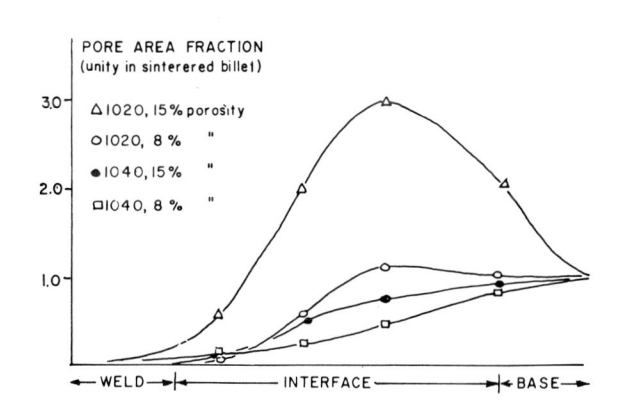

FIGURE 3 Normalized Pore Area Fraction From Sintered Base Alloy to Weld Zone.

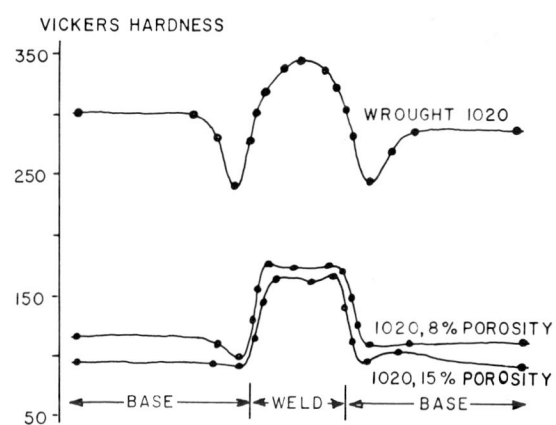

FIGURE 4 Vickers Hardness of Sintered Base Material and Weld Zones in 1020 Steel.

Source: *International Journal of Powder Metallurgy and Powder Technology*, Oct 1977

the thermal conditions are not sufficient to prevent pore extension. Hardness is maximized in the weld zone for both wrought and sintered materials. For the wrought samples, microscopic examination indicated that a greatly reduced grain size was the source of weld zone strengthening. As grain size did not change appreciably for the sintered materials, the near-elimination of porosity would account for the strength increase. These strength changes are wholly consistent with variations in porosity previously studied in similar alloys (3).

The fracture resistance of base materials (sintered only) was compared to that of the material in the weld zone by determination of the plane strain fracture toughness K_{IC}. Fracture toughness values were computed from the bend test specimen geometry and maximum load using the appropriate stress intensity formula (4). The results are given in Table 1. The fracture resistance of the welds is clearly superior to that of the sintered and still porous form of the alloys. Although increasing yield stress often leads to a lessened fracture toughness in conventional wrought-density alloys, through reduced plastic zone formation at the crack tip (5), the reduction in porosity in the weld area evidenced in the present circumstances overshadows this tendency and results in increased fracture resistance (6,7).

TABLE I: Fracture Toughness (K_{IC}) for Sintered Alloys and Weld Zones

Composition AISI	Compacting Pressure N/m^2	K_{IC} (sintered) $MN/m^{-3/2}$	K_{IC} (Weld) $MNm^{-3/2}$
1020	345	8.80	23.7
1020	690	8.91	24.9
1040	345	13.5	31.1
1040	690	13.7	31.6

Summary and Conclusions

Sintered billets of AISI 1020 and 1040 steel were prepared at densities of about 85% and 92% of the theoretical (full) density via two different compacting pressures. Both compositions and both densities were successfully inertia welded to form weldments with vanishing porosity and improved hardness (yield strength) and fracture resistance evidenced by the welds.

Clearly, the added flexibility of product design inherent in the application of welding technology is available to the designer utilizing powdered metals. The use of inertia welding eliminates porosity in the weld zone and greater mechanical strength can be expected when the degrading influence of pores is eliminated, a situation not as clearly achieved when conventional weld joining, using filler material, is employed. The application of inertia welding, or its close relative fraction welding, demands a greater empirical study of the parameters that are variable in the process, such as bearing pressure, energy input, and energy input rate if welds are to be uniformly and routinely successful. Extensive compilations of such data are available for the case of fully dense metals (8), but the authors found that these data provided only a rough scheme for fixing the parameters in the welding of sintered and porous metals.

Acknowledgements

The authors wish to acknowledge the generosity of the American Welding Society whose grant-in-aid partially supported the present study. In addition, helpful technical assistance was given by Dr. R. E. Keith, Associate Professor of Industrial Engineering, The Pennsylvania State University, and P. W. Ramsey, Manager of Welding Research, A. O. Smith Corporation.

References

1. W. V. Knopp, Int. J. Pow. Met. and Pow. Tech., II, 63 (1975).
2. J. P. Hinrichs, P. W. Ramsey, and M. W. Zimmerman, Weld. J., *50*, 242 (1971).
3. M. Eudler, Powder Met., *9*, 265 (1962).
4. H. Tada, P. Paris, and G. Irwin, Stress Analysis of Cracks Handbook, Del. Research Corporation, Hellertown, PA 27.2 (1973).
5. G. Irwin, Proceedings of Seventh Sagamore Conference, Syracuse University, Syracuse, NY, IV-63 (1960).
6. J. T. Barnby, D. C. Ghosh, and K. Dinsdale, Powder Met., *16*, 55 (1973).
7. G. A. Clarke and R. A. Queency, Int. J. of Powder Met., *8*, 81 (1972).
8. "Friction Welding", Metals Handbook, 8th edition, Vol. 6, ASM, Metals Park, Ohio, 507 (1971).

Orbital-motion technique for friction welding non-circular components

This article is based on a paper presented at a recent meeting of the Friction Welding Technical Group of The Welding Institute by Mr J Searle, Bsc, CEng, MIMechE, technical director, The Friction Welding Co Ltd, Chasetown Industrial Estate, Ring Road, Chasetown, near Walsall, Staffs WS7 8XD, which was concerned with a new type of friction welding technique and machines developed by his company. Friction welding is now well established as a means of joining metal components. Among the advantages of the technique are the production of high quality welds, reliability, adaptability of the process to automation, and the elimination of 'consumables' such as fluxes and filler metal.

However, there is a limitation to the process as currently employed, whereby heat is generated by rotating one component while it is pressed firmly against the other. Although this arrangement provides a very simple method of producing the necessary heat, it cannot be used where the interface of the components is not substantially circular. Furthermore, the current practice of stopping rotation by the use of a friction brake or by allowing the rotating part to stop as its energy is dissipated at the weld interface, does not permit specific angular alignment of the components after welding.

A further disadvantage is that the rate of heat generation is not uniform over the interface. The result is a heat-affected zone of non-uniform thickness, which, in consequence, can be neither the minimum nor the optimum thickness at all points across the interface.

These limitations and disadvantages can be avoided by using some other form of relative motion.

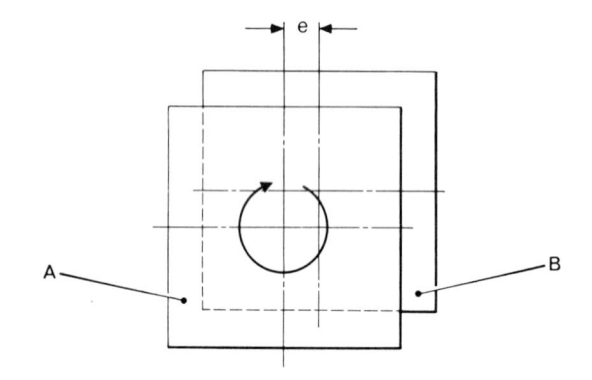

Fig 1

Diagram illustrating the principle of orbital friction welding. The orbiting component is indicated at A, the stationary component at B and the orbiting radius at e

Fig 2

During the first stage of orbital friction welding, the two components are rotated with their axes offset, and force is applied axially to generate frictional heat. For the second, forging stage, the components are brought into axial alignment and the endwise force is increased

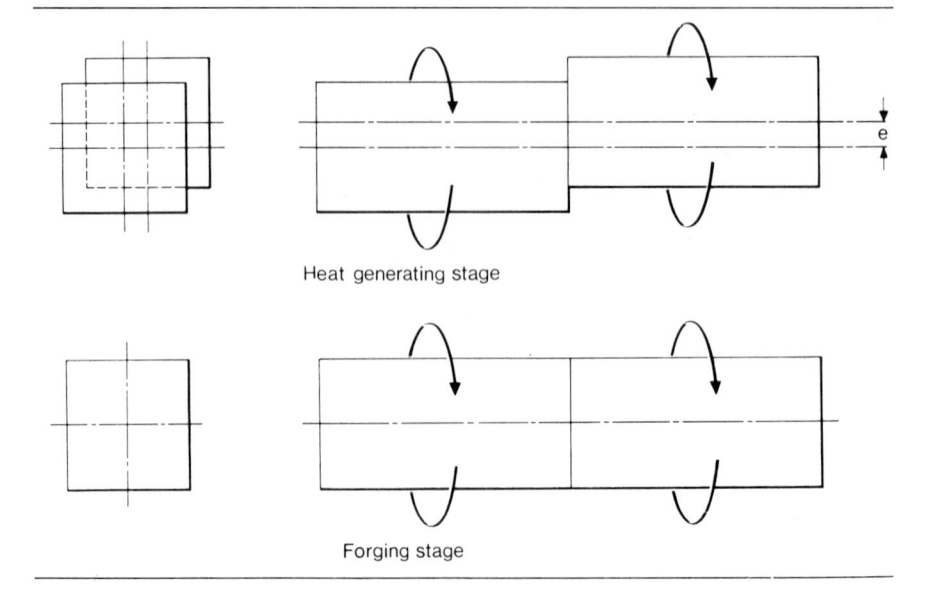

Heat generating stage

Forging stage

Orbital welding as a mathematical solution

Mathematical analysis has suggested that the most convenient of the possible modes of motion for producing heat is orbital movement. In this mode, one component is stationary and the other is caused to pursue a circular path without rotating about its own axis, as indicated in Fig 1. Here, the orbiting component A moves in a circular path of radius e about the centre of the stationary component B.

When sufficient heat has been generated it is necessary only to reduce the radius of orbit e to zero – that is, the moving component is centralized with the stationary part, thereby achieving the required alignment.

Three important advantages are claimed for this orbital welding technique. (1) A component can be of any cross-sectional shape, subject only to the limitation that it must have sufficient strength to withstand the lateral forces imposed during the heating phase. (2) More than one pair of components can be welded at a time, thus permitting very high production rates. (3) The relative motion between the interfaces is identical over those parts that are in continuous contact. The rate of heat generation is thus uniform over the greater part of the interface, so that a heat-affected zone of uniform thickness across virtually the full width of the interface is produced.

Problems of a practical machine

The idea of utilizing this form of motion led to the conception of the Friction Welding B-type orbital machine, in which the components are welded as outlined above. The advantages can be appreciated readily, but there are considerable difficulties in producing a practical machine. These difficulties arise from the fact that the orbiting component must be held in a substantial work-holder which, therefore, must be heavy. This work-holder must be able to orbit and then return to a zero-offset position. While it is in orbit, the work-holder will develop a very considerable centrifugal force, the magnitude of this force depending on the square of the welding speed.

If the work-holder and component together weigh 200 lb, then a welding speed of 2 ft/sec with an orbiting radius of $\frac{1}{8}$ in will produce a centrifugal force of about 1 ton; if it is necessary to use a welding speed of 4 or 6 ft/sec, this force increases to 4 or 9 tons, respectively. The corresponding figures for orbits/min are approximately 2,000, 4,000 and 6,000.

Clearly a machine of this type would require extensive development and, therefore, it was considered to be advantageous if a simple machine could be built which would enable orbital friction welding to be investigated, without having to develop a machine of the complex design originally envisaged.

The Friction Welding A-type orbital machine

Simpler equipment was designed and built in the form of the Friction Welding A-type orbital machine. In this unit, both components are rotated with their rotational axes offset. The relative motion between the components is orbital, but the high inertia forces and the balancing problem are avoided. The arrangement is shown diagrammatically in Fig 2.

Force is applied to the components with their axes offset. When sufficient heat has been generated as a result of friction, the axes are realigned and a greater 'forging' force is applied.

Such a machine appeared to be simple to design and develop, and a prototype was built with the following objectives in mind: (a) To prove that orbital friction welding gave satisfactory results; and (b) to provide the data necessary for the design and construction of the more complex B-type machine, as conceived originally.

Both objectives have been achieved as can be seen from the etched specimen, Fig 3. It will be observed that the weld is free from defects, with good flash formation, and it is expected that a prototype B-type orbital machine will be operational in September.

Although the A-type orbital friction welder which currently is in use was designed as a test rig and not as a production machine, it is considered that its low cost, wide capability and simplicity should allow it to be employed advantageously in industry.

Fig 4 is a general view of the A-type welding machine, and it has two heads, each of which has a spindle fitted with a collet for holding the work. Conventional rolling bearings are used to carry the spindle. Since the spindle speed is not dependent on the size of the weld interface, there is no necessity for a change in speed as the machine is developed for larger components. When larger thrust forces are required, hydrodynamic bearings may be employed since both the speed and operating cycle facilitate the application of such bearings. There are no balancing problems of any magnitude, and it is necessary for each spindle to be balanced only about its own axis of rotation.

The heads are driven in unison, and such a drive could be effected through a lay shaft connected independently to each head by gears or chains or some similar arrangement. A system of this type may be desirable in some applications, but it is expensive and the number of parts required adds considerably to the problem of eliminating backlash and relative movement between the two spindles. In the solution adopted, one spindle is driven directly from the motor and the second spindle is driven from the

Fig 3

The sound, defect-free interface, extending beyond the edges of the workpieces, as produced by orbital friction welding, is indicated by this etched specimen (by kind permission of The Welding Institute)

Fig 4

Friction Welding A-type orbital welding machine. A forging force of 6-tons can be applied, and workpieces of $\frac{7}{8}$-in square cross-section can be welded

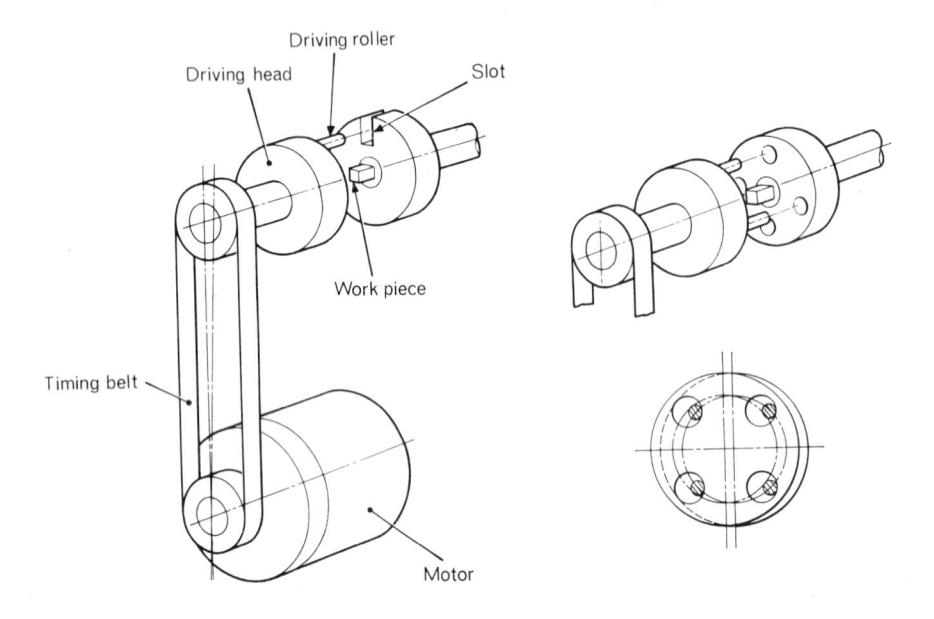

Fig 5

Drive systems as employed on the Friction Welding A-type machine. A roller-and-slot arrangement used for low-speed applications is seen at the left, and a roller-and-hole system for high-speed drive, at the right

Fig 6

Close-up view of the Friction Welding A-type machine showing the force frame, also the roller-and-hole drive system

first through a driving roller and a slot, as indicated in Fig 5.

This arrangement gives good alignment in the forge position, but rotation of the driven spindle is non-uniform. The arrangement proved satisfactory for speeds up to about 2,000 rev/min, and it has the advantage that no adjustment is required when the orbital radius is altered. For operation at high speeds, a drive-roller and hole combination is employed which gives uniform motion at the throw for which it is designed, as well as accurate final alignment. A close-up view of the machine showing the latter arrangement is given in Fig 6.

A force frame for applying thrust to the workpieces is incorporated in the machine and is shown schematically in Fig 7. This arrangement has to permit lateral movement of the head while load is applied, and, therefore, tie bars with spherical bearings are provided. The tie bar axes are in the same plane as the centre-line of the components, thus relieving the machine bed of any bending moments. The cross bar is pivoted on the mounting for the thrust cylinder which is free to move along the bed. In this manner, the tie bars provide all the reaction to the thrust cylinder.

The machine bed is subjected only to the forces generated in the plane of the weld interface. It is relatively free from stress, and since there are no welding forces when alignment between the components has been achieved, the bed is stress free when freezing takes place. Component inaccuracies due to machine deflections are thus minimized. A linkage and cylinder serves for moving the non-driven head to orbit radius and returning it to zero-offset position.

The machine based on these ideas is not large for a unit which, at this stage in its development, welds ¼-in square mild steel bar satisfactorily. As may be seen from Fig 3 the bond area extends well beyond the periphery of the bar. The machine has also allowed the multiple welding of components to be demonstrated and eight pieces of ⅛-in diameter bar have been welded simultaneously to produce four components.

It is considered safe to assume that orbital friction welding is capable of carrying out any operations that are now performed by rotational friction welding, with the possible exception of joining cylindrical sections with walls that are too thin to withstand the lateral loads imposed by the process. In this connection, however, it may be noted that the company have welded pipe of 0·080-in wall thickness, and no problems due to bending of the pipe wall have been experienced.

There was reason to suppose that the exposure of some of the weld interface during the orbital motion would lead to the formation of oxides and that the orbital movement might not have the efficacious cleaning action associated with rotational motion. It is now apparent from practical tests that these doubts were not justified.

It is anticipated that the full exploitation of the orbital principle may not be realized until a B-type orbital machine is available. The proposed design is shown schematically in Fig 8. As can be seen, the machine has two platens, one of which provides the orbital motion, while the other serves for application of axial thrust. Tooling to suit the work to be undertaken will be mounted on the platens and provision for manual or automatic loading will be possible.

The machine should lend itself to automatic loading. For example, batches of 8 components made from ⅛-in diameter bar could be welded simultaneously, and as mentioned earlier, the A-type orbital machine has been employed in this manner, using hand loading and produc-

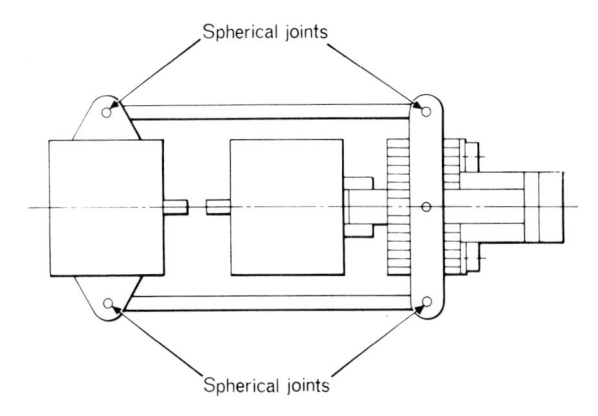

Fig 7

Diagrammatic plan view of the force frame arrangement for a Friction Welding A-type orbital welding machine. The frame has a floating cross-bar support

Fig 8

Schematic layout of a Friction Welding B-type machine for orbital welding, showing the platens which provide flat obstruction-free surfaces for mounting a wide range of tooling

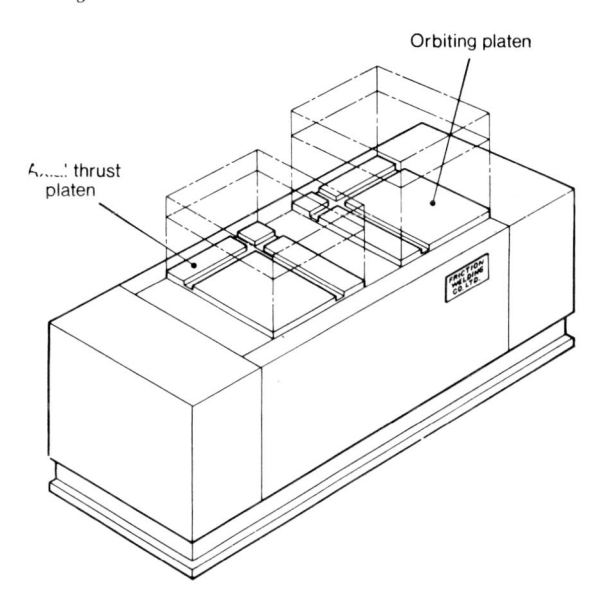

ing batches of four components. A walking-beam arrangement could be used to load and unload the work since it could pass between the work-holders, even while the orbital movement is taking place, since the orbital radius is only a fraction of an inch. The anticipated production rate would be of the order of 2,000 components an hour, allowing 14 s for each cycle. Orbital motion machines, it is considered, could be used for welding strip for continuous rolling, welding rails or pipe, and similar applications.

The choice between A-type and B-type orbital friction welding machines will depend upon the output required. Where quantities are large, the B-type will be more economical, since its higher cost will be offset by the greater output rate. Where quantities are smaller, the cheaper A-type orbital machine will be more economical.

Source: *Machinery and Production Engineering*, Sept 1, 1971

ADVANCES IN FRICTION WELD MONITORING AND CONTROL

By

Gerald S. Ellsworth
Advanced Technologies, Inc.

ABSTRACT

The friction welding process is being used increasingly in the United States for production of automotive and aircraft components having critical requirements meriting 100% quality control; an improved "Monitor/controller" has been developed for use with friction welders for quality control. The unit is also helpful in machine control, parameter development, and part design. The features of this Monitor/Control are discussed.

INTRODUCTION

Friction welding was developed primarily in Russia and England during the 1950's; the process is more widely used in Europe and Japan today with the United States lagging in general acceptance for production, partly because of confusion and difficulty in consistently obtaining reliable and repeatable results.

All friction welding machines inherently include control devices for machine operation; as requirements of the bonded joints become more severe, and as the process is used more extensively, several comprehensive units have been developed and evolved to a sophisticated degree. The monitor/control discussed, shows an added degree of convenience and reliability.

In most welding applications quality control is of utmost importance, regardless of the welding method. Generally used test methods include destructive bend, tensile, or rupture tests, or non-destructive ultrasonic, x-ray, or magnaflux tests. In general, these are expensive and time consuming tests that inhibit high production.

The monitor/controller is an in-process device that is reliable, easy to setup and provides instant substantiation of the quality of each and every weld joint made.

WELD PROCESS DESCRIPTION

Friction welding is a joining method which uses the heat generated at the interfaces by rubbing (rotating) one piece of material against another to form a metallurgical bond.

Usually the process involves one round part (tube or solid) which is spun at a particular speed, under pressure, against a stationary member. At the completion of the heat phase, the rotating member is stopped and the parts are forged together. The results include a bond having a narrow heat affected zone. (That is, the temperature does not reach the melt point of either part, since the metallurgical bond is achieved by diffusion, rather than fusion, and the heat does not migrate back far from the interface).

(The specific parameters for any given assembly depend upon the cross section of the joint as well as the physical and metallurgical properties of the materials.)

The friction produced heat first elevates the joint temperature to a degree that plastic deformation occurs, expelling oxides and contaminants from the joint area.

With the joint properly preheated, frictioning continues to further raise the interface temperature. During this heat phase, material is displaced from the joint forming the characteristic curl associated with friction welding.

When the interface reaches bonding temperature, rotation is stopped and forge pressure is applied.

Increased pressure during the forge phase produces a smaller grain structure in the interface and heat affected zones, promoting greater weld strength.

High integrity friction weld joints result with adherence to the required variables, as follows:

    A. Burn off rate (axial shorting with respect to time).

    B. Burn off distance.

    C. Forge distance.

For many applications, the values of these variables are not extremely critical; a wide range of values can be used to produce satisfactory results.

For illustration, the following graphs and motor load traces indicate stages of the friction weld process:
    (See Figure 1 & Figure 2)

MONITOR HISTORY

Several monitor or controller devices have been described (1,2) and marketed in the past.

Some of the inadequacies of existing unit are listed:

    A. Analog device (LVDT), stroke limited, wear problems.

    B. Required complicated adjustments for setup and "zeroing".

    C. Transducer problems, mounting of components, package ruggedness.

    D. Overall system accuracy, .1mm best, with temperature variation, ageing, and drift.

With the above inadequacies in mind, it was decided that an improved system would be developed that would monitor/control the friction weld process. In addition, the unit would include metric and English capabilities.

FRICTION WELD MONITOR

The A.T.I. Friction Weld Monitor was designed to provide concise control and monitoring of each stage of the friction weld process.

To insure quality and uniformity the monitor must:

    A. Control heat distance or heat time. (Some people use the word burn in place of heat).

This insures surface contaminants have been burned away and the interface has been raised to forging temperature.

    B. Monitor forge distance.

This insures that correct forge pressure is present at the right time.

The monitor is physically composed of three units.
(See Figure 3)

    The first is a motor load meter.

The second is a rugged, well sealed glass scale digital transducer.

The third is a self contained electronics package.

ITEM A - MOTOR LOAD DEVICE

In the past, several different methods of sensing part contact have been employed. Traditionally, pressure switch contacts, inertia switches (3) or limit switches have been used.

These were hard to adjust or have time delays that are difficult to overcome.

By using a motor load meter and its associated torroid coil, from one leg of the motor primary side, we can see the instant a demand is placed on the motor, indicating the parts have contacted and the system is functioning. By blanking off the clutch pick up load spike at the start of the spindle rotation, we have a consistant and non variable starting point for the process even when parts vary in length.

Experience has shown that badly deformed parts (cupped ends; nonsquare cut ends) or parts with lubrication contaminents will not affect the motor load enough to trip the set point until a sufficient cross section of the interfacial joint is reached.

This in effect allows you to wear away the deformed ends before the monitor starts to operate.

ITEM B - GLASS SCALE LINEAR ENCODER

The Linear Voltage Displacement Transformer (LVDT) has been around a long time and served as a very useful tool, but by going to a Glass Scale Linear Encoder, we have solved the listed limitations and provided other useful information. Since ram or tail stock travel may be up to two feet for loading and automation clearances, we can now know where our ram is at all times, if this is required. This information can be used to tell us if we have short parts, or not loaded one of the components (ram will travel too far before it sees a motor load). This could prevent welding an empty chuck to the tooling.

The Glass Scale gives us a resolution of .01mm or .001 inch for accuracy.

This is a typical unit; the same as employed on the glass scale readouts on many machine tools.

The device gives us an up/down count so we can use it for backing out the ram or telling us if we have a welded part extracted from the holding device.

ITEM C - ELECTRONICS PACKAGE

This group of items consist of a Nema 1 housing with a built in 110V power supply. It has a front panel with thumb wheel presets for the three phases of the welding process, LED output displays and setup instructions.

WELDING STAGES

PREHEAT

During this stage, the surfaces are squared up and contaminants are thrown away from the weld area. The surfaces continue to gall and stable friction is developed.

HEAT

Once stable friction is established by the pre-heat stage, weld pressure is increased to raise the interface temperature to the materials forging range.

DECLUTCH AND BRAKE

When forging temperature has been reached, the drive system is de-clutched and the spindle is stopped.

FORGE

Upon application of braking, pressure is raised and held until the interface cools, forming the weld.

MONITOR OPERATION

When the components to be welded make contact, the monitor self zeros and begins its cycle. (See Figure 4 & 5)

The electronics package processes the transducer information, thus keeping track of the weld rams exact position at all times.

The pre-heat timer is started by the motor load meter and is set only long enough to get the parts to a consistant cross section, usually in the order of 1-2 seconds.

The distance the ram travels during this stage will be displayed on the front panel.

At the completion of pre-heat time, a second clock is started to keep track of heat time. As material is consumed, the monitor compares the distance the ram travels to the setting on the heat distance tumb wheels. When this preset is reached, the clock timer is stopped and a machine interlock is provided to stop the chuck and begin forge. The amount of time required to reach this distance is displayed on a 4 digit readout on the front panel. This display is held until the next part cycle.

Upon receiving this interlock, the monitor compares the distance traveled with two thumb wheel presets; minimum and maximum forge distance.

Forge distance must fall within these limits for the weld to be considered acceptable.

Next, the monitor looks at the amount of time taken during burnoff and compares it to the minimum and maximum heat time thumb wheel presets on the front panel. The amount of time required to reach heat distance must also fall within these limits for the part to be acceptable.

Thus, it can be seen that the monitor not only provides constant control of heat distance but also performs a go - no go check of every part welded.

TOTAL DISTANCE

While total distance moved during the welding process may or may not be important, we have the capability to read this distance, after weld completion.

Source: SME Technical Paper No. AD78-746, 1978

By pressing a total distance display button, the unit will read the travel from start of weld to completion. This distance will be more than the sum of the pre-heat distance, plus heat distance, plus forge distance by the amount of travel during the arrest time. This is quite consistant for each part and is a good check on clutch and brake performance.

METALLURGICAL CROSS CHECK

Of course, the above explanation of the weld monitor/controller does not relieve the requirement of a periodic sampling of the welded product for metallurgical evaluation. What it does do is resolve some of the "doubt factor" that is always present in a fabrication process and brings the friction weld process closer to a science than a "black art".

MONITOR FEATURES

1. All displays are long life 7 segment LED type.

2. All displays are plug in for quick replacement.

3. All connections to the monitor are via quick release cannon plugs.

4. Indicator lamps are replaceable from the front panel.

5. All logic is solid state integrated circuit for highest noise immunity and reliability.

6. Fault logic uses fail safe approach and is power failure proof.

7. Metric inch conversion is standard on all monitors.

8. All distances are measured with a bi-directional (up-down counted) to guarantee positional accuracy.

9. System is entirely digital; therefore, no adjustments are ever required.

PRINTOUT OPTION

A paper tape printer is available as a plug addition. It can be added at any time this feature is desired; as all wiring is connected through a single cannon plug on the monitor.

Forge distance and burn time are printed in black for a good part and red for a bad part.

SUMMARY

This paper reviews the friction weld process and describes a new tool to give an "in-process" answer to the quality control question, "Is it a good weld?".

The equipment described has been proven in production and has been shown to be easy to install, set up and operate. With such a tool, an unskilled operator can run a friction welder and the product assurance question is virtually resolved.

The device is also useful in inertia welding, to monitor distance traveled.

REFERENCES

1. Ellis, Nicholds, The Welding Institute.

2. Electro Test, Harlow Ltd.

3. Inertia Switch by Blacks, Metalurking Production, June 74.

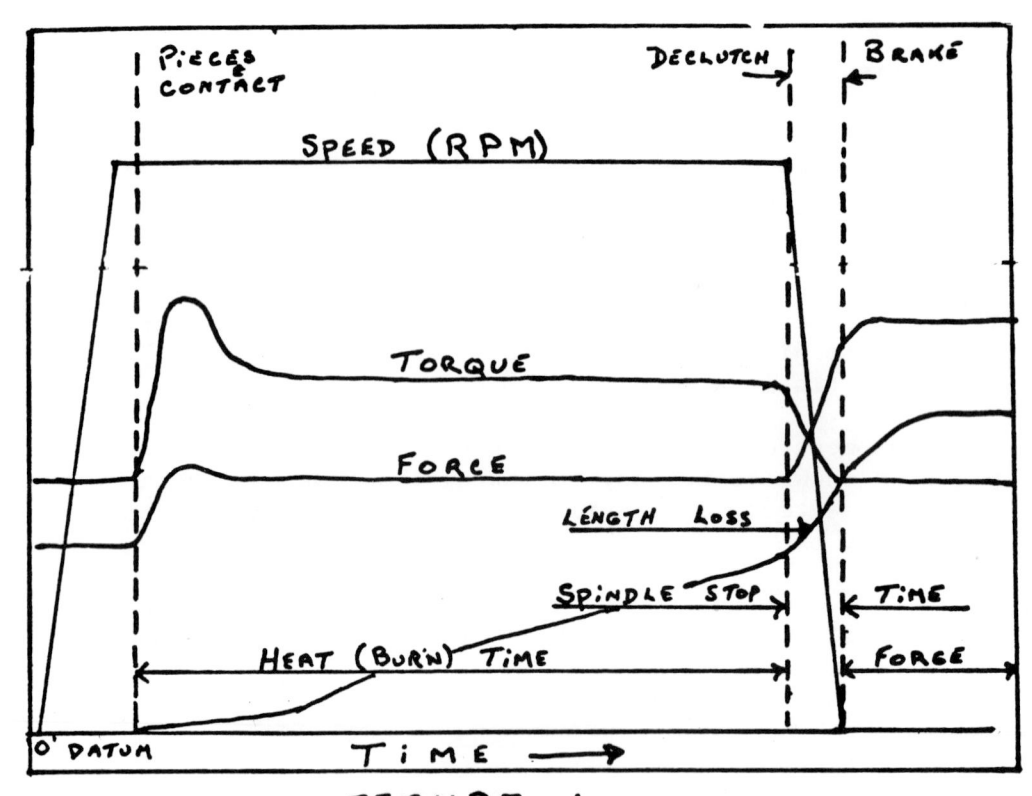

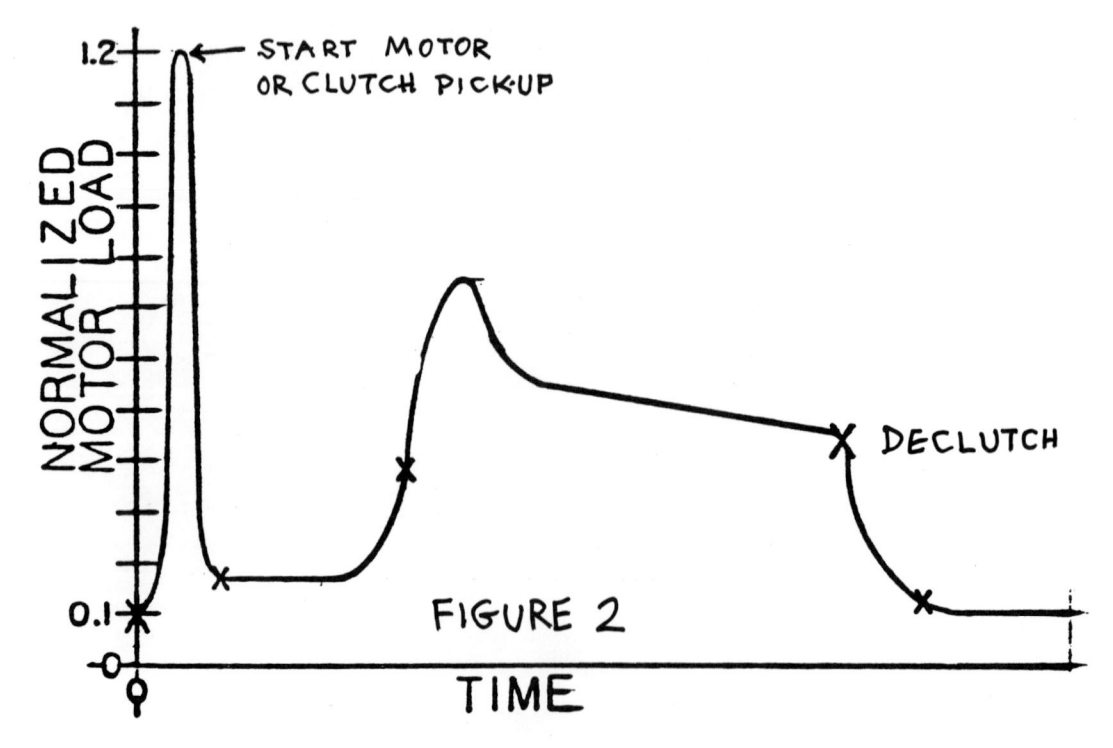

FIGURE 1

FIGURE 2

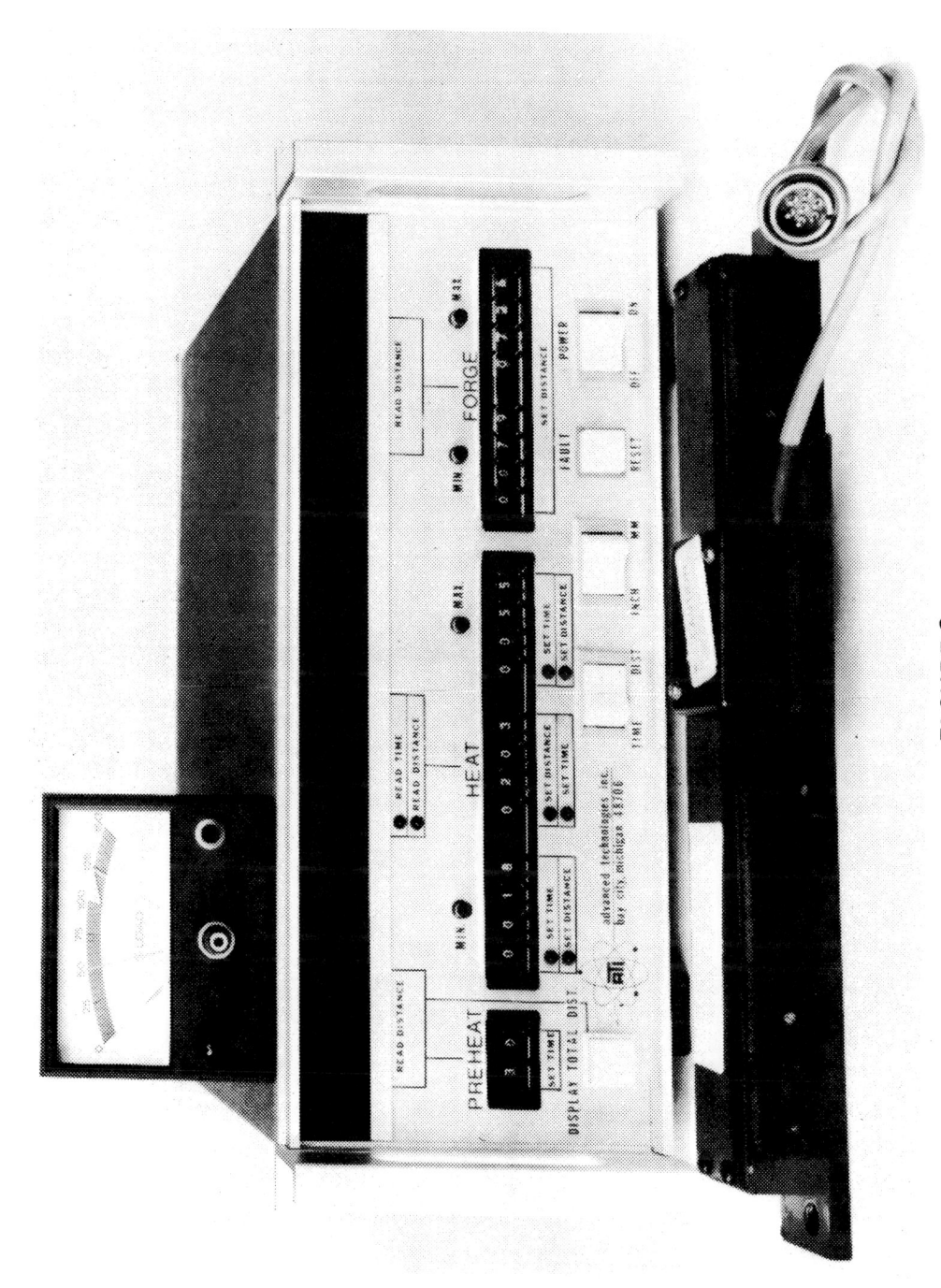

FIGURE 3

Source: SME Technical Paper No. AD78-746, 1978

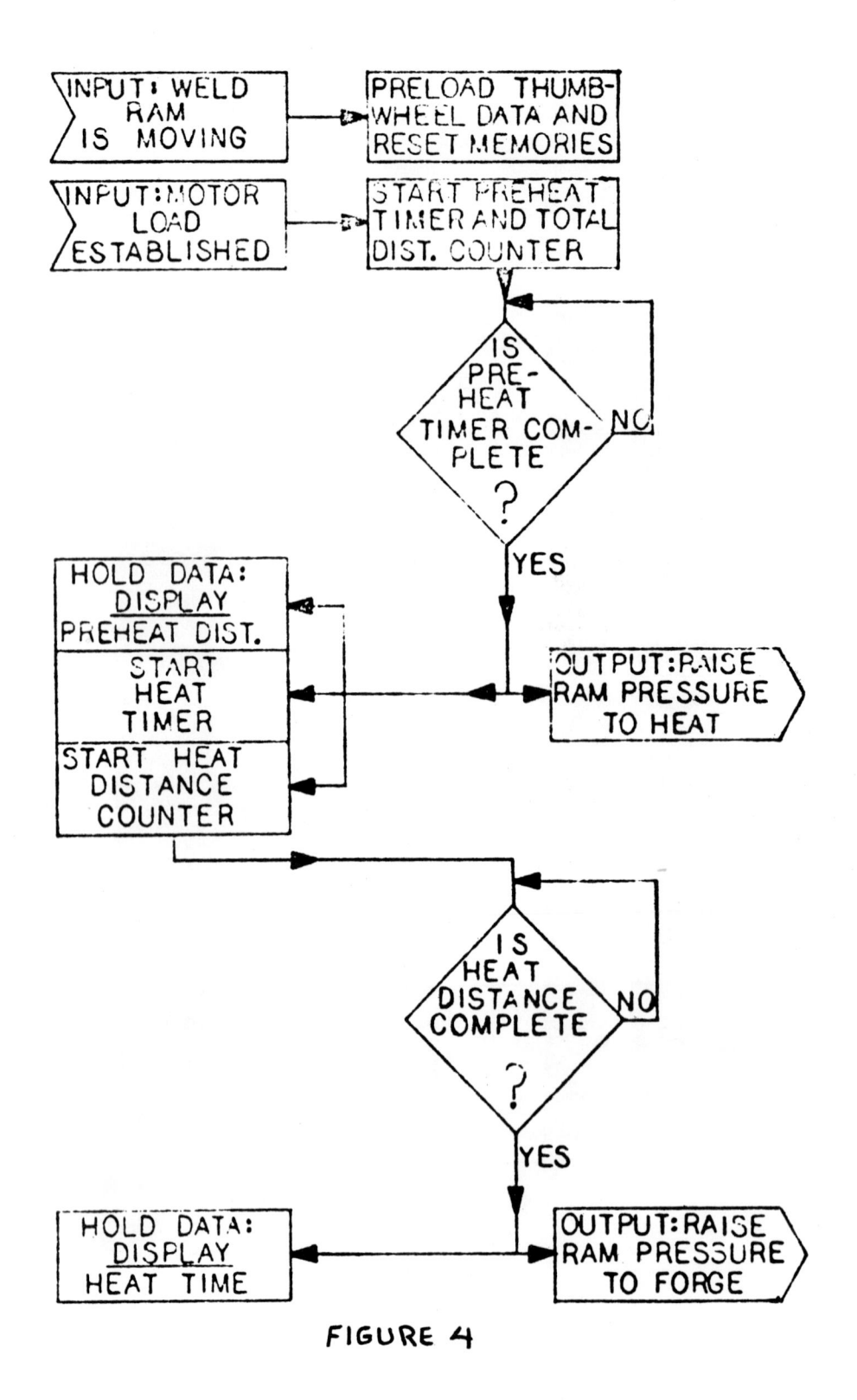

FIGURE 4

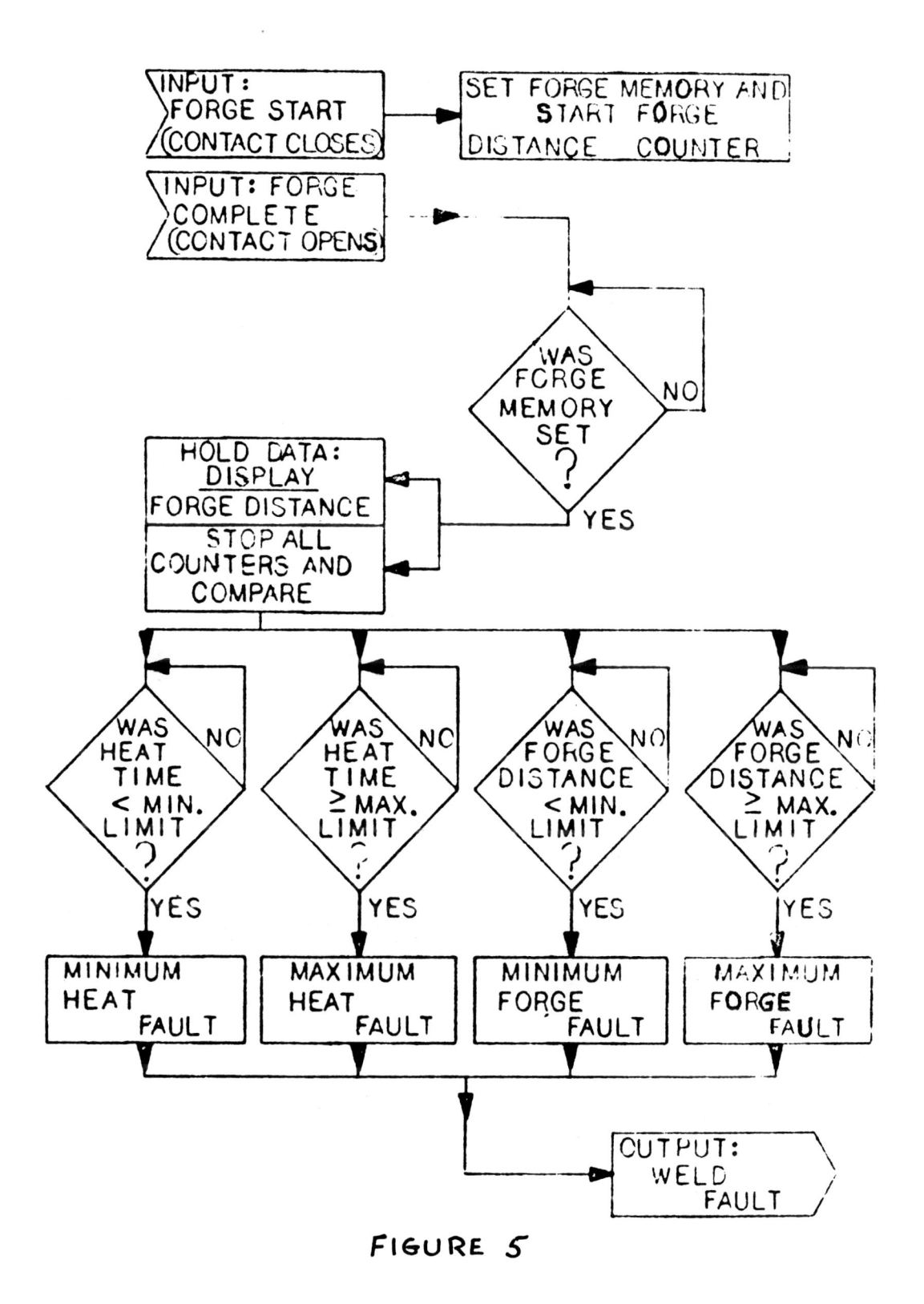

FIGURE 5

FRICTION WELDING OF MARAGING STEEL FOR
SMALL MISSILE-SYSTEMS APPLICATIONS

D. A. Seifert and K. E. Meiners
BATTELLE
Columbus Laboratories
505 King Avenue
Columbus, Ohio

ABSTRACT

A study of the application of friction welding to the fabrication of small diameter rocket-motor cases was made, and several areas where friction welding could prove cost effective were identified. Experimental investigations of the factors affecting the quality of friction welds between thin walled 18Ni(250) maraging-steel tubes were conducted. An empirical model relating weld strength to rotational velocity, axial heating and forging forces, and axial upset or shortening was developed from these investigations. It was found that this model could, with a fair degree of reliability, be used to predict the conditions necessary for obtaining optimum-quality continuous-drive friction welds between maraging steel tubes from 2.8 to 5.8 in. in diameter having diameter-to-wall thickness ratios of from 30:1 to 85:1.

1. INTRODUCTION

The basic principle of friction welding involves the simultaneous application of pressure and relative motion, generally in a rotational mode, between the components to be joined. The frictional heat thus generated raises the interface temperature of the components to very nearly their melting points, while the applied pressure perpendicular to the plane of motion serves to extrude the heated material including any dirt and oxide films from the interface, bringing the components to be joined into intimate contact. Termination of the relative motion while maintaining or even increasing the applied pressure then serves to produce a sound metallurgical bond between the two components. In the case where the components to be joined are metallurgically similar, the properties of the weld zone can approach, or even exceed, those of the parent material. Microstructurally friction welds are often typical of fine-grained wrought material, with the actual interface being indistinguishable from the surrounding material.

Continuing advances in the power and efficiency of solid-rocket propellants have created a need for stronger and more reliable materials of construction for the various missile-system components, especially the motor cases. In the case of small-diameter ordnance missiles, the use of more exotic and consequently more expensive materials for structural components brought about a need for improved production efficiency and decreased material waste in order to keep manufacturing economy at an acceptable level. Friction joining offers a promising solution to these problems. Not only is it adaptable to high speed automated production, it offers highly reliable joints with properties approaching those of the parent material and the potential for significant material and manufacturing savings through greatly reduced machining losses and the possible elimination of costly and difficult forming operations.

It was, then, the purpose of this investigation to determine possible areas where friction joining may be successfully applied to ordnace missile fabrication and to develop the necessary technology for its implementation. A parametric study was thus conducted to define the effects of the independent friction-welding variables on the quality of specified weld-joint configurations using materials of primary interest for missile-systems fabrication. The findings for one such joint configuration between maraging-steel components are the topic of this discussion.

2. EXPERIMENTAL

The main technical effort was aimed at optimizing the conditions for achieving satisfactory friction welds between tubular sections having both the same and different wall thicknesses and between tubular and plate-type elements. The materials of principal interest included the 18 percent nickel maraging steels and a high-strength aluminum alloy. This discussion, however, will be limited to the results obtained for simple butt welds between thin-walled maraging tubes having similar cross sections.

Continuous drive (or conventional) friction-welding experiments were carried out to determine the effects of heating pressure, relative rotational velocity, axial shortening (upset), and forging pressure on butt welds between 18Ni(250) maraging-steel tube segments ranging in diameter from 2.8 to 5.8 in. and having diameter-to-wall thickness ratios ranging from 30:1 to 75:1. Microstructural examination and mechanical-property evaluation, including tensile and simple bend tests, were used to evaluate the integrity of the welds. Multiple correlation/regression techniques were then employed to evaluate and weigh the effects of the various independent weld-cycle variables on the resultant weld properties, principally tensile strength. This statistical treatment of the data yielded an empirical model from which optimum friction-welding conditions can be chosen for tube-to-tube or similar weld-joint configurations between maraging-steel components.

3. RESULTS AND DISCUSSIONS

3.1 APPLICATION STUDY

Several potential areas for the application of friction welding to missile-systems hardware fabrication were determined by careful study of designs presented in the Rocket Motor Manual.[1] In most instances, the motor cases for these missiles systems were fabricated by deep drawing or spin forging, followed by machining of a suitable tool or alloy steel. Examination of the design sketches of each of the systems showed the motor cases to consist essentially of long thin-walled tubes with thickened end sections to permit attachment by either welding or mechanical fastening of end caps, nozzle assemblies, etc. One design even incorporated an intermediate thick-walled section for the support of internal hardware. Unless extremely sophisticated forming tooling and techniques were employed, fabrication of these motor cases would require a significant amount of costly and time consuming internal machining. Even the simplest design reviewed required the attachment of the head end by welding.

The application of friction-welding technology to the fabrication of these rocket-motor systems could greatly increase the cost effectiveness of their production by eliminating machining steps, speeding up production, and possibly by simplifying forming operations. Neither would the mechanical integrity of motor cases fabricated by this technique be sacrificed as friction-welded joints can be produced whose mechanical properties are as good as and sometimes better than those of the parent material. With the proper selection of joint configuration, essentially all internal machining could be eliminated from the production of wall-thickness transitions. Thick- and thin-walled tubular sections could simply be friction welded together in a matter of seconds using a half-lapped joint configuration. The need for deep drawing closed end tubes could also be eliminated by friction welding tubular case body sections to more easily formed separate end caps. Capital equipment expenditures might also be reduced somewhat by replacing deep-drawing equipment, including associated heat-treat furnaces, with high-production-rate friction welders. Tubing of the proper diameters and wall thicknesses could then be purchased in quantity from a commercial vendor, cut to the proper lengths, and friction welded together to form rocket motor-case bodies.

The assembly of other missile-system components might also be facilitated through the use of friction welding. Many dissimilar as well as similar materials have been successfully joined by various investigators.[2-11] Thus, the probability exists that guidance packages, warheads, guidance fins, and nozzle assemblies might be quickly and reliably joined to rocket-motor cases by friction welding. These components, not required to withstand the extreme internal pressure of rocket engines, could be fabricated from aluminum and other lightweight alloys.

Notwithstanding the above cited advantages to friction-welding missile-system components, there are several materials- and configuration-oriented variables which must be studied when considering the application of friction-welding technology to a specific missile-system production process. Thermal conductivity and particularly thermal-expansion differences between the two components to be joined greatly influence friction weldability. Strength could also be an important factor. First, the strength of a friction-welded joint between dissimilar materials is not likely to be much stronger than the weaker of the two. Second, and possibly more important, residual stresses generated on cooling from the welding temperature by differences in thermal expansion could seriously weaken or even cause failure in some systems. Rigorous investigation of these factors may be, to some extent, bypassed if friction weldability of the two component materials in question has been previously established.

The physical design and shape of components to be friction welded could also influence the decision whether to incorporate friction welding in a production process, particularly where tubular components are involved. Such factors as diameter-to-wall-thickness ratio (D/T), area available for gripping the components, minimum proximity of faying surfaces to chucking devices, and allowable flash formation during welding must be considered.

Thermal considerations such as temperatures required at joint interfaces and allowable temperature distributions as a function of distance from joint interfaces may also be important for some applications. Several investigators[12-18] have provided mathematical models by which these factors may be estimated with reasonable accuracy once certain experimental information regarding heat-input rates, coefficient of friction, etc., are determined.

Information required for specification of friction-welding equipment for a particular joint configuration and materials combination can generally be determined only through experimental investigation. Power requirements are a prime example. The amount of torque generated at faying surfaces is highly dependent on coefficient of friction, a temperature- and axial-pressure-dependent function, and configurations; e.g., for a given pair of materials and a fixed axial pressure, the torque generated during frictional heating is obviously greater for a tubular configuration than for a solid rod configuration having the same cross-sectional area. A Russian investigator[17] has suggested that a set of nomograms might be constructed for specifying the various friction-welding-cycle variables based on a knowledge of the temperature dependence of the shear strength of the material being welded. The interrelationship between shear strengths, speed of rotation, axial pressure, and time to reach equilibrium torque must still be experimentally determined for the configuration in question.

Based on the above review of potential applications of friction welding to missile-system hardware production, three friction-welding joint configurations were chosen for experimental investigation. A simple butt joint between thin-walled tubes was given primary consideration, with tube-to-plate and "flashless" tube-to-tube configurations also being studies. The material of specific interest was 18Ni(250) maraging steel.

3.2 PRELIMINARY INVESTIGATIONS ON MARAGING STEEL

A series of some 30 friction-welding experiments were carried out using maraging-steel tubes having a mean diameter of approximately 2.8 in. and a diameter-to-wall thickness ratio (D/T) of 30 to 34. Examination of the microstructures of these preliminary welds indicated that bonding is readily achieved over a wide range of welding conditions, but that optimum microstructures, e.g., minimum perturbation of the parent metal structures, were considerably more difficult to produce. Since the parent metal had a very fine grain structure (about ASTM #7 as shown in Figure 1, it was found that any

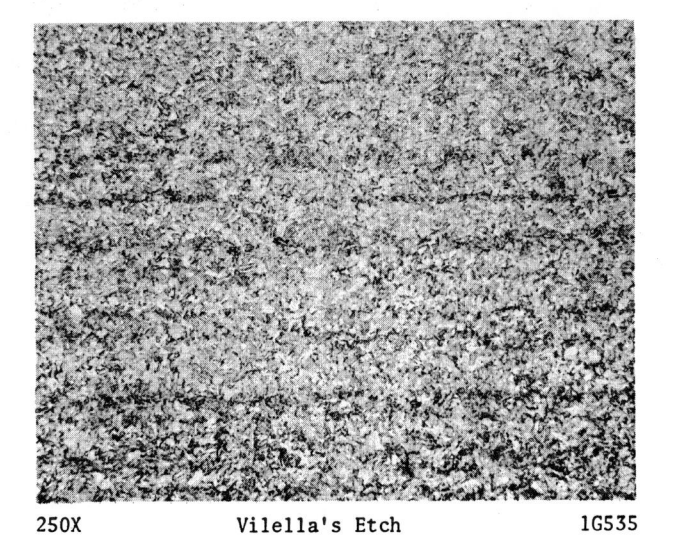

250X   Vilella's Etch   1G535

FIGURE 1. PARENT METAL STRUCTURE OF 18Ni(250) MARAGING-STEEL TUBE SPECIMENS Aged 3 hr at 900 F.

set of welding conditions which generated more than sufficient frictional heating tended to cause significant grain growth in the areas immediately adjacent to the weld interface. This was particularly true where heating rates were low and/or frictional heat generation times were long. This is illustrated in Figure 2, which shows the effects of (a) prolonged heating at moderate pressures and (b) heating at low contact pressures which increased the times required to achieve reasonable axial displacements. Excellent microstructures were achieved, on the other hand, at high heating and forging pressures and moderate speeds and at low heating pressures and speeds as shown in Figure 3. Unfortunately, however, those conditions which produced good microstructures at low axial pressure and rotational speeds did not achieve axial displacements sufficient to insure bonding over the entire faying surface areas.

Preliminary studies on the effects of forging pressure indicated that, in general, increasing the forging pressure above the level of the heating pressure tended to reduce the width of heat-affected zones and to produce some grain refinement at the weld interfaces, particularly where relatively low (3000 to 5000 psi) heating pressures were used. This is illustrated in Figure 4, where, after heating for approximately 3.7 sec. at an axial pressure of 4800 psi and a rotational speed of 2000 rpm, axial pressure (a) was not significantly increased and (b) was approximately doubled at ermination of the weld cycle. Some disadvantages to the use of high axial forging pressures were also found, as the possiblity of axial misalignment in the welded specimens was increased, particularly where heat inputs were high and prolonged and/or specimen support was relatively distant from the faying surfaces.

One of the objectives of this investigation was to determine the feasibility of friction welding together missile-systems components which contained certain viscoelastic materials. In the case of maraging steel this would mean friction welding together previously maraged components. A single

100X   20H$_2$O, 10HNO$_3$, 20HCl, 1/2 g FeCl$_3$ 5F433   100X   6F939

(a) Welding conditions: 1000 rpm; 5840 psi heating pressure; 6360 forging pressure; approximately 0.4 in. upset; 21 sec (as welded).

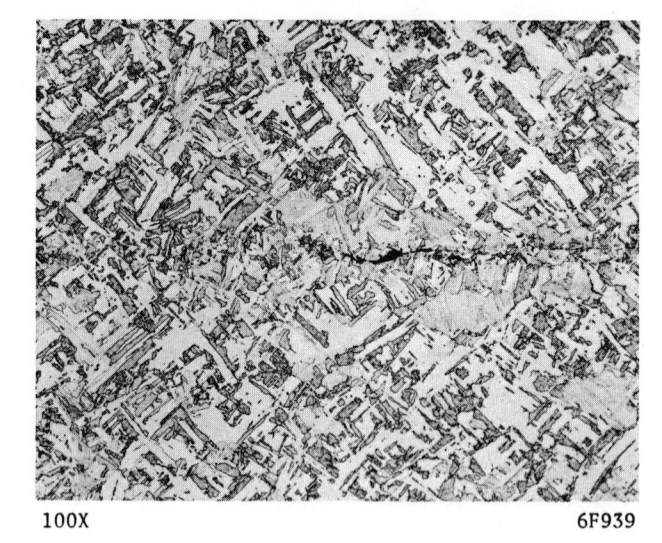

(b) Welding conditions: 1500 rpm; 3880 psi heating pressure; 6700 psi forging pressure; 0.044 in. upset; 5.1 sec (as welded).

FIGURE 2. ILLUSTRATION OF GRAIN GROWTH ADJACENT TO FRICTION WELDS CAUSED BY PROLONGED HEATING AND/OR LOW HEAT-GENERATION RATES

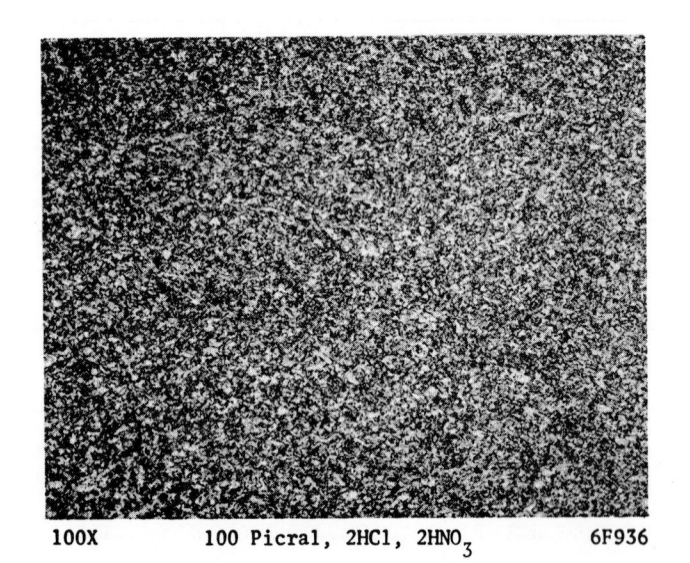

100X   100 Picral, 2HCl, 2HNO$_3$ 6F936   100X   20H$_2$O, 10HNO$_3$, 20HCl, 1/2 g FeCl$_3$ 5F431

(a) Welding conditions: 2000 rpm; 11,960 psi heating pressure; 21,290 psi forging pressure; 0.045 in. upset; 0.88 sec (as welded).

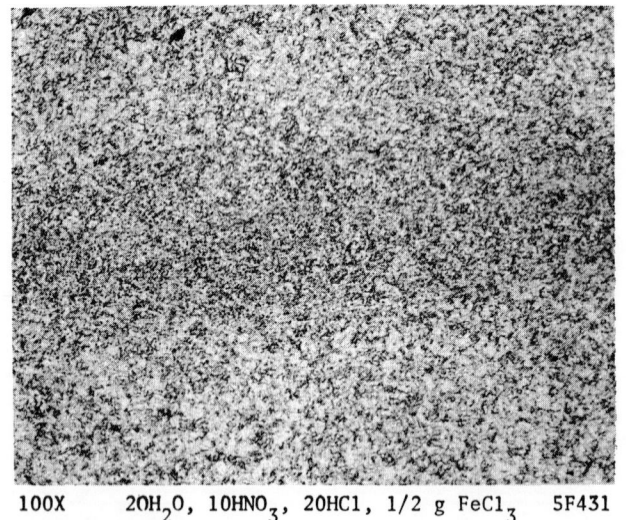

(b) Welding conditions: 1000 rpm; 3880 psi heating pressure; 3880 psi forging pressure; 0.008 in. upset; approximately 1 sec (as welded).

FIGURE 3. ILLUSTRATION OF RETENTION OF FINE-GRAIN STRUCTURE AND GRAIN REFINEMENT PRODUCED BY SHORT WELD CYCLES AT BOTH HIGH AND LOW HEATING PRESSURES

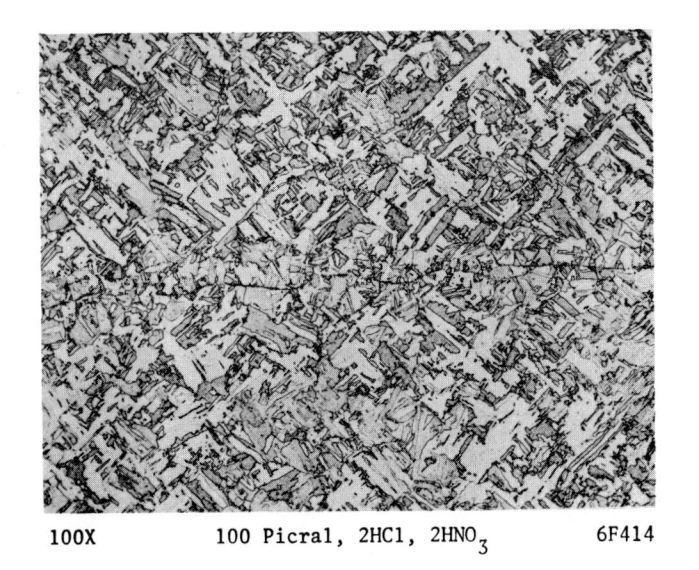

100X 100 Picral, 2HCl, 2HNO$_3$ 6F414	100X 100 Picral, 2HCl, 2HNO$_3$ 6F413
(a) Welding conditions: 2000 rpm; 4830 psi heating pressure; 5160 pis forging pressure; 0.033 in. upset; 3.73 sec.	(b) Welding conditions: 2000 rpm; 4740 psi heating pressure; 9360 psi forging pressure; 0.038 in. upset; 3.72 sec.

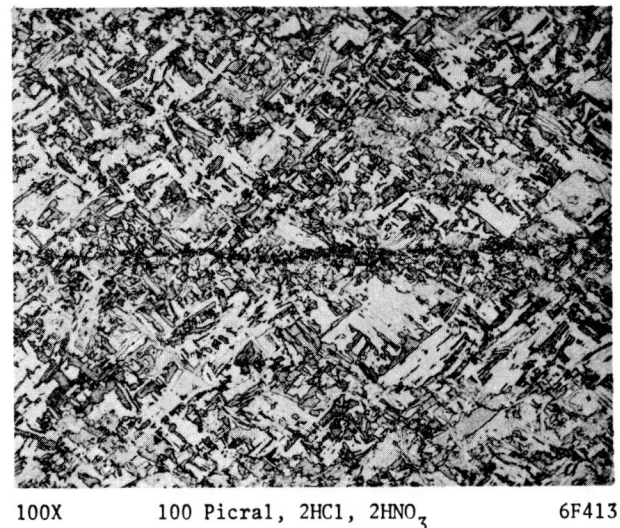

FIGURE 4. ILLUSTRATION OF THE EFFECT OF INCREASED FORGING PRESSURE ON THE MICROSTRUCTURE OF MARAGING-STEEL FRICTION WELDS

experiment was conducted to investigate this possibility by welding together two tube segments which had been previously maraged for 3 hr at 900 F. Welding conditons which had previously been shown to yield satisfactory weld microstructures were used for this experiment. After welding, the specimen was examined metallographically and a microhardness trace across the weld interface made. As shown in Figure 5, a distinct loss of hardness was evident at the weld interface. This, as was

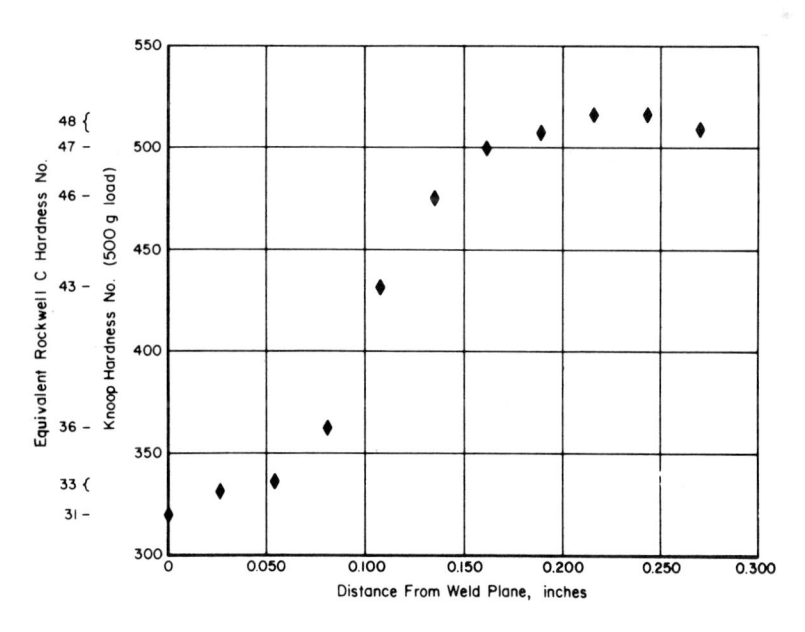

FIGURE 5. MICROHARDNESS OF FRICTION WELD BETWEEN MARAGING-STEEL TUBES AGED AT 900 F FOR 3 HOURS PRIOR TO WELDING

Source: *Materials & Processes for the 70's*, SAMPE, 1973

not unexpected, was probably caused by rapid re-solutioning of the Ni_3Ti and Fe_2Mo precipitates which give maraging steel its extraordinary strength and toughness.[18,19] A band of decreased strength commensurate with the band of reduced hardness was probably therefore generated. This would have to be removed by an additional maraging treatment to realize the optimum properties of maraging steel. Such a heat treatment would, in the very least, be detrimental to any contained viscoelastic material as well as requiring two maraging heat treatments for each component produced. In view of these results it was decided that further friction-welding studies involving previously maraged specimens would be unwarranted. This and a previous experiment had already indicated that the conditions needed to weld maraged tubes did not differ significantly from those needed to weld maraging-steel tubes in the solution annealed condition.

3.3 PARAMETRIC INVESTIGATIONS OF FRICTION BUTT WELDS BETWEEN MARAGING-STEEL TUBES

It was the primary purpose of this investigation to determine the interrelationships among the independent (controllable) friction-welding cycle variables, e.g., rotational speed, axial heating pressure, axial forging pressure, and axial displacement (or, alternatively, heating time), and their effects on resultant weld quality. The bulk of the welding experiments were carried out using tubular specimens having a mean diameter of about 2.8 in. and a diameter-to-wall thickness ratio (D/T) of about 30:1. Additional experiments, using specimens having mean diameters of about 2.8 in. and D/T's of 45 to 49 and specimens having mean diameters of about 5.8 in. and D/T's of 75 to 85 were performed to investigate the scalability of those weld-cycle variables proven by mechanical-property tests to have produced acceptable welds in the 2.8 in. in diameter, 30:1 D/T specimens.

The mechanical properties, specifically tensile strengths, of the welded and subquently maraged specimens were first examined with respect to each of the independent variables individually. This was done graphically as shown in Figures 6 through 8. It was immediately apparent from these representations of the data that no single independent variable was particularly predominant in controlling the tensile strength of friction-butt-welded maraging-steel tubes. Similar correlations between ductility (elongation at rupture) and the independent cycle variables were considerably less conclusive while those involving either strength or ductility with rotational speed were decidedly inconclusive. It was therefore decided that a relatively complex relationship in which each of the independent variables contributes to the attainment of the resultant weld properties must exist.

Multiple correlation/regression techniques using a relatively broad but powerful statistical analysis computer program[20] were applied to the data generated in order to determine the influence of each of the independent weld-cycle variables on the strength of friction-welded maraging-steel tubes. Examination of the data as presented in Figures 6, 7, and 8 suggested that axial displacement and heating pressure had a stronger influence on mechanical properties than did forging pressure or relative rotational velocity. Further, the trends suggested by these data indicated that the functional relationship between tensile strength and both upset and heating pressure should be either exponential, logarithmic, or parabolic in form and that forging pressure should have a weak parabolic relationship with tensile strength. All combinations of these functional relationships, along with linear and logarithmic functions of surface velocity, were fit to the data and evaluated statistically. The most satisfactory statistics, e.g., multiple correlation coefficient, standard error, and F-ratio were obtained for a polynomial describing tensile strength as dependent on the exponential of axial displacement, U_H, the square root of heating pressure, P_H, the square root of the ratio of forging pressure, P_F, to heating pressure, and on linear surface velocity. The relationship was statistically improved by normalizing the axial displacement values based on specimen

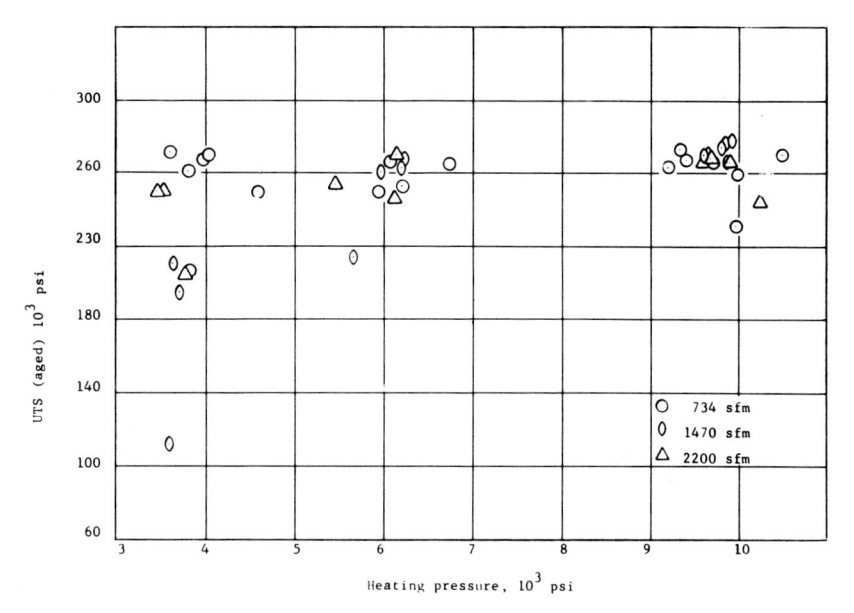

FIGURE 6. EFFECT OF HEATING PRESSURE ON TENSILE STRENGTH OF FRICTION-BUTT-WELDED MARAGING-STEEL TUBES

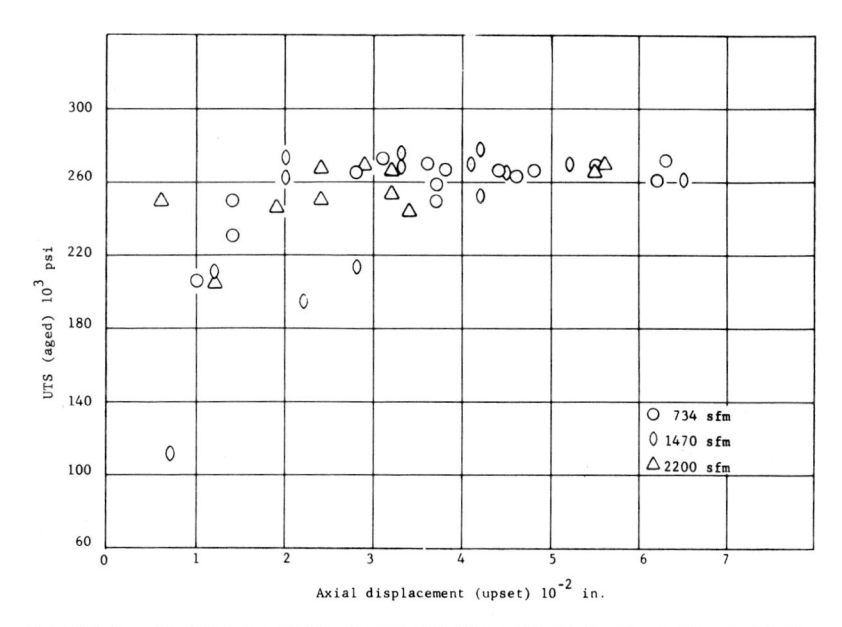

FIGURE 7. EFFECT OF AXIAL DISPLACEMENT DURING HEATING ON TENSILE STRENGTH OF FRICTION-BUTT-WELDED MARAGING-STEEL TUBES

Source: *Materials & Processes for the 70's*, SAMPE, 1973

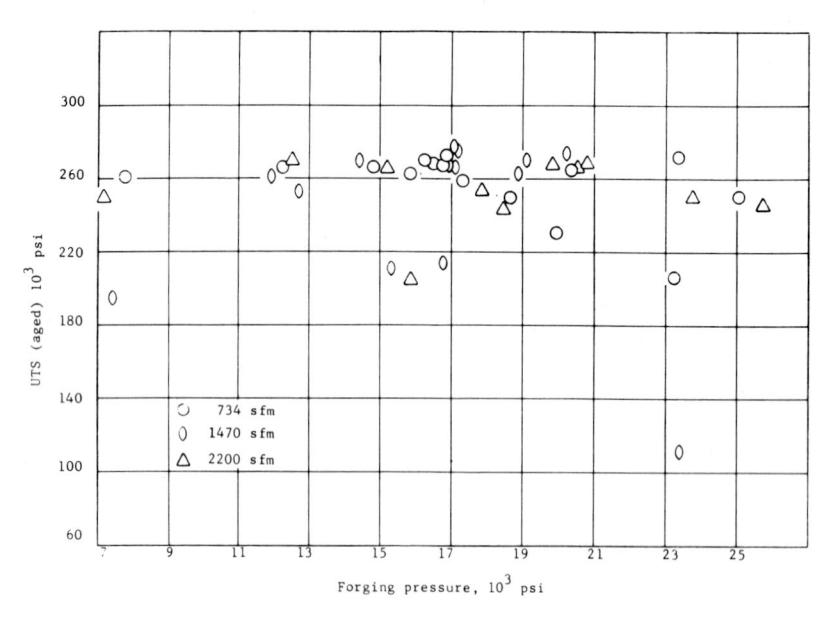

FIGURE 8. EFFECT OF FORGING PRESSURE ON TENSILE STRENGTH OF FRICTION-BUTT-WELDED MARAGING-STEEL TUBES

cross-sectional areas in terms of volume of material displaced from the faying surfaces. The relationship thus obtained is presented in Table 1, along with the relevant statistics. This relation is taken to be valid for butt-welded maraging-steel tubes having mean diameters between 2.8 and 5.8 in. with D/T ratios between 30:1 and 85:1. According to the statistics presented in the table, the regression model accounts for some 55 percent of the variance observed in the tensile strengths, and strength data predicted from the model should not be in error by more than 10 percent of levels greater than 218,000 psi. A comparison of actual tensile strengths with those predicted from the regression equation is presented in Figure 9. It can be seen from this relation that most of the experimentally determined strength data fall within the standard error limits of the model, except at the lower strength levels where the density of experimental data points is much lower. The solid symbols represent values for experiments using specimens having larger mean diameters and higher D/T's and whose welding condtions were derived by scaling based on specimen cross-sectional area. This indicates, then, that the model derived here is adequate for use by designers in specifying the conditions for friction-butt-welding thin-walled maraging-steel tubes.

According to the model presented above, acceptable friction-butt welds can be obtained over a relatively wide range of speeds, heating pressures, axial displacements, and forging pressures. In order to determine what combinations of these variables might provide the most cost-effective welding conditions, the relationship between equilibrium axial-displacement rates, axial heating pressure, and rotational velocities was studied. Equilibrium displacement velocities for welds between 2.8 in. in diameter, 30:1 D/T tubes, when reached prior to weld-cycle termination, were determined from the displacement versus time curves on the weld-cycle records. Equilibrium volumetric displacement velocities are presented in Figure 10 as a function of heating pressure for three rotational speeds. Several interesting points can be derived from these relationships. First, it would appear that there is little effect of rotational velocity on displacement rates at low heating pressures, and that displacement velocity increases with increasing axial pressure at a rate that is greater at higher rotational speeds. Also, the pressure-versus-displacement velocity relationships

TABLE 1. EMPIRICAL RELATIONSHIP BETWEEN STRENGTH AND INDEPENDENT
WELD-CYCLE VARIABLES FOR FRICTION BUTT WELDS BETWEEN
THIN-WALLED MARAGING-STEEL TUBES

$$UTS = 168509 - 140703 \exp(-100\, V_H) + 783\sqrt{P_H} + 22861\sqrt{P_F/P_H} + 1.53\, S_L$$

UTS = Tensile strength, psi

V_H = Volume of material expelled from faying surface, in.^{3}

P_H = Axial heating pressure, psi

P_F = Axial forging pressure, psi

S_L = Relative surface velocity of sliding components taken at the mean tube diameter, standard ft per min.

Statistics:

Multiple correlation coefficient, R	0.74470
R squared (portion of variance in dependent variable accounted for by regression eqn.)	0.55458
Standard error	21,879
Degrees of freedom in regression	4
Degrees of freedom in residual	35
F ratio	10.89453

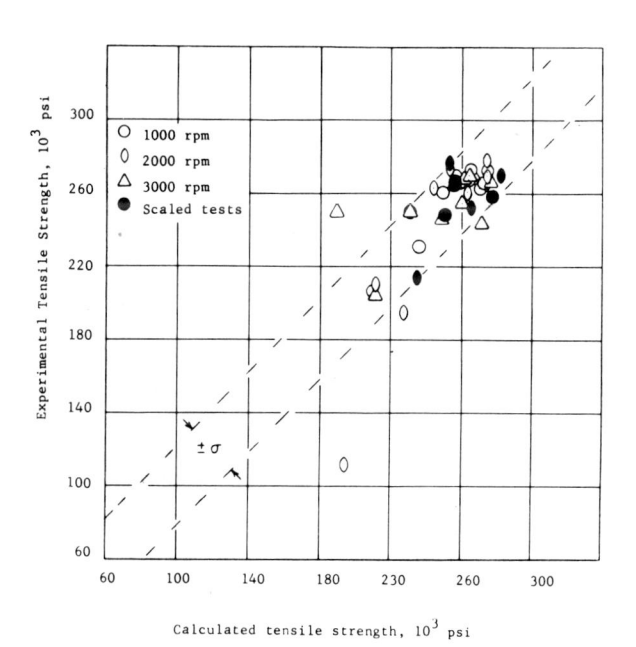

FIGURE 9. COMPARISON OF STRENGTHS OF FRICTION-BUTT-WELDED MARAGING-STEEL TUBES WITH THOSE DERIVED FROM $168509 - 140703 \exp(-100\, V_H) + 783\sqrt{P_H} + 22861\sqrt{P_F/P_H} + 1.53\, S_L$

Source: *Materials & Processes for the 70's*, SAMPE, 1973

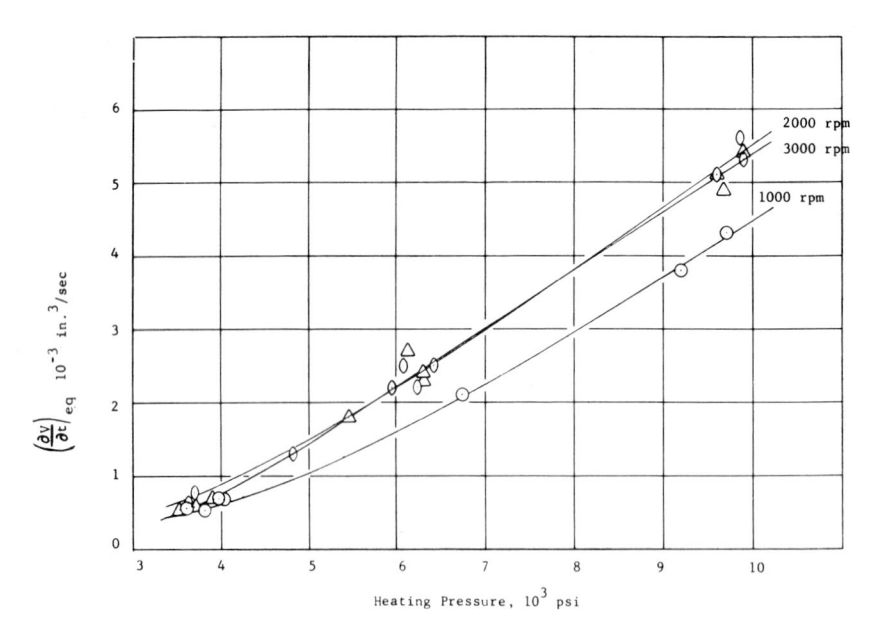

FIGURE 10. EFFECT OF HEATING PRESSURE ON EQUILIBRIUM VOLUMETRIC DISPLACEMENT RATE FOR FRICTION-BUTT-WELDS BETWEEN 2.8-IN.-DIAMETER STEEL TUBES

appear to be nonlinear at 1000 and 2000 rpm but approach linearity at intermediate pressures at 3000 rpm. This is somewhat contrary to the findings of Ellis[9] for similar investigations using mild steel bars and may, for the most part, be due to differences in specimen geometry. It would appear from Figure 10 that while higher displacement rates, and therefore welding rates, are attained at increased heating pressure, there is little effect of differences in speed above 2000 rpm. Graphical differentiation of these curves with respect to heating pressure, shown in Figure 11 as a function of rotational speed, would indicate that welding is more efficient at 2000 rpm than at 3000 rpm at the higher heating pressures. Furthermore, the figure suggests that peak welding efficiency occurs at lower rotational speeds as heating pressure is increased within the range of these investigations.

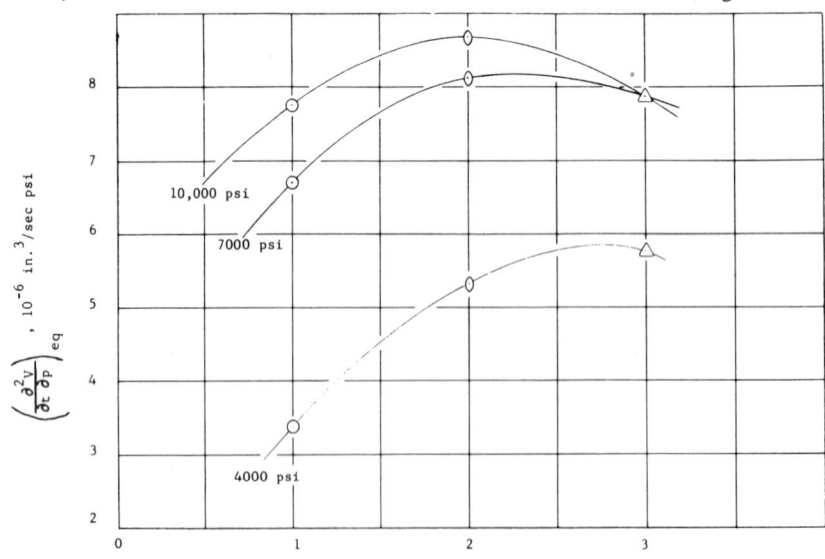

FIGURE 11. EFFECT OF ROTATIONAL VELOCITY ON CHANGE IN EQUILIBRIUM VOLUME DISPLACEMENT RATE WITH AXIAL HEATING PRESSURE FOR FRICTION-BUTT WELDS BETWEEN 2.8-IN.-DIAMETER MARAGING-STEEL TUBES

Microstructural changes in the maraging-steel friction welds due to variations in welding conditions during the parametric studies were considerably less obvious than those discussed above for the preliminary welding study. Differences in grain size of the heat affected zone and character of the weld interfaces were discernible, however, as shown in Figure 12, which essentially represents the extrmes of the welding conditions studied. Changes which might be associated with differences in mechanical properties were usually less obvious and occasionally are not discernible at all.

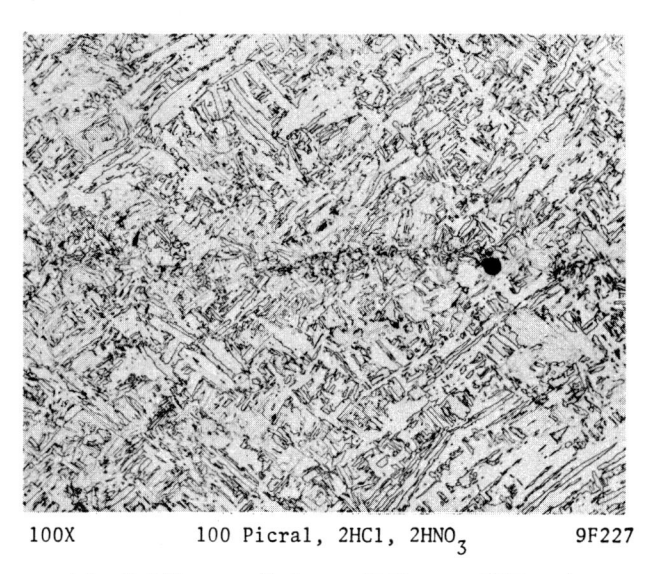

100X   100 Picral, 2HCl, 2HNO$_3$ 9F227

(a) Welding conditions: 1000 rpm; 3800 psi heating pressure; 7730 psi forging pressure; 0.091 in. upset; 9.83 sec.

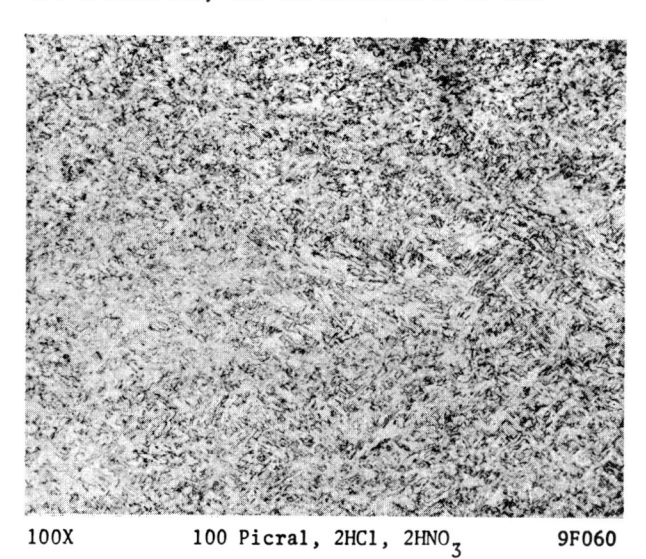

100X   100 Picral, 2HCl, 2HNO$_3$ 9F060

(b) Welding conditions: 3000 rpm; 9680 psi heating pressure; 19,850 psi forging pressure; 0.065 in. upset; 1.23 sec.

FIGURE 12. MICROSTRUCTURAL EFFECTS OF THE GREATEST DIFFERENCES IN CYCLE VARIABLES FOR FRICTION-BUTT WELDS BETWEEN MARAGING-STEEL TUBES

The effects of the maraging heat treatment (900 F for 3 hr) on both microstructure and hardness of maraging steel friction welds were studied for a number of welding conditons which caused significant differences in mechanical properties. Again, variations due to differing welding conditions were not particularly discernible. The maraging heat treatment did have a significant effect on both microstructure and hardness, as shown in Figure 13. Longitudinal hardness and structure variations across the interfaces as a result of welding can probably be attributed to both heating and mechanical-working effects. The narrow region of increased hardness at the weld interface, as shown in Figure 13, was probably caused by mechanical work introduced by the applied forging pressure. The adjacent areas were probably softened by the high temperatures generated during frictional heating while the regions of high hardness toward the outer edges of the heat affected zone probably result from a partial aging effect on the solution annealed workpieces. Maraging subsequent to welding not only served to increase overall hardness and to flatten the hardness profile as shown by the figure, but also accentuated the texture of the material by preferential precipitation of what appeared to be carbides in longitudinal bands which flare outward toward the specimen surfaces at the weld. This longitudinal banded structure is thought to have been introduced during fabrication of the material from which the specimens were made. Because of the axial shortening during welding, some of these carbides have a tendency to become concentrated at the weld interface, and may have contributed to lack of strength in some specimens. Scanning electron micrography

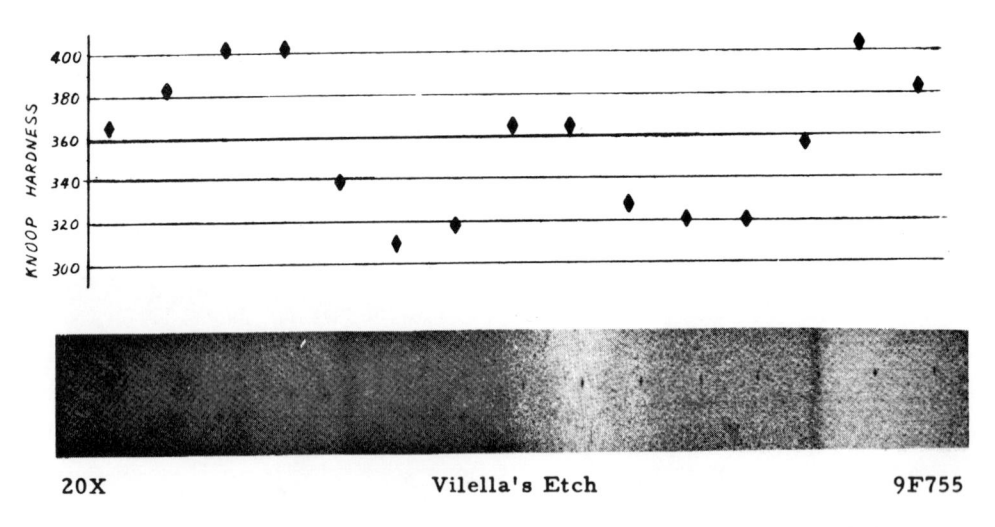

a. Knoop Hardness and Macrostructure as Friction Welded

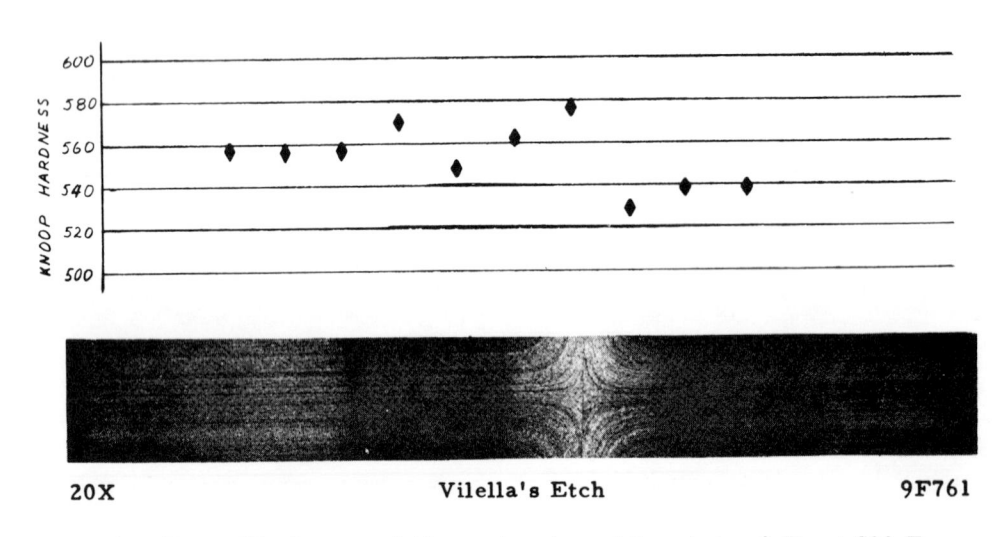

b. Knoop Hardness and Macrostructure After Aging 3 Hr at 900 F

FIGURE 13. EFFECT OF AGING ON MACROSTRUCTURE AND HARDNESS OF MARAGING-STEEL FRICTION-BUTT WELDS

Reduced 23 percent in reproduction.

of the tensile-specimen fracture surfaces from two of the welds, as shown in Figure 14, indicated that failure was ductile in welds exhibiting both high and low tensile strengths. Those areas of Figure 14b which are not representative of fracture surfaces, either ductile or brittle, are throught to be a result of trapped carbides or oxides. Solution treatment of the weld prior to aging tended to reduce the carbide-precipitate concentration at the interface as well as to produce some grain refinement, as shown in Figure 15. Hardness profiles were not significantly affected by this

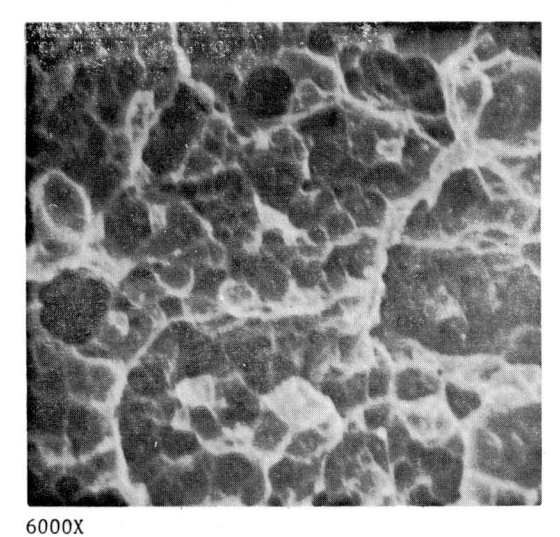

6000X
(a) Weld No. 95 Fracture Surface

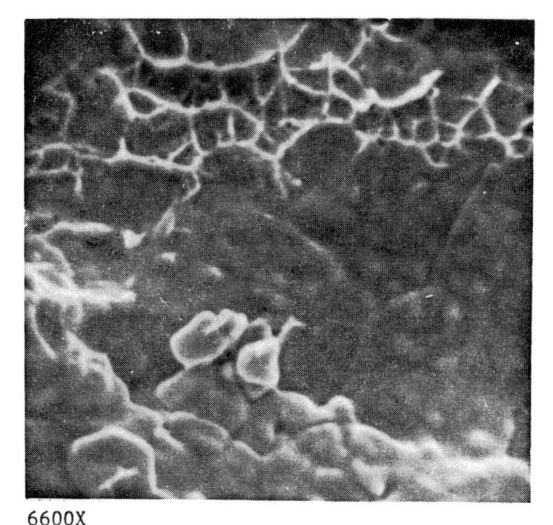

6600X
(b) Weld No. 92 Fracture Surface

FIGURE 14. SCANNING ELECTRON MICROGRAPHS OF TENSILE FRACTURE SURFACES OF FRICTION-WELDED MARAGING-STEEL TUBES

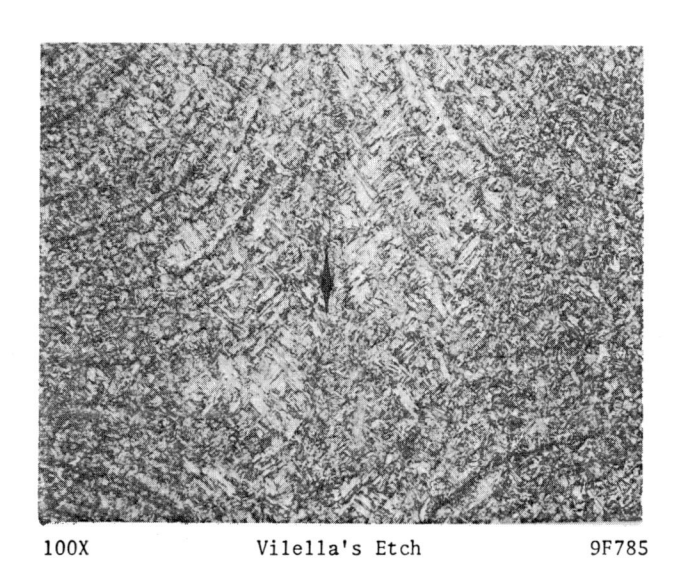

100X   Vilella's Etch   9F785
(a) Maraged 3 Hr at 900 F After Welding

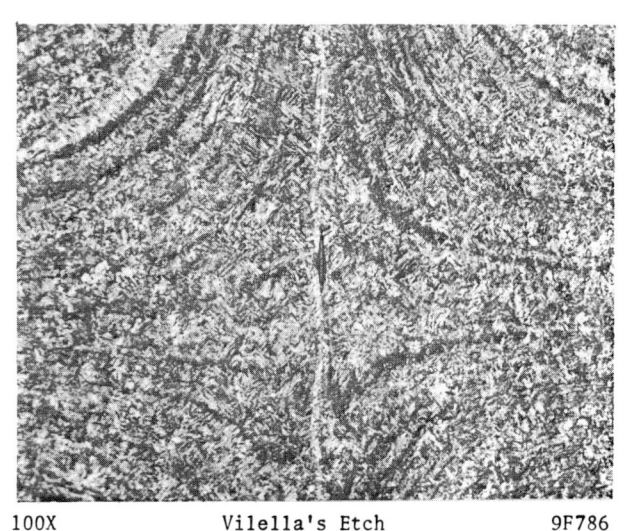

100X   Vilella's Etch   9F786
(b) Solution Treated at 1500 F for 15 Min After Wleidng, Then Aged 3 Hr at 900 F

FIGURE 15. MICROSTRUCTURAL EFFECT OF INTERMEDIATE ANNEAL PRIOR TO AGING OF MARAGING-STEEL FRICTION WELDS

treatment. It should be noted here that while this practice is usually undesirable after conventional welding because of possible distortion due to uneven stress fields, no such distortion would be expected in friction welds because of the completely uniform heating and welding which occurs simultaneously over the entire weld cross section.

4. CONCLUSIONS

(1) In general, it can be concluded from this study that friction welding can be effective in decreasing costs and increasing efficiency in missile-systems production operations. Its incorporation would result in simplification

of component forming operations, reduction in the number of machining steps, and increased speed and reliability in joining operations, which would thereby increase the cost effectiveness of missile-systems production.

(2) An empirical model has been derived which will permit specification of the conditions of rotational speed, axial heating and forging pressures, and axial displacement necessary to achieve high-quality friction butt welds between 18Ni(250) maraging-steel tubular components having mean diameters between 2.8 and 5.8 in. and diameter-to-wall thickness ratios between 30 and 80. The greatest welding efficiencies were found to occur at rotational velocities equivalent to 1470 sfm and axial heating pressures between 9000 and 10,000 psi.

(3) It was found from this study that the quality of friction welded joints between maraging-steel tubes has, in terms of welding variables, a greater dependence on axial pressure and displacement during the heating phase of the weld cycle than on either forging pressure or relative rotational velocity. Microstructural investigations have indicated that high joint integrity is most likely when frictional heating is evenly distributed between the components to be joined, is limited to the immediate vicinity of the weld interface, and is sufficient to permit plastic deformation of the interfacial material but not prolonged enough to cause significant grain growth in the areas adjacent to the weld. Tensile failure of friction welds was found to be ductile in nature with weakness being at least partially due to included carbides and/or oxides which act to reduce the effective bond area.

(4) Auxiliary braking is desirable for conventional friction welds between thin-walled components because it serves to decrease the probability of radial distortion in the welded joint and reduces the torque peak generated during deceleration of the rotating component.

(5) Maraging-steel components should be friction welded in the solution-annealed condition, as the heat generated during welding is sufficient to cause re-solution of the strengthening precipitates, thus requiring additional aging to achieve full maraged properties.

(6) Solution treatment of friction-welded maraging-steel components prior to aging was found to be beneficial to joint quality by irreversibly dissolving carbides concentrated at the weld interface and causing refinement of the grain structure in the bond zone.

(7) It is not feasible to friction weld maraging-steel components containing viscoelastic materials because of the subsequent heat treatment required to achieve optimum properties. Maraging-steel components containing viscoelastic materials could, however, be friction welded to dissimilar materials where re-solution would not affect the overall joint strength.

4.1 ACKNOWLEDGMENT

Research sponsored by Research and Engineering Directorate, Army Missile Command, Department of the Army, Contract No. DAAH01-71-C-0142. The views presented are not to be construed as an official Department of the Army position. Complete documentation of the research on which this paper was based may be obtained from the Defense Documentation Center, Alexandria, Virginia, by requesting Report No. AD747838. The authors wish to thank Mr. E. J. Wheelahan, AMSMI-RSM, of the Army Missile Command for the impetus and leadership which made the work possible.

4.2 REFERENCES

(1) Rocket Motor Manual (U), Chemical Propulsion Information Agency, The Johns Hopkins University Applied Physics Laboratory, Silver Springs, Maryland, CPIA/MI, March, 1969 (CRD).

4.2 REFERENCES
(Continued)

(2) Hazlett, T. H., "Properties of Friction Welds Between Dissimilar Metals", Welding Journal, 41 (10), 448-505 (1962).

(3) Hazlett, T. H., and Gupta, K. K., "Friction Welding of High-Strength Structural Aluminum Alloys", Welding Journal, 42, 4905-4945 (1963).

(4) "Friction Welding Spins Its Way Onto the Production Floor", Iron Age, 194 (14), 50-61 (1964)

(5) Hodge, E. S., "Friction Joining", Battelle Technical Review, 14 (8-9), 10-13 (1965).

(6) Rao, M., and Hazlett, T. H., "A Study of the Mechanisms Involved in Friction Welding of Aluminum Alloys", Welding Journal, 49 (4), 1815-1885 (1970).

(7) Nessler, G. G., et al, "Friction Welding of Titanium Alloys", Welding Journal, 50 (9), 3795-3855 (1971).

(8) Moore, T. J., "Friction Welding", Welding Journal, 51 (4), 253-262 (1972).

(9) Ellis, C.R.G., "Continuous Drive Friction Welding of Mild Steel", Welding Journal, 51 (4), 1835-1975 (1972).

(10) Metals Handbook, Vol. 6, 8th Ed., American Society for Metals, Metals Park, Ohio (1971), p. 511.

(11) Paprocki, S. J., et al, "Joining Zircaloy-Stainless Steel and SAP Alloys by Friction, Rolling, and Explosive Techniques", BMI-1594, Battelle Memorial Institute (September 4, 1962), pp 8, 20.

(12) Cheng, C. J., "Thermal Aspects of the Friction Welding Process (Part I)", American Machine and Foundry Company, CRL TR 296 (July 7, 1961).

(13) Cheng, C. J., "Thermal Aspects of the Friction Welding Process (Part II)", American Machine and Foundry Company, CRL TR 334 (August 15, 1962).

(14) Cheng, C. J., "Transietn Temperature Distribution During Friction Welding of Two Similar Materials in Tubular Form", Welding Journal, 41 (12), 5425-5505 (1962).

(15) Wang, K., and Nazappan, P., "Transient Temperature Distribution in Inertia Welding of Steels", Welding Journal, 49 (9), 4195-4265 (1970).

(16) Rich, T., and Roberts, R., "Thermal Analysis for Basic Friction Welding", Metal Const. and Brit. Weld. Jour., 3 (3), 93-98 (1971).

(17) Voinov, V. G., "Mechanism of Joint Formation in Friction Welding", Svarochnoe Proizvodstvo (Welding Production), 15 (1), 8-13 (1968) (Translation).

(18) Miller, G. P., and Mitchell, W. I., "Structure and Hardening Mechanisms of 18 Percent Nickel-Cobalt-Molybdenum Maraging Steels", J. Iron and Steel Inst., 203, 899-904 (1965).

(19) Marcus, H., Schwartz, L. H., and Fine, M. E., "A Study of Precipitation in Stainless and Maraging Steels Using the Mossbauer Effect", Trans. ASM, 59 468-478 (1966).

(20) Nie, N., Bent, D. H., and Hull, C. H., Statistical Package for the Social Sciences, McGraw-Hill, New York (1970).

Friction-Welded Heat Pipes Stabilize Arctic Soils

By Robert R. Reinking

THE TRANS-ALASKA pipeline is buried when it encounters rock or dry, stable permafrost. Roughly half of its 789 mi (1279 km) route, however, is over fine-grained, poorly drained sediments and potentially unstable, wet permafrost strata. Here, burying is precluded. Instead, the pipe is laid on elevated support assemblies consisting of 18 in. (455 mm) in diameter steel vertical support members (VSM's) ranging in length from about 25 to 55 ft (7.6 to 16.8 m). The VSM's are driven into the ground (tundra and permafrost) to a depth between one-half and three-quarters of their length.

Without the addition of the soil stabilizers discussed here, the VSM's would be subject to pile settling — one of two major problems peculiar to construction in these arctic soils. Pole jacking is the other.

Damage Potential: Pole jacking occurs when the active, wet layer of the permafrost freezes from the surface down, bonding the soil to the piling more tightly near ground level. Expansion accompanying freezing lifts the member up. This frost heaving can raise lightly loaded structures as much as 18 in. (455 mm) a year.

When the active layer thaws in the arctic summer, supports partially settle, usually to different degrees, causing damage to the supported structure.

The cycle is weighted in favor of pole jacking. Consequently, lightly loaded poles can eventually rise completely out of the ground and topple.

Because VSM's are relatively heavily loaded, pole jacking is not expected to be a serious problem.

The presence of surface vegetation and a thin layer of topsoil normally helps insulate the permafrost and maintain a stable permafrost table. Construction activities, however, commonly disturb this insulating layer, hastening permafrost thawing and causing table instability. Settling of imbedded support piles result. The gravel work pads added under the Trans-Alaska pipeline's elevated sections aren't sufficiently insulating to counter thawing and settling.

The potential for both pole jacking and pile settling is eliminated with the McDonnell Douglas Cryo-Anchor soil stabilizing heat pipes.

Operation: The heat pipe is a passive, natural-convection, two-phase heat transfer device. It has no moving parts and does not require any power for operation. Its basic component is a pipe containing a low-temperature-evaporating working fluid with a radiator pressed onto one end.

In the Trans-Alaska pipeline application, most of the heat pipe below the radiator is imbedded inside the VSM. The radiator is exposed to the atmosphere.

Heat from the soil vaporizes the working fluid in the pipe's lower, evaporator section. The vapor rises, transferring heat into the condenser section. As heat is lost to the condenser wall and dissipated to the atmosphere by the radiator, the vapor condenses and drains back down the pipe wall. A special internal structure permits thin-film evaporation and cooling along the evaporator's entire length and circumference.

Operation is automatic whenever the temperature of the air is below that of the heat pipe's imbedded evaporator section. The cycle is interrupted when the air is warmer than the soil, preventing heat transfer back into the ground (except for a small amount contributed by conduction). Test results indicate that heat transfer rates are several hundred times higher than those obtainable by thermal conductivity or conventional liquid convection cycles.

The heat pipes work well. They minimize permafrost degradation — provide soil stabilization — by cooling it substantially below the level attained with natural freezing. This more than offsets any increases in thermal transmission into the permafrost during summer via the VSM's, gravel overlays, or disturbed surface insulating layers.

An additional important benefit is an increase in the adfreeze bond between the permafrost and VSM which ups the piling's load-bearing capacity and further enhances pole jacking resistance. Test data show that the permafrost remains frozen at a temperature below normal during the entire thaw season. Soil studies indicate that this permafrost subcooling can increase adfreeze bond strength by a factor of three. This, in turn, indicates a potential for reducing the pile length needed to support a given load.

Heat Pipe Material and Process Selection

The Cryo-Anchor heat pipe is the brainchild of the Donald W. Douglas Laboratories component of McDonnell Douglas Corp., Richland, Wash. It was designed, tested, and qualified to Alyeska Pipeline Service Co. specifications.

Above ground sections of the Trans-Alaska pipeline rely on ammonia-filled heat pipes to prevent damage caused by seasonally induced changes in permafrost consistency. The devices are already in place in the center vertical support members of 150 ft (45 m) long pipeline and block valve test section.

Results of an extensive qualification program, which included fabrication and accelerated life testing of full-size heat pipes, insure that service life will exceed the 30 year life design goal.

Heat pipes were fabricated at the McDonnell Douglas-Tulsa facility, Tulsa, Okla., on an automated production line turning out a heat pipe every 90 s. The only manual operations: monitoring of the line to insure a smooth work flow, detecting equipment malfunctions, and banding completed pipes into shipping bundles.

The heat pipe's evaporator section is 2 in. (50 mm) OD, 0.25 in. (6.4 mm) wall mechanical tubing. Length ranges from 25 to 58 ft (7.6 to 17.7 m).

The condenser section is 3 in. (75 mm) OD, 0.75 in. (19 mm) wall mechanical tubing. Length is 6 ft (1.8 m) for evaporators up to 31 ft (7.6 m) long, and 8 ft (2.4 m) for longer evaporators.

Larger tubing was specified for the above ground condenser section simply to provide additional protection

How Heat Pipes Prevent Pole Jacking and Pile Settling

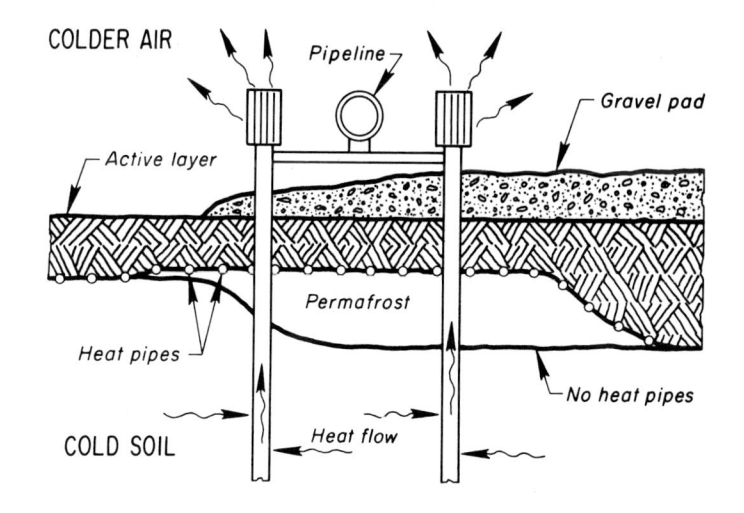

Source: *Metal Progress*, Feb 1976

against damage by stray bullets, vandalism, or accident.

Most of the evaporator-to-condenser tube joints were made by friction welding.

The working fluid is contained by hexagonal steel caps (blanked from plate) friction welded to each end of the pipe.

Low-carbon grades were selected for all steel components for reasons of cost, availability, and compatibility with high-speed, automatic welding methods.

The radiator is a 4 or 6 ft (1.2 or 1.8 m) long, 20-fin aluminum alloy extrusion. The OD envelope measures about 11 in. (280 mm), and each fin is 3.75 in. (95 mm) deep. The longer radiators are used with the longer condensers.

Liquid anhydrous ammonia (Federal Specification O-A-445) is the working fluid. Selection was based on the results of thermal and thin-film flow and evaporation studies, and on considerations of heat transfer efficiency, possible operational failure modes, fluid contaminants and their long-term chemical reactions, thermal and chemical stability, and compatibility with the heat pipe material and its possible contaminants.

Joining: Four welding processes — friction, gas tungsten-arc (GTAW), gas metal-arc (GMAW), and flash welding — were evaluated in the qualification program.

Acceptance criteria were: helium leak tightness to 1×10^{-9} cm^3/s; the ability to withstand a 2000 psi (14 MPa) proof test; and a maximum internal flash or drop-through of 0.125 in. (3.2 mm).

The last criterion applied to the evaporator-to-condenser joint and was necessary to meet Alyeska's requirement that a failed heat pipe be salvageable by cutting off the portion above the VSM and inserting a 1.25 in. (32 mm) OD, sealed, ammonia-filled tube with its own radiator into the remaining portion of the heat pipe. The retrofit pipe would provide adequate soil stabilization.

Flash welding was eliminated because joint impact re-

A simulated heat pipe installation is about to begin at McDonnell Douglas-Tulsa. Note the special handling fixtures. Each VSM gets a pair of heat pipes.

sistance was too low and excessive inner surface flash jeopardized meeting the retrofit salvage requirement.

The other three qualified although it was recognized that GTAW was slower by a substantial margin, and that joint preparation, spacing and alignment, and machine settings were much more critical for GMAW.

Friction welding was consequently selected as the production method because of its high speed, low cost, unskilled operator requirement, high reliability and reproducibility, and the very low criticality of joint preparation and tolerances. The dependence on electrical controls and current uniformity, and the need for electrodes, filler metals, shielding gas, and extensive machining (for joint preparation) were also eliminated. In addition, joint low-temperature impact resistance proved slightly better.

The leak-tight welds could be easily inspected by simply measuring the external flash.

Production welds had to meet these requirements:
1. Be helium leak-tight to 1×10^{-8} cm^3/s.
2. Withstand an internal proof pressure of 1700 psi (11.7 MPa) or higher without failure.
3. Withstand an internal proof pressure of 850 psi (5.9 MPa) without permanent dimensional change.
4. Exhibit 100% penetration of wall thickness (determined by metallurgical section).
5. Exhibit no more than 0.1 in. (2.5 mm) interior or exterior flash, drop-through, or build-up.
6. Be free of surface flaws exceeding ⅛ in. (3.2 mm) in length at right angles to the weld zone.
7. Be free of porosity or inclusions exceeding ⅛ in. (3.2 mm) in diameter or length in the fusion zone transverse to the wall thickness.

Friction welding beat out GTAW, GMAW, and flash welding for making heat pipe joints. Shown here are interior views of, left to right, typical condenser-end cap, condenser-evaporator tube, and evaporator-end cap joints.

Unfortunately, friction welding couldn't be used for the first 18 000 condenser-evaporator joints because of the welding machine's delivery schedule. Automatic GTA welding with four orbital-head machines was used instead. These welders held the two pipe sections in position while the weld head rotated around the joint. Two passes were required; the root pass was done without filler. One joint took about 6 min. to complete.

Friction welding — used for the remainder of the 112 000 heat pipes — was roughly eight times faster.

Plant Tour: Here's an outline of how the heat pipes were fabricated.

The 2 in. (50 mm) evaporator tubing enters on one side of the building and is cut to length. The special internal structure that provides thin-film evaporation is then inserted and the assembly is cleaned and dried. End cap friction welding follows.

The 3 in. (75 mm) condenser tubing enters on the other side of the building. After being cut to length, one end is faced for cap welding and an exterior lead taper for radiator installation is machined. Next step is facing of the opposite end of the tube for joining to the evaporator section. The condenser tube is then cleaned and dried.

Evaporator and condenser sections are conveyed to a second, larger friction welder where the condenser section is clamped in the welder's rotating jaws. The evaporator is butted against the end of the condenser and also clamped. The condenser is then rotated and forced against the evaporator section to complete the friction weld. Total welding time is approximately 45 s.

The assembly is then charged with ammonia through the condenser's open end and a temporary plug is inserted. A third friction welder is then called on to weld the other end cap to the condenser over the seal plug.

End caps and plugs are processed in another area of the plant. They're cleaned and packed in metal containers. Containers are purged with dry nitrogen to prevent rusting and contamination, and delivered to the charging station or to the appropriate end cap friction welding station.

After the application of identifying marks, the condenser section of the completed heat pipe rolls over a pad saturated with a corrosion preventive. This precaution is taken to insure that excessive corrosion will not hinder radiator installation in the field.

Pipe welds are checked for leaks on a sample basis using an ammonia-sensitive detecting solution.

Tubing, end cap, and plug cleaning solutions are checked daily.

Ammonia quality is closely monitored. Each truckload is tested before transfer to bulk storage, and each batch is further purified by a Douglas Lab-developed process before heat pipe charging. Each batch is also tested for moisture content by the nickel catalyst, furnace decomposition technique. The measurement is made with a dew pointer.

Completed heat pipes are randomly sampled and the ammonia charge is checked by lifting the pipe into a vertical position and measuring the liquid level with an ultrasonic tester.

One of every 200 pipes is also functionally tested to gage operating efficiency. An accelerated life test is performed on one out of every ten of these pipes.

Bundles of 24 heat pipes are shipped by rail or truck to Seattle for transshipment to Alaska.

Special installation handling fixtures designed and fabricated by McDonnell Douglas-Tulsa are used at the pipeline site.

One, a trailer-mounted hydraulic press, installs the extruded radiator. Another, a sling-like device, is used with a crane to lift the heat pipe and insert it into the VSM.

Source: *Metal Progress*, Feb 1976

SECTION II:
Explosive Welding

Fundamentals of Explosive Welding 99
Review of the Present State-of-the-
   Art in Explosive Welding 116
Explosive Welding and Cladding —
   Overview of the Process and
   Selected Applications 129
Mechanical Properties of
   Explosively-Cladded Plates 145
Recent Developments in the Theory and
   Application of Explosion Welding 151

Fundamentals of Explosive Welding

B. CROSSLAND and A. S. BAHRANI

Mechanical Engineering Department of The Queen's University of Belfast

SUMMARY. An account is given of the basic mechanism of adhesion in all welding processes, and a brief review of fusion and pressure welding techniques is given. A completely new method of welding which employs high explosives is briefly described and explained in terms of the principle used in the hollow charge, which was developed during days of war to defeat heavy armour plate.

The results of experiments carried out to find the mechanical properties of explosively welded joints are discussed and the application of the principle to the problem of welding tubes into end plates is described.

Results of metallurgical examination of the welded joints are presented and discussed and the mechanism of the formation of the interfacial waves is briefly mentioned.

1. Introduction

Welding can be defined as the process of joining two or more materials, often metallic, by localized coalescence or union across the interface. The essential conditions for any form of welding are that the two surfaces prior to welding should be absolutely clean and uncontaminated and that these surfaces should be brought into intimate contact with one another. It is impossible to produce such surfaces by normal mechanical or chemical cleaning processes. However, under conditions of high vacuum it has been reported by Bowden and Tabor (1954) and Keller (1963) that nearly perfectly clean surfaces can be produced.

If surfaces produced under and maintained at high vacuum are brought into contact then adhesion will occur between the asperities. The adhesion of harder and less ductile materials will be considerably reduced due to the rupturing of the junctions by elastic recovery when the load is removed. With softer and more ductile materials the junction will probably not be ruptured when the load is removed, and the adhesion measured will be appreciable. If in addition to the normal force a tangential force is applied which is not sufficiently great to produce macroscopic sliding, then it is found that the adhesion is greatly increased even with the harder metals and it is found that the adhesion force can reach values well in excess of the normal load applied. The effect of a tangential force has been satisfactorily explained by Bowden and Tabor (1950) as being caused by the increased area of contact at the asperities as a consequence of the reduction of normal load required to cause yielding in the presence of a tangential stress.

Even though the importance of the presence of an oxide film is demonstrated by the work carried out under high vacuum, nevertheless this still does not provide a basic understanding of the nature of the adhesion force. If two perfectly clean and atomically flat surfaces of the same metal are brought together, interatomic repulsive and attractive forces will come into play and equilibrium will be reached at the equilibrium interatomic distance for the metal concerned when the potential energy of the system is a minimum as shown in fig. 1. The strength of the bond will depend on the crystallographic misorientation across the interface and the diffusion and recrystallization which occurs, both of which are dependent on temperature. The situation in relation to the adhesion of different metals which may not only have a different atomic spacing but also a different crystal structure is obviously a problem of great complexity, but nevertheless the adhesion which can still occur is a consequence of the interatomic forces. There is no doubt that the mismatching, which is even more serious than with adhesion between the same metal, must influence the adhesion force.

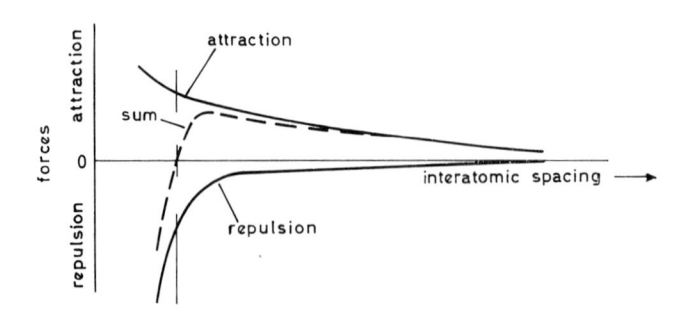

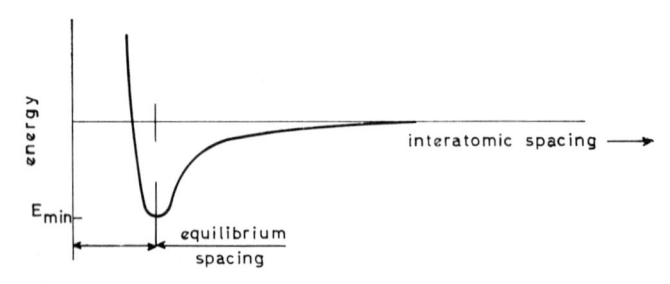

Fig. 1. Forces and potential energy of a pair of atoms.

There are basically two welding processes and there are many variants of each. Firstly there is fusion welding in which the surface of the two metals are melted by the application of heat and the contaminant surface layers are brought to the surface of the melt pool. Additional metal in the form of a filler rod may be added to the melt pool. Heat for fusion welding can be provided by several methods which have been developed mainly during this century. A very usual source of heat is that produced by an acetylene oxygen flame formed by a fine gas jet. Another method which is used extensively is to employ an electric arc which is formed between an electrode and the metals being welded. Frequently this arc is formed in an inert gas shield, such as in the argon arc process in which a shield of argon is used, to prevent excessive oxidation in the region of the weld. More recently electron-beam welding has been introduced in which the work-piece is placed in a high vacuum system and bombarded with a dense and focused stream of high velocity electrons. Even more recently focused laser (light amplification by stimulated emission of radiation) beams have been employed for welding.

The second basic welding process is pressure welding, in which the surfaces to be joined do not attain their melting temperature. A fine example is provided by the age-old process of forge welding which up to the twentieth century was the only welding process in general use. In forge welding the work-pieces are heated to an appropriate temperature then quickly superimposed and then hammered together. During this process the mating surfaces are plastically deformed, breaking up the contaminate surface films and creating fresh uncontaminated areas where adhesion can occur. The work-pieces are heated so that it is easier to plastically deform them under a forging hammer. Cold pressure welding is possible if the two metals to be welded are squeezed together between indenting dies or rolls, and this method is extensively used to clad stainless steel on to steel. Yet another example is provided by friction welding, in which frictional heat is generated at the contacting surfaces between a stationary and rotating work-piece which are pressed together. When they reach a suitable temperature the rotation is stopped while the end load is maintained or increased while the member is allowed to cool.

Generally with fusion welding it is difficult to weld metals with appreciably different melting-point temperatures and impossible if the boiling point of one metal is higher than the melting point of the other. Even with some metals with a similar melting point there may be serious metallurgical problems such as the formation of intermetallic compounds with undesirable properties. With pressure welding it is impossible to weld metals of appreciably different hardness as one would plastically deform before the other. These problems do not arise in explosive welding, and though explosive welding is seriously limited in its field of application, nevertheless it does allow greatly dissimilar metals to be welded together and provide joints of great strength.

2. Explosive welding

During the First World War it was observed that fragments of the steel shells of bombs occasionally stuck to metallic objects in the vicinity of the explosion. This, had it been realized, was an example of explosive welding.

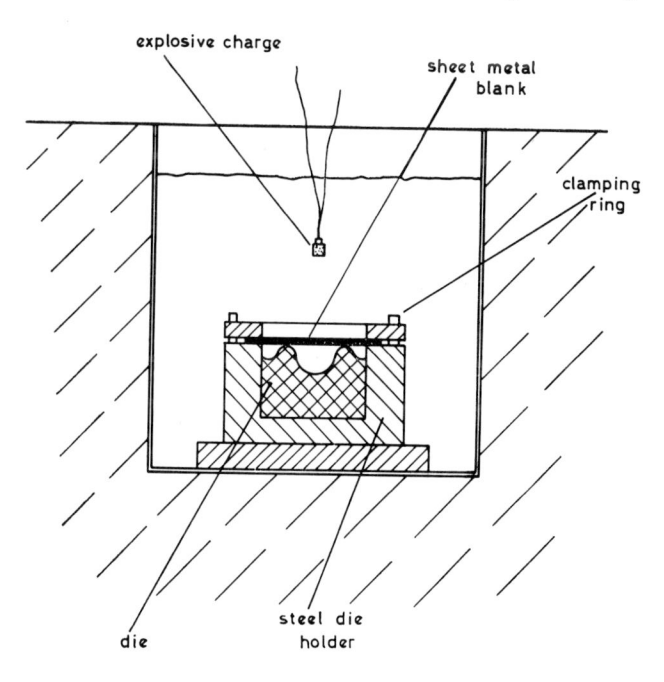

Fig. 2. Explosive forming.

In 1954 Allen, Mapes and Wilson found when right circular cylinders were fired obliquely into thin plane lead targets, that above a certain critical angle which depended on the velocity the surface of the front of the cylinder was marked by a series of ridges or 'waves'. Abrahamson (1961) who discussed the formation of these waves, gave some photomicrographs of the junctions between a steel bullet and a copper target and a steel bullet and a steel target which showed that welding had occurred between them.

Welding between metals was observed in explosive-forming experiments in 1957. In explosive-forming an explosive charge is detonated under water (see fig. 2) and the shock wave generated imparts momentum to a sheet-metal blank clamped above a die, thus forcing the sheet metal into the die. It was noted that if a metal die was employed and an excessive explosive charge was used, then the metal sheet became welded to the die. This discovery gave rise to an interest in the possibility of using high explosives for welding processes and during the last few years such processes have been actively examined in many countries and industrial use is now beginning to be made of such processes. In 1963 the first serious work on explosive welding to be disclosed in Great Britain was started in the Mechanical Engineering Department of the Queen's University of Belfast.

3. Mechanism of explosive welding

The mechanism of explosive welding is based on the analysis of the hollow charge given by Birkhoff, MacDougall, Pugh and Taylor (1948). Fig. 3 (a) shows the general arrangement of a hollow charge. It will be seen that a conical cavity in the explosive charge is lined with a thin metal liner. When

(a) Hollow charge with metal liner before explosion

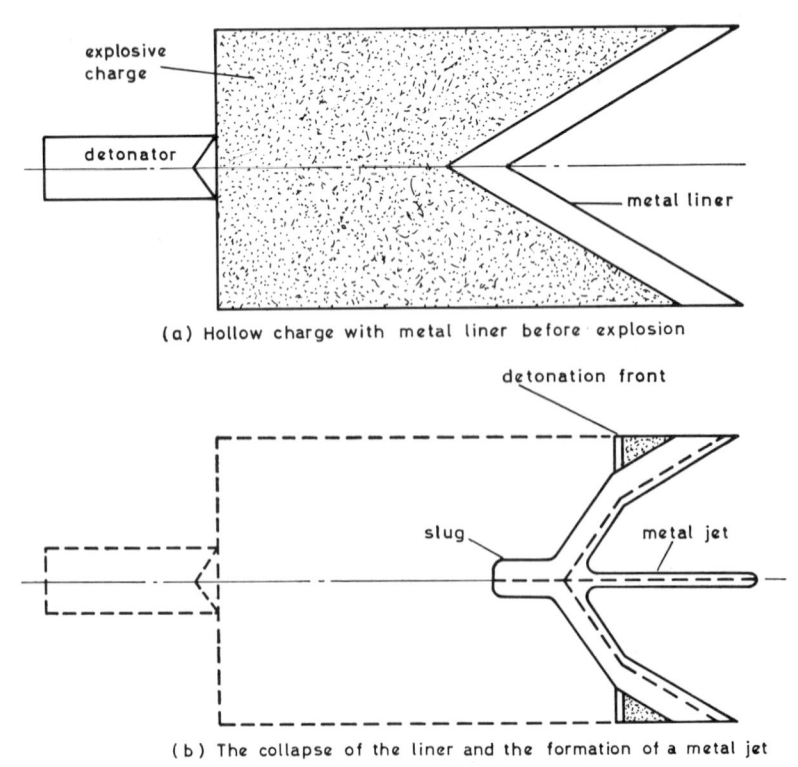

(b) The collapse of the liner and the formation of a metal jet

Fig. 3 (a) and (b). Hollow charge with metal liner.

the charge is detonated the detonation wave moves down the explosive charge, and when it reaches the apex of the liner it subjects the outer surface of the cone to a very high pressure which causes its walls to collapse. The pressure produced in the metal in the region where the walls of the liner collide (see fig. 3 (b)) is extremely high, probably of the order of 10^5 atmospheres which is much higher than the shear strength of the metal. Consequently the material in the region of impact behaves as an inviscid fluid and the laws of fluid mechanics can be applied to the situation. It can be shown that the liner material divides into a high velocity 'metallic jet' and a slower moving slug as shown in fig. 3 (b). This high velocity metallic jet has remarkable penetrating powers, and the hollow charge in fact forms the basis of the bazooka weapon used so successfully in the last war against heavy armour plates on tanks.

In explosive welding we shall see that the forming of a metallic jet is essential to achieve welding. Fig. 4 illustrates the set-up commonly used for cladding; the top or 'flyer plate' is supported with a minimum of constraint at a small angle of incidence relative to the stationary or 'parent plate' which is supported on a relatively massive anvil plate. The top surface of the flyer plate is covered

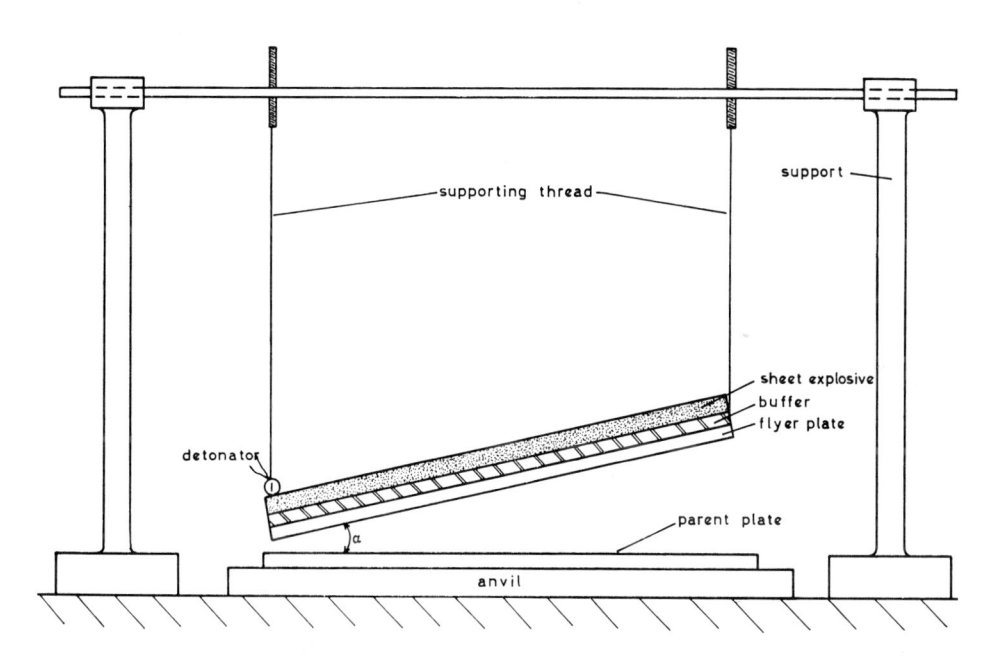

Fig. 4. Set-up for explosive cladding.

with a protective buffer such as rubber or polystyrene, and above that the sheet of high explosive is laid which is detonated from the lower edge. As shown in fig. 5 the detonation of the explosive imparts a velocity V to the flyer plate which collides with the parent plate at an increased angle of incidence β. It will be seen from fig. 5 that

$$V = U \sin \phi = U \sin (\beta - \alpha) \tag{1}$$

where U is the detonation velocity for the explosive sheet used and α is the initial angle of incidence. It will also be noted that as the flyer plate collapses on to the parent plate it moves at every instant as though it is hinged at S.

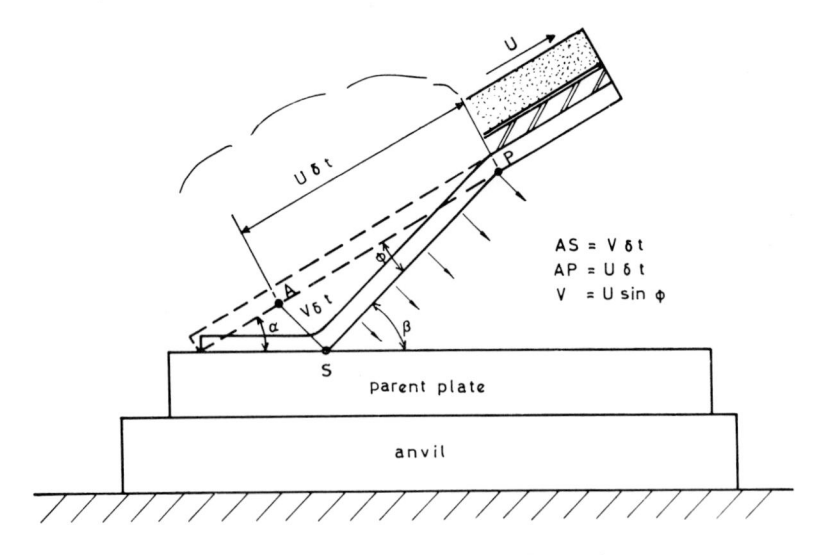

Fig. 5. Mode of collapse of flyer plate.

As the flyer plate collides with the parent plate it suffers a rapid retardation and an extremely high pressure is generated in the region of impact at S. As in the hollow charge, the pressure is so high compared with the shear strength of the materials involved that they behave for a very short interval of time in a similar manner to inviscid fluids.

It is convenient to change the coordinates of the system shown in fig. 5 so as to bring point S to rest. To do this it is necessary to apply a backward velocity of $V/\sin \beta$ to the parent plate as shown in fig. 6. If this backward

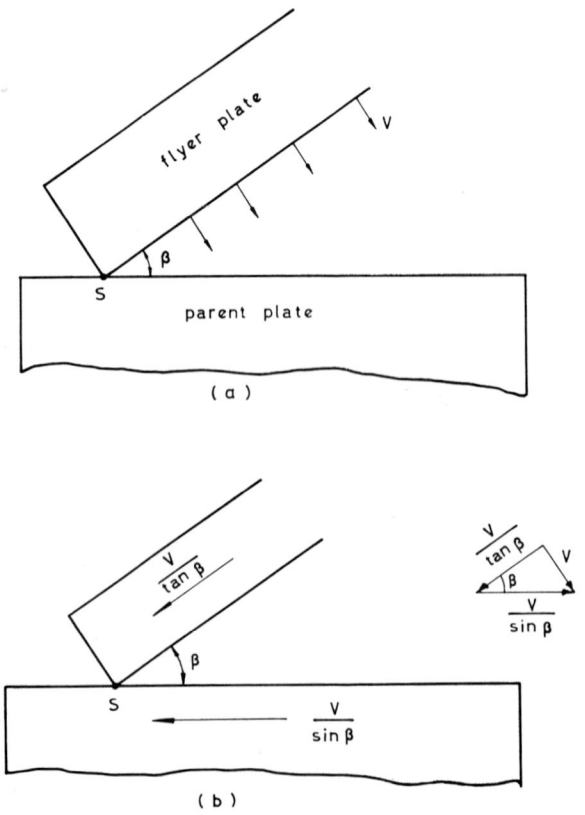

Fig. 6 (a) and (b). Mechanics of collision; changing the coordinates.

velocity is applied to the system, then the velocity of the flyer plate is $V/\tan\beta$ towards S. As the metals at the point of impact are behaving as inviscid fluids, the system becomes equivalent to a liquid jet of velocity $V/\tan\beta$ impinging on a stream travelling at a velocity of $V/\sin\beta$ at an angle of incidence β.

The liquid jet impinging the stream at S is deflected into a horizontal direction still travelling at the same velocity, but this implies that the conservation of momentum in the horizontal plane has not been satisfied.

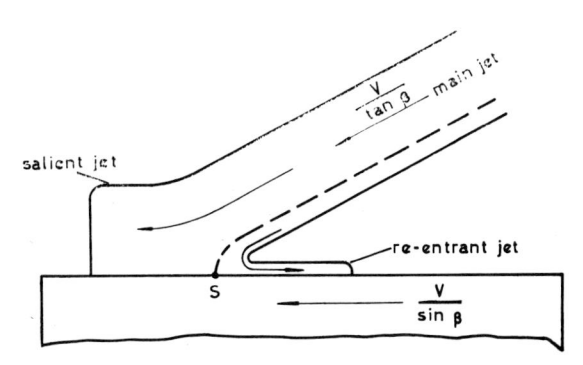

Fig. 7. Mechanics of collision; division of the main jet.

Consequently it must be concluded that the jet divides into salient and re-entrant jets as shown in fig. 7.

Applying the conservation of momentum to the configuration in fig. 7 gives

$$m\frac{V}{\tan\beta}\cos\beta = m_s\frac{V}{\tan\beta} - m_r\frac{V}{\tan\beta}$$

or

$$m\cos\beta = m_s - m_r \qquad (2)$$

where m is the mass of the jet, m_s is the mass which is diverted into the salient jet, m_r is the mass diverted into the re-entrant jet and $V/\tan\beta$ is the velocity of the jet. It will be noted that

$$m = m_s + m_r \qquad (3)$$

and this, coupled with eqn. (2), gives

$$m_r = \frac{m}{2}(1-\cos\beta) \qquad (4)$$

$$m_s = \frac{m}{2}(1+\cos\beta). \qquad (5)$$

The absolute velocity of the re-entrant jet will be

$$\frac{V}{\tan\beta} + \frac{V}{\sin\beta} = \frac{V}{\sin\beta}(1+\cos\beta). \qquad (6)$$

This theory is accurate as long as the velocity of the flyer plate relative to the point of impact, S, is subsonic or $V/\tan\beta$ is less than the velocity of sound in the material being welded. However, at higher velocities the simple hydrodynamic theory is not applicable, as the role of shock waves within the flow cannot be neglected. For example, if the sheet explosive is Metabel with a detonation velocity of 6500 m/sec and if the weight of charge is such as to give a velocity to the flyer plate of 500 m/sec, then from eqn. (1)

$$\sin(\beta - \alpha) = \frac{500}{6500}$$

or

$$\beta - \alpha = 4° 25'.$$

If α is 10° then β is 14° 25′, and from eqn. (4)

$$m_r = 0.01568 \text{ m},$$

or 1·5 per cent of the flyer plate is diverted into the re-entrant jet. So for a flyer plate thickness of 5 mm, a layer 0·078 mm thick will be 'peeled' off to form the re-entrant jet, which will have a velocity from eqn. (6) of 3950 m/sec or a Mach number of 11·5. The velocity of the point of impact is $V/\sin \beta$ or 2000 m/sec which is considerably lower than the velocity of sound in the parent plate, so that deformation will spread into regions of the parent plate ahead of the point of impact. Consequently a hump develops in the parent plate ahead of the point of impact due to the action of the jet.

It is now possible to explain the mechanism of welding. The high velocity re-entrant jet sweeps the surface of the parent plate and it picks up by surface traction a thin layer from the top surface of the parent plate. Effectively the lower surface of the flyer plate has been peeled off and the top surface of the parent plate has been scraped off, leaving two absolutely clean and uncon-

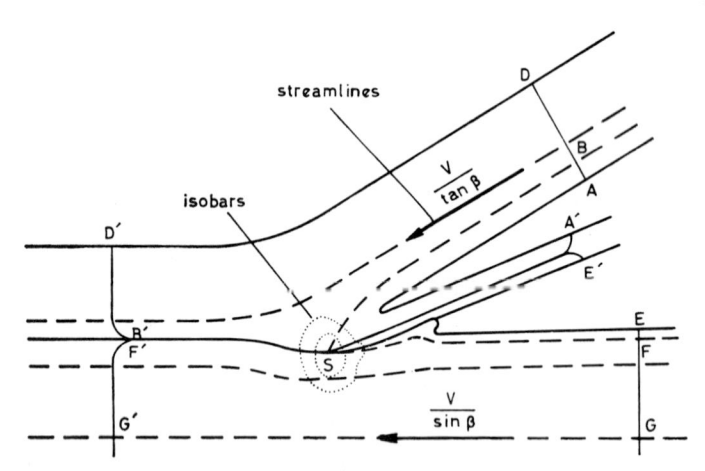

Fig. 8. Flow configuration in region of collision.

taminated surfaces which are brought together under pressure at the point of impact. The essential conditions for welding are achieved without melting or the need to plastically deform the metals to break up the oxide or contaminant film on the surfaces. As a consequence the limitations imposed by fusion or pressure welding do not apply to explosive welding. Fig. 8 shows the flow configuration in the region of impact. If sections DBA and EFG are considered, then layers BA and EF are removed and points F and B will be brought together as shown by D′B′F′G′.

The inclined plate technique shown in fig. 4 is necessary in order to get the velocity of the point of impact less than the speed of sound in the material of the parent plate, even though the detonation velocity of the explosive may be higher than the velocity of sound, and in order that a fluid metal jet is created. In most metals the velocity of sound lies within the range 4000–5500 m/sec, while most explosives have a detonation velocity in excess of this range. However, there are a few powder explosives such as Trimonite which have a much lower detonating velocity, and it is then possible to use a parallel-plate technique such as that shown in fig. 9. It will be seen that the velocity of the point of impact is equal to the detonation velocity of the explosive and an oblique collision is achieved where the effective angle of incidence is

$$\beta = \sin^{-1}\frac{V}{U}.$$

If the velocity of the plate is 500 m/sec and U for Trimonite is 3900 m/sec, then $\beta = 7° 22'$. The spacing between the flyer and parent plates is unimportant as long as it is sufficient to allow the flyer plate to accelerate to its terminal velocity.

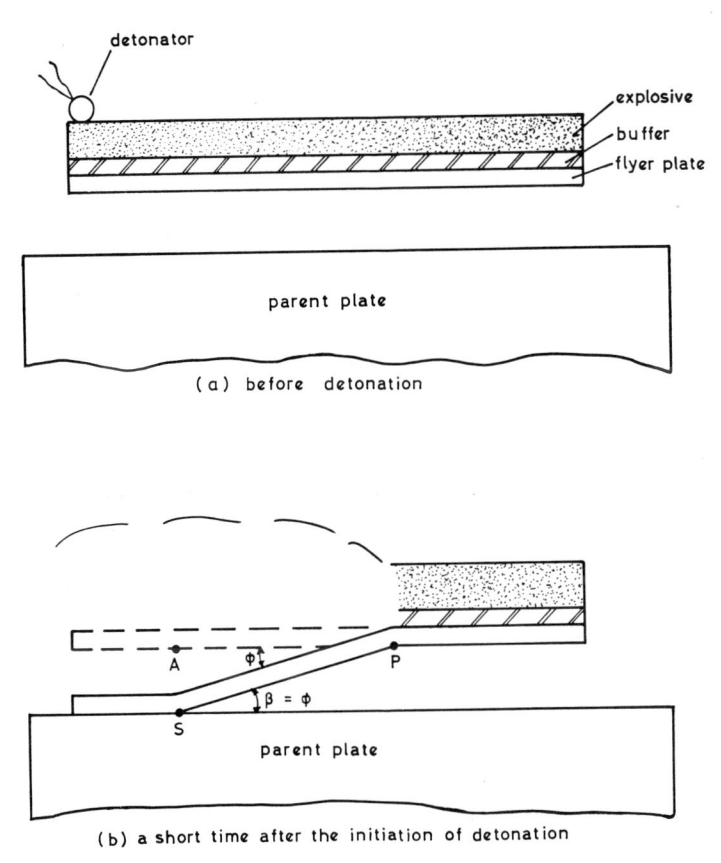

Fig. 9 (a) and (b). Arrangements for explosively welding parallel plates.

4. Experimental investigation

Fig. 4 shows the general arrangement of the inclined plate technique and the actual set-up is shown in fig. 10. In the particular experiment shown the flyer plate is stainless steel and the parent plate is mild steel, the area of the plate is 2 ft × 1 ft and the weight of explosive sheet is about 3 lb. The detonation of this charge is shown in fig. 11. The parallel-plate technique has also been used with Trimonite powder which is more difficult to handle than Metabel sheet explosive, but the experimental set-up is otherwise simpler and it would

Fig. 10. Sheet explosive being laid on top of flyer plate.

ease the task of handling very large plates. There is no reason to prevent very large areas of cladding being done by this method, except that the weight of explosive charge will be correspondingly very large and there are very few sites sufficiently far away from human habitation and yet close to centres of industry where such large charges could be detonated.

Up to date it has proved possible to explosively weld stainless steel to steel, copper to stainless steel, copper to steel, brass to stainless steel, brass to steel, titanium to steel, aluminium to steel, etc. The mechanical strength of these various combinations has been exhaustively tested, and it has been found possible with correct adjustment of the angle of inclination and weight of explosive charge to obtain welds of great strength. Frequently the weld is stronger than the weaker of the two materials being welded, and failure occurs well away from the joint in the weaker of the two materials. For optimum strength it has generally been found that the initial angle of inclination should be 5° or thereabouts.

Fig. 12 shows how the parallel-plate technique can be applied to welding tubes into end plates, which is a problem of very great practical value in engineering; for example, in the construction of condensers, feed-heaters, boilers, etc. In order to locate the tube, the tube plate is bored to the outside diameter of the tube, and then for most of the thickness counterbored to give a clearance of 1 mm. The 'explosive pack' is made up out of dental wax which also forms the buffer, and ends are fitted with wood plugs with a wood spigot which can be varied in diameter to suit the weight of charge needed. A detonator can be inserted in a hole provided in one of the wood end plugs. Such a pack is shown being inserted in a tube in fig. 13.

Fig. 11. The explosive is detonated.

So far it has proved possible to weld cupro-nickel and aluminium brass tubes to brass tube plates; and stainless steel, mild steel and titanium tubes to mild steel tube plates. The weld achieved is typical of an explosive weld and its strength is expected to be similar to that found for the weld between one plate and another. There remains the question of how many tubes can be welded simultaneously into a tube plate and if there are any problems of interaction between the shock waves originating at the same instant from various centres; this problem is being investigated at the present time.

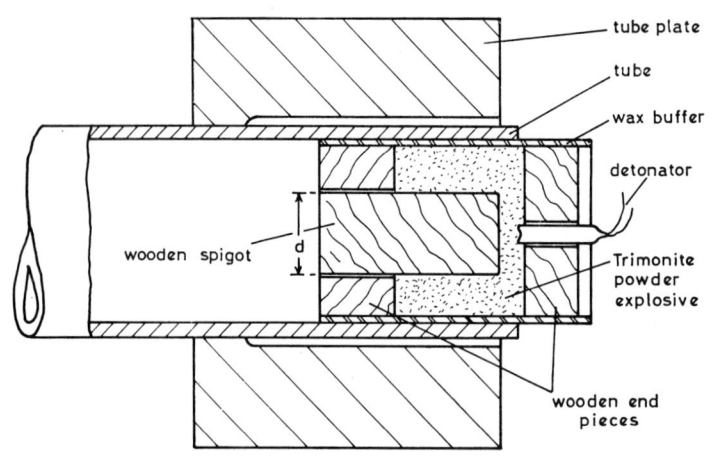

Fig. 12. Experimental set-up for welding tube to tube plates.

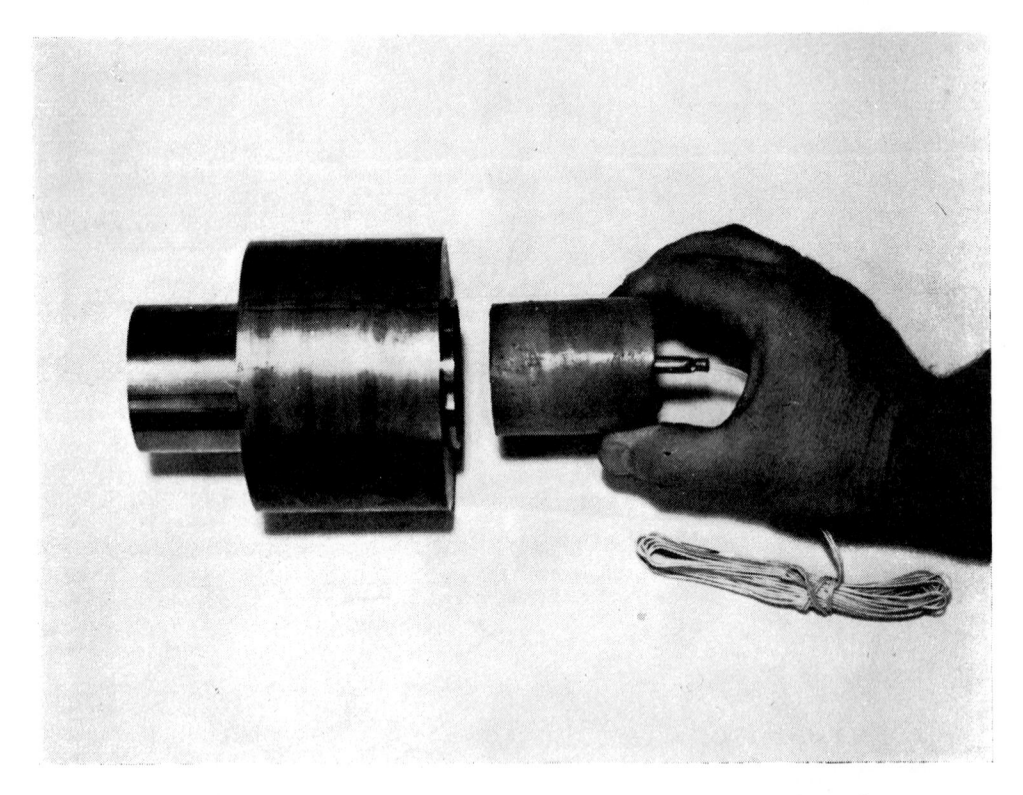

Fig. 13. The explosive pack ready for insertion into the tube.

5. Metallurgy of explosive welds

Photomicrographs of the weld interface are shown in fig. 14, which show the characteristic waves which are formed in the explosive welding process. These waves are produced at the rate of about 10^6/sec and it can be seen that there are areas of high vorticity in which the rate of shearing will be extremely high. In some cases the rate of shear has been so high that the frictional heating has caused melting in this region, but this melted zone is immediately quenched by the surrounding bulk of metal resulting in a cast structure, and sometimes cooling cavities caused by contraction on cooling can be observed. In these areas there is a mixture of the two metals which can lead to hitherto unknown intermetallic compounds. These new parameters of high rates of shear, high rates of cooling and high instantaneous pressure have opened up a completely new field of metallurgy.

Fig. 14 (*a*) shows the wave formation extremely clearly, and the vortex in front of the wave and the tail behind can be seen in great detail. The flow of the two surfaces are clearly indicated by the distortion of the crystal structure of the two steels. Traces of the flyer plate material can usually be found in the front vortex, though without a micro-probe analyser it is a little difficult to detect these traces even when very dissimilar metals are being welded together. Fig. 14 (*c*) shows stainless steel welded to mild steel, and the dark zone close to the surface of the mild steel is a 'shocked zone' which is indicative of the very high pressure instantaneously applied during impact. This shocked zone is a consequence of the pressure applied and it can be eliminated by using less explosive and a greater angle of obliquity. Fig. 14 (*d*) shows the interface between titanium and steel, and the white layer is an intermetallic compound of titanium and steel. The micro-diamond identations which can be seen are an indication of hardness and it will be noted that the intermetallic compound is extremely hard, in fact much harder than the hardest of steels. This layer is not only hard but also extremely brittle and consequently the weld is weak. However, with the correct welding conditions the intermetallic compound is not produced except in the centre of the front vortex as can be seen in fig. 14 (*e*).

There is still an immense amount of work to be done on the metallurgical aspect of this work, including electron microscopy and x-ray examination.

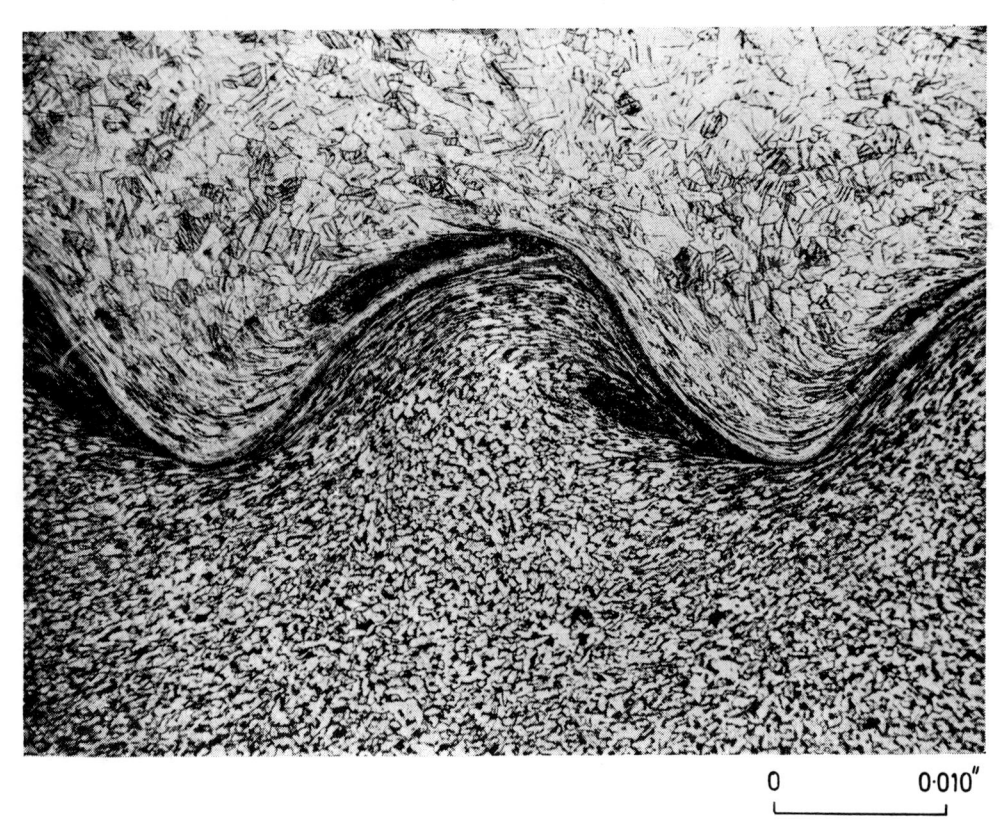

Fig. 14. Photomicrographs of the interface of explosively welded metals.
(*a*) Mild steel welded to mild steel.

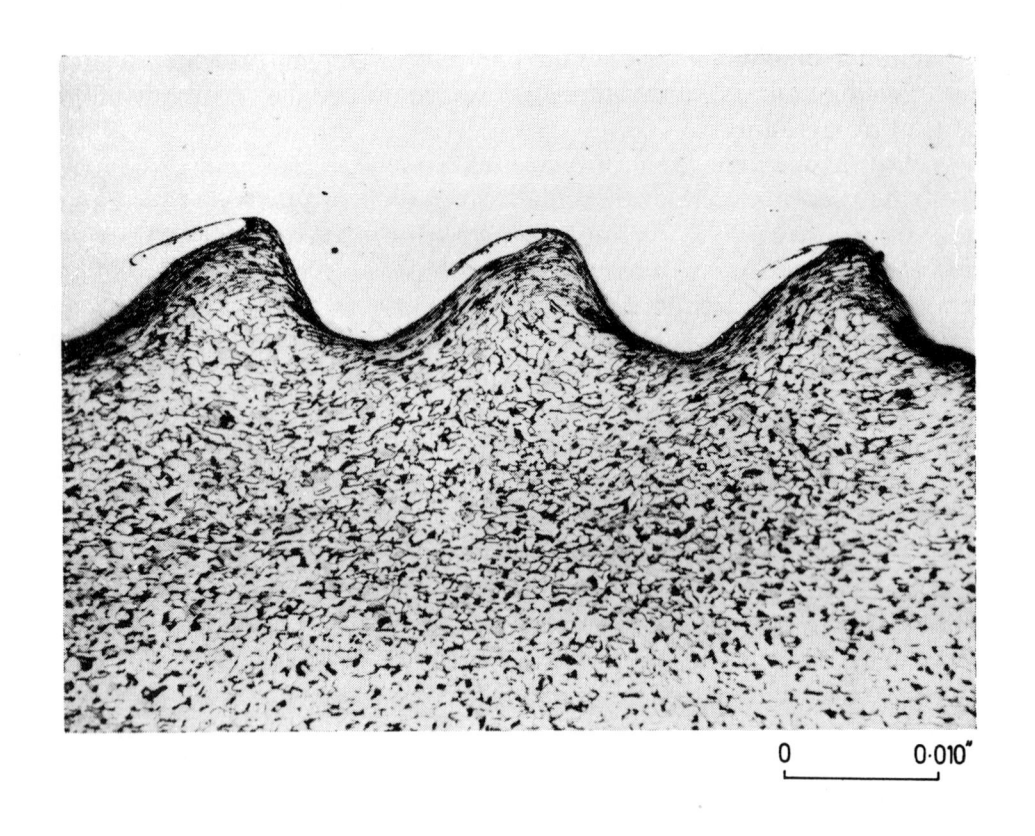

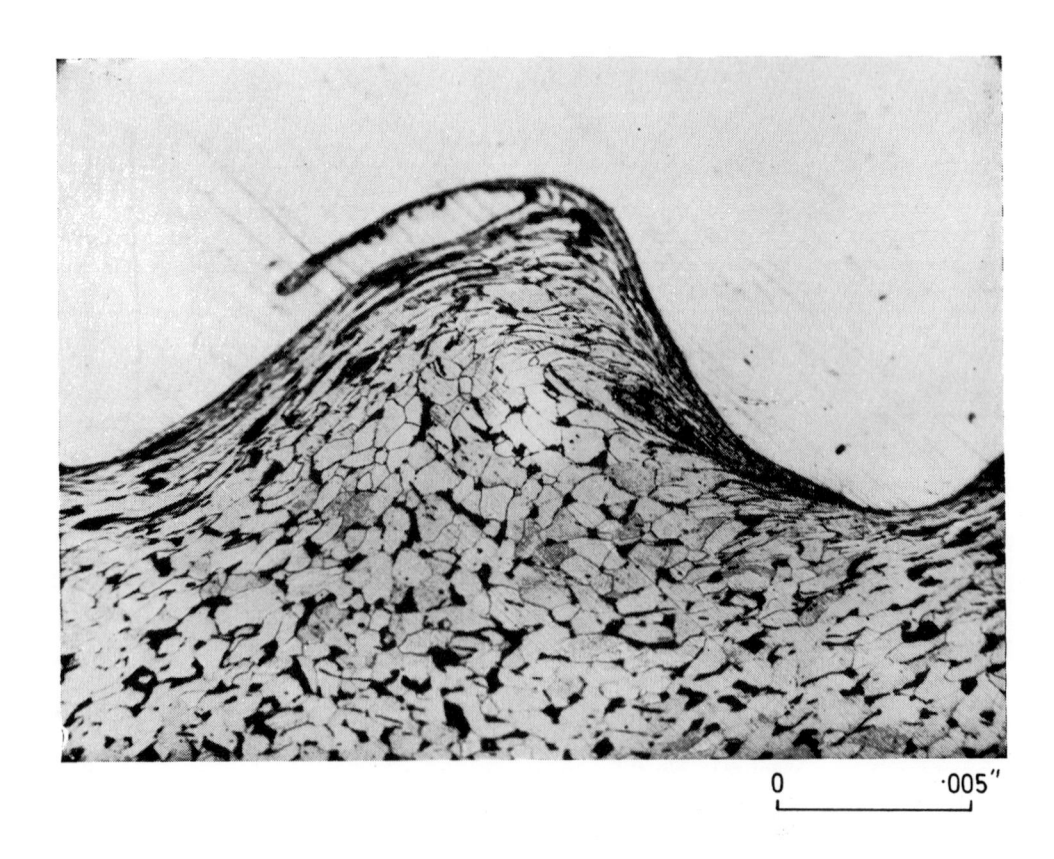

(b) Copper welded to mild steel.

0 0.010"

(c) Stainless steel welded to mild steel.

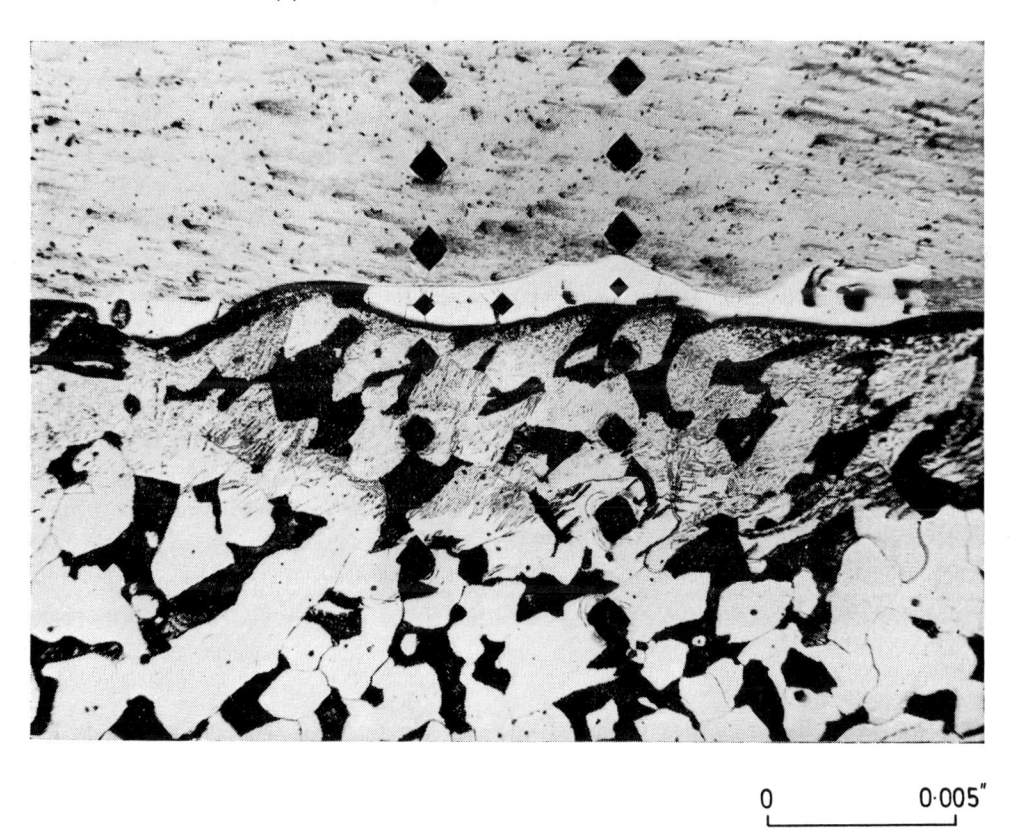

0 0.005"

(d) Titanium welded to mild steel when the welding conditions are not properly adjusted leading to the formation of hard brittle intermetallic compound at the interface.

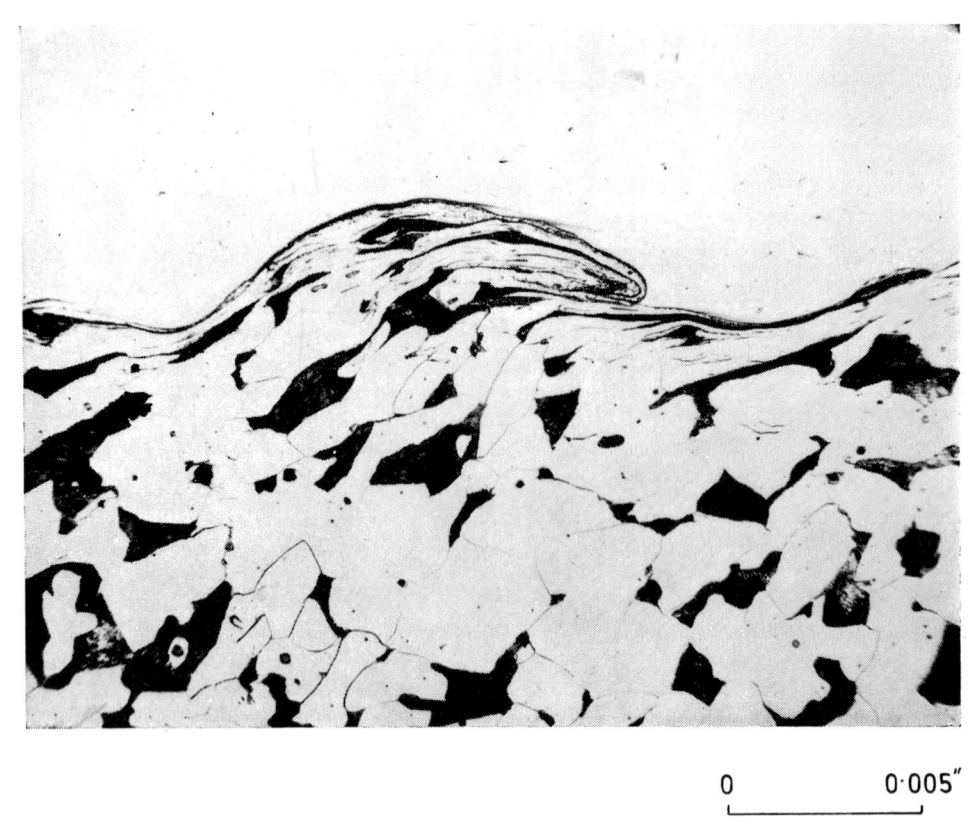

(e) Titanium welded to mild steel when the correct welding conditions are used.

The waves noted at the weld interface are similar to those noted on the nose of right circular cylinders fired at lead targets reported by Allen, Mapes and Wilson (1954). Abrahamson (1961) suggested a mechanism of wave generation which depended on the salient jet (see fig. 7). From the analysis it will be seen that the salient jet has a thickness very nearly equal to the thickness of the flyer plate, and yet waves with a wavelength of 0·05 mm have been noted in an experiment where the flyer plate was 2 mm thick. It is hardly conceivable that waves could be laid down at about 10^6 waves/sec by a mechanism involving the entire thickness and momentum of the salient jet. Bahrani, Black and Crossland (1967) have suggested a mechanism of wave generation which is dependent on the re-entrant jet. For the numerical case given in § 3 the re-entrant jet had a thickness of 1·568 per cent of the flyer plate, or for a 2 mm thick flyer plate a thickness of 0·03 mm which could conceivably give waves having a wavelength of 0·05 mm. They have also provided experimental evidence which provides strong support for the mechanism they propose.

6. Conclusions

The generation of a high velocity fluid metal jet in the explosive welding process provides a satisfactory explanation of the mechanism involved in removing the contaminant surface layers, which is an essential of any welding process. This process enables strong welds to be produced between metals which it would be difficult or impossible to weld by any other means. It can also be applied in situations where conventional welding methods are not applicable. In particular the process can be applied to the manufacture of cladded plate, which is of great value in the chemical industry, or to the age-old problem of fixing tubes into tube plates so that no leakage can occur.

One of the major practical problems is one of noise and of suitable sites for such work. At present, experiments are being carried out to investigate the possibility of carrying out this process in a large vacuum chamber to overcome this problem. Preliminary tests indicate that this idea holds great promise.

The metallurgical problems which arise from this work are considerable, as several new and important parameters are involved in this process. Further work involving electron microscopy and x-ray examination is being carried out to investigate some of the new materials found at the interface in an explosively welded joint.

References

ABRAHAMSON, G. R., 1961, *J. Appl. Mech., Trans. Amer. Soc. Mech. Engrs.*, **83** (series E), 519.
ALLEN, W. A., MAPES, J. M., and WILSON, W. G., 1954, *J. Appl. Phys.*, **25,** 675.
BAHRANI, A. S., BLACK, T. J., and CROSSLAND, B., 1967, *Proc. Roy. Soc.* A, **296,** 123.
BIRKHOFF, G., MACDOUGALL, P. P., PUGH, E. M., and TAYLOR, G. I., 1948, *J. Appl. Phys.*, **19,** 563.
BOWDEN, F. P., and TABOR, D., 1954, *The Friction and Lubrication of Solids* (Oxford University Press).
KELLER, D. V., 1963, *Wear*, **6,** 353.

The Authors:

B. Crossland served an engineering apprenticeship with Messrs. Rolls Royce Ltd. He graduated in Mechanical Engineering from Nottingham University. Subsequently he became an Assistant Lecturer, then Lecturer and finally Senior Lecturer in the University of Bristol. He was appointed to the Chair of Mechanical Engineering at the Queen's University of Belfast in 1959.

A. S. Bahrani was born in Baghdad and received his primary and secondary education there. He came to the United Kingdom in 1948 to study Mechanical Engineering at the Queen's University of Belfast where he graduated B.Sc. with honours in 1952 and M.Sc. in 1953. From 1954 to 1962 he was employed by the Iraqi Ports Administration, Basrah, in various capacities. He returned to the University of Belfast in 1962 to carry out research on high rate forming and welding of metals. He was awarded a Ph.D. in 1965 and was appointed Lecturer in Mechanical Engineering the same year.

Review of the present state-of-the-art in explosive welding

B. Crossland

A brief review of the history and simplified mechanism of explosive welding is presented. Consideration is then given to the present state of understanding of the parameters which control the explosive-welding process, and the implications of these parameters in relation to thin- and thick-plate cladding. The problem of cladding large thick plate is then considered and experimental data are reported which support the welding parameters proposed, and which allow definite conclusions to be drawn of the set-up required to achieve welding over the whole plate and close to the edges of the parent plate. Applications of explosive welding to cylindrical surfaces are reported including tube-to-tube welding, plugging of heat exchangers, welding of collars to tubes, etc. In particular, the problem of ligament distortion in tube-to-tubeplate welding and plugging is considered. The alternative possibility of placing bungs in the holes adjoining that in which a tube is being welded is examined. Other applications and potential applications of explosive welding are considered. MT/219S

©*1976 The Metals Society. The paper was presented at the Metals Society/Centro Nacional de Investigaciones Metalurgicas meeting on 'New aspects of metal forming and heat treatment', held in Madrid on 22–28 September 1974. The author is Professor of Mechanical and Industrial Engineering at the Queen's University of Belfast.*

It is nearly 20 years since Pearson in the USA probably first recognized the potential usefulness of explosive welding which had first been noted by Carl[1] in 1944, though Philipchuk[2] is credited with its first public recognition as a potential industrial process in 1957. Since then, considerable work has been reported in many countries, notably the USA, the USSR, West Germany, Czechoslovakia, Japan, and the UK, and this work has been the subject of reviews by Crossland and Williams[3] and Crossland.[4,5]

It is now clearly recognized that the application of explosive welding is seriously restricted, but there are a few applications where explosive welding is ideally suitable and commercially viable; for example, in large-plate cladding of one metal on another, tube-to-tubeplate welding, and plugging of heat exchangers. There are also potentially useful applications of explosive welding such as tube-to-tube connections, manufacture of duplex and triplex tubing, fabrication of wire-reinforced metals, line welding of metals, etc.

Mechanism

Though the simplified mechanism has been adequately described on many occasions, it is so relevant to the discussion of welding parameters that it is desirable to state it briefly yet again.

The basic arrangement is shown in Fig.1a in which there is a buffer above the flyer plate which may be of rubber, cardboard, or similar material to protect the top surface of the flyer plate from damage from the detonation of the explosive charge. Above the buffer is a layer of explosive which is detonated from the lower edge. Figure 1b shows the state of affairs when the detonation front has reached B and the collision point between the flyer plate has got to S. The portion of the flyer plate BS is moving at a velocity V_P which is dependent on the explosive loading (i.e. the ratio of mass of explosive to the mass of the flyer plate plus buffer). The exact direction of V_P is not obvious, and its precise computation is complicated and of doubtful value, so it is necessary to make an assumption, and fortunately the various assumptions which have been made lead to very similar conclusions. The assumption made here is that as the flyer plate is deformed it is slightly stretched, as shown in Fig.1b, as $AS+SB>AB$, but after stretching its length remains constant so that $SB=SB'$. From the trigonometry of this assumption, as described in Fig.1b, it can be seen that

$$\frac{V_P}{\sin(\beta-\alpha)} = \frac{V_D}{\cos\beta/2} \qquad (1)$$

from which the dynamic angle of obliquity β can be

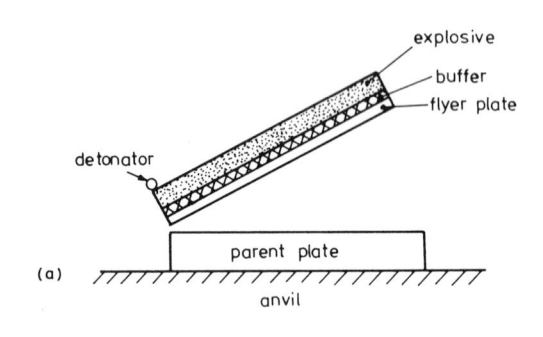

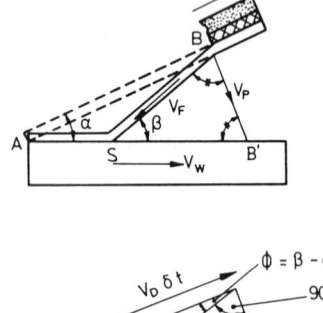

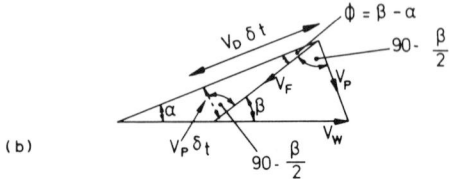

1 Inclined arrangement for explosive welding

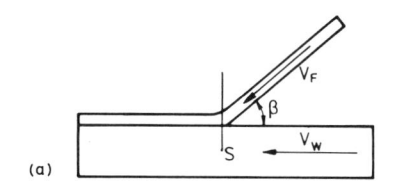

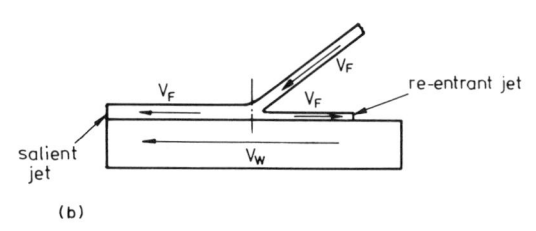

2 Formation of re-entrant jet

determined for given values of V_P, V_D, and α. At the point of impact S, the stagnation pressure is extremely high, and as with the treatment of the hollow charge by Birkhoff *et al.*[6] it will be assumed that the shear strength of the metals at S is small compared with the pressure, and consequently the two materials can be considered as inviscid fluids. Although this is an over-simplification, nevertheless, this assumption leads to worthwhile and apparently valid conclusions.

For convenience, the co-ordinate system so far used is changed by applying a backward velocity equal to the velocity of weld propagation to the whole system, which brings point S to rest. It will be seen from the velocity diagram in Fig.1*b* that the flyer-plate velocity relative to S is V_F and that

$$V_F = V_W = \frac{V_P/2}{\sin \beta/2} \quad \ldots \ldots \ldots (2)$$

Consequently, the system reduces to that depicted in Fig.2*a* where, to an observer at S, the horizontal stream travelling at velocity V_W is met by an oblique stream travelling at velocity V_F. If the parent plate was to be considered as rigid then the oblique stream will be deflected into the horizontal plane as shown in Fig.2*a*. However, this does not satisfy the conditions for conservation of momentum in the horizontal direction, so it must be concluded that the oblique jet is split into two at S to form salient and re-entrant jets, as shown in Fig.2*b*. Conservation of mass requires that

$$m = m_s + m_r \quad \ldots \ldots \ldots (3)$$

where m is the mass of the flyer plate m_s is the mass deflected into the salient jet, and m_r is the mass deflected into the re-entrant jet.

Conservation of momentum dictates that

$$mV_F \cos \beta = m_s V_F - m_r V_F \quad \ldots \ldots \ldots (4)$$

From (3) and (4),

$$m_s = \frac{m}{2}(1 + \cos \beta) \quad \ldots \ldots \ldots (5)$$

$$m_r = \frac{m}{2}(1 - \cos \beta) \quad \ldots \ldots \ldots (6)$$

Obviously, this is a gross over-simplification because, first, the parent plate is not rigid, any more than the flyer plate, so the re-entrant jet will, in fact, consist of material from both the flyer and parent plates. The actual amount of each material present in the re-entrant jet will depend on

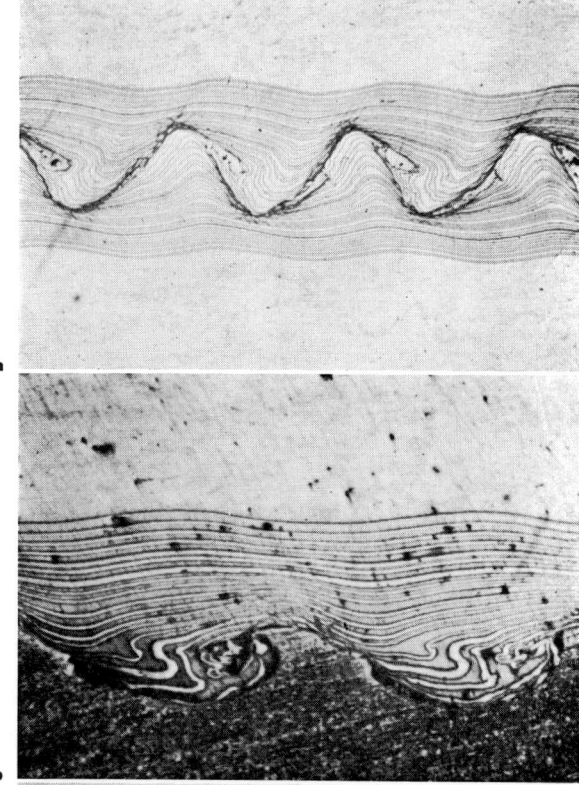

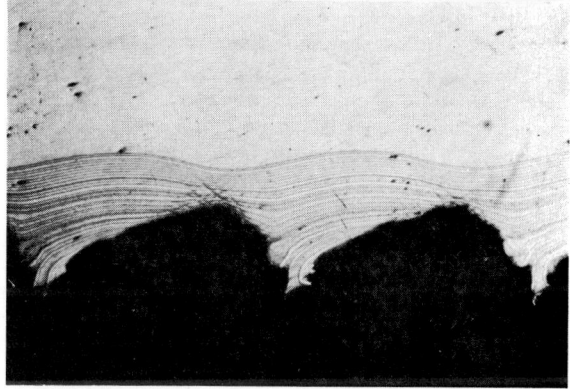

a copper to copper, ×60; *b* copper to titanium, ×100; *c* copper to lead, ×60

3 Effect of relative density on wave formation at weld interface

the relative densities and other relative properties. For instance, with a soft aluminium flyer plate impacting a medium-strength steel, it would appear reasonable to assume that the aluminium will jet more readily than the steel, not only because of their relative densities but also because aluminium is of much lower strength. Secondly, the whole situation is made more complicated by the typical wavy interface formed at the small values of β, which are for various reasons experienced in practice. Figure 3 shows such wavy interfaces between materials of equal densities and materials of widely different densities, in which one or both the surfaces being welded are electro-plated with alternate layers of copper and nickel. It is found that under these conditions the jet is largely or completely trapped in the vortices formed, and electron-probe analysis shows that material from both plates is present in these vortices. Finally, it has been assumed in

Source: *Metals Technology*, Jan 1976

this very simple treatment that the flow can be treated as subsonic, whereas welding can also take place under supersonic flow conditions, though the limited evidence available suggests that welding does not occur if the velocities are much above sonic conditions. Under most practical welding conditions the flows are clearly subsonic, so the above treatment is not too inadequate. Despite the considerable complexity of the problem it seems reasonable to assume that the ability to remove the contaminant surface layers from both surfaces, which is the essential requirement for welding, is dependent upon the impact velocity being sufficiently great to induce fluid-like properties, and on the dynamic angle β. From equation (7) it can be seen that as β increases, the amount of material removed from the surfaces increases, and it appears, everything else being equal, that there is a minimum value of β required to achieve the requisite cleaning action for welding to occur.

Recent experiments with a compressed-air gun in which a flyer plate can be fired at an oblique target have shown that some material escapes from the system, probably in the form of a spray of material. It appears that welding only occurs under conditions where there is a loss of material. The maximum loss of material, which probably gives the best conditions for removing the contaminant surface layers, occurs at the upper value of β at which waves are generated. This work will shortly be published in the proceedings of the 5th International Conference of the Centre for High Energy Forming.

Welding parameters

It is necessary to establish the parameters which control the welding process, but before one can start to investigate welding parameters it is necessary to define what is a satisfactory weld and to attempt to quantify it so that one weld can be compared with another. Bahrani and Crossland[7, 8] have described several tests, such as tensile test across the weld interface, tensile shear test, side-shear test, and bend test; however, these tests are not entirely satisfactory as adequately high results can be obtained even when there is some unfavourable metallurgical condition at the interface such as a brittle intermetallic compound. A dynamic test might well indicate the presence of such an unfavourable metallurgical condition, and Wylie et al.[9] have devised an impact test across the plane of the weld. They proposed a figure of merit, which is the ratio of the impact energy to fail the weld interface to the impact energy to shear the material of the parent or flyer plate, and a value of unity indicates an entirely satisfactory weld. Having reached an arbitrary decision on what is a satisfactory weld, the next problem is to decide on the independent variables which are likely to influence the weld. Perhaps the more obvious variables are the detonation velocity of the explosive V_D, the flyer-plate velocity V_P, the sonic velocity of the parent-plate material V_{SP}, the sonic velocity of the flyer-plate material V_{SF}, some mechanical property of the two materials being welded such as yield strength and the initial angle of obliquity α, if an inclined process is used. It might be considered in the parallel arrangement of cladding that the stand-off distance, that is, the initial clearance between the flyer and parent plates, is another independent variable, but in fact experiment indicates that the terminal velocity of the flyer plate under explosive loading is achieved if the stand-off is in the range $0.5\,t$–$1.0\,t$, where t is the flyer-plate thickness. There are no doubt other independent variables, such as the solid solubility of the two metals being welded, which could affect the metallurgical state at the weld interface. The more obvious and important dependent variables are the dynamic angle of obliquity β, and the velocity of the flyer and parent plates relative to the point of impact S. Wylie et al.[9] and Williams et al.[10] have carried out a considerable programme of experimental work to establish welding parameters, but although they have reached some tentative conclusions, nevertheless, it will be realized that the task is a formidable one in view of the number of variables. It is also difficult with explosive experiments to vary independently one parameter at a time, such as the angle β which appears to be an important parameter. Currently, experiments are being carried out with a gas gun to fire small, flat plates against target plates, which allows better control and measurement of β and V_P which are two of the more important variables. From these explosive welding experiments it has been concluded that the parameters to be satisfied are:

(i) V_W/V_{SP} and V_F/V_{SF} should be less than unity though some welds have in fact been achieved up to values of 1·25

(ii) the impact angle β should exceed a limiting value below which welding will not occur regardless of V_P. If this value of β is related to the thickness of the contaminant surface layers is not yet known, but experiments are planned using a gas gun in which the thickness of the contaminated surface layers will be varied

(iii) a minimum value of contact pressure which is related to the impact velocity V_P, must be achieved to obtain a satisfactory weld

(iv) above this minimum impact velocity the conditions for an acceptable weld appear to be associated with the kinetic energy imparted to the flyer plate by the explosive charge. The kinetic energy for a satisfactory weld may be related to the strain energy necessary to cause dynamic yield, and it also appears to be related to the stronger of the two materials being welded

(v) for metal combinations where brittle intermetallic can be formed there is perhaps an upper limit of impact energy beyond which the weld is weakened by the formation of excessive intermetallic. Even where brittle intermetallics are not a problem there will still be an upper limit because if there is excessive melting at the weld interface, then the weld created may be torn apart by the reflected tension wave from one of the free surfaces before the metal at the weld interface has developed sufficient strength

(vi) the stand-off, or clearance between the flyer and parent plate in a parallel welding arrangement should exceed half the flyer-plate thickness to allow the flyer plate to achieve its maximum velocity

(vii) in general, the use of high detonation-velocity explosive should be avoided as the very high pressure pulse produced in the flyer plate can, while the flyer plate is in flight, produce a reflected tension wave which in turn can cause spalling of the flyer plate while it is still in flight

(viii) there appears to be no effect of anvil characteristics on weld quality, but in practice this could well affect the damage imparted to the product.

These welding parameters indicate that a parallel arrangement, such as that shown in Fig.4, is possible as it can be seen from equations (1) and (2) that

$V_P = 2V_D \sin \beta/2$

and

$V_F = V_W = V_D$

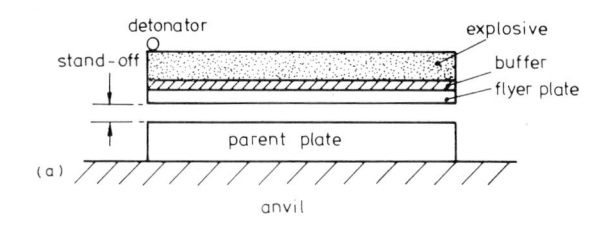

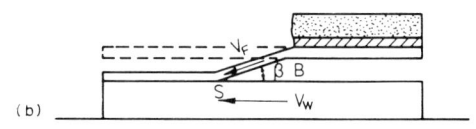

4 Parallel arrangement for explosive welding

so, if V_D is less than V_{SP} and V_{SF}, the first parameter is satisfied. However, it is also necessary to satisfy the rest of the parameters. To demonstrate the limitations imposed by these parameters, the welding of brass to mild steel will be considered where experiment suggests that the KE should exceed about 150 J/cm² and V_P should exceed 200 m/s while β should be greater than 7°. It should perhaps be noted that the $KE/V_P/\beta$ plane has not been fully explored and it is probably unjustified to assume no inter-relation between these parameters; however, it does allow a few interesting facts to emerge. In the first instance, the use of an explosive such as Trimonite 1, which is an ICI explosive consisting of ammonium nitrate, TNT, and atomized aluminium will be considered. Data on the detonation velocity of this explosive as a function of layer thickness, and the velocity imparted to a flyer plate as a function of explosive loading are well established, for instance in the work of Shribman and Crossland.[11] Table 1 gives the essential information as the thickness of the flyer plate is increased. Up to 8 mm the KE is kept at the minimum value but above 8 mm the KE must be raised in order to achieve the required value of V_P, so V_P is kept constant above 8 mm.

From Table 1 it will be seen that even for a 4 mm plate thickness the angle β is slightly less than the specified 7°, and though welding probably could be achieved, nevertheless, for a plate thickness of 6 mm β is probably too small. This perhaps indicates that an inclined procedure is required with an angle α of a few degrees, but this is not satisfactory for large plates as the stand-off at the edge remote from detonation is too large, and, as will be seen in the section below on 'Large-plate cladding', this leads to insuperable problems concerned with the dynamics of the displacement of the flyer plate. An alternative solution is to increase the explosive loading and hence V_P so that β is increased, but if the KE is excessive then the weld quality deteriorates and ultimately excessive damage to the product can occur. Wylie et al.[9] suggest that the KE should not exceed 450 J/cm² for this combination though this is perhaps a little conservative. If this value of KE is accepted, however, it leads to the data given in Table 2. From this table it can be seen that for a flyer-plate thickness of 10 mm it is impossible to achieve a sufficiently large angle β without getting excessive kinetic energy. At 8 mm the angle β is just in excess of 7° for the maximum kinetic energy so it would be possible to weld an 8 mm thick flyer plate. However, there is considerable interest in thicker flyer plates and the only practical solution would appear to be to use an explosive with a lower detonation velocity. The only suitable explosive with a lower detonation velocity which is readily available is ammonium nitrate with 6% diesel oil or mixtures of ammonium nitrate, diesel oil, and sawdust which, in thick layers, gives a detonation velocity of about 2·5 km/s. This explosive will not detonate in small layer thicknesses and it needs a booster charge of high detonation-velocity explosive in order to achieve stable detonation. The data in Table 3 are obtained with this explosive working within the upper bound for kinetic energy and assuming that V_D is constant at 2·5 km/s. Here, it is shown that it is now possible to weld up to at least 12 mm thick flyer plates with a low detonation-velocity explosive; and bearing in mind that the limiting upper bound for the kinetic energy and the lower bound for β are probably on the conservative side, it would be possible to exceed 12 mm. Obviously, if the detonation velocity could be further reduced, then even thicker flyer plates could be welded.

This type of approach, based on the proposed welding parameters, can be extended to examine the lower limit of flyer-plate thickness. With Trimonite 1 explosive it is concluded that even at $t = 0.5$ mm, the angle β is probably unacceptably large. If an explosive with a detonation velocity close to V_{SF} or V_{SP}, whichever is lower, is used, then a 0·5 mm thick flyer plate and perhaps a 0·2 mm plate can be welded. Below this thickness it appears that a driver plate may be necessary to keep β within reasonable limits.

Table 1 Explosive welding parameters using Trimonite 1

Flyer-plate thickness t, mm	Flyer-plate velocity V_P, m/s	Kinetic energy KE, J/cm²	Explosive loading R	Thickness of explosive, mm	Explosive velocity V_D, km/s	$\beta°$
2	440	150	0·77	11·0	2·61	9·7
4	310	150	0·4	11·4	2·63	6·8
6	250	150	0·24	12·6	2·69	5·4
8	218	150	0·22	14·2	2·81	4·4
10	200	158	0·21	15·0	2·88	4·0
12	200	190	0·21	18·1	3·04	3·8
16	200	252	0·21	24·1	3·23	3·6
20	200	316	0·21	30·1	3·38	3·4
24	200	380	0·21	36·1	3·50	3·3
28	200	442	0·21	42·0	3·60	3·2

Table 2 Further experimental welding parameters

Flyer-plate thickness t, mm	Explosive loading R	Flyer-plate velocity V_P, m/s	Kinetic energy KE, J/cm²	Thickness of explosive, mm	Explosive velocity V_D, km/s	$\beta°$
10	0·4	310	380	28·5	3·37	5·3
10	0·5	375	555	36·0	3·6	6·0
10	0·6	418	685	43·0	3·61	6·6
10	0·7	450	800	50·0	3·7	7·0
8	0·4	310	305	22·7	3·21	5·6
8	0·5	375	445	28·8	3·37	6·4
8	0·6	418	550	24·5	3·48	6·9
8	0·7	450	640	40·0	3·58	7·2
6	0·4	310	227	17·1	3·03	5·9
6	0·5	375	333	21·5	3·18	6·8
6	0·6	418	410	25·8	3·29	7·3
6	0·7	450	480	30·0	3·38	7·7

Table 3 Low detonation-velocity explosive parameters

Flyer-plate thickness t, mm	Flyer-plate velocity V_P, m/s	$\beta°$	Flyer-plate thickness t, mm	Flyer-plate velocity V_P, m/s	$\beta°$
10	340	7·8	20	240	5·6
12	310	7·2	24	228	5·3
16	269	6·2	28	202	4·7

Source: *Metals Technology*, Jan 1976

Large-plate cladding

Large thick plates of various metals cladded to thicker parent plates of different metals, such as stainless steel to steel, naval brass to steel, or Monel metal to steel, are of considerable commercial interest. Though roll cladding, continuous strip welding, or weld overlay are practised, nevertheless, the technological problems involved are considerable and the combinations which can be produced are limited. With explosive welding, an essentially solid-phase weld is produced if the correct welding conditions are achieved, and it is possible to form a satisfactory metallurgical bond between materials of widely different physical and mechanical properties.

The commercial exploitation of the cladding of large, thick plate was introduced by Du Pont in 1963–64 and it appears that the commercial organizations producing explosively clad plate in the West are mainly or entirely licensees of Du Pont. It may be as a result of the commercial interest in large-plate cladding, particularly by the large chemical companies, that little useful information on the process is available in the literature. Much work on small-scale testing using thin flyer plates has been reported in the literature, but none on cladding large, thick plate until the work of Crossland et al.[12] and more importantly Wylie and Crossland.[13]

The basic problem in large-plate cladding lies in the end effects which can be influenced by the position of the initiation of detonation, stand-off distance, the use of a booster, and the mode of propagation of detonation. Wylie and Crossland[13] have reported many tests and they concluded that the ideas on flyer-plate motion described by Ruppin[14] were essentially correct. Ruppin considered that a flyer plate would not collapse in the ideal manners shown in Figs.1 and 4, and he suggested a mechanism of unstable edge motion which is illustrated diagrammatically in Fig.5. At the edge of the flyer plate the pressure will fall to atmospheric so the momentum imparted to the edges will be less than at the centre, and as a result the displacement of the edges will lag behind the centre. There is an additional effect due to the reduction of the detonation velocity at the free boundaries of the explosive. The net result is what might be described as a 'whip-lash effect' which leads to an increase in the angle of obliquity resulting in large-amplitude waves, sometimes excessive melting, and even shearing of the flyer-plate material. The effect can be reduced by extending the charge beyond the edge of the flyer plate or increasing the thickness of the charge at the edges, but the most certain way is to keep the stand-off to a minimum which reduces the magnitude of the whip-lash effect. It is for this reason that the inclined configuration of Fig.1 is impracticable for large-plate cladding.

After an extensive experimental programme reported by Wylie and Crossland[13] it was concluded that the arrangement for cladding shown in Fig.6 was satisfactory. It can be seen that the explosive extends beyond the edge of the flyer plate, which is achieved by adopting the Z-section frame which is attached to the edge of the flyer plate to form a tray to hold the explosive. The flyer plate is supported by screws in lugs welded to the plate, which bear on four corresponding lugs welded to the parent plate. A buffer in the form of cardboard or linoleum off-cuts is stuck down to the top surface of the flyer plate to prevent air bubbles being trapped between the buffer and flyer plate. Explosive in the form of powder is spread to a uniform thickness to give the required explosive loading, and a booster charge of some high detonation velocity explosive together with a detonator is placed at one corner. Using this set-up, as shown in Fig.6, the Queen's University of Belfast have cladded a 9·5 mm thick by 1·07 m square stainless steel plate to a 63·5 mm by 1·05 m square mild steel plate. The detonation velocity was measured at four stations using the Dautriche

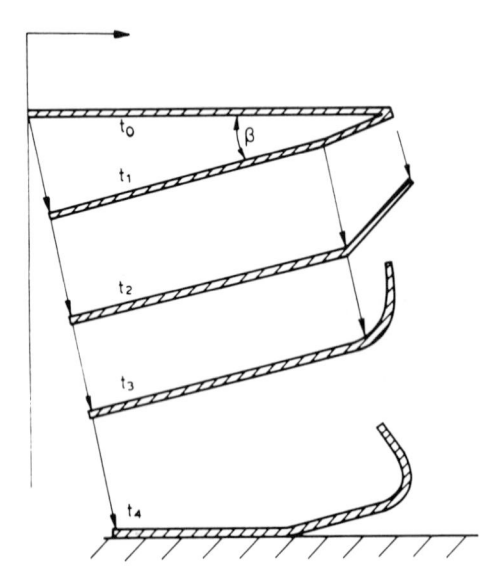

5 Development of edge damage due to unstable plate movement

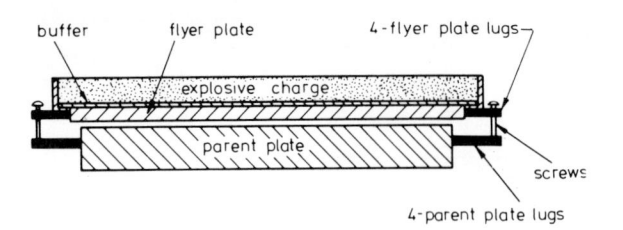

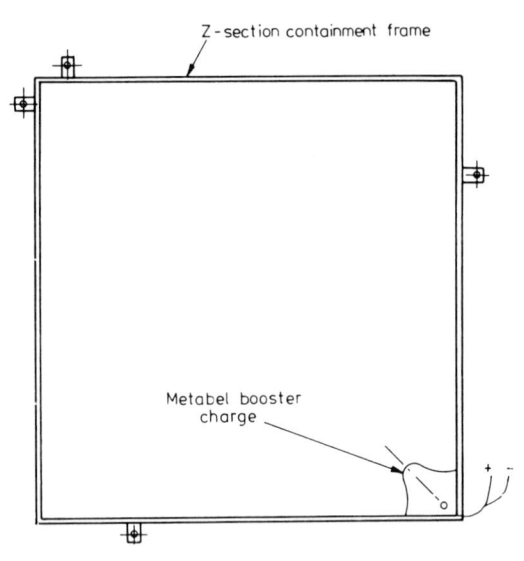

6 Arrangement for cladding large areas with thick flyer plates

7 3·2 m dia. cladded plate: 15 mm of aluminium bronze welded to 115 mm of steel

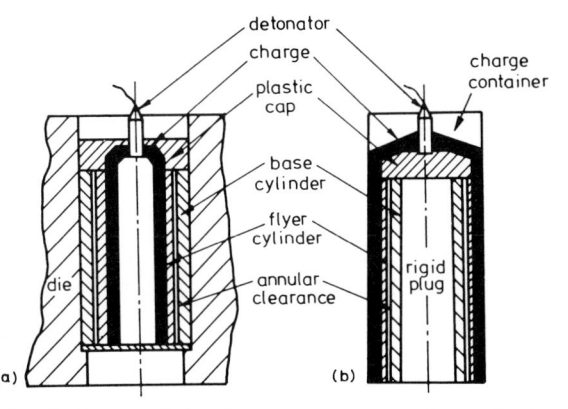

9 Explosive and implosive systems of producing duplex tubing

method.[11] A layer of Trimonite 1 explosive was used which gave an explosive loading of about 0·32. The stand-off was varied from 3·2 mm at the corner at which detonation was initiated, to 9·5 mm at the furthermost corner. An entirely satisfactory weld was achieved with a uniform wave amplitude close up to the edges of the plate, except immediately in the vicinity of the booster charge where a small non-bonded area was observed. However, when welding a smaller 12·7 mm thick manganese bronze flyer plate to mild steel with Trimonite 1 explosive, no welding was achieved, while with the same explosive loading of ammonium nitrate–diesel oil explosive, a perfect weld was achieved. The essential difference is that the detonation velocity of ammonium nitrate–diesel oil is appreciably less, while the velocity imparted to the flyer plate is somewhat reduced, but the net result is a significant increase of β from its previous value of 4·33°. A 1·07 m square by 12·7 mm thick flyer plate of naval brass has also been successfully welded to mild steel using an ammonium nitrate–diesel oil explosive charge.

In the UK, explosive welding is carried out commercially by Nobel's Explosive Company Ltd at Stevenston in Scotland. Figure 7 shows a 3·2 m dia. cladded plate consisting of 15 mm thickness of aluminium bronze to ASTM B171 alloy 614 welded to 115 mm of steel to BS 1501–161, while Fig.8 shows the completed heat exchanger which is destined for use in a Swedish district heating scheme. Anderson[15] describes other clads produced by Nobel's Explosive Co.

Ltd, and also methods of fabricating vessels from cladded material.

Welding of cylindrical surfaces

The possibility of welding cylindrical surfaces was recognized early on by Holtzman and Rudershausen[16] who mention implosive and explosive systems for producing duplex tubing, though they would appear to have only patented the explosive system. Doherty and Knop[17] used an implosive system to produce a stainless steel tube cladded to a tantalum inner tube. The o.d. of the stainless steel was 20·3 mm with a wall thickness of 2·4 mm while the inner tube had a wall thickness of 0·5 mm. Those authors report having produced lengths of up to 10·6 m of such tubing. Blazynski and Dara[18] have reported implosive fabrication of duplex tubes which are subsequently to be reduced by conventional tube drawing. Figure 9 shows the basic arrangement of the implosive and explosive systems.

The first practical application of explosive welding of cylindrical surfaces was probably that of tube-to-tubeplate welding which was probably first explictly recognized by Crossland et al.,[19] though at about the same time several groups in the UK and the USA were involved in developing such processes. The first and perhaps only truly commercial process of tube-to-tubeplate welding is the YIMpact process described by Cairns and Hardwick[20] and Cairns et al.[21]

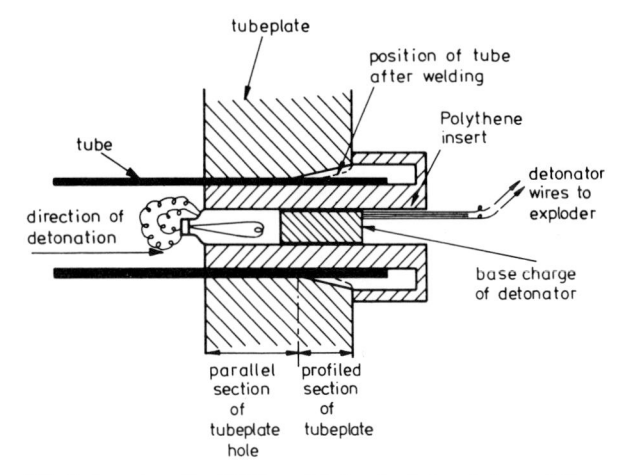

8 Cladded plate of Fig.7 used as a tubeplate for a heat exchanger district heating scheme

10 Set-up for the YIMpact tube-to-tubeplate welding process

Source: *Metals Technology*, Jan 1976

11 Superheater for CEGB Eggborough power station with 12·7 mm o.d.×0·92 mm stainless steel tubes welded to mild steel

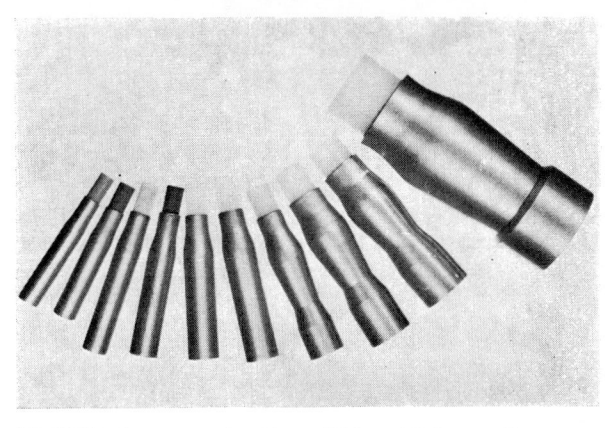

13 YIM plugs ranging from 12·2 to 50·8 mm dia.

Figure 10 shows the basic arrangement of the YIMpact process which employs a tapered arrangement with a high detonation-velocity explosive. The explosive can be provided by the charge in a conventional detonator, which is sufficient to weld a small thin-walled tube, and supplemented by ring charges of PETN explosive if larger thicker-walled tubes are to be welded. The tube before welding projects beyond the front surface of the tubeplate and the excess material is sheared off during the welding process. The plastic insert positions the explosive charge axially so that welding is achieved up to the front surface of the tubeplate, and no significant crevice can be detected. Figures 11 and 12 show two applications of the YIMpact tube-to-tubeplate welding process. In other applications the tubes have been welded to the cladding of explosively cladded tubeplates.

Another important application is that of tube plugging. In modern atomic or conventional power stations, the steam generator may contain more than 10000 heat exchanger tubes and in the lifetime of the plant leakage from some of the tubes cannot be ignored. In many situations, and especially in nuclear power plant, there are no possibilities of replacing leaking or damaged tubes as the cost in down-time would be prohibitive. The only alternative is to plug each end of the defective tube, but it is essential to achieve a high-quality weld and frequently it is necessary to have no crevice between the plug and the front surface of the tubeplate in order to avoid the possibility of crevice corrosion. Yorkshire Imperial Metals Ltd have developed a series of plugs ranging from 12·2 to 50·8 mm dia. which are shown in Fig.13. They have adopted a tapered technique with a high detonation-velocity explosive but with the taper on the plug so that it is not necessary to have a tapered hole in the tubeplate, as in Fig.10. Bahrani et al.[22] have made use of the parallel method of welding tubes to tubeplates (see Fig.14) using a low detonation-velocity explosive, which was proposed by Crossland et al.,[19] to plug 41·3 and 44·5 mm dia. holes in the heat exchanger of the advanced gas-cooled reactor shown in Fig.15. Figure 16 shows the arrangement of the plug and the charge used in this application, while Fig.17 shows a photograph of the plug being inserted in a mock-up tubeplate. Yet another

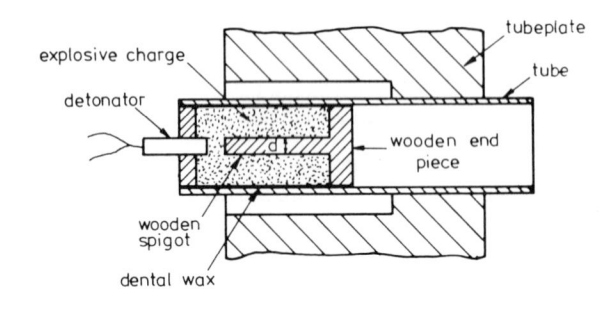

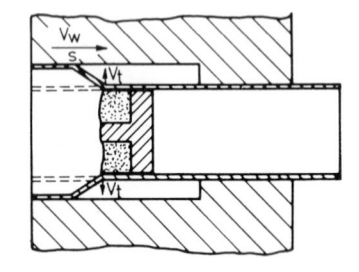

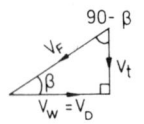

12 Clark liquor calandria with titanium tubes 50·8 mm o.d.×1·63 mm welded to Type 321 stainless steel

14 Parallel arrangement for tube-to-tubeplate explosive welding

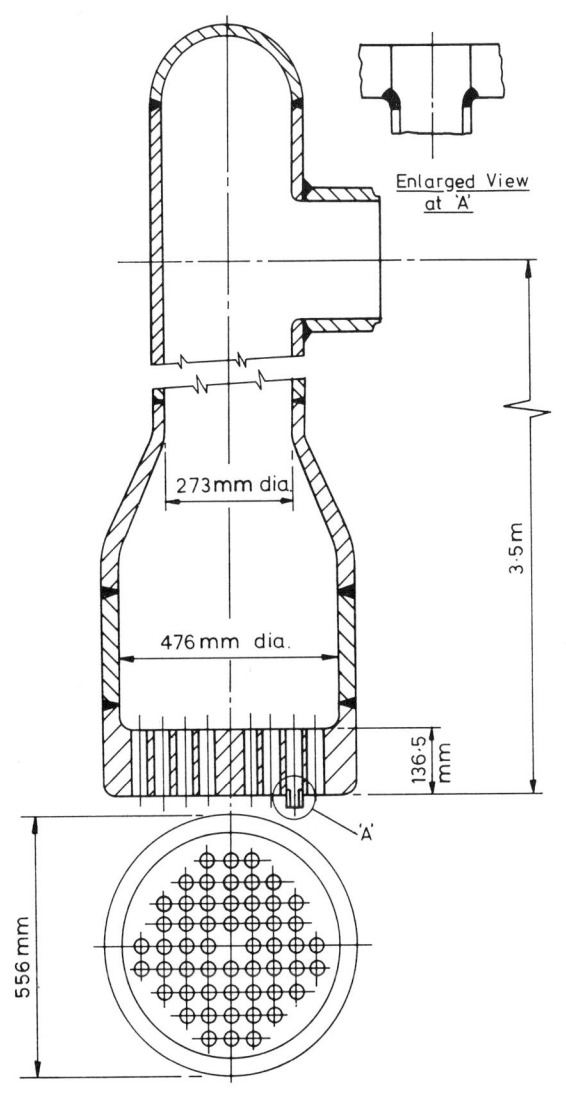

15 Reheater tubeplate and header of the advanced gas-cooled reactor

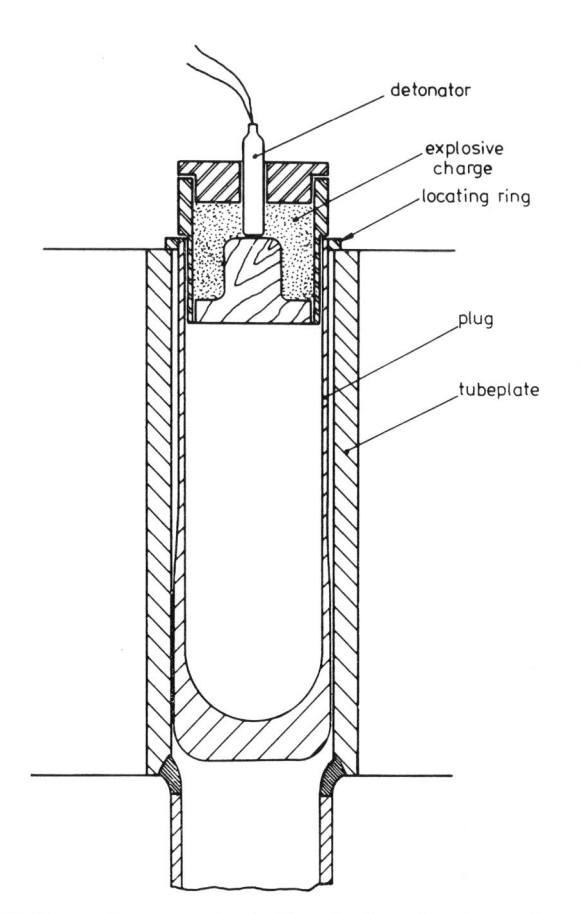

16 General arrangement of explosive plugging set-up

plug has been developed by Bahrani et al.[23] for plugging the 21·1 mm dia. hole in the toroidal superheater and reheater exchangers of the prototype fast reactor in which there is little headroom and the access is difficult. In this case, a radically different technique was used which is shown in Fig.18. Basically, a point charge is used which causes a spherical bulge to form in the wall of the plug, so that oblique impact is experienced each side of the centre line of the bulge. Consequently, two regions of weld are formed, one each side of the centre line of the bulge with the interfacial waves running in opposite directions. If the centre of the charge is correctly positioned axially the excess length of the plug is sheared off and welding can occur up to the front surface of the tubeplate. Figure 19 shows the form of the plug while Fig.20 shows the access to the tubeplate. This process has been used successfully on the superheater and reheater during the construction phase, and, at the time of writing, it was being applied on site to two of the superheaters.

For tube-to-tubeplate welding and plugging, the welding parameters have been explored by Williams et al.[10] and Crossland and Williams,[24] who found that they were similar to those proposed by Wylie et al.[9] for explosive cladding. However, there are serious additional problems, and in particular the distortion of the ligaments adjoining a hole in which a tube has been welded. Crossland and Williams[24, 25] have considered this problem in detail, and for a given explosive loading, which depends on the combination of materials and thickness of the tube, and a given tube and tubeplate configuration, they have established by dimensional analysis a relationship which allows an estimation to be made of the ligament distortion. Using this, they have proposed a strategy for establishing the permissible ligament thickness and consequently the centre-line spacing. The ligament thickness and centre-line spacing for a

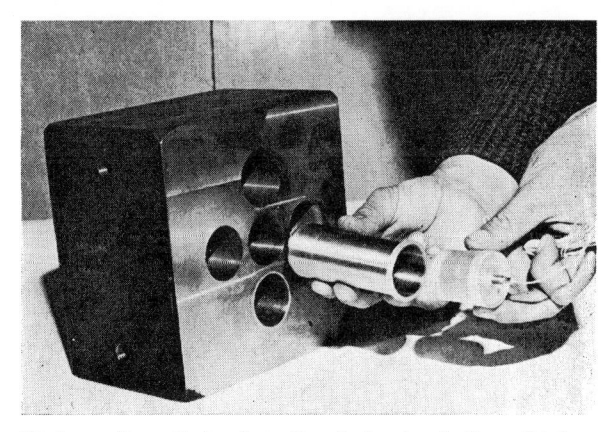

17 Insertion of plug into five-hole simulation of tubeplate

Source: *Metals Technology*, Jan 1976

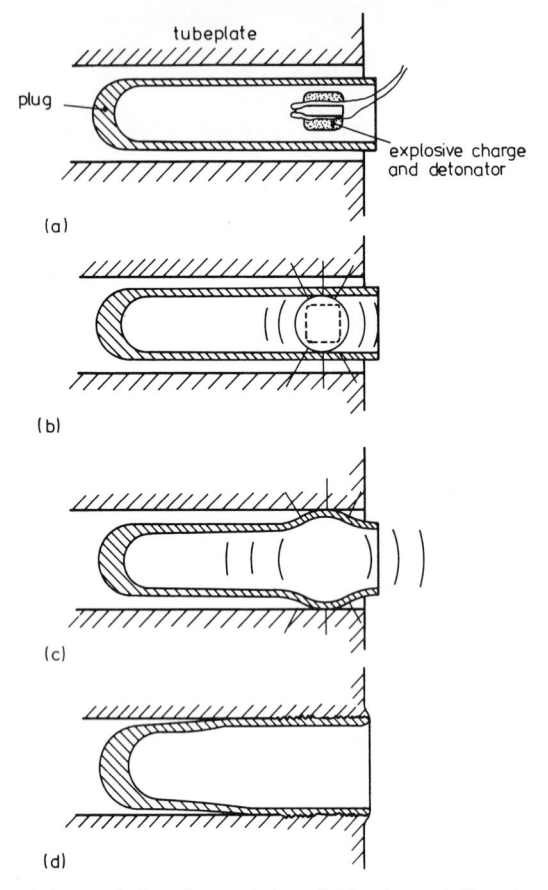

a before exploding charge; *b* immediately after exploding charge; *c* collision between walls of plug and tubeplate; *d* welded plug with end sheared off

18 Schematic illustration of welding process between plug and tubeplate

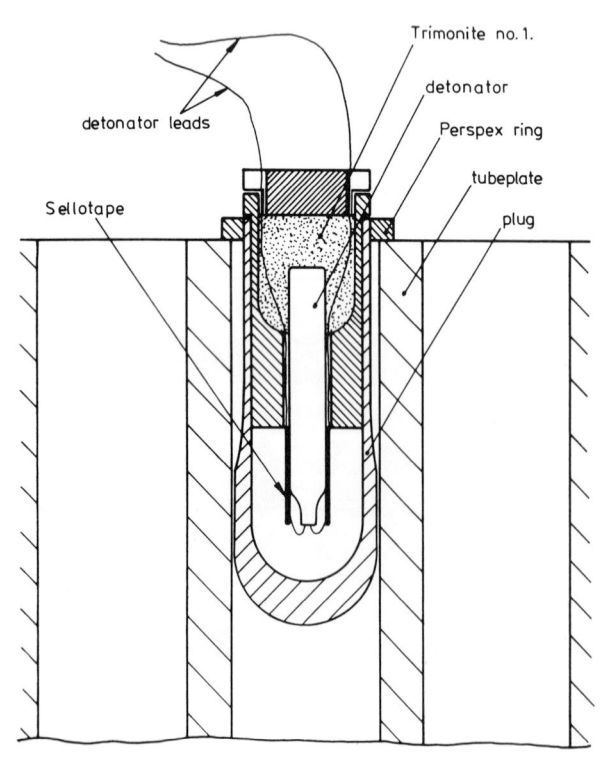

19 Explosive plugging set-up for heat exchanger of prototype fast reactor

20 Access to tubeplate of prototype fast reactor

particular case of welding a 30 mm o.d. annealed copper tube to a mild steel tubeplate is shown in Fig.21. From this diagram it will be seen that there is little to be gained in centre-line spacing by increasing the stand-off beyond $1t$, which, for a 2 mm wall thickness, means an initial radial clearance between the tube and the hole in the tubeplate, of 2 mm. For this analysis it was assumed that the minimum stand-off which can be tolerated is $0.5t$, so if the initial stand-off is qt the distortion which could be permitted from welding a tube in an adjoining hole would be $qt - 0.5t$, which, if q is 1 reduces to $0.5t$ or 1 mm for a t of 2 mm. However, it is found that if the ligament is very small in relation to the hole diameter then the distortion becomes unacceptably large, and in this case the adjoining holes must be bunged with metal bungs before welding a tube in position in order to keep the distortion within acceptable bounds. Both Yorkshire Imperial Metals Ltd and the Queen's University of Belfast have developed satisfactory bunging techniques which have permitted satisfactory welds to be achieved with a very small distortion. For example, it has been possible to bung a 44·5 mm dia. hole with a 6·4 mm ligament using bungs, whereas without bungs the distortion was unacceptably large.

Yet another potentially important application of the welding of cylindrical surfaces is provided by the joining of pipes. Willis and Murdie[26] used an explosive welding technique to expand the butted ends of two pipes against a common sleeve to form a weld. Chadwick and Evans[27] have carried out an extensive programme of pipe welding using an internal charge system such as that shown in

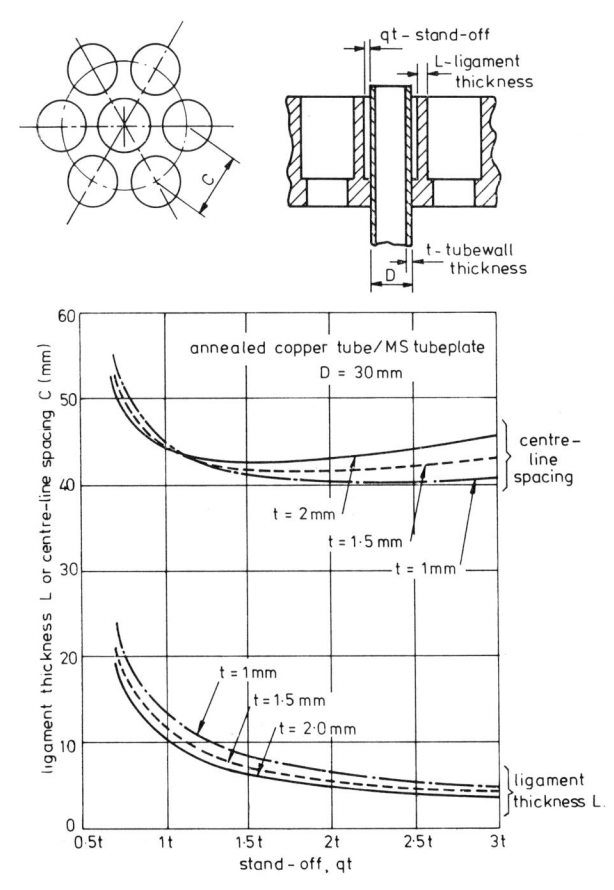

21 Calculated values of minimum ligament thicknesses *L* and centre-line spacing *C* for welding annealed copper tubes to mild steel tubeplates

Fig.22, in which either the external sleeve is sufficiently massive to take the reaction or an external anvil is used so that a thinner sleeve can be employed. Several advantages are claimed for this arrangement over the external charge system with an internal expanding anvil, which has been developed by the Denver Research Institute and the Alcan Research Laboratory, Canada, such as reduction of noise, containment of the debris, and the ease of fitting an external anvil compared with inserting an expanding mandrel. Figure 23 shows the external anvil system for a small pipe and Fig.24 shows a length of welded 89 mm dia. by 3·2 mm wall-thickness tube which was pressure tested to 1·03 MN/m². Chadwick and Evans[27] also report welding much larger diameter tubes using the arrangement shown in Fig.25 and an example of two tubes welded by this method is shown in Fig.26. The International Research and Development Co. Ltd have successfully extended this work to welding pipes under water, which is potentially of great value in the North Sea. Completely independent of this work, Wylie and Crossland[28] have used a similar system to weld collars to tubes to form a transition joint between a tube and a thermal sleeve of a different and incompatible material. However, in this case it was important to achieve a weld between the tube and sleeve without a crevice, and

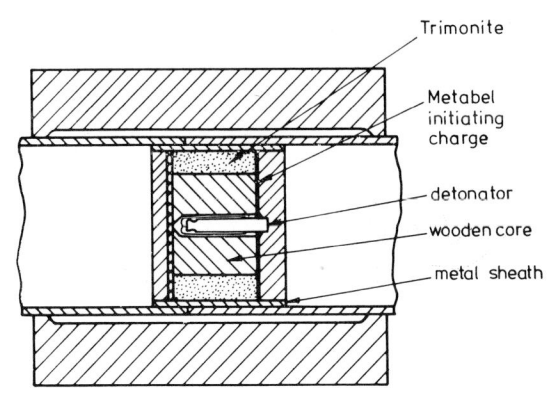

22 Schematic arrangement of pipe welding with an internal charge and external sleeve

23 External anvil system with a 25·4 mm o.d. pipe

24 Explosively welded trial length of 89 mm o.d. × 3·2 mm wall-thickness pipe

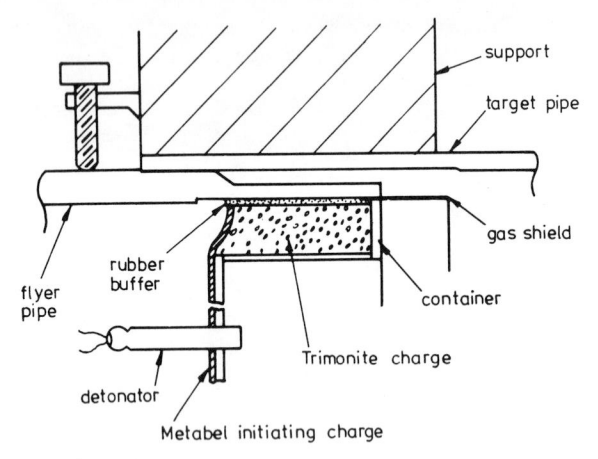

25 Set-up for welding large-diameter pipes

26 Explosive weld between a 762 mm o.d.×12·7 mm wall tube and a 812 mm o.d.×19·1 mm wall tube

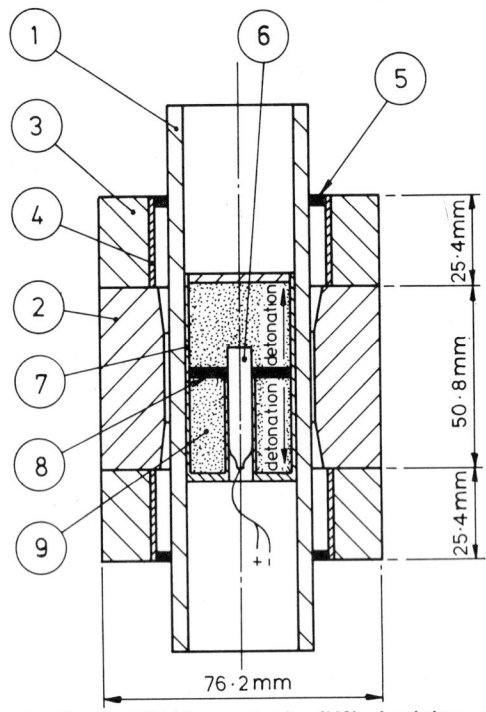

1 tube; 2 collar (SS); 3 support collar (MS); 4 polythene sleeves; 5 polythene location rings; 6 electric detonator; 7 polythene charge holder; 8 Metabel sheet explosive; 9 Trimonite 1 powder explosive

27 Arrangement for welding of a tube to a collar

a tube-to-collar weld; *b* section through tube-to-collar

28 Welding of a tube to a collar with central initiation of the explosive charge as in Fig.27

it was necessary to develop a satisfactory support system each side of the sleeve to allow welding up to the end of the sleeve without a crevice being formed, while preventing the tube being sheared. The arrangement adopted is shown in Fig.27 while Fig.28 illustrates such a weld.

Other welding processes

A few other applications have been investigated but they do not appear so far to have found commercial application.

The explosive fabrication of fibre-reinforced material was probably first described by Jarvis and Slate,[29] while more recently Wylie *et al.*[30,31] have produced larger samples of high-tensile wire-reinforced aluminium with volume concentrations of over 20% wire mesh. However, it would appear that this process is only really applicable to metal-wire reinforcement of soft metal matrix materials.

Another region of great interest is in lap welding using

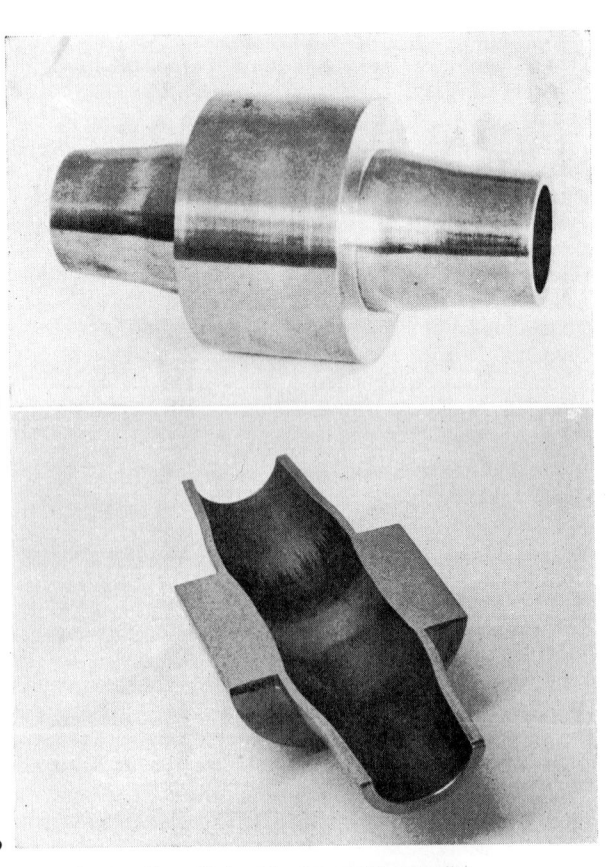

high detonation-velocity line charges. Shribman et al.[32] have briefly mentioned lap welding and more recently Bement[33] has demonstrated its applicability in fabricating large aluminium structures out of aluminium and aluminium alloy sheet. Currently, attempts are being made to exploit line welding in the manufacture of large containers.

Metallurgy of explosive welds

Space does not permit a detailed discussion of the metallurgical problems of explosive welding. In general, it can be stated that as explosive welding is essentially a solid-state welding process, the metallurgical problems are much less severe than with a fusion-welding process as demonstrated by Lucas et al.[34] However, in some cases it is possible, under incorrect welding conditions, to get the formation of a continuous molten interlayer and this can sometimes lead to the formation of a brittle intermetallic compound, as in the case of titanium welded to steel. In the area of the vortices in the interfacial welds there is always, as far as can be ascertained, a molten zone in which there is a mixture of the two metals being welded. This molten zone is surrounded by solid metal at a relatively low temperature, so the rate of cooling is extremely high, and Williams et al.[35] and Dhir et al.[36] have observed metastable compounds in these zones which are basically like those found in splat cooling. Away from the weld interface, metallurgical defects associated with shock loading are observed, such as twinning, and if a high detonation-velocity explosive is used, then spalling failure of the flyer plate may occur during its free flight. Normally, if the formation of harmful intermetallics or of a continuous-cast interlayer are avoided, then the strength of the weld will equal the strength of the weaker of the two materials being welded, as determined by the side-shear test or some form of tension test across the weld interface. Fatigue tests on explosively fabricated composite material give surprisingly high values, though the fatigue failure frequently originates at the weld interface. However, it should be noted that the fatigue properties of the materials being welded are normally enhanced by the hardening they experience due to the shock loading to which they are subjected in the welding process, and on the basis of these enhanced properties the fatigue strength of the composite material is perhaps somewhat reduced. In general, the materials after welding are appreciably harder and their ductility reduced; it is also probable that the fracture toughness is somewhat reduced, and with many material combinations this reduction cannot be eliminated by subsequent heat treatment for metallurgical reasons; any reduction in toughness must therefore be accepted.

Conclusions

The author has been concerned mainly with a review of the developments in the application of explosive welding technology in the UK and has made no attempt to review the extensive contribution to this field by groups working in other countries. It can be seen that explosive welding has a very limited, albeit useful, application. In general, it can be used to weld very simple, flat, and cylindrical surfaces, but it cannot be used for welding surfaces with sharp transitions in section as the reflection of shock waves from these discontinuities would cause failure. However, the process is already being exploited successfully for large-plate cladding of one metal on another, particularly for large tubeplates of heat exchangers and for producing cladded plate for pressure-vessel construction. It has also found application in the welding of tubes to tubeplates for heat exchangers, and for the plugging of faulty tubes in heat exchangers particularly in conventional and nuclear power plants.

Metallurgically, the problems with explosive welding are much less acute than with fusion welding as the process is essentially a solid-state welding process. However, there are metallurgical aspects of explosive welds which give rise to considerable interest, particularly the metastable compounds formed in the vortex areas of the weld.

Acknowledgments

Thanks are due to the following people and firms for supplying information and photographs: D. K. L. Anderson of Nobel's Explosives Co. Ltd, R. Hardwick of Yorkshire Imperial Metals Ltd, N. H. Evans of the British Gas Corporation Engineering Research Station, and M. D. Chadwick of International Research and Development Co. Ltd. Thanks are due also to the author's many collaborators over the years: Dr A. S. Bahrani, Dr V. Shribman, Dr J. D. Williams, Dr W. Lucas, S. K. Banerjee, Dr P. Dhir, Dr H. K. Wylie, Dr P. E. G. Williams and Dr A. Cave. Acknowledgment is made for the assistance so readily given by the workshop and technical staff under R. H. Agnew and R. J. Harvey. This work would not have been possible without very generous grants from the Science Research Council over many years, and research contracts from Babcock and Wilcox Ltd.

References

1. L. R. CARL: *Met. Progr.*, 1944, **46**, 102.
2. V. PHILIPCHUK: 'Explosive welding status'; 1965, ASTME Creative Manufacturing Seminar, Paper no.SP65-100.
3. B. CROSSLAND and J. D. WILLIAMS: *Metallurgical Rev.*, 1970, **15**, 79.
4. B. CROSSLAND: 'Review of the state-of-the-art in explosive welding'; 1971, Internat. Inst. of Welding, document 4/71/71.
5. B. CROSSLAND: *Metals and Materials*, 1971, **5**, 401.
6. G. BIRKHOFF et al.: *J. Appl. Phys.*, 1948, **19**, 563.
7. A. S. BAHRANI and B. CROSSLAND: *Proc. Inst. Mech. Eng.*, 1964, **179**, 264.
8. A. S. BAHRANI and B. CROSSLAND: *ibid.*, 1966, **180**, 31.
9. H. K. WYLIE et al.: Proc. 1st Internat. Symposium on 'Use of explosive energy in manufacturing metallic materials of new properties and possibilities of application thereof in chemical industry', 1970, Marianske Lazne.
   See also Proc. 3rd Internat. Conf. of Centre for High Energy Forming, 1971.
10. P. E. G. WILLIAMS et al.: Proc. conf. on 'Welding and fabrication of non-ferrous metals', 1; 1972, Welding Institute.
11. V. SHRIBMAN and B. CROSSLAND: Proc. 2nd Internat. Conf. of Centre of High Energy Forming, 1969.
12. B. CROSSLAND et al.: 'Explosive cladding of large plates', Proc. select conf. on 'Explosive welding', The Welding Institute, 1968.
13. H. K. WYLIE and B. CROSSLAND: Proc. internat. conf. on 'Uses of high-energy rate methods of forming, welding, and compaction', Leeds, 1973.
14. D. RUPPIN: Paper presented to Internat. Inst. of Welding meeting, 1968, Warsaw.
15. D. K. C. ANDERSON: *Sheet Metal Ind.*, 1974, **51**, 36.
16. A. H. HOLTZMAN and C. G. RUDERSHAUSEN: *ibid.*, 1962, **39**, 399.
17. A. E. DOHERTY and L. H. KNOP: Proc. 2nd Internat. Conf. of the Centre for High Energy Forming, 1969.
18. T. Z. BLAZYNSKI and A. R. DARA: Proc. 3rd Internat. Conf. of the Centre of High Energy Forming, 1971.
19. B. CROSSLAND et al.: *Welding*, 1967, **35**, 88.

20. J. H. CAIRNS and R. HARDWICK: Proc. select conf. on 'Explosive welding', The Welding Institute, 1969.
21. J. H. CAIRNS et al.: Proc. 3rd Internat. Conf. of Centre for High Energy Forming, 1971.
22. A. S. BAHRANI et al.: *Internat. J. Pressure Vessels and Piping*, 1973, **1**, 17.
23. A. S. BAHRANI et al.: Proc. internat. conf. on 'Welding research related to power plant', Southampton, 1972, CEGB.
24. B. CROSSLAND and P. E. G. WILLIAMS: Proc. 2nd Internat. Conf. on 'Pressure vessel technology', 1973.
25. P. E. G. WILLIAMS and B. CROSSLAND: Proc. 3rd Internat. Conf. on 'Advances in welding processes', Harrogate, 1974.
26. J. WILLIS and D. C. MURDIE: *Sheet Metal Ind.*, 1962, **39**, 811.
27. M. D. CHADWICK and N. H. EVANS: *Metal Construction*, 1973, **5**, 285.
28. H. K. WYLIE and B. CROSSLAND: Proc. 4th Internat. Conf. of Centre for High Energy Forming, 1973.
29. C. V. JARVIS and P. M. B. SLATE: *Nature*, 1968, **220**, 782.
30. H. K. WYLIE et al.: *Welding*, 1971, **39**, 214.
31. H. K. WYLIE et al.: Proc. 3rd Internat. Conf of Centre for High Energy Forming, 1971.
32. V. SHRIBMAN et al.: *Prodn. Engineer*, 1969, **48**, 69.
33. L. J. BEMENT: 'Small-scale explosive welding of aluminium'; 1971, NASA, Langley Research Centre.
34. W. LUCAS et al.: Proc. 2nd Internat. Conf. of Centre for High Energy Forming, 1969.
35. J. D. WILLIAMS et al.: Proc. 3rd Internat. Conf. of Centre for High Energy Forming, 1971.
36. P. DHIR et al.: Proc. 4th Internat. Conf. of the Centre for High Energy Forming, 1973.

3. EXPLOSIVE WELDING AND CLADDING — OVERVIEW OF THE PROCESS AND SELECTED APPLICATIONS

J. A. YOBLIN, J. D. MOTE, and L. E. JENSEN
E. F. Industries, Incorporated
Louisville, Colorado

Explosive welding is a viable and rapidly growing manufacturing technology. It has established itself as an important and basic process which is generating a wide variety of critical production parts. The technology is not dependent on either large pools of skilled labor or massive amounts of capital equipment; it has an outstanding safety record.

Products manufactured by explosive welding include key components for chemical process plants, oil refineries, heat exchangers, large-diameter pipeline joining, cryogenic systems, ship construction, etc.

More than fifty U.S. patents have been issued since E. F. Industries' fundamental first-in-time patent, "Explosive Welding," which issued in March, 1962. Most of these patents are owned by either E. F. Industries, Inc., of Louisville, Colorado, or by E. I. du Pont of Wilmington, Delaware. As a result of patent litigation between the two companies, EFI has licensed du Pont under its patents and du Pont has cross licensed EFI under its patents.

The science of explosive welding is highly developed; literally millions of dollars have been spent by private industry and the government to develop this technology. Although explosive welding was invented in 1958, large commercial growth has just started in the past few years and is increasing very rapidly.

INTRODUCTION

Explosive welding (DYNAWELD*tm*) and explosive cladding (DYNACLAD*tm*) is used by E. F. Industries, Inc., on a commercial scale to manufacture a wide variety of metallurgically bonded similar and dissimilar metals for many basic industries including:

Refineries
Chemical Processing Plants
Power Generation (fossil fuel and nuclear powered)
Cryogenics
Shipbuilding
Pipelines (oil and gas)
Electronics
General Industrial-Machinery
Heat Exchangers
Aerospace and Aircraft (commercial and military)
Desalinization Plants

The basic process and resulting products were first invented in 1958, and the resulting first U.S. patent (titled "Explosive Welding") was issued in 1962. This original pioneer patent, which is owned by E. F. Industries, Inc., is the first of more than 50 U.S. patents which have been issued. The process is not an art, but rather a precise science. Appropriate products resulting from the process are technically superior to those produced by competitive processes such as fusion welding or roll cladding. In many cases unique combinations, such as titanium to steel, have resulted in new products for which no competitive process currently exists.

In many modern industries, three metal systems commonly occur (often in the same system). These include:

1. Steel (and steel alloys)
2. Titanium (and titanium alloys)
3. Aluminum (and aluminum alloys).

No two of these three systems can be welded together by conventional techniques. Yet any combination can be explosively welded together in either flat or cylindrical configurations.

At the present time, explosive welding and cladding is a rapidly growing multimillion-dollar annual business which is also being practiced commercially in several foreign countries.

BACKGROUND

Explosive welding was invented in early 1958 by Mr. Vasil Philipchuk (now deceased) and Mr. Franklin LeRoy Bois. While Philipchuk and Bois were doing some explosive forming work they observed instances of welding between the explosively formed product and the steel die or tool. They investigated this phenomenon and determined that parts of the explosively formed product were welded to the steel die with a substantially continuous metallurgical bond. By April 1, 1958, the inventors retained a member of the Metallurgy Department of M.I.T. as a consultant to analyze samples of welds obtained by explosive techniques. Very shortly thereafter, relatively sophisticated combinations of metal (such as aluminum to Inconel and titanium to steel) were being explosively welded and clad, and relatively sophisticated products and applications were being articulated, such as flat plate cladding, welding liners inside pipes and cylinders, explosively welding tubes and pipes to headers or tubesheets, etc.

E. F. Industries, Inc. acquired several operations (including their respective patents) involved in explosive welding and cladding. These included: Explosive Fabricators, Inc. Division of Tyco Laboratories; Explosive Metalworking Operations, Martin Marietta Corporation, Denver Division; and Explosive Welding Operation, Hexcel. In addition, Martin Marietta Corporation received a seven-year multimillion-dollar ARPA research and development contract in explosive metalworking in which they issued a subcontract to Denver University's Denver Research Institute. The last phases of this program were run while Martin Marietta's Explosive Metalworking Operation was part of EFI. E. F. Industries, Inc., has more than 21 corporate years in the field of explosive welding and cladding, and this extensive experience is enabling the technology and applications of explosive welding and cladding to be developed at an increasing rate.

Figure 1 is a pictorial schematic of the DYNACLAD^{tm} explosion cladding process. This figure illustrates the same principles which are involved in explosive welding (*e.g.*, pipelines). The main features of this process are:

1. A specific standoff between the cladder metal and the backer metal.

2. Progressive detonation of the explosive over the metal to be explosively clad or welded.

3. A specific bend angle by the cladder metal as it is forced to weld to the backer metal.

4. The formation of a jet within the included angle between the cladder metal and the backer metal. This jet actually causes the surface metal between the cladder and backer to behave as a fluid — these surfaces are scoured clean by the jet.

5. A characteristic (and visible) wavy pattern at the weld interface caused by the swirling jet as it freezes.

The cladder and backer metal are truly metallurgically joined. In many cases the strength of the weld is greater than the weaker of the two metals explosively welded together.

The list of metals which can be explosively welded is almost limitless. "Incompatible" systems such as titanium to steel, aluminum to

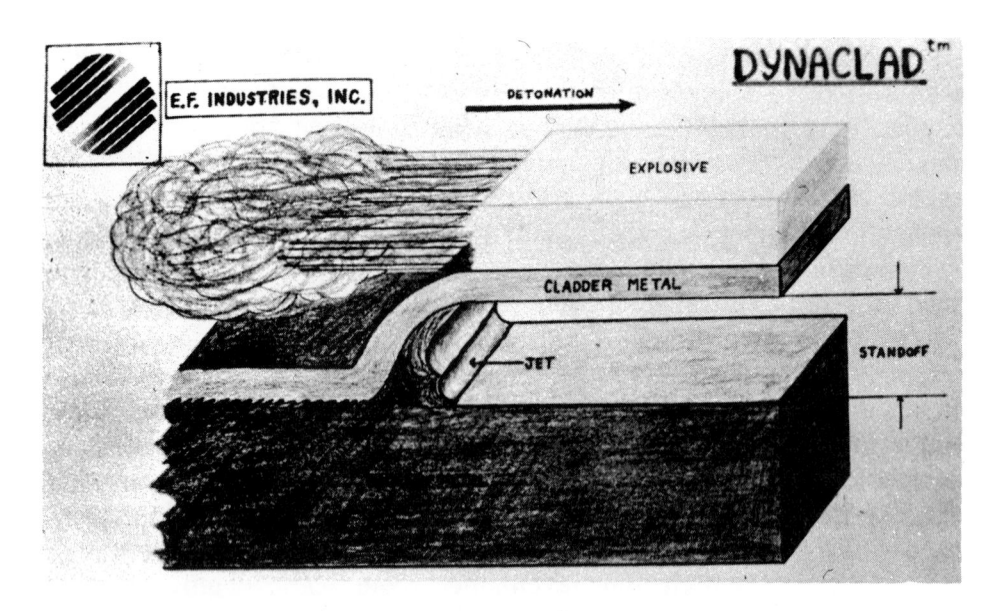

Figure 1. Explosion cladding process.

steel, and aluminum to titanium have been successfully explosion clad and welded on a commercial scale. Table 1 is a partial listing of metals which have been explosively clad. Many other combinations, including multiple clads of more than two different metals, are feasible and have been produced.

Competitive processes to explosive welding or cladding are shown in Table 2.

Certain metal systems are metallurgically incompatible to fusion welding, roll cladding, or diffusion bonding. In these cases, explosive welding and cladding offer a viable manufacturing alternative. In other cases where weld overlaying or roll cladding are feasible, large section sizes might indicate the explosive welding process to be either more cost effective, or to yield a metallurgically superior material.

Table 3 is a partial listing of products which have been manufactured using the DYNAWELD^{tm} or DYNACLAD^{tm} process.

PROCESS

Figure 2 is an engineering schematic of the explosion welding and cladding process. The schematic focuses at the collision point and shows the Collision Angle B, which is determined by the standoff distance, the explosive used (amount and detonation velocity), and the thickness and

Table 1. Partial Listing of Metals Which Have Been Explosion Welded

Steels to	Stainless Steels	Copper
	Aluminum	Nickel
	Titanium	Columbium
	Monels	Tantalum
	Copper-Nickels	Hastelloys
	Brasses	Stellites
	Bronzes	Magnesium
	Silver	Gold
Aluminum to	Steels	Titanium
	Stainless Steels	Nickels
	Copper	Silver
	Copper Alloys	Magnesium
Titanium to	Steels	Copper Alloys
	Stainless Steels	Nickel Alloys
	Aluminum	Tantalum
	Copper	Columbium

Table 2. Competitive Processes to Explosive Welding and Cladding

Fusion Welding (including weld overlaying)
Roll Cladding
Diffusion Bonding

Table 3. Partial Listing of Explosion Welded and Clad Products

DYNACLAD[tm]	Tube Sheets (for heat exchangers)
DYNACLAD[tm]	Plates (for pressure vessels and tanks)
DYNACLAD[tm]	Bi-metal Tubing (for chemical plants, heat exchanger tubing, desalinization plants, etc.)
DYNACLAD[tm]	Nozzles (for high-pressure vessel attachments)
DYNACLAD[tm]	Cu clad Al, and Al clad Steel Strip (for electrical transition joints)
DYNACLAD[tm]	Al clad Steel Plate and Strip (for structural transitions — marine, automotive, etc.)
DYNABOND[tm]	Tubular Transition Joints (to join dissimilar metal tubes and pipes)
DYNAWELD[tm]	Explosion Welded Pipelines
DYNAWELD[tm]	Overlays (for erosion/corrosion resistant surfaces; for building up worn surfaces)

Source: *Advances in Joining Technology*, Brook Hill Publishing Co., 1976

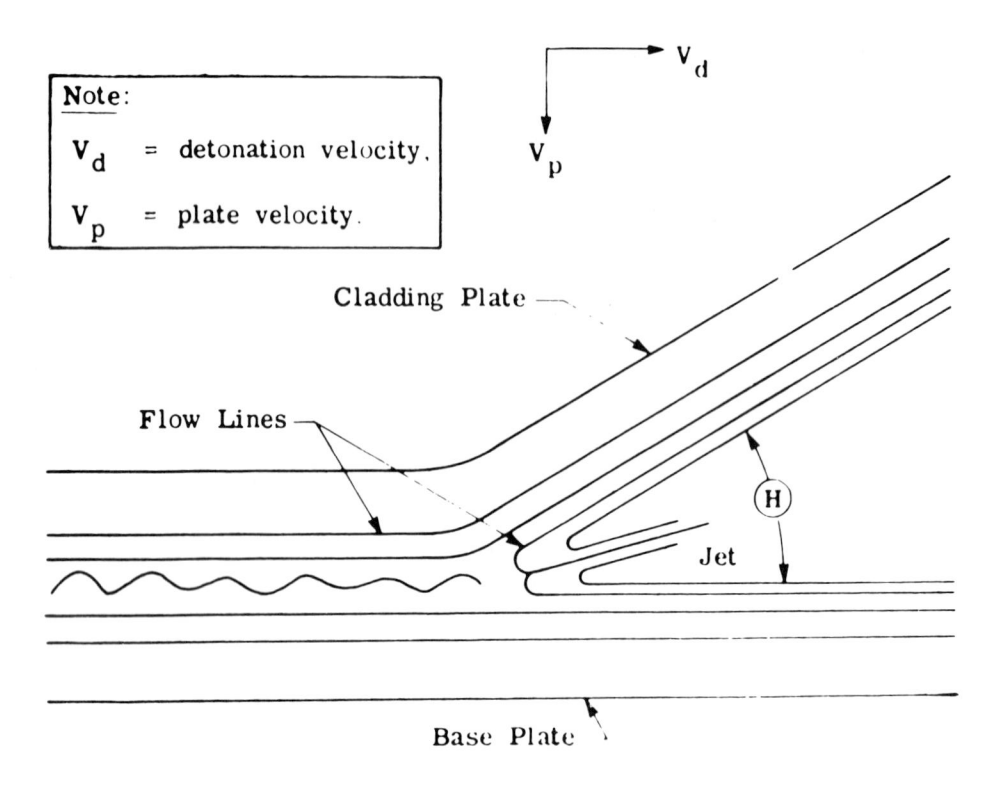

Figure 2. Engineering schematic of explosion welding process.

Note:
V_d = detonation velocity.
V_p = plate velocity.

strength of the cladding plate. The schematic also illustrates a few imaginary flow lines in the cladding plate and in the base or backer plate. The flow lines nearest respective inside surfaces of the cladder and backer are incorporated in the jet at the collision point and turn into the jet. As the detonation front and the collision point advance, the swirling action of the jet freezes and the wavy pattern which is characteristic of the explosion welding and cladding process is formed. If too much energy (excessive amounts of explosive) is coupled with a large standoff, sufficient heat is generated in the jet to actually melt the interfacial surface of the cladder and/or the backer. This manifests itself as a melt condition, either discontinuous pockets or continuous layers, at the interface.

In situations where an undesirable compound can be formed (for example titanium/steel where a brittle iron-titanium intermetallic compound will result if melting is permitted at the interface), the explosion welding parameters are adjusted so that no melting occurs. Photomicrographs will then show a clean wavy interface free of melt pockets or zones, and tensile shear tests at the interface will confirm high strengths in excess of ASTM code requirements. For critical applications, EFI will run shear tests on each clad plate to verify bond-strength integrity.

Figure 3 is a photomicrograph (original magnification @ 100X) of titanium (SB-265-II) explosion welded to steel (SA-516-70).

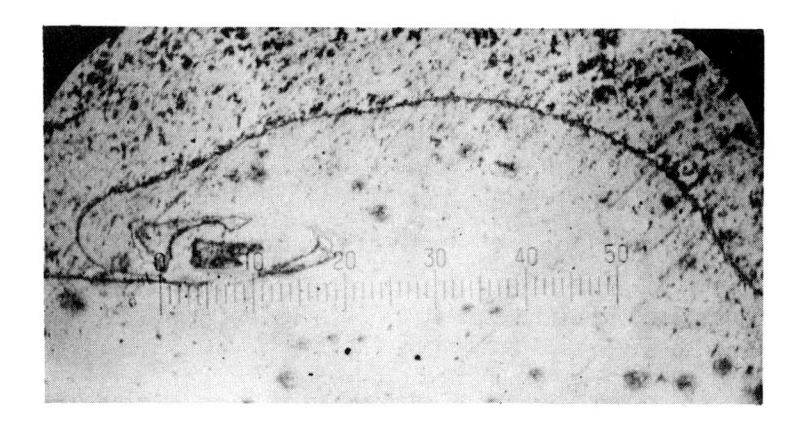

Figure 3. Photomicrograph (100X) of titanium (SB-265-II) explosion welded to steel (SA-516-70).

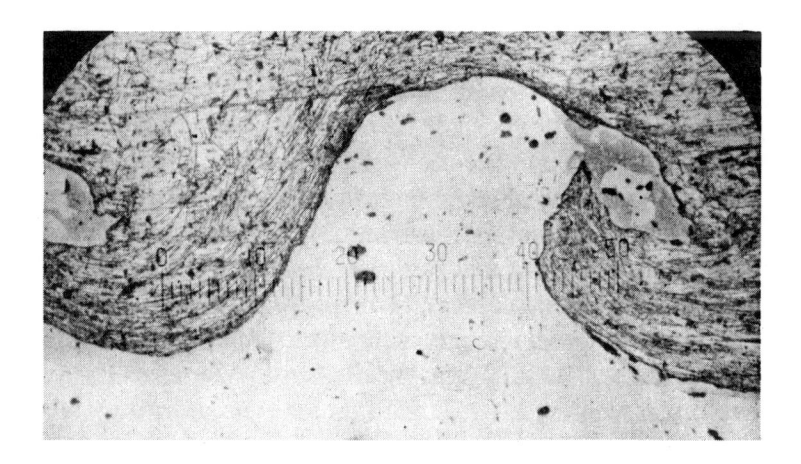

Figure 4 is a photomicrograph (100X) of copper-nickel (80-20) explosion welded to steel (SA-516-70).

Figure 5 is a photomicrograph (100X) of stainless steel (type 304L) explosion welded to carbon steel (SA-516-70).

Figure 4. Photomicrograph (100X) of copper-nickel (80-20) explosion welded to steel (SA-516-70).

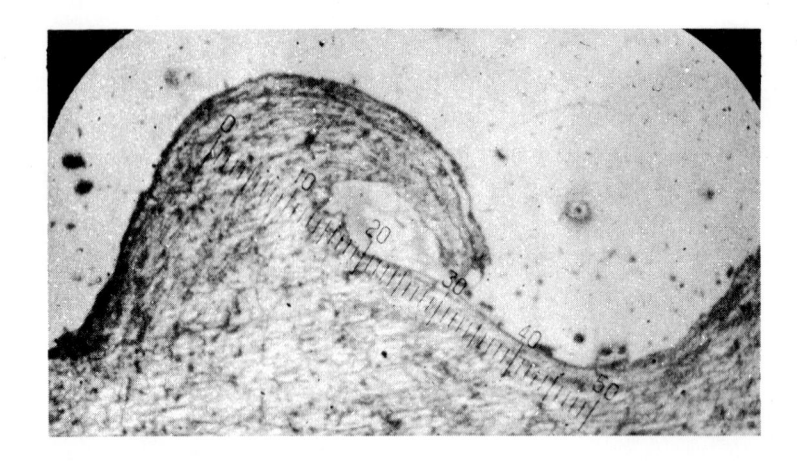

Figure 5. Photomicrograph (100X) of stainless steel (Type 304L) explosion welded to carbon steel (SA-516-70).

The character of the waves (including amplitude and frequency) is controlled by such variables as the detonation velocity of the explosive, the standoff distance, the collision angle, the thickness of the cladder, and other properties of the cladder and the backer. In a given system (*e.g.*, a given thickness of metal A explosion welded to metal B), the character of the wave is reproducibly controlled by the appropriate manufacturing procedure. Literally thousands of explosion clad and welded products have been manufactured; once the manufacturing procedure has been established, the scrap rate is generally less than 1%.

Ultrasonic testing has been developed to the point where it is a fine nondestructive testing technique which permits 100% inspection of a clad surface. In certain cases X-ray has been used to show the frequency of the waves. In cases where the explosion welded product is machined across an interface (*e.g.*, the O.D. of a DYNACLAD^{tm} tubesheet) dye penetrant inspection is an outstanding tool.

The following series of photographs illustrates a typical sequence for explosion cladding bimetal tubesheets at E. F. Industries, Inc.

Figure 6 is a truckload of carbon-steel backer plate arriving at EFI.

Figure 7 shows each plate is inspected dimensionally and is metal stamped with the appropriate identification. Mill certifications are reviewed. In certain cases EFI may reultrasonic the steel backer plates prior to explosion welding.

Figure 8 shows a plate having one surface sand blasted to remove mill scale and other surface contaminants.

Figure 9 shows a steel backer plate being surface conditioned with disc grinders. Typically the surfaces to be clad are conditioned to the range of RMS 125.

Figure 6. Steel backer plate arriving at EFI.

Figure 7. Incoming plates being inspected and identified.

Figure 8. Sandblasting of plate surface to remove mill scale and other contaminants.

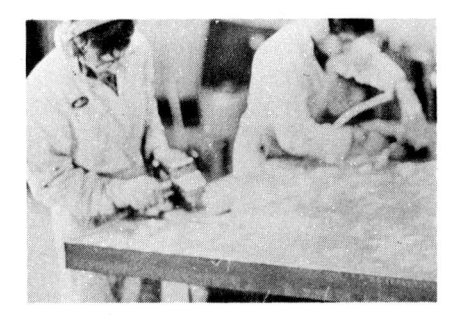

Figure 9. Backer plates being surface conditioned to approximately 125 rms.

Figure 10 is a cladder plate being surface conditioned with a disc grinder.

Figure 11 is a stack of steel backer plates whose surfaces are ready for explosion cladding.

Figure 12 is the surface of a 10-inch-thick aluminum slab which has been prepared to have a 1/4-inch-thick copper plate explosion welded to it.

Figure 13 shows the prepared interfacial surfaces of a cladder and a backer which are about to be assembled for explosive cladding.

Figure 14 shows the cladder resting on the backer, and separated from the backer by the appropriate standoff distance using small aluminum blocks (which have been machined to the correct height) as spacers (these can be seen at the corners). The production technicians are taping the edges to keep the interface space clean during transportation to the field and during field setup.

Figure 15 shows a plate being loaded onto a trailer for transportation to one of EFI's explosion welding sites.

Source: *Advances in Joining Technology*, Brook Hill Publishing Co., 1976

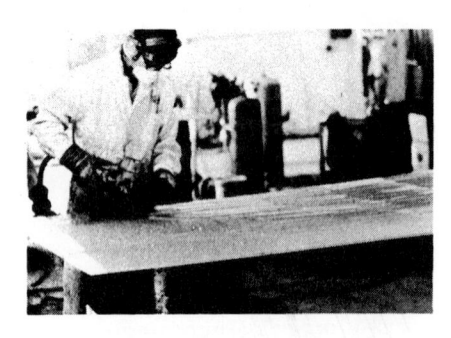

Figure 10. Cladder plates being surface conditioned to approximately 125 rms.

Figure 11. Prepared backer plates ready for assembly.

Figure 12. Ten-in.-thick aluminum backer (to be clad with 1/4-in.-thick copper).

Figure 13. Cladder and backer ready for assembly.

Figure 14. Cladder and backer showing standoffs.

Figure 15. Loading of assembled plates onto trailer.

Figure 16 shows a trailer of assembled plates leaving for the field.

Figure 17 shows a plate being lifted from the trailer. It will be placed on the ground and the explosive will be added.

Figure 18 shows the explosive powder being placed on top of the cladder. The explosives are premixed and weighed for each plate. As a double check, the side rails are cut to a calculated height such that the proper amount of explosive will just fill the space inside the side rails.

Figure 19 shows a final assembly ready for shooting. The initiator is buried in the middle of one edge.

Figure 20 is a photograph of the shot taken from a distance.

Figure 21 shows how the force of the explosive has driven the now explosively welded bimetal plate into the ground.

Figure 22 is a photograph of the thin copper explosively welded to the 10-inch-thick aluminum slab, just after the explosion welding event.

Figure 23 shows an explosion clad plate being lifted to the trailer for return to the EFI plant.

Figure 16. Trailer load of assemblies leaves for shooting site.

Figure 17. Unloading of assembly at shooting sight.

Figure 18. Explosive powder mix being placed on top of cladder.

Figure 19. Final assembly ready for shooting.

Source: *Advances in Joining Technology*, Brook Hill Publishing Co., 1976

Figure 20. Explosive cladding shot.

Figure 21. Explosion-clad plate after detonation.

Figure 22. Explosion-clad copper on 10-in.-thick aluminum.

Figure 23. DYNACLAD*tm* plate being lifted onto trailer for return to EFI.

Figure 24 shows the copper clad thick aluminum after the explosion welded plates have been cleaned.

Figure 25 shows a circular bimetal tubesheet being cut from the explosion clad plate.

Figure 26 is a close-up of EFI's new powder torch. This cutting torch is a new Japanese invention which overcomes the problems of earlier powder torches, and is actually superior to a plasma torch in most respects. The kerf is about 3/8-inch wide and is truly vertical (not curved). Carbide precipitation is only about 0.002 to 0.003 inches, and the cut surface of the steel is readily machinable.

Figure 27 is a photograph of a batch of explosively clad tubesheet circles ready for shipment of EFI.

Figure 28 is a photograph of two sections of large-diameter steel pipe which has been explosively welded together using a collar which slips over both sections of pipe which are to be joined.

Figure 29 shows a long piece of test pipe with explosion welds joining the individual sections.

Figure 24. Copper 1/4 in. thick clad to 10-in.-thick aluminum plates.

Figure 25. Bimetal tube sheet being cut from DYNACLAD*tm* plate.

Figure 26. Special powder torch for cutting tube sheets.

Figure 27. Explosion-clad tube sheets ready for shipment from EFI.

Figure 30 shows a section of explosion clad bimetal pipe. The inside of the pipe is copper base alloy, and the outside of the pipe is steel. The two metals are metallurgically bonded over the entire interfacial area.

Figure 31 shows a slice from the as-explosively clad bimetal pipe, as well as a room-temperature twist test which proves the quality of the bond. This pipe has been tube reduced and cold drawn to a reduction of approximately 45:1. The resulting tubing is approximately 0.75 inches O.D by 0.040-inch wall thickness. Each 10-foot-long explosion clad pipe will yield approximately 400 feet of perfectly clad bimetal tubing.

Figure 32 shows a columbium alloy cover sheet which was explosively welded to each of the columbium alloy ribs which were machined into solid columbium backer. The application for this fin-tube construction is a very high-temperature heat exchanger.

Figure 33 is a photograph of six H-14 steel billets (each approximately 3 inches in diameter ×18 inches long) and each has had a 5/8-inch-thick copper sleeve explosion welded to its O.D. From these explosion clad composites an injector nozzle (for molten zinc) was machined. The injector nozzles needed a method of rapid heat transfer to maintain proper temperature control of the machined injector and the metallurgically bonded thick copper layer provided that.

Figure 34 shows a group of DYNABOND*tm* tubular transition joints which were machined from explosively welded 5083 aluminum to 304L stainless steel, with a 0.030-inch interlayer of silver. These transition joints are used typically in cryogenic systems piping where an all welded system is desired. The aluminum end of the DYNABOND*tm* joint can be conventionally welded to the aluminum piping, and the stainless steel end of the joint can be conventionally welded to the stainless piping or structure. Other DYNABOND*tm* tubular transition joints include titanium directly to aluminum, and titanium to steel.

Figure 35 is a photograph of a titanium-faced steel blind flange. The titanium was explosion welded to the steel to provide a complete metallurgical bond to maximize the heat transfer characteristics of the flange (in addition to the obvious corrosion protection provided by the titanium face).

Figure 36 shows a high-pressure carbon steel nozzle which has had the inside surface explosion clad overlaid with 1/4 inch of stainless steel. In certain sizes, explosion clad overlaying of stainless steel is less expensive and yields a superior product compared to weld overlaying.

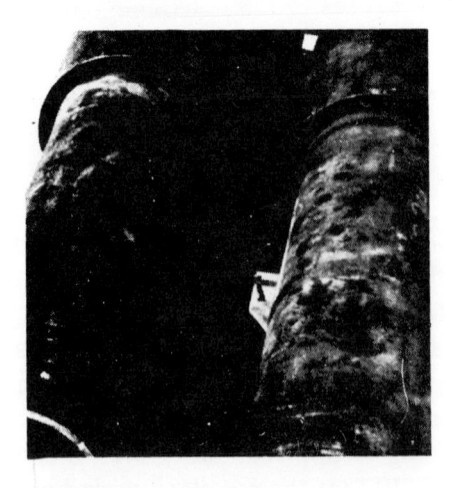

Figure 28. Explosion-welded steel pipe (text section).

Figure 29. Explosion-welded steel pipe (long section).

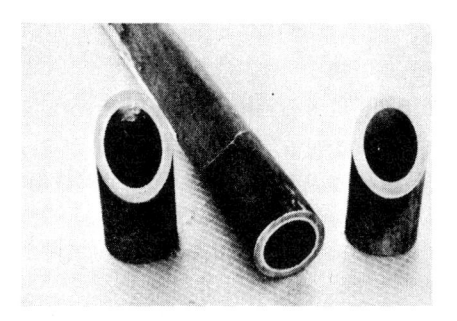

Figure 30. Explosion-clad bimetal tubing (pipe).

Figure 31. DYNACLAD^{tm} bimetal pipe and twist specimen.

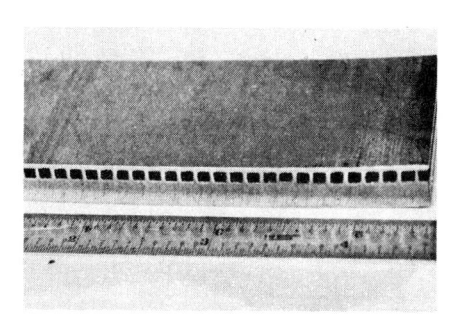

Figure 32. Columbium-alloy cover sheet explosively welded to columbium alloy ribs.

Figure 33. DYNACLAD^{tm} 5/8-in.-thick copper sleeves clad to 3-in.-diameter H-13 steel billets.

Figure 34. DYNACLAD^{tm} 5083 aluminum to 304L stainless steel transition joints.

Figure 35. DYNACLAD^{tm} titanium-faced steel blind flange.

Figure 36. Carbon steel nozzle with 1/4-in. stainless steel DYNACLAD^{tm} to inside diameter.

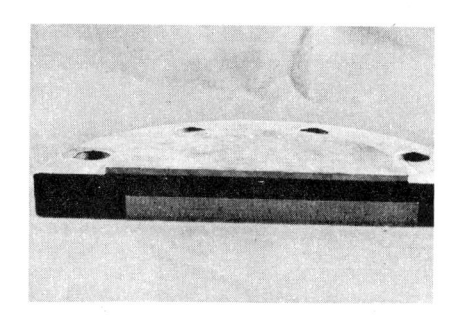

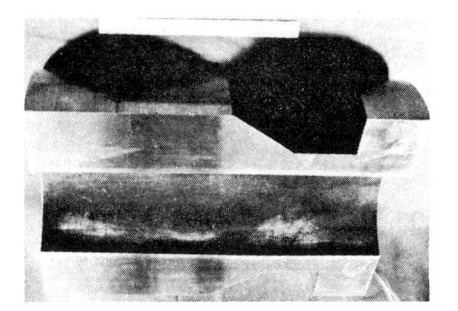

SUMMARY

Explosion welding and explosion cladding is a basic viable manufacturing technology which lends some new dimensions to the design engineers. Certain compositions or geometries which do not lend themselves to fusion welding or roll cladding can be explosion welded or clad. In other cases, where the systems are metallurgically compatible, explosion welding or cladding may offer a more cost-effective alternative.

REFERENCES

1. Cook, M. A., *The Science of High Explosives*, rev. ed. Huntington, N.Y.: Robert E. Krieger Publishing Co., Inc. (1971).
2. Ezra, A. A., *Principles and Practice of Explosive Metalworking*. London: Industrial Newspapers Ltd. (1973).
3. Johansson, C. H. and Persson, P. A., *Detonics of High Explosives*. New York: Academic Press (1970).
4. Rinehart, J. S. and Pearson, J., *Explosive Working of Metals*. London and New York: Pergamon Press (1963).

Mechanical properties of explosively-cladded plates

by S. K. Banerjee and B. Crossland

This article reports on the strength of explosively produced composites. Experimental work was carried out on stainless steel/mild steel and brass/mild steel composites. The ultimate tensile strength of the composites was higher than that of either of the component materials, due to shock hardening, and the percentage elongation of the composites was appreciably lower. The fatigue strength of the composites was found to be slightly greater than the stronger of the two component materials. Tensile shear tests on the composites after they had been subjected to 10 cycles of thermal fatigue showed that even for a temperature range of 20°-700°C there was no significant reduction in strength.

Explosive welding was probably first recognised as a practical proposition by Phillipchuk[1] who observed welding when explosively forming an aluminium U-channel on a steel die. Much earlier Carl[2] had observed solid phase welding between two metallic discs in contact with a detonator. Deribas[3] records that explosive welding had been noted in 1946 in the USSR by Professor M. A. Lavrentiev. Since 1957 the process has been considerably developed for plate cladding, tube welding, etc, and the number of papers published has been considerable. A general review of the subject is given by Crossland and Williams.[4]

Essentially, in explosive welding, a high velocity oblique impact of the two members to be welded is arranged, which generates a high velocity metal jet from the surface of one of the plates. This jet sweeps the surface of the other plate and removes the contaminant surface by surface traction. The two resulting virgin clean surfaces are then brought into intimate contact, providing the conditions for solid phase welding. In actual fact the process is more complicated due to the interfacial waves which are frequently observed, and which are discussed by Bahrani et al.[5]

It is obvious that the strength of the bond developed

Mr Banerjee is a scientist in the MMT Division of the National Metallurgical Laboratory, Jamshedpur, India. Professor Crossland is Head of Department of Mechanical Engineering, The Queen's University of Belfast.

in an explosive weld is of importance in relation to its application in service. In general non-destructive testing, and in particular ultrasonic testing, is employed to test the integrity of the weld, and destructive testing is required to establish the bond strength for design purposes.

Addison et al.[6] used a transmission ultrasonic test in which the specimens were submerged in water between the transmitter and receiver with a transmitter having a frequency of 10 Mc/s. They established that the amplitude of the signal was related to the bond quality though surface conditions and plate curvature affected the recorded amplitude. However, there was no indication that it was possible to differentiate between a fusion and an adhesive bond. It might be expected that a cast structure would be more 'lossy' than a solid phase weld, but perhaps the sensitivity is insufficient.

Many experimenters have reported tests to establish the bond strength of explosive welds but most of them have given little detail except to state that the bond is strong. Bahrani and Crossland[7,8] have given details of tensile shear, side shear, bend and tensile tests and also hardness surveys for stainless steel/mild steel, mild steel/mild steel and titanium/mild steel welds. They showed the effect of explosive loading, initial angle of obliquity when using a high detonation velocity explosive and an inclined technique, and the harmful effect of the intermetallic which can form between mild steel and titanium.

Rowden[9] gave results of tension and charpy tests carried out on the flyer and parent plate materials, and tensile shear tests carried out to establish the strength of the weld. He examined welds between austenitic stainless steel/carbon–manganese steel and titanium 115/carbon–manganese steel, and he showed the increase in the strength of the parent and flyer plate materials caused by shock hardening, and the effects of a stress relieving heat-treatment.

The major contribution to the subject of strength of explosively cladded plate is reported by De Maris and Pocalyko[10] who carried out tension, shear, bend, impact and bending fatigue tests. These were carried out on 304-L stainless steel, TMCA 35-A aluminium, Hastelloy C, Inconel 600, 1100H14 aluminium and DHP copper all welded to ASTM A–212–B FbQ steel, and stainless steel welded to ASTM A–387–D FbQ low alloy steel. The tensile tests showed increased strength and slightly reduced ductility in all cases compared with the parent plate material properties, and in all cases the ASTM requirements for the material of the parent plate were satisfied. With the exception of the titanium clad material all the composites survived a 180° bend test with the flyer plate material on either side of the bend. The impact strength close to the weld interface is reduced by welding process but in the parent material the effect is negligible at 13 mm ($\frac{1}{2}$ in.) from the weld interface. However, a stress-relieving heat treatment restores the impact strength to the pre-clad value. It is difficult to comment briefly upon the fatigue properties because of the differences in Young's modulus between the parent and flyer plate materials. In the case of Hastelloy and stainless steel, however, Young's modulus was similar to that of the backing steel. In both cases the fatigue results showed a lower fatigue strength than that of the backing steel. In the other composites there was a significant reduction in fatigue strength compared with that of the parent plate material, except in the case of Inconel welded to A–302–B steel. In all cases it is claimed that the fatigue failures did not start at the interface.

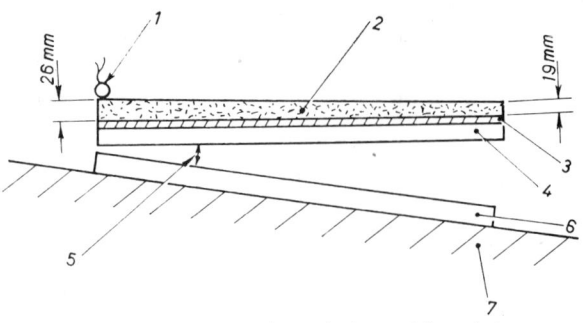

1 Diagrammatic arrangement for explosive welding: 1 detonator, 2 explosive charge, 3 buffer, 4 flyer plate, 5 small angle, 6 parent plate, 7 anvil

2 Explosive weld interface, mild steel/stainless steel composite (× 100)

3 Explosive weld interface, mild steel/brass composite (× 100)

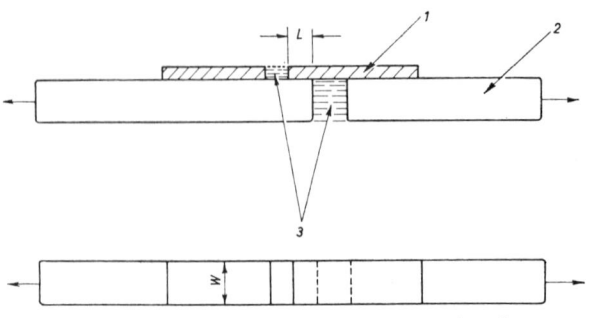

4 Tensile shear specimen: 1 flyer plate, 2 parent plate, 3 grooves milled after welding

Gelman et al.[11] have also reported on symmetrical bending fatigue tests on two different steels cladded to 22K steel. In their work they cladded each side of a 50 mm thick plate with 3 mm thick flyer plates. For 1 Kh 18 N9T steel cladded to 22K steel there was a significant reduction in fatigue strength, particularly without tempering. With 0 Kh 13 steel cladded to 22K steel, the reduction in fatigue strength was less pronounced and after a triple heat treatment the fatigue limit of the clad and unclad parent material were the same. The differences in the two composites were attributed to the differences in the linear coefficients of expansion.

This article reports tests on two explosively produced composites carried out in the Department of Mechanical Engineering of the Queen's University of Belfast.

Material specification

Two combinations were tested: mild steel clad with austenitic stainless steel and mild steel clad with 70/30 brass. The chemical analyses are given in Tables 1 and 2 and the mechanical properties in the as received state are given in Table 3. Metallurgical examination revealed that the three metals were in either the hot rolled or annealed condition.

Explosive welding of the composite plates

A clad plate of about 305 mm × 610 mm (12 in. × 24 in.) was required for the specimens it was planned to test. Plates of 3·2 mm ($\frac{1}{8}$ in.) thick mild steel and stainless steel were prepared by surface grinding on the surfaces to be welded and then removing any oil or grease prior to welding. The explosive charge was put in a perspex box with a linoleum base which served as a buffer between the explosive and the top or flyer plate. Trimonite No. 3 was used, an ICI explosive which has a maximum detonation velocity of 3900 m/s, but a lower detonation velocity below 25 mm thickness as shown by Shribman and Crossland.[12] A charge of 7 kg (16 lb) of explosive was estimated to be necessary, but from previous experience it had been concluded that a uniform thickness of charge did not give a uniform flyer plate velocity, but one which increased with the distance from the point of detonation. This led to a poor bond near the point of initiation of detonation and excessive melting and either a poor bond or no bond remote from the

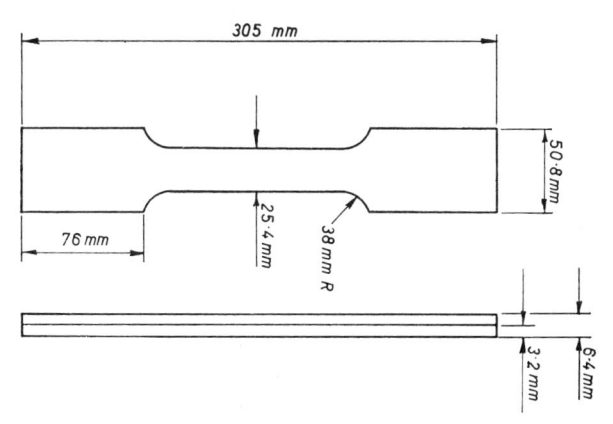

5 Static tensile and tensile fatigue specimen

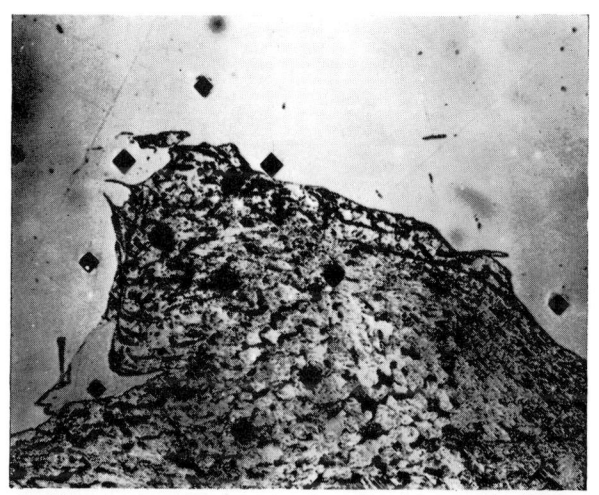

6 Micro hardness survey at interface of stainless steel/mild steel weld (× 250)

7 Micro hardness survey at interface of brass/mild steel weld (× 400)

TABLE 1

Material	Chemical composition %						
	C	S	P	Si	Mn	Cr	Ni
Mild steel EN3A	0·18	0·038	0·020	0·04	0·36	—	—
Stainless steel EN58C	0·05	—	—	—	—	17·9	10·5

TABLE 2

Material	Chemical composition %			
	Zn	Pb, Fe and Ni	Sn and P	Cu
Brass	35	Nil	Nil	Balance

TABLE 3

Material	Yield stress hbar	Ultimate tensile stress hbar	Elongation %	Hardness VPN	Fatigue limit[1] hbar
Mild steel EN3A	29·6	43·3	41·5	136	±11·6
Stainless steel EN 58C	—	58·0	55	160	±13·2
Brass	—	34·0	50	92	± 5·8

[1]Carried out in a Losenhausen UHS 40 fatigue machine with a mean stress of 21·6 hbar (14 tonf/in²).

detonator. To overcome this the charge was tapered from 26 mm thickness near the detonator to 19 mm remote from the detonator, as shown in Fig. 1. A small angle of incidence between the plates was obtained by using wax spacers of different thickness along the edge of the plate. The whole assembly was mounted on a thick steel anvil, and the charge was detonated with a No. 8 star detonator.

A similar procedure was followed for welding a 3·2 mm brass plate to mild steel but no initial angle of inclination was used. A parallel arrangement of the flyer and parent plate was achieved by separating the two plates by polystyrene particles to give an initial clearance of 1·6 mm ($\frac{1}{16}$ in.). With polystyrene particles there is no evidence of any harmful effect such as that noted by Crossland and Bahrani when using copper particles.

When the plates had been welded they were subjected to ultrasonic inspection using a 6 Mc/s probe and a light lubricating oil to ensure effective coupling between the probe and the plate. Except close to the end of the brass/steel composite plate no reflection could be detected from the weld interface, which confirmed that there was a bond between the two component materials. However, an ultrasonic test of this type certainly cannot establish the quality of the bond, and in particular if there is excessive melting. Metallurgical examination was consequently carried out to check on the interface. Figures 2 and 3 show photomicrographs of the stainless steel/ steel and brass/steel composites. It will be seen that in the stainless steel/steel weld there is obviously a small melted zone in the front vortex and in the brass/steel weld there are melted zones in both the front and rear vortex regions. However, there is not a continuous melted zone in either case, though the amount of melting in the brass/steel weld is probably greater than desirable.

Mechanical tests

One of the main objects of the present work was to examine the fatigue and thermal fatigue properties of composite plate produced by explosive welding. However, it is necessary to examine the strength of the bond under static loading to establish that it is representative of explosive welding. Previous workers have used a range of tests, for example, Bahrani and Crossland[7,8] have reported results of tensile shear tests, side shear tests, bend tests and tension tests across the weld interface. If the welding conditions are correct then the bond is frequently stronger than the weaker of the two component materials and none of these tests is very informative. Consequently in this investigation only tensile shear and bend tests have been carried out.

The tensile shear specimen is shown in Fig. 4; the lightly hatched regions are removed by machining and the specimen is then pulled in a tensile testing machine. As a result of bending of the specimen when it is pulled the weld zone, which is being tested, is subjected to combined peel and shear, which is a more demanding test than simple shear. Bend tests were carried out on full thickness specimens 4·8 mm ($\frac{3}{16}$ in.) wide which were bent through 180° round a 12·7 mm ($\frac{1}{2}$ in.) diameter former. Tests were carried out with the clad metal on the inside, outside and on the side of the bend.

During explosive welding there is some general hardening of the metal due to shock hardening, in addition the material in the melted zones or the heavily worked material near the interface may be appreciably harder. The general increase in hardness represents an increase in the tensile strength which is obviously of great importance. Consequently a micro-hardness survey was carried out on a cross-section of the composites.

The general increase of hardness may also be accompanied by a reduction in ductility. Consequently tensile tests were carried out on the component metals and the composite plate using a specimen shown in Fig. 5.

Fatigue tests were carried out on specimens of the component metal plates and the composite plates using

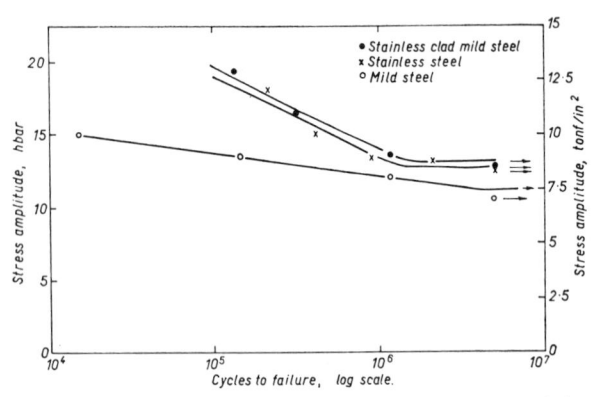

8 Fatigue of stainless steel/mild steel composite and of the component metals

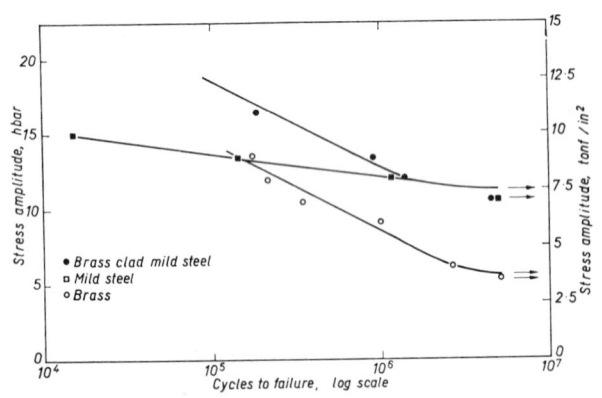

9 Fatigue of brass/mild steel composite and of the component metals

10 Point of initiation of fatigue fracture in composite material (mild steel/stainless steel) (×13)

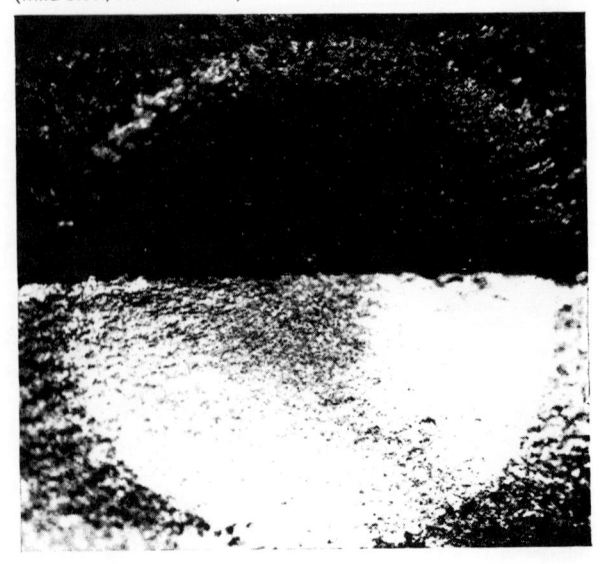

specimens of the form shown in Fig. 5. The specimens were carefully polished finishing with a grade 600 emery cloth. They were tested in a Losenhausen UHS 40 fatigue machine and the mean stress based on the cross-sectional area was maintained constant at 21·6 hbar (14 tonf/in.²) while the amplitude was varied from test to test. Tests were run to failure or 5×10^6 cycles.

Thermal stresses can occur in metals due to temperature differences; if sufficiently severe they can lead to fatigue failure. However, in bimetal composites in which the component materials have a different coefficient of expansion thermal stresses will occur when the temperature is changed, even in the absence of a temperature gradient. One of the main problems in thermal fatigue testing is the length of time required to heat up the specimen and cool it and to maintain the testing cycle constant. Consequently for this preliminary investigation it was decided to apply 10 cycles of heating and cooling at several temperature levels, and then carry out a tensile shear test to establish if there was a reduction in strength as a result of the cycling.

The thermal fatigue specimens were either 6·4 mm × 6·4 mm × 152 mm or 6·4 mm × 12·8 mm × 152 mm. They were heated in an electric furnace for 30 minutes which was sufficiently long to have closely approached the furnace temperature, and they were then quenched in water at 20°C. With this procedure there was a thermal stress at the maximum temperature caused by the difference in the coefficient of expansion, and also immediately on quenching in cold water due to the resulting temperature gradient. Tests were carried out at 400°, 500°, 600° and 700°C.

Results

The results of the tensile shear tests are given in Table 4.

In the bend tests there was no failure at the weld interface for any of the bend tests carried out, even with the cladding on the side, which is a severe test.

Figures 6 and 7 are micrographs showing microhardness indentations which give some idea of the hardness variations, given in Table 5.

Tensile tests were carried out on the composite plates and the data in Table 6 should be compared with the properties of the component metal plates before welding which are given in Table 3.

The results of the fatigue tests on the component materials and the composite plates are given in Figs. 8 and 9. In the composite plate the failure was initiated at the interface, as shown in Fig. 10, where the circular region is centred on the interface, though it is not apparent if this initiated the final fracture.

The results of tensile shear tests after 10 cycles of thermal fatigue are shown in Figs. 11 and 12.

Discussion and conclusions

The use of a tapered charge for welding the composites for this work gave a more uniform weld quality than that given by a uniform charge. This aspect of explosive welding is being examined in greater detail to ascertain the conditions to give a uniform flyer plate velocity.

Metallurgical examination has shown the presence of mechanical twinning and Neumann bands in the mild steel. In the stainless steel there is an irreversible change of austenite to martensite in the region of the interface and this may lower the corrosion resistance in this region. In the brass there was no significant metallurgical change though the grains had become elongated as in cold worked material. Appreciable increases in hardness of the component materials were observed especially in

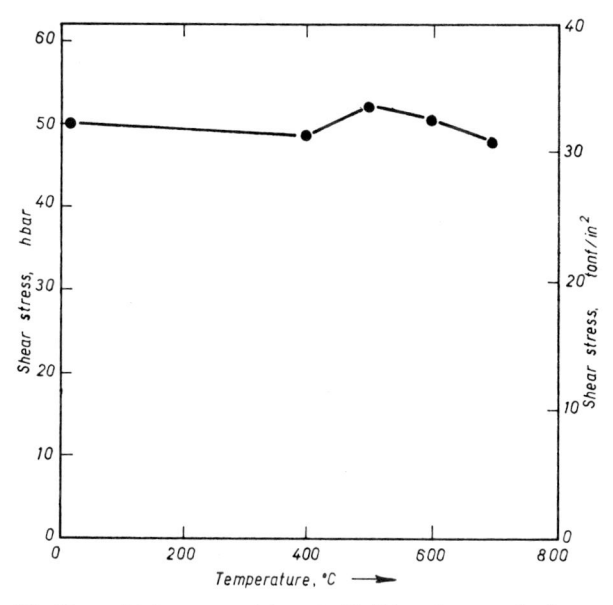

11 Thermal fatigue on stainless steel/mild steel composite (note there was no separation of the weld interface)

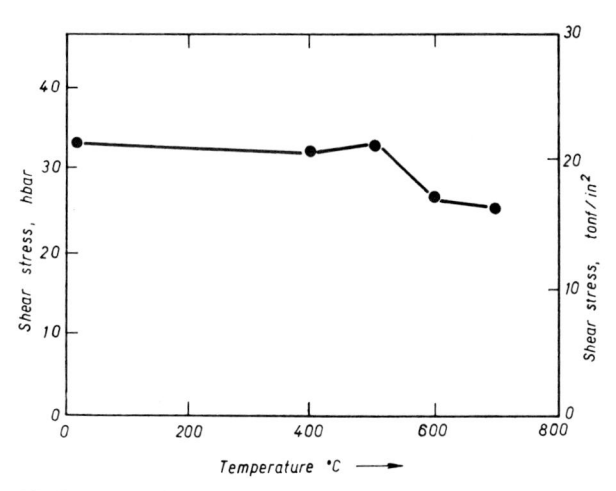

12 Thermal fatigue on brass/mild steel composite (note there was no separation at the welded joint)

TABLE 4

Material	No. of tests	Mean shear strength hbar	Remarks
Stainless steel —mild steel	3	49·8	Failures occurred in mild steel
Brass—mild steel	3	33·2	Failures occurred in brass

TABLE 5

Materials of composite	Original hardness of components UPN	Hardness after welding UPN
Stainless steel —mild steel	160 136	302 195
Brass—mild steel	92 136	128 169

Source: *Metal Construction and British Welding Journal*, July 1971

the region of the interface and the top surface of the flyer plate.

The ultimate tensile strength of the composite plates is greater than the ultimate tensile strength of either of the component materials before welding; the elongation is appreciably less than that of either component material. This increase in the ultimate tensile strength is a result of shock hardening. The percentage elongation which is a measure of ductility is reduced but it is not known if this gives rise to a corresponding reduction of impact strength or fracture toughness. The effect of heat treatment on the ductility has not been investigated though it might be expected to give rise to metallurgical problems, such as the diffusion of carbon from the mild steel into the stainless steel.

Under fatigue loading the two composite plates tested showed a slight improvement of fatigue properties compared with the fatigue properties of either of the component materials. In the case of the mild steel/stainless steel composite the stress conditions in each component material are identical except that there may have been some residual stresses created by the welding process especially as the coefficients of thermal expansion are different. No attempt was made to investigate the state of residual stress. It will be seen from Fig. 8 that the fatigue properties of the mild steel component have been raised a little but there appears to have been little effect on the fatigue properties of the stainless steel.

With the mild steel/brass composite the difference in the values of the Young's modulus of the component materials leads to a very different situation from that of the mild steel/stainless steel composite. Assuming that there are no residual stresses and that there is compatibility of strains, then the stresses in each of the component materials at the fatigue limit can be calculated. These stresses are given in Table 7 and it can be seen that in the mild steel component the mean stress is not $21 \cdot 6$ hbar (14 tonf/in.^{2}) but $29 \cdot 5$ hbar (19·1 tonf/in.^{2}). As the mean stress has increased it might have been expected that the cyclic stress at the fatigue limit, according to the Goodman diagram, should have been lowered, whereas it has been raised from $11 \cdot 6$ hbar ($7 \cdot 5$ tonf/in.^{2}) given in Table 3 to $15 \cdot 8$ hbar (10·2 tonf/in.^{2}). Both the mean stress and the stress amplitude in the brass has been significantly reduced and comparing them with the data given in Table 3 for brass it would be expected that fatigue would not be initiated in the brass.

TABLE 6

Materials of composite	Ultimate tensile stress hbar	Elongation %
Stainless steel—mild steel	64·3	23·9
Brass—mild steel	45·6	18·1

TABLE 7

Material	Mean stress hbar	Stress amplitude at the fatigue limit hbar
Mild steel/brass Mild steel	21·6	11·6
	29·5	15·8
Brass	13·7	7·4

The results of the fatigue tests demonstrate the increase in strength caused by shock hardening of mild steel. However, with mild steel/stainless steel composite the increase in the strength of mild steel is less marked than with the mild steel/brass composite; this may be due to the intermetallic compounds formed at the interface in the mild steel/stainless steel composite being a source of weakness, or to carbon migration. That the interface is a cause of weakness is clearly demonstrated by the origin of the circular fatigue cracks shown in Fig. 10 which are centred on the interface.

Whatever the cause the fact remains that for the two composites investigated the fatigue strength is, if anything, raised above the higher of the two fatigue curves of the component metals prior to welding, and this must be considered very satisfactory.

The thermal fatigue tests show no significant weakening of the strength of the weld, but many more cycles would be necessary before reaching any firm conclusions. Again, however, the results so far must be considered as very satisfactory.

All the mechanical tests seem to indicate that the composite plates produced by explosive welding have satisfactory mechanical properties except for some reduction of ductility.

Acknowledgements

The work was carried out in the Department of Mechanical Engineering of the Queen's University of Belfast. The authors are indebted to the Science Research Council who gave a generous grant and the Council of Scientific and Industrial Research, India, and the Ministry of Overseas Development, UK, for providing a grant for one of us. The advice of Dr J. D. Williams was greatly appreciated and the assistance of the workshop staff under Mr R. H. Agnew and the laboratory staff under Mr. R. J. Harvey.

References

1 Phillipchuk, V. 'Explosive welding status' ASTME Creative Manufacturing Seminar Paper No. SP65–100, 1965.
2 Carl, L. R. 'Brass welds: made by detonation impulse' *Metal Progress*, **46** 7, 1944, 102
3 Deribas, A. A. 'Explosive welding', published Academy of Sciences, USSR, Siberian Branch 1967.
4 Crossland, B. and Williams, J. D. 'A review of explosive welding', to be published by the *Metallurgical Review*.
5 Bahrani, A. S., Black, T. J. and Crossland, B. 'The mechanics of wave formation in explosive welding' Proc of the Royal Society, A, 1967, **296**, 123.
6 Addison, H., Fogg, W. E., Betz, I. G. and Hussey, F. W. 'Explosive welding of aluminium alloys', Weld. Research Supplement, Proc. AWS 44th Annual Meeting 1963, 359.
7 Bahrani, A. S. and Crossland, B. 'Explosive welding and cladding: an introductory survey and preliminary results'. Proc. of the Institution of Mechanical Engineers, **179**, 1964, 264.
8 Bahrani, A. S. and Crossland, B. 'Further experiments on explosive welding and cladding with particular reference to the strength of the bond'. Proc. of the Institution of Mechanical Engineers, **180**, 1966, 31.
9 Rowden, G. 'The effects of explosive welding on the properties of BS1501-161-28A, EN 58J and Titanium 115'. Proc. Select Conference on Explosive Welding, Weld. Institute, 1969, 31.
10 DeMaris, J. L. and Pocalyko, A. 'Mechanical properties of detaclad explosion-bonded clad metal composites'. ASTME, Engineering Conference paper AD66–113, 1966.
11 Gelman, A. S., Zaitsev, G. Z., Ponomarev, V. Ya. and Tsemakhovich, B. D. 'Fatigue strength of two layer steel, clad by explosion bonding' Svar Proiz. 10, 1966, 4.
12 Shribman, V. and Crossland, B. 'An experimental investigation of the velocity of the flyer plate in explosive welding'. Proc. Second Internat. Conference of the Center for High Energy Forming, Denver 1969, 7.3.1.

RECENT DEVELOPMENTS IN THE THEORY AND APPLICATION OF EXPLOSION WELDING

by:

Steve H. Carpenter

and

Robert H. Wittman
Denver Research Institute
University of Denver

ABSTRACT

Recent investigations have shown that four concepts seem to provide the necessary boundary conditions to describe the optimum explosion conditions. The four concepts are:

1. The critical angle for jet formation,
2. The critical impact pressure for jet formation,
3. The critical flow-transition velocity, and
4. The kinetic energy of the flyer plate and the heat dissipation characteristics of the collision region.

From these concepts it is possible to describe an explosion weldability parameter for any metal or alloy. Current applications of the explosion welding process are also discussed.

INTRODUCTION

Explosion welding is basically a solid phase welding process, in which the bonding occurs as a consequence of an oblique high velocity collision between the components to be welded. The process was discovered quite by accident when ordnance specialists found shell fragments bonded to metal target plates against which they had been impacted. The first reported explosion welding was by Philipchuk[1], in 1957, who stated that he observed the effect when explosively forming an aluminum U-channel on a steel die. He found that the U-channel could not be removed from the die because a circular shaped area of the channel had come welded to the steel die. From this unlikely beginning applications of the explosion welding process have had mixed successes. In too many cases applications were attempted with little or no knowledge of the basic problems involved. This lack of information resulted in many programs being conducted on a trial and error basis which proved costly and resulted in very little to show for the investment of time and money. However, during the past five to six years significant progress has been made in developing a comprehensive understanding of the explosion welding process. This has in turn provided the information and techniques necessary for the application of the explosion welding process to meaningful joining and/or bonding problems. The purpose of this paper is to present and review the progress of the past few years in both understanding the basic principles and actual application of the explosion welding process. For a more complete and historical discussion of explosion welding with an extensive bibliography the reader is referred to reference 2.

THE EXPLOSION WELDING PROCESS

The explosion welding process depends on the use of explosives to accelerate the parts to be joined into a high velocity oblique collision. A majority of investigators believes that a fundamental condition necessary for bonding is that the oblique collision take place in such a way as to give what is known as a jetting collision. A jetting collision is defined as an oblique collision in which a jet or spray of metal is formed at the apex of the collision and is forced outward from between the colliding plates at very high velocities.

In the explosion welding process the jet is the mechanism which produces a break-up and effacement of the surface films. During the oblique collision the jet is formed, causing a scarfing or cleaning of the metal surfaces as they move toward the collision point. At the collision point the virgin or clean metal surfaces are forced into intimate contact by the high pressure from the explosion, effecting a metallurgical bonding across the interface. The explosion welding process can be considered as a two-step process: first, the jet breaks up and cleans the inhibiting surface layers and, second, the high pressure forces the clean metal surfaces into such intimate contact that interatomic forces can be established across the bond interface.

When all of the parameters are properly controlled in the explosion welding process, a weld interface of the type shown below in Fig. 1 is obtained. The ripple or wave-like appearance of the explosion welded interface provides explosion welding with its most unique characteristic. The novel appearance of the weld interface has by itself evoked and stimulated much interest in explosion welding. The phenomena and mechanisms responsible for wave formation at the explosion weld interface have been the subject of numerous investigations.[3],[4],[5] These investigations, for the most part, however, have not contributed substantially to a better technique of application or a greater ability to predict the precise conditions that will result in optimum weld quality.

Figure 1 Typical wavy interface found in explosion welding. 4130 steel explosion welded to 4130 steel.

WELDING PARAMETERS

In order to predict the conditions that will produce an optimum quality weld it is necessary to understand how the welding parameters control weld quality. In the explosion welding process the welding engineer basically has two parameters which can be varied at his disposal. These parameters are (1) the type and amount of explosive and (2) the geometry or welding configuration. It is very difficult to relate these parameters directly to weld quality. A closer examination of the welding process can, however, provide information in dynamic variables which can be related directly to weld quality. Explosion welding operations are carried out in either a constant standoff configuration. These are discussed in detail below.

The constant standoff configuration with the dynamic parameters labeled is shown below in Fig. 2.

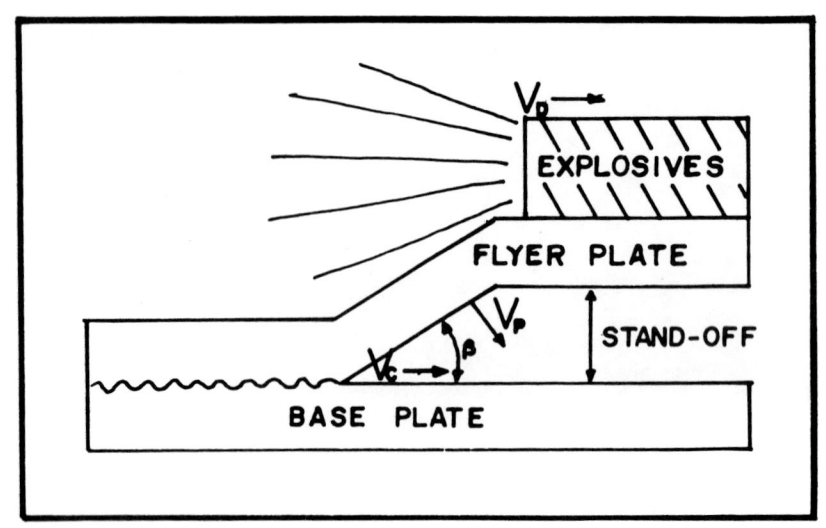

Figure 2 Schematic of the constant standoff welding configuration.

where

V_D = detonation velocity of explosive

V_p = flyer plate velocity

V_c = collision point velocity

β = dynamic bend angle

It can easily be shown that the following relations hold for the constant standoff configuration:

$$V_p = 2 V_c \sin \beta/2 \qquad (1)$$

$$V_c = V_D \qquad (2)$$

The preset angle standoff configuration is slightly different as shown below in Fig. 3.

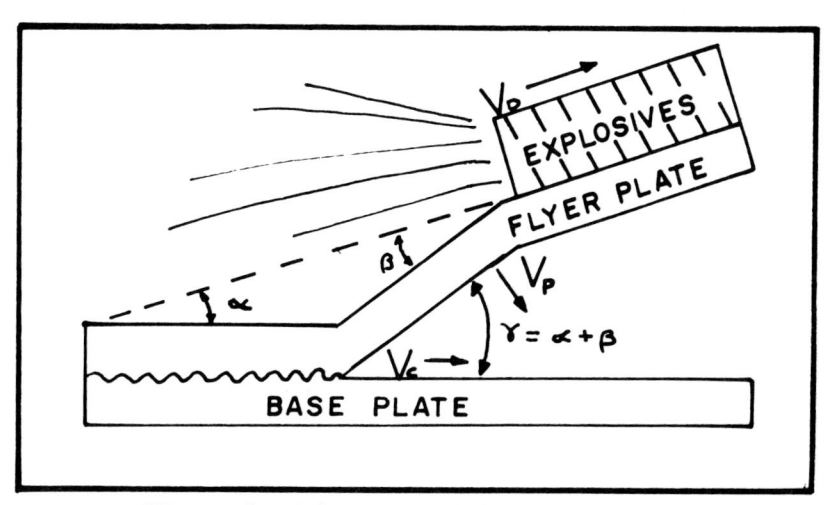

Figure 3 Schematic of the preset angle standoff welding configuration.

where

V_D = detonation velocity of explosive

V_p = flyer plate velocity

V_c = collision point velocity

β = dynamic bend angle

α = preset angle

γ = collision angle

For the preset angle configuration we have the following relationships:

$$V_p = V_D \sin \beta \qquad (3)$$

$$V_c = \frac{V_D \sin \beta}{\sin (\alpha + \beta)} \qquad (4)$$

The problem until fairly recently has been the difficulty in measuring these dynamic parameters for different explosives.

The use of two pressure-actuated, continuous-writing velocity probes (one on the flyer plate and one on the base plate) has made it possible to determine the position of the detonation front and the collision

Source: SME Technical Paper No. MF74-819, 1974

point at the same instant of time. Using this principle it is possible to reconstruct the collision profile along the length of weld and determine collision angle (γ), collision point velocity (V_c), plate velocity (V_p), and the detonation velocity (V_D) at any point. The collision point velocity, V_c, can be varied by explosive selection and by use of the preset angle geometry. The impact velocity, V_p, can be changed by adjusting the explosive loading and the standoff, S. The velocity probe technique was developed by the U. S. Bureau of Mines Explosives Research Center between 1963 and 1967 and described in the literature in 1968.[6] The probe consists of an aluminum tube of 0.058 cm. diameter and 0.0038 cm. wall, having a skip insulated resistance wire of 0.0076 cm. diameter on the tube axis as shown below in Fig. 4. A constant, battery-generated, direct current is passed through tube and wire. A pressure wave of relatively small magnitude propagating along the tube collapses the tube onto the wire, shortening the electrical path and decreasing the resistance. The voltage drop is monitored by an oscilloscope and is proportional to the distance along the probe. In this way the position-time profile of an advancing pressure front can be recorded.

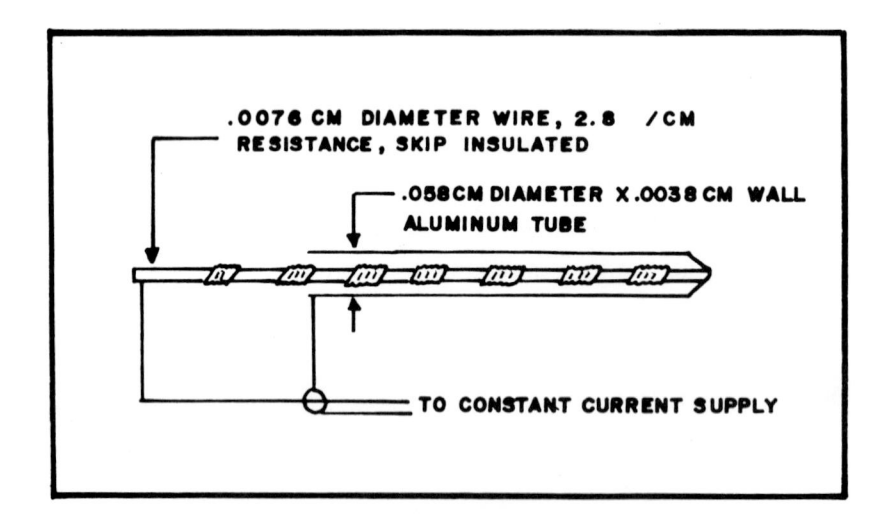

Figure 4 Schematic of the continuous velocity probe.

OPTIMUM CONDITIONS FOR WELDING

As shown by Wittman[7], four concepts seem to provide the necessary boundary conditions to describe optimum explosion welding characteristics. They are:

    A. The critical angle for jet formation,

    B. The critical impact pressure for jet formation,

    C. The critical flow-transition velocity, and

    D. The kinetic energy of the flyer plate and the heat dissipation characteristics of the collision region.

A. The Critical Angle for Jet Formation

The concept of a critical angle for jet formation has been described by Walsh et al.[8] and Cowan and Holtzman[9]. It provides an upper flow velocity boundary and the related impact velocity or angle. At flow velocities higher than critical, the collision is jetless and no welding is presumed to occur. At flow velocities below the boundary, jetting occurs and welding is said to result.

The critical angle for jetting can be calculated as described in Reference (8).

B. The Critical Impact Pressure

Clearly the pressure at the collision point must be sufficient to produce jetting of the metal surfaces. Since the metal at the interface must flow in a fluid manner to produce jetting, the impact pressure for explosion welding must exceed the dynamic yield stress by a considerable margin. The impact pressure will be determined by the flyer plate velocity.

Experience has shown that a good estimation of the minimum flyer impact velocity can be obtained by the use of the expression

$$V_p = \left(\frac{\sigma_{TS}}{\rho}\right)^{1/2} \quad (5)$$

where σ_{TS} is the ultimate tensile strength in dynes/cm^2 and the density.

C. The Critical Flow Transition Velocity

The existence of a critical flow (collision point) velocity for transition to turbulent flow (wavy bond zone) has been noted or discussed by Deribas et al.[10], Burkhardt et al.[11], and Cowen et al.[12]. Cowan et al. appear to have developed a means for predicting the critical flow velocity that seems both accurate and useful. They used flyer and base plate density (ρ_F and ρ_B) and static hardness (Hd_F and Hd_B) as the yield criteria in an expression defining the Reynolds number, R_T, at the flow transition velocity, V_T. Their expression

$$R_T = \frac{(\rho_F + \rho_B) V_T^2}{2(Hd_F + Hd_B)} \tag{6}$$

gives values of R_T ranging from 8.1 to 13.1 for both symmetric and asymmetric cladding configurations, the average R_T being 10.6.

D. The Kinetic Energy of the Flyer Plate and the Heat Dissipation Characteristics of the Collision Region

The range of optimum impact velocities, V_p, beyond the minimum seems to be controlled by the thermophysical properties and the melting point of the metals being welded. A fraction of the impact kinetic energy of the flyer plate will be trapped within the system. The trapped energy will be converted into heat by flow deformation and friction. The generated heat will then give rise to a rapid rise in temperature, the magnitude being governed by the thermophysical properties of the metals being welded. For optimum weld strength to be realized the interface behind the collision point must be sufficiently cooled and maintained free of melt-induced defects so that tensile wave passage may occur without complete or partial rupture.

Using the above concept, it is possible to express the relationship of impact parameters to the physical properties of the metal that lead to heat dissipation without melting during tensile wave passage. This problem has been discussed in detail by Wittman[7]. He has determined an expression for the maximum flyer plate velocity as

$$V_p = \frac{1}{N} \frac{(T_{MP} C_B)^{1/2}}{V_c} \frac{(KCC_B)^{1/4}}{(\rho h)^{1/4}} \tag{7}$$

where

T_{MP} = melting point temperature

C_B = bulk sound speed

K = thermal conductivity

C = specific heat

V_c = collision point velocity

ρ = density of flyer plate

h = thickness of flyer plate

N = constant

EXPLOSION WELDABILITY

All of the boundary conditions for optimum welding discussed in the preceding section can be simultaneously represented in a plot of

$$V_p = 2V_c \sin \frac{\gamma}{2} \qquad (8)$$

This relationship is valid for either the preset angle or parallel explosion welding geometry as it deals only with the collision parameters, not initial or intermediate states. This relationship is plotted below in Fig. 5 with the four boundary conditions we have discussed schematically represented. Optimum welding occurs when dynamic parameters fall within the crosshatched area.

It is interesting to note that since we have expressions for the maximum and minimum flyer plate velocity in terms of material properties an explosion weldability ranking of metals can be made. Explosion weldability is chosen to be proportional to the ratio of max to min flyer plate velocities that will result in a parent metal strength weld, i.e.

$$\text{Weldability} = \frac{V_{p\,max}}{V_{p\,min}} \qquad (9)$$

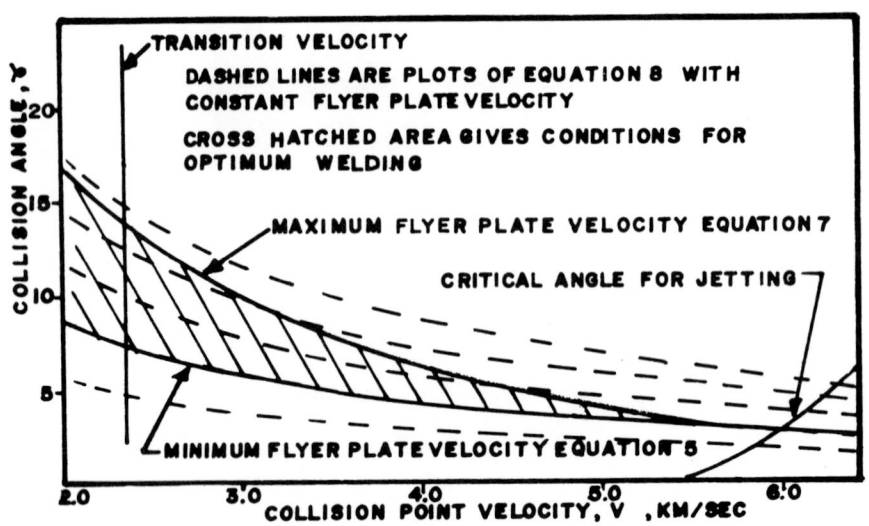

Figure 5 Schematic of Equation 8 with the four boundary conditions giving the optimum conditions for welding.

Table 1 shows data on the weldability of several metals and alloys. The data are for flyer plate thickness equal to 0.635 cm and a collision point velocity equal to 1/2 the buck sound speed.

It is now possible with some degree of certainty to design a successful explosion weld configuration and determine by analysis which applications are feasible and which are not. It is hoped that through these methods more successful applications will result.

CURRENT APPLICATIONS

By far the most important commercial application of explosion welding is the cladding of flat steel plate with a variety of corrosion-resistant metals and alloys. Because the explosion welding process is readily capable of joining dissimilar metals, the most competitive applications involve the cladding of such combinations as titanium/steel, tantalum/steel, aluminum/steel, and aluminum/copper. Many other combinations can be and are commercially joined by explosion welding.

Explosion clad plate is subjected to a wide variety of metal-work operations before end use in most instances. The 3-layer coin stock produced by duPont during the 1960's illustrates an interesting process concept. A 3-layer billet approximately 6 inches thick was explosion clad and then rolled to the final sheet thickness required for coining. It is possible to extend the bond area of other explosion welded combinations not only by rolling but by extrusion, drawing and swaging. We have hot rolled Ti/steel explosion bonded strip to a 20:1 reduction[13] and cold rolled explosion bonded copper/aluminum strip to reductions of over

Table I. Mechanical-Physical Properties of Selected Metals and Alloys and Their Weldability Rating

Metal	ρ g/cm^2	σ_{TS} dynes/cm^2	C_B cm/sec	T_{MP} °C	K ergs-cm/cm^{2}sec°C	C erg/g°C	Weldability Index $\frac{1}{N_p}(V_p\text{max}/V_p\text{min})$	Relative to Silver
Silver (1/2 hard)	10.49	20 × 10^8	2.8 × 10^5	960	4.18 × 10^7	.234 × 10^7	.359	1
Aluminum 1100-0	2.71	9	5.1	657	2.22	.961	.330	.92
Copper (1/2 hard)	8.91	31	3.94	1082	3.62	.385	.295	.82
Columbium (annealed)	8.57	36.5	4.44	2468	.44	.272	.214	.596
Aluminum (5052-H32)	2.68	22.8	5.3	649	1.38	.96	.180	.50
AISI 1020 Steel	7.86	45	4.59	1516	.52	.447	.170	.47
Aluminum 6061-T6	2.70	31	5.35	652	1.54	.961	.159	.44
Titanium 50A	4.50	35	5.22	1704	.16	.52	.136	.38
304 Stainless Steel	7.90	58	4.57	1454	.16	.50	.115	.32
Aluminum 7075-T6	2.80	57.3	5.1	635	1.21	.96	.111	.31
Inconel 600	8.43	62	4.98	1427	.15	.443	.109	.30
T16AT-4V (annealed)	4.42	95	5.22	1649	.07	.564	.06	.167

Source: SME Technical Paper No. MF74-819, 1974

200:1. The potential advantage of such processing is the independence from special reduction procedures (i.e., large reductions per pass, elevated temperatures) requiring much greater machine forces and usually greater operator skill.

Depending on the individual material properties, most explosion clad plate is amenable to conventional fabrication techniques such as bending, pressing, flanging, and spinning either hot or cold. Cutting and trimming can be accomplished by sawing, flame cutting, or shearing.

Several important product applications of explosion welding are derived by explosion cladding large plates and then cutting and/or machining the plate into small individual components. Detacouple transition joints by duPont are made in this way. The most popular transition combinations of metals appear to be aluminum/steel and aluminum/copper. Aluminum/steel transition inserts are useful for both structural and electrical applications while copper/aluminum transition inserts are primarily for electrical application. Examples are shown in Fig. 6 and 7.

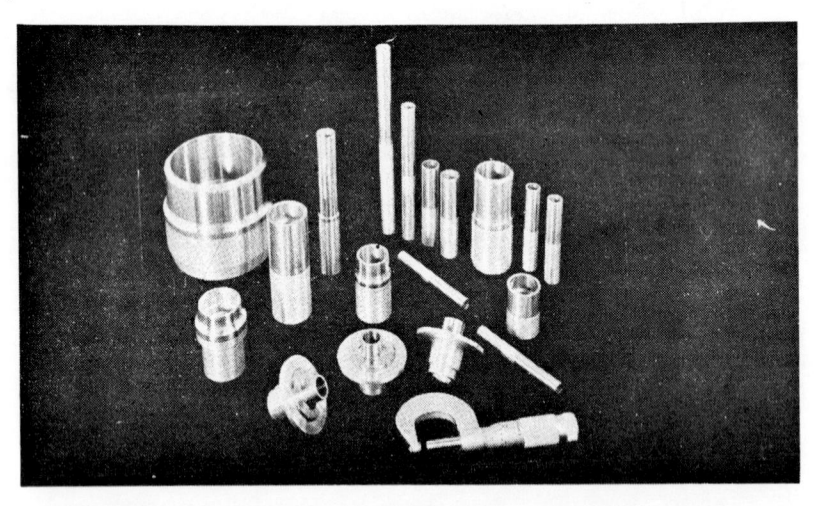

Figure 6   Examples of explosion boned transition joints-fabricated by duPont Co.

One special problem often encountered in the fabrication of explosion clad dissimilar alloys into pressure vessels or other welded structures is the procedure for electric fusion welding of one clad plate to another. For example, when the cladding plate is a refractory metal such as tantalum and the base metal is steel, co-fusion of these two metals in a conventional weld joint would produce the brittle TaFe intermetallic compound. Special procedures must be used to prevent mixing of the clad and base metal and de-bonding of the clad due to thermal expansion stresses. Techniques now exist for separately fusion welding clad to clad and base to base, maintaining structural integrity and continuity of cladding for combinations such as tantalum/steel, titanium/ steel, and the more conventional combinations such as stainless steel/ steel and Inconel/steel. Typical joint configurations are described in a recent paper by Richter[14] and in the Welding Handbook[15].

Figure 7 Examples of transition joints fabricated using tubular explosion welding techniques--2-1/2 in and 4 in transition tubes and sections. (Ref. 22)

Explosion welding is not limited to flat plate cladding for pressure vessels, tube sheets and transition joints. Explosion welding techniques have been adapted for specialty field welds in several instances. Two of the most notable applications are the welding of tubes in tube plates[16] and the remote plugging of leaking heat exchanger tubes in pressurized water reactor steam generators[17], [18]. The techniques of both processes are illustrated in Figures 8 and 9. The advantage of tube in tube sheet explosion welding is the ability to weld dissimilar metal combinations and thin wall tubes. In addition, welding skills are not required as the weld parameters are contained in the manufactured explosive cartridge and tube sheet preparation. The tube plugging process illustrates the ability to weld remotely since only mechanical placement and surface preparation are the weld controlling factors.

Other applications of explosion welding recently reported at a symposium in Czechoslovakia[19] included the field welding of over 2000 Km of aluminum communication conduit and the production of aluminum-tin alloy explosion bonded to a steel backing for bearing inserts in diesel engines. Probably one of the most significant developments reported was the use of an explosion welding machine installed in a conventional factory environment for butt joining of hydrostatically extruded copper clad aluminum rods. The rods are explosion joined end to end to create a continuous rod for wire drawing. Use of blast and noise suppression chambers for explosion welding and other explosive metalworking operations were also reported in Czechoslovakia and Russia.

Figure 8 Group of 19 tubes explosion welded in a hexagonal arrangement on 1.3 inch centers (Courtesy Williams et al., Ref. 16).

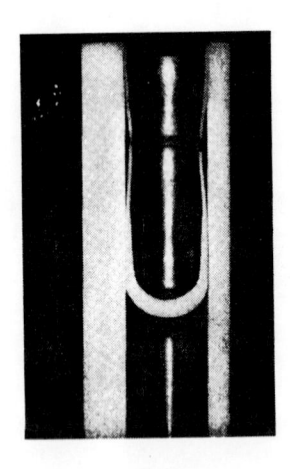

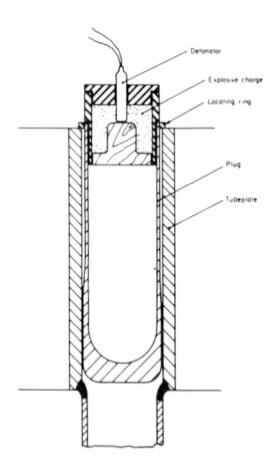

Figure 9 Example of closing tube using explosion welding techniques (Ref. 18).

One final development that may have commercial significance is the demonstration of explosion welding without using an explosive energy source. It has been shown by Wittman[20] and M. Meyer[21] of Sandia Laboratories that welding by oblique impact can be achieved using electromagnetic impulse in place of an explosion generated impulse. Aluminum to aluminum and copper to copper have been welded by this method at the present time. However, it should be possible to weld a wide variety of dissimilar metal combinations since the "impact" welding principles are identical to those of explosion welding. The process will be limited to localized spot or lap welds and relatively thin sections because of the practical energy storage limitations of capacitor systems, but nevertheless could still be an attractive nonexplosive process for welding where dissimilar or other difficult to fusion weld combinations are encountered.

Other applications of the explosion welding process are discussed in detail in Reference 2.

BIBLIOGRAPHY

1. V. Philipchuk, "Creative Manufacturing Seminar 1965," Paper SP 65-100, New York (Amer. Soc. Tool Manuf. Eng.)

2. "Principles and Practices of Explosive Metal Work," Edited by A. A. Ezra, Industrial Newspapers Ltd., 1973, Chapt. 10, pp. 173-228.

3. Abrahamson, G. R., "Permanent Periodic Surface Deformations Due to a Travelling Jet," J. Appl. Mech. $\underline{28}$, Dec. 1961.

4. A. S. Bahrani, T. J. Black, and B. Crossland, "The Mechanics of Wave Formation in Explosive Welding," Proc. of the Royal Society, Series A, Vol. 296, Jan. 1967.

5. R. F. Rolsten, H. H. Hunt, W. A. Dean, and A. K. Hopkins, "Phenomena Associated with Impulsive Loading: Welding, Rippling, Hardening," Tech. Report AFML-TR-67-15 under Air Force Contract No. AF 33 (615)-2040, Feb. 1967.

6. J. Ribovich, R. W. Watson, and F. C. Gibson: AIAA Jour. $\underline{6}$, pp. 1260-1263 (1968).

7. R. H. Wittman, "The Influence of Collision Parameters on the Strength and Microstructure of an Explosion Welded Aluminum Alloy," 2nd Int. Sym. Use of Explosive Energy in Manufacturing Metallic Materials, Marianske Lazne, Czech, 9-12 Oct. 1973.

8. J. M. Walsh, R. G. Shreffler, and F. J. Willig, J. Appl. Phys. $\underline{24}$, pp. 349-359 (1953).

9. G. R. Cowan and A. H. Holtzman, J. Appl. Phys. $\underline{34}$, pp. 328-339 (1963).

10. A. A. Deribas, V. M. Kudinov and F. I. Matveenkov, Combustion, Explosions and Shock Waves, Vol. 3, pp. 344-348 (1967). (English Translation)

11. A. Burkhardt, E. Hornbogen, and K. Keller, Z. Metallk. $\underline{58}$, pp. 410-415 (1967).

12. G. R. Cowan, O. R. Bergmann, and A. H. Holtzman, Met. Trans. $\underline{2}$, pp. 3145-3155 (1971).

13. R. H. Wittman and L. Trueb, Z. Metallkunde, 64 (1973) 9, pp. 613-618.

14. U. Richter, Schweissen and Schneiden, 25 (1973) 6, pp. 1-3.

15. Welding Handbook, Sixth Ed., Section 5, Chapter 93, American Welding Society, Miami, Florida 33125 (1973).

16. J. D. Williams and B. Crossland, "Explosive Welding of Hard Drawn Copper Tubes to Tube-Plates," Welding and Metal Fabrication, Jan. 1969.

17. W. Johnson, "Explosive Welding Plugs into Heat Exchanger Tubes," Welding Journal 50, p. 22 (1971).

18. B. Crossland, R. F. Halliburton, and A. S. Bahrani, "The Explosive Plugging of Heat Exchangers," 4th Int. Conf. of the Center for High Energy Forming, Vail, Colo., July 9-13, 1973.

19. 2nd Int. Sym. Use of Explosive Energy in Manufacturing Metallic Materials, Marianske Lazne, Czech, 9-12 Oct. 1973.

20. R. H. Wittman, unpublished data.

21. M. D. Meyers, "Impact Welding Using Magnetically Driven Flyer Plates," 4th Int. Conf. of the Center for High Energy Forming, Vail, Colo., July 9-13, 1973.

22. R. Grollo, "Explosive Joining of Dissimilar (Transition) Metal Tubes," Proc. of the 3rd Int. Conf. of the Center for High Energy Forming, Vail, Colo., July 1971.

SECTION III:
High-Frequency Welding

A High Speed Welding System for the Production of Custom Designed HSLA Structural Sections 171
High-Frequency Welding of Pipe and Tubing 199
HF Contact or Induction Tube Welding: Which Is Better? 233
High-Frequency "Bar-Butt" Welding 236
How To Make a Wheel Rim — in 4 Seconds 240
Melt Welding — a New High Frequency Process 243

A HIGH SPEED WELDING SYSTEM FOR THE PRODUCTION OF CUSTOM DESIGNED HSLA STRUCTURAL SECTIONS

H. N. Udall
  Director of Research & Development
  THERMATOOL CORP., Stamford, CT. U.S.A.

J. T. Berry
  Consultant to THERMATOOL CORP.
  Professor, School of Mechanical Engineering
  The Georgia Institute of Technology, Atlanta, GA.
    U.S.A.

E. D. Oppenheimer
  Consulting Engineer
  Mamaroneck, NY., U.S.A.

INTRODUCTION

The high frequency electric resistance welding process for the production of structural beams is well established; however, up to now its use has been largely confined to the welding of conventional grades of structural steel with yield strengths of the order of 250 MPa (36 ksi). The more recently available HSLA (microalloyed) steels have now opened up the possibility of making welded structural beams with considerably higher yield strengths which in turn will lead to significant weight savings. This process appears to have great potential for welding structural beams from this class of material. The high weld rates, typically 15 - 45 m/min (50 - 150 ft/min), the low heat input required, the very narrow heat affected zone and the forge pressure applied during welding combine to produce welds which retain the fine grain character and the excellent mechanical properties of the parent material. This paper will describe both the equipment used to produce continuous welded HSLA structural beams and certain aspects of the metallurgy of welds produced by this process in a number of different HSLA steels with nominal yield strengths of 550 MPa (80 ksi).

BACKGROUND OF THE PROCESS

This structural welding system uses a continuous electric resistance welding process* and is based on the use of electric current at a nominal frequency of 400,000 Hertz. At this frequency the current characteristically flows in the surface of the material to a depth of approximately 0.8 mm. (0.03 in.) in carbon steels at their welding temperature. This enables a very high power density to be concentrated at the welding surfaces and enables the process to weld at high speeds with low heat input. The joining is also enhanced by the forging action which occurs as the heated surfaces are brought together. Consequently the resulting joint is characterized by a distinct lack of the grain coarsening which is frequently seen in fusion type welds. Joint preparation or cleaning are normally not required, no fluxes or filler metals are used and no inert gas atmosphere is required.

To produce an I-beam section three strips of steel, two for the flanges and one for the web, are fed from conventional uncoiling equipment. Immediately upstream of the weld station each edge of the incoming web strip is cold upset to locally increase its thickness. This enables the subsequent weld to be typically from 120% - 150% of the original web thickness. This significantly enhances the competence of the welded assembly. Figure 1 is a schematic diagram of the process which shows the placement of the lower contact shoes on the web and flange upstream of the weld point. A similar set of contacts (not shown) provides current to the upper weld. Both welds are made simultaneously, one directly above the other at the common weld pressure point.

In addition to the usual I-beam configuration a wide range of other products can be made using this system. It can be readily adjusted to weld sections with flanges of different widths and/or thicknesses and either one or both flanges may be offset relative to the centerline of the web as shown by Fig. 2. Additionally different grades of steel may be used for the web and flanges.

The system is continuously adjustable throughout its size range so that beams may be designed for optimum performance rather than having to be selected from the nearest standard size available. This of course leads to the maximum economy in the use of material.

---

*Patented in the USA and other countries by THERMATOOL CORP.

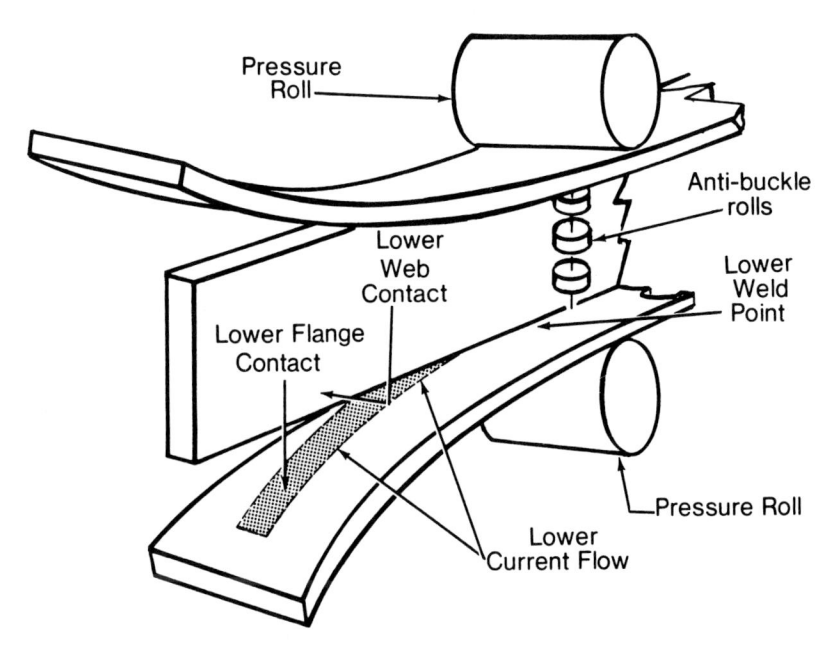

Simplified Representation of High Frequency Welding of an I Beam
Figure 1

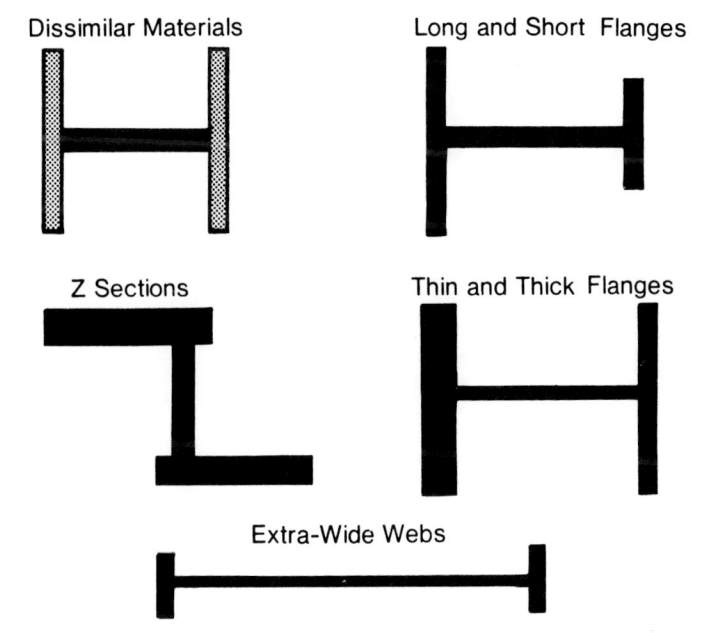

Some Typical Sections Which Can Be High Frequency Welded
Figure 2

PRODUCTION SYSTEMS DESIGN FOR WELDING I-BEAMS

The flow diagram for a typical structural beam production system is illustrated in Fig. 3. It consists of the THERMATOOL welding section and the associated material handling equipment and drives. The exact layout of any actual

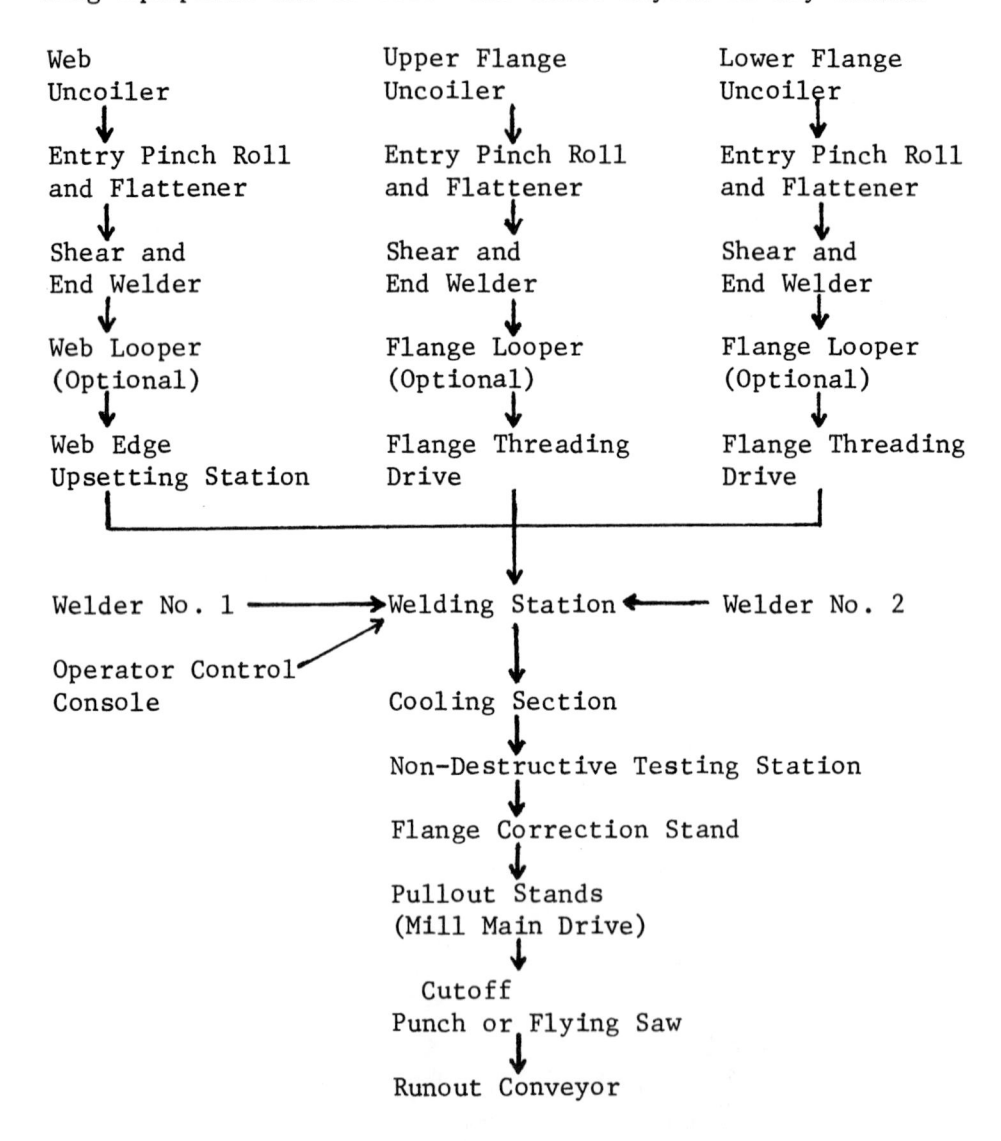

```
Web                    Upper Flange           Lower Flange
Uncoiler               Uncoiler               Uncoiler
   ↓                      ↓                      ↓
Entry Pinch Roll       Entry Pinch Roll       Entry Pinch Roll
and Flattener          and Flattener          and Flattener
   ↓                      ↓                      ↓
Shear and              Shear and              Shear and
End Welder             End Welder             End Welder
   ↓                      ↓                      ↓
Web Looper             Flange Looper          Flange Looper
(Optional)             (Optional)             (Optional)
   ↓                      ↓                      ↓
Web Edge               Flange Threading       Flange Threading
Upsetting Station      Drive                  Drive
   └──────────────────────┬──────────────────────┘
                          ↓
Welder No. 1  ──→  Welding Station  ←──  Welder No. 2
Operator Control  ↗
Console               ↓
                   Cooling Section
                      ↓
                   Non-Destructive Testing Station
                      ↓
                   Flange Correction Stand
                      ↓
                   Pullout Stands
                   (Mill Main Drive)
                      ↓
                   Cutoff
                   Punch or Flying Saw
                      ↓
                   Runout Conveyor
```

FLOW DIAGRAM FOR TYPICAL I-BEAM MILL USING THE
HIGH FREQUENCY ELECTRIC RESISTANCE WELDING PROCESS

Fig. 3

plant is designed to suit the needs of the individual user. In addition to the items shown in the flow diagram, facilities for coil slitting to prepare coils of the correct widths must also be available, together with the necessary handling equipment for the incoming coil stock and the outgoing finished beams. Beam painting facilities may also be added if this is a requirement.

Each system uses two high frequency power supplies usually either, 160 KW or 300 KW output each. Other sizes are available to meet individual plant requirements. The choice of size depends upon the maximum web thickness for which the line is designed and the maximum line speed required. In addition to being conservative of material and energy the system is also one of inherently high productivity. Table I shows typical production weld rates for various web thicknesses.

TABLE I

Typical Production Speed Meters Per Minute			Typical Production Speed Feet Per Minute		
Web Thickness (mm.)	Power Supplies 160KW	300KW	Web Thickness (in.)	Power Supplies 160KW	300KW
3.0	22	46	0.125	70	150
6.0	8	22	0.250	25	70
9.5	-	11	0.375	-	35

LABORATORY AND PILOT SCALE TESTING

A. Scope Of Tests

From 1960 to 1971 high frequency I-beam welding was done using steels with yield strengths of up to 280 MPa (40 ksi). In 1971 a series of Tee weld tests was made at the THERMATOOL Corp. labs using steels with yield strengths of up to 380 MPa (55 ksi). These tests indicated that in order to prevent the formation of an undesirable brittle weld heat affected zone, the carbon equivalent of such steels should be kept below 0.40 using the formula given below which was developed by THERMATOOL Corp. at that time.

$$CEq\% = C\% + \frac{Mn\% + Si\% + Cr\% + Mo\% + V\% + Cb\%}{5} + \frac{Ni\% + Cu\%}{15}$$

In 1976 another series of Tee weld tests was run using the more recently developed high strength low alloy steels with yield strengths up to 550 MPa (80 ksi). These tests included a pilot run on a commercial I-beam line using two 550 MPa (80 ksi) yield steels, one for the web and the other for the two flanges. These are identified as 80CI and 80V respectively. Beams were also made from a 310 MPa (45 ksi) steel at this time. Concurrently a series of tests was made at the THERMATOOL Corp. labs using an experimental fixture in which the same two steels together with three additional 550 MPa (80 ksi) yield steels were used both individually and in various combinations to make welded tees. Two further test samples were made. In both tests the flange was of 550 MPa (80 ksi) steel. In one the web was of 310 MPa (45 ksi) steel and in the other it was of 250 MPa (36 ksi) steel. The chemical composition and mechanical properties of the 550 MPa (80 ksi) steels which are the subject of this paper are given in Table 2.

B. Equipment Used In Preparing Welds

Pilot Or Commercial Scale Runs. The mill which was used for the pilot production runs has been in operation welding I-beams from commercial grades of structural steel since 1971. It is equipped with two 160 KW welders. Although it was not designed for use with HSLA steels, it was sufficiently adaptable to enable this pilot run to be made. These beams were subsequently incorporated into the structure of two experimental road transport vehicles which were built for test purposes.

Laboratory Scale Runs. The fixture consisted of a weld station designed to weld Tees using 64 mm. (2.5") wide strip for both the flange and the web. The Tees were welded from strip lengths between 3 m. and 6 m. (10 ft. and 20 ft.). A drawbench was used to pull the strips through the weld area tooling. Weld power was supplied by a THERMATOOL high frequency welding generator. In some tests a 160 KW unit was used and in others a 300 KW unit.

TABLE 2

Details of HSLA Steels Involved in Current Tests

Steel Designations		80C1	80C2	80C3	80T	80V
	C	0.04	0.06	0.05	0.08	0.11
	Mn	0.52	1.35	1.19	0.43	1.09
	Si	0.10	0.55	0.036	0.03	0.13
	P	0.008	0.007	0.008	0.003	0.008
Composition	S	0.015	0.004	0.010	0.019	0.014
Wt. Percent	Cb	0.10*	0.10	0.13	0.007	–
	V	–	–	–	0.012	0.13
	Al	–	0.06	0.10	0.084	0.042
	Cu	–	–	0.06	–	0.02
	Ni	0.004	–	0.04	–	0.03
	Mo	–	–	0.02	–	0.01
	Cr	0.020	–	0.04	–	0.03
	Ti	–	–	–	0.29	–
	Ce	–	0.025	0.005	–	–
	N_2	–	0.010	0.007	0.0055	0.019
Yield Strength	ksi	80.5	80.5	86.7	79.5	85.2
	MPa	555	555	598	549	588
Ultimate Tensile Strength	ksi	87.0	89.8	95.2	104.2	100.6
	MPa	600	620	657	719	694
Elongation	%	23.0	25.5	23.3	17.4	21.0
Thickness	in.	0.087	0.255	0.149	0.109	0.135
	mm.	2.21	6.48	3.78	2.77	3.43

Notes: – indicates no information available
 * Not determined, specification is 0.08 – 0.12

Source: *Welding of HSLA (Microalloyed) Structural Steels*, ASM, 1978

C. Test Procedures

Qualification Test. The initial test used to qualify any welded sample is a peel test in which a short section of beam or Tee to be tested is securely clamped and the web is peeled away from the flange using a special peeling tool. This is shown in Figure 4. To pass this test the beam must fail in the parent metal of either the web or the flange but not on or immediately adjacent to the bond plane.

Metallographic Examination Of Weld And Heat Affected Zone. A sample weld from each combination of web material and flange material that was welded and which had satisfactorily passed the peel test was taken for metallurgical examination. Details of these runs are given in Table 3. Run number 3698 was from the pilot scale commercial run. All other run numbers refer to welds made on the laboratory fixture.

A substantial number of sections was made and subsequently examined using standard light microscopy as well as micro- and standard hardness testing procedures. The sample sections were representative of all of the combinations of materials that were run.

It was the intention of the metallographic and hardness surveys to reveal any occurrence of weld area or heat-affected zone cracking or unusually high or low hardness areas in the vicinity of the weld. The sections prepared were taken in a plane perpendicular to the welding direction and included material about 13 mm. (0.5 in.) from the bond plane in the web and both sides of the flange. The samples were cut using a water cooled abrasive cut-off wheel and ground and polished using accepted metallographic preparation procedures. After lightly etching in nital, sections were scanned at low magnification (X46) for any evidence of heat affected zone cracking, excessive grain growth, lack of fusion and bond-line defects. This was undertaken no sooner than about eight hours from the time of welding.

Immediately after the low power examination, sections were examined at higher magnifications for microcracks, fissures or hot-tears. Finally, representative photomicrographs were taken of the structures in the heat affected zones near the bond plane and in the unaffected parent material.

Note: A long tubular lever arm (not shown) is used on the end of the peeling tool to provide the necessary peeling force

Figure 4

Source: *Welding of HSLA (Microalloyed) Structural Steels*, ASM, 1978

TABLE 3

Summary of Metallographic And Hardness Data And Observations

Run	3698	24	86	60	44	74	105	107	108	109
Flange										
Material	80V	80V	80V	80V	80C1	80C2	80T	80C3	80V	80C3
Parent(1)Hv	221	220	213	218	213	193	225	211	223	220
HAZ (2)Hv	412	428	412	355	302	213	345	302	382	272
Structure(3)	A	A	A	A	E	A	E	A	A	A
Bond (4) Hv	262	241	285	246a	216	266b	264	264	288	243
Web										
Material	80C1	80C1	80C1	80V	80C1	80C2	80T	80T	80T	80C3
Parent(1)Hv	225	220	220	223	220	197	222	220	220	220
HAZ (2)Hv	328	310	355	470	328	175	355	372	345	345
Structure(3)	E	E	E	A	E	A	E	E	E	A
Heat Input										
KJ/mm.	0.31	0.31	0.19	0.33	0.23	0.51	0.25	0.26	0.26	0.30
KJ/in.	7.8	7.9	4.8	8.5	5.8	13.0	6.3	6.5	6.5	7.5
Speed m/min.	24	20	46	20	20	17	26	26	26	23
ft/min.	80	67	150	67	67	55	86	86	86	75

Notes: (1) Parent Material Hardness determined as Rockwell 30-T
(2) Maximum Micro-Hardness in Heat Affected Zone
(3) HAZ Structure Type A - Acicular   E - Equiaxed
(4) Hardness at Bond Plane determined as Rockwell 15-N except where marked a for 30T, b for 15-T.

Microhardness Testing. Microhardness surveys were undertaken using a Knoop indentor employing a 25 or 50 gm. load. The survey commenced in unaffected parent material in the web, ran across the weld zone and usually into the unaffected parent material in the flange. The impressions were generally spaced 0.64 mm. (.025 in.) apart on the ends of each hardness run and 0.13 mm. (.005 in.) apart in the vicinity of the bond plane. Usually 18 to 30 micro-hardness indentations were made in each traverse along a centrally located axis.

The Knoop hardness values were subsequently converted to the more common Vickers hardness values for convenience.

Light Load Hardness Testing. In view of the well known effects of impression size upon "averaging" hardness level, impressions were also made in the immediate vicinity of the bond plane and in the unaffected parent metal using the Rockwell Superficial hardness tester. Either 15 or 30 Kg. loads were used together with the indentor (diamond or ball) appropriate to the general hardness level experienced.

To render the interpretation of the above data more convenient, the hardness figures obtained were converted to Vickers hardness values, although in all tabular material presented, the method of taking hardness is indicated.

Tee Tensile Impact Tests. A limited series of tensile impact tests was conducted on various welded specimens to explore the resistance of the weld area to shock loading at normal and sub zero temperatures. It was necessary, because of the unusual shape of the region to be tested, to design a test specimen, Fig. 5 and appropriate apparatus, Fig. 6, which could be mounted on a standard Tinius Olsen Universal Impact Tester.

A small number of specimens of each of six materials and material combinations were tested at room temperature, 19C (66F), and at one arbitrarily selected low temperature, -45C (-50F).

The purpose of the Tee tensile impact tests was to simulate service shock loading of welded structural joints. This was considered necessary because of the possible use of welded HSLA beams in vehicles where random loading and low temperature operation can occur.

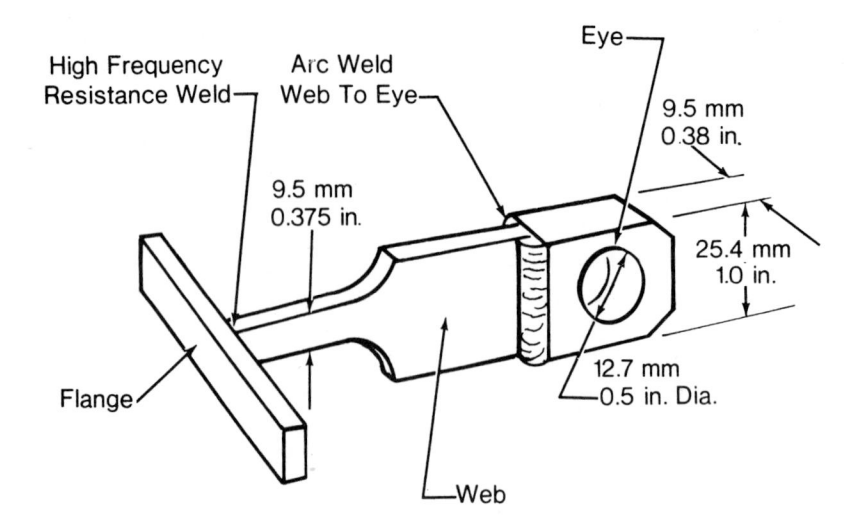

Tee Tensile Test Specimen
Figure 5

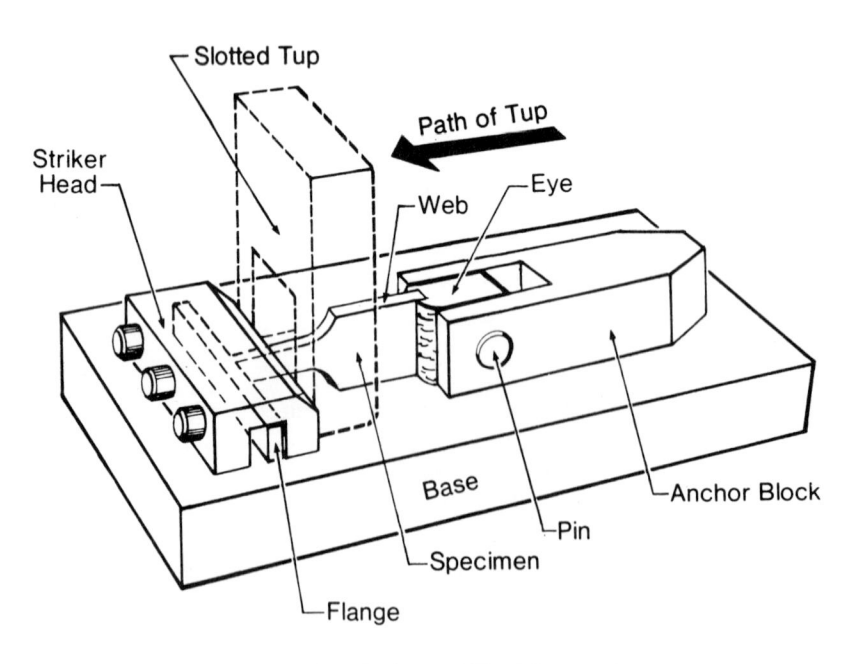

Tee Tensile Impact Test Apparatus
Figure 6

A practical problem in the testing of welded joints is to achieve high local loading in the joint region without causing premature failure in other parts of the welded assembly. Loading should however occur in a manner which will simulate actual service conditions. Intentional notches can serve to place stress concentrations in locations where initial failure is desired. Natural stress concentrations due to the shape of the assembly, the presence of defects in the material or the joint, or notches and crevices resulting from the joining process are, however, of greater interest since they will always be present in the assembly when it is in service and may initiate failures.

The current test has been designed, to our best ability, to produce failure in the joint if it is likely to occur there in service but to allow failure in neighboring regions if the joint is stronger, consequently no artificial notches were added to the test specimens. This investigation was limited to two test temperatures 19C (66F) and -45C (-50F) which are typical of the range for testing high strength low alloy steels for general service.

RESULTS OF TESTS CONDUCTED

A. Metallographic and Hardness Tests

The salient features of the examination and testing program outlined above were:

a) No cracking was observed in any of the microsections examined;

b) No micro-fissures could be ascertained at a magnification of 1060 diameters;

c) Although individual microhardness values of the order of 420 - 470 Hv (43 - 47 Hrc) were sometimes seen in several welds (see Table 3) in the heat affected zone adjacent to rather than actually on the bond plane, the Rockwell Superficial hardness tests conducted upon the same welds yielded values in the general area of the bond line no greater than about 300 Hv (30 Hrc) and generally between 240 and 285 Hv (20 - 28 Hrc).

d) The parent plate micro-structures examined varied considerably; however, the heat affected zone structures appeared to fall into two principal categories:

(i) Those made up of more lenticular transformation products (lower bainites, martensite and/or acicular ferrite);

(ii) Those made up of more equiaxed transformation products.

Figures 7, 8, and 9 illustrate these points. One of the parent plate materials exhibited some segregation. This became particularly noticeable upon examining HAZ material (Fig. 10).

X1060

Cross Section showing lenticular transformation products in heat affected zone of Run 77 flange side. 80C2 Steel

Figure 7

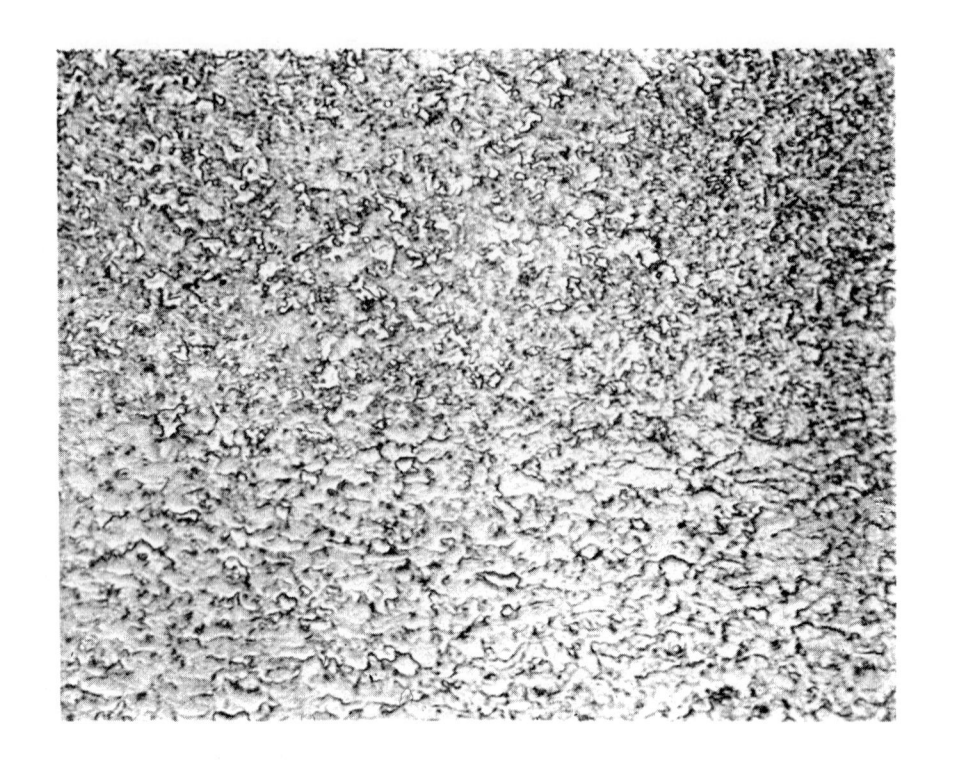

X500

Cross Section showing more equiaxed products
of transformation at bond line of Run 44

Flange and Web 80C1 Steel

Figure 8

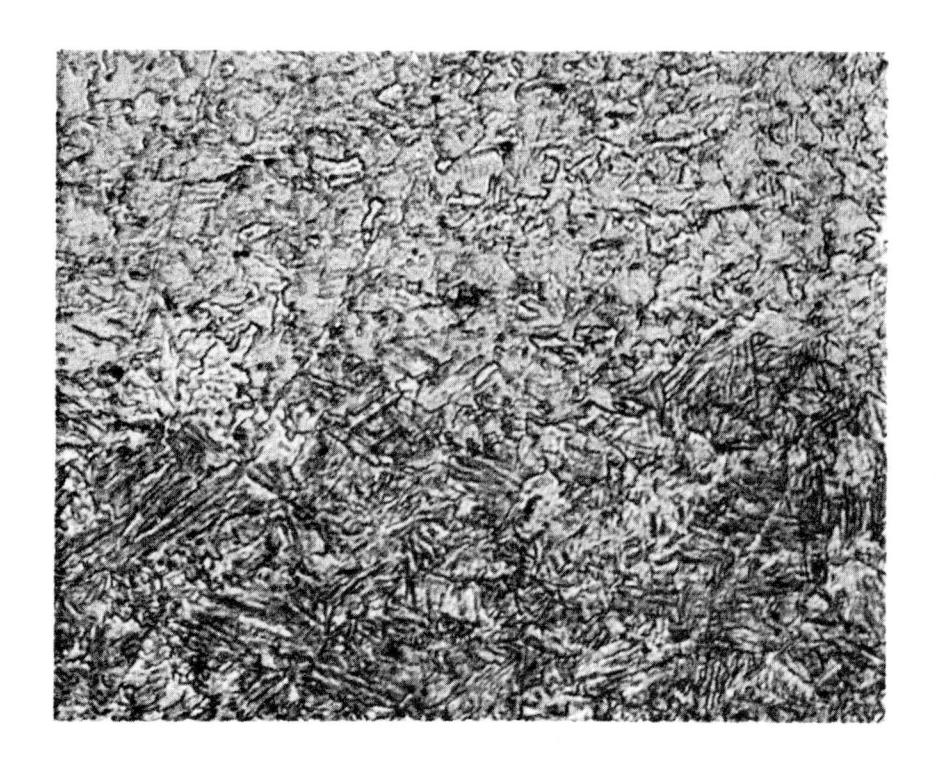

X1060

Cross Section showing transition from more lenticular to more equiaxed products of transformation at bond line in Run 86

Flange 80 V Steel   Web 80C 1 Steel

Figure 9

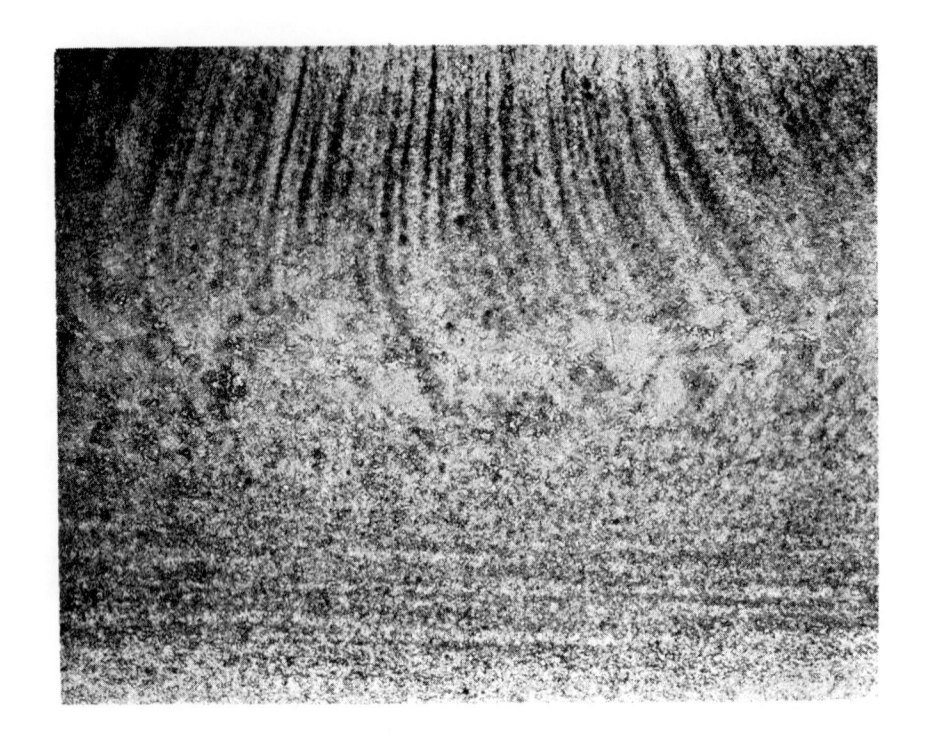

Cross Section showing segregation bands
accentuated by partial austenization
during welding

    Flange and Web 80C2 Steel

        Figure 10

B. Tee Tensile Impact Tests

In most cases failure of the welded assembly occurred in the web material well away from the heat affected zone. Failure energies recorded for these tests were essentially the same for any given material in both the room temperature (19C) and sub-zero (-45C) tests.

In the case of Run 105 with flange and web of material 80T, failure at low temperature (-45C) occurred in the flange material at an energy of about 20% of the room temperature (19C) web failure energy; however, in Run 107 with a web of 80T material but a flange of 80C3 the typical web tensile failure occurred at both temperatures examined, at a similar energy level to the room temperature test of Run 105.

Table 4 presents the results of the full range of tests including fracture energies and the location and appearance of fracture surfaces. Since each material used was a different thickness, the energy values have been expressed on a per unit web cross sectional area basis. Some important interrelational aspects of this test are contained in the discussion and conclusions of this paper.

TABLE 4

Summary of Tee Tensile Impact Test Results

Run	Flange Material	Web Material	19C (66F) Nature Of Fracture	Energy* J/sq.cm.	ft.lb/sq.in.	-45C (-50F) Nature Of Fracture	Energy* J/sq.cm.	ft.lb/sq.in.
3698	80V	80C1	Web Ductile	231	1107	Web Ductile	225	1077
34	80C1	80C1	Flange Single Shear & HAZ Tensile (1)	157	753	Flange (2) Double Shear	184	878
60	80V	80V	Web Ductile	329	1574	Web Ductile	352	1685
105	80T	80T	Web Ductile	306	1463	Flange Tensile Lamellar (3)	61	291
107	80C3	80T	Web Ductile	313	1498	Web Ductile	315	1507
109	80C3	80C3	Web Ductile	363	1736	Web Ductile	348	1664

* Average of two test results of each run at each temperature. Expressed on a per unit area of web basis to compensate for different web thickness.
(1) See. Fig. 11 (2) See Fig. 12 (3) See Fig. 13

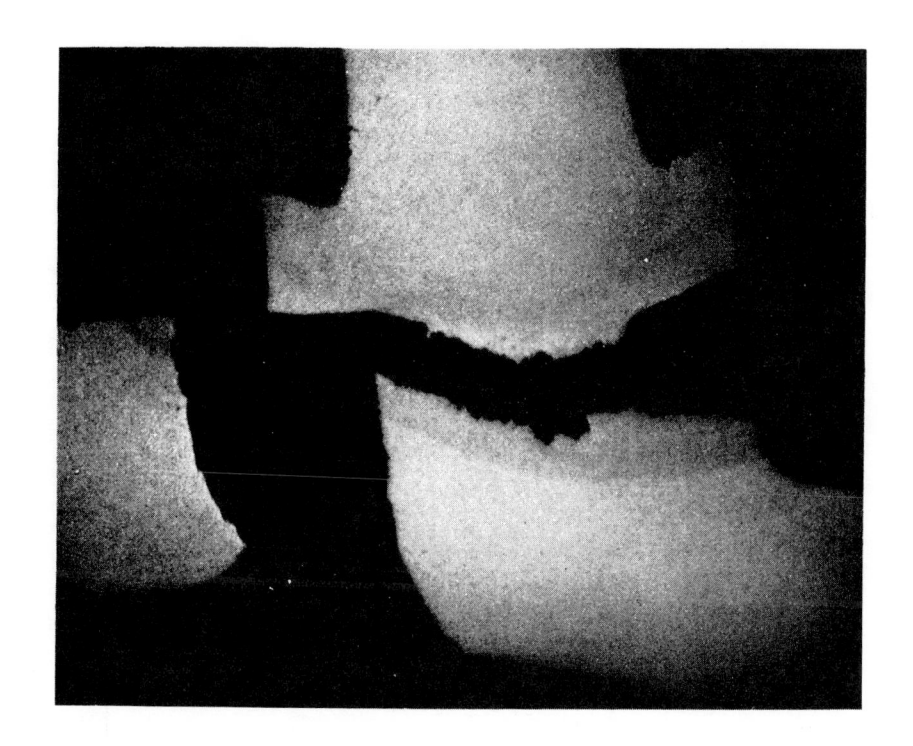

Cross section of Tee Tensile impact test of Run 34 at 19C showing shear failure in one side of the flange followed by a peeling type failure between the web and the other side of the flange.

Figure 11

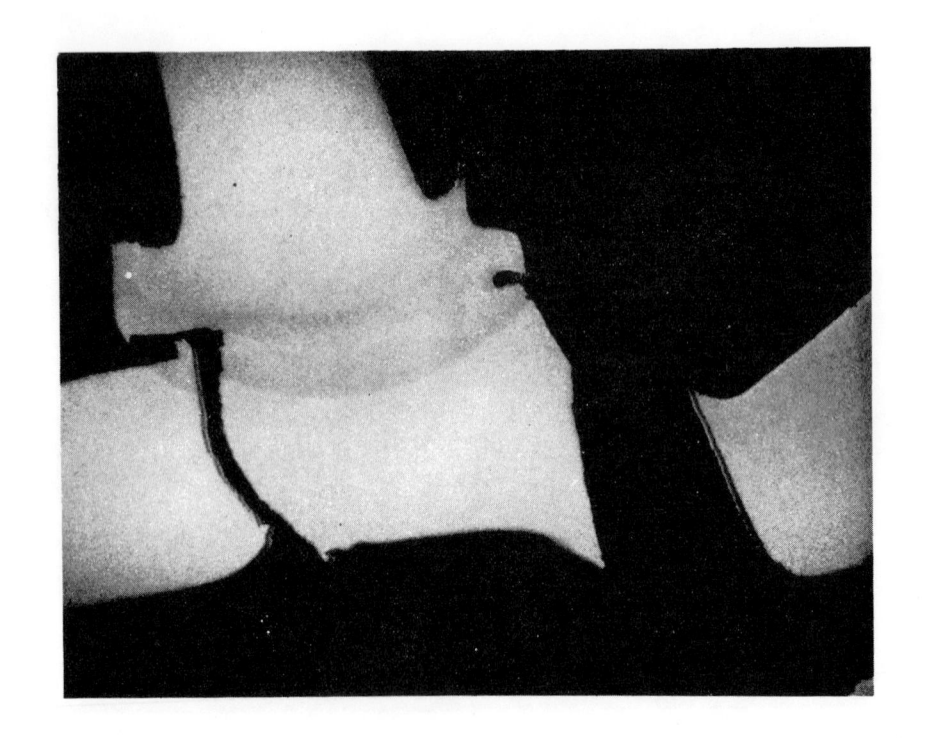

Cross Section of Tee Tensile impact
test of Run 34 at -45C showing double
shear failure of the flange on either
side of the web.

Figure 12

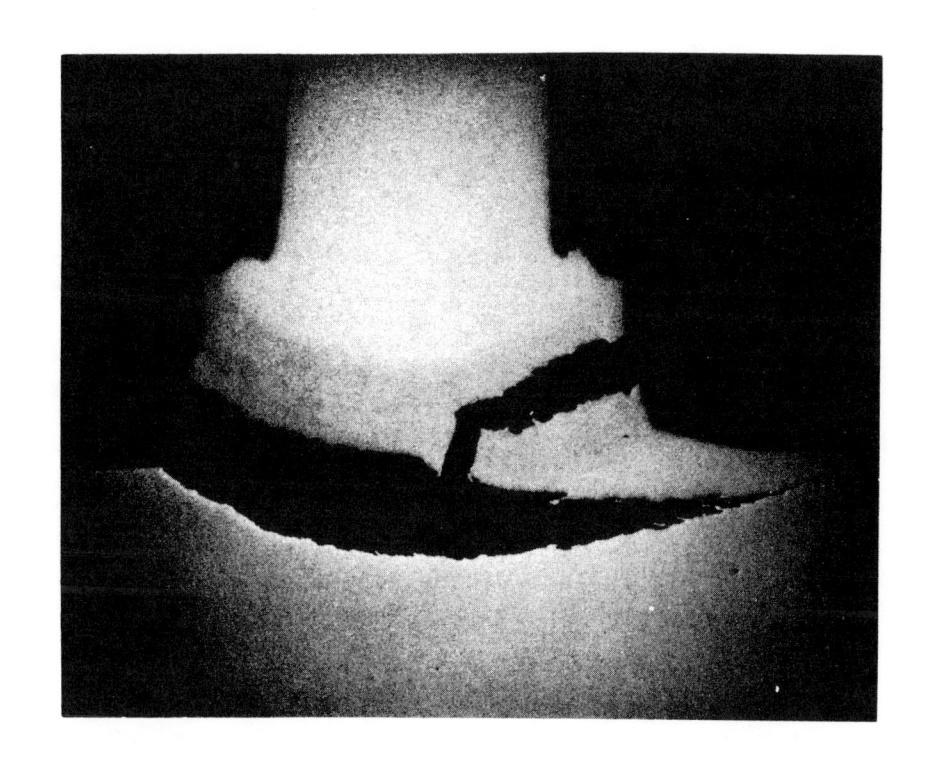

Cross Section of Fracture in Run 105
Tested at −45C (−50F)

Figure 13

DISCUSSION OF TEST RESULTS

Metallographic and Hardness Tests

Two striking observations became apparent upon examining the metallographic sections together with the accompanying hardness tests:

a) Absence of heat affected zone cracking, fissures or bond line defects;

b) Absence of significant grain coarsening in the welded zone;

Both of the above features undoubtedly contribute to the superior performance that the majority of the joints exhibited in the Tee-tensile tests.

A particular aspect of the structure-property combinations involved is that they would seem to be typical of those produced by the inherently low heat inputs seen in the high frequency resistance welding process. The hardness level increases observed are parallel with the effects small heat source-large heat sink combinations are seen to have upon cooling rates experienced in fusion welds; however, at the same time such hardness increases in welds of the latter type have often lead to concern regarding heat affected zone toughness.

Reference to the HSLA* welding literature (References 1, 2,3,4,5,6) indicates frequent instances of the ductile-brittle fracture transition temperature (Tc) being raised in the heat affected zone of fusion welds, or in specimens subjected to fusion-weld thermal cycles. Although increased hardness will play a part in raising Tc, the mean grain diameters of the transformation products and the disposition of carbo-nitride preciptates reforming during cooling of the heat affected zone will also influence Tc.

Both the austenite coarsening and carbo-nitride dissolution which will occur prior to the above precipitate reformation in these steels appear to be minimized with rapid heating

*Vanadium and Niobium/Columbium containing varieties

and cooling in the austenite range. In particular it seems likely that rapid cooling through the 800-500C region referred to by Hannerz (Ref. 6) is important in suppressing the re-precipitation phase. Such rapid cooling is typical of the high frequency resistance welding process. Furthermore, on examining Hannerz results (Ref. 6) one is led into believing that the grain coarsening experienced in many fusion welds may in itself be as important as the re-precipitation effects with respect to transition temperature.

In essence then for the Cb/V treated steels and with the low heat inputs experienced in high frequency resistance welding, one would anticipate considerably less trouble than with fusion welds with respect to HAZ brittleness for the following reasons:

- a) Short austenitizing times, therefore minimum dissolution of pre-existing carbo-nitrides, which will result in minimum austenite grain-growth as a consequence.

- b) Rapid cooling rates which will further limit austenite grain growth and also re-precipitation of carbo-nitrides subsequent to above dissolution.

- c) Further beneficial effects due to the thermomechanical working which occurs in this type of welding.

The above would seem to be consistent with our observations for these steels.

The embrittlement effect partly evidenced by the titanium treated steel, while it may not be typical of all heats of such steel, gives cause for concern and warrants further work to determine the underlying causes. Referring to the Tee tensile impact test results the presence of a more equiaxed type structure would seem to be a possible factor contributing to its low temperature failure mode; however, the role of precipitation in this instance needs further clarification.

Nonetheless, the remaining steels do seem to be relatively problem-free and carry with them considerable promise for the applications currently in mind.

Tee Tensile Impact Tests

Interpretation of test results requires that one concede that if, despite efforts to the load joint area, failure occurs elsewhere, then one may assume that the joint is at least as strong as the rest of the assembly in the proposed application.

Failure in this type of test will start where local stresses first exceed the yield stress. As the specimen shape changes due to plastic deformation, failure may continue where yielding started or the stress distribution may change so that further failure occurs elsewhere.

When testing an assembly where numerous modes of failure are possible, several of which may be of nearly equal static strength, failure energies can be dramatically different for very small variations of local strength dependent on which mode of failure takes place.

The clamping of the tee flange can vary in ways which result in greatly varying bending stresses in the flange. Bending deformation of the flange can produce additional tensile stress concentration at the corners of the joint. The span between flange supports, degree of rounding of the corners of these supports, and tightness of clamping can all have visible effects on the results. Although various spans were tested, the only results considered useful here are those for tests where the support span was as small as possible.

The following conclusions can be drawn from these tests:

1. The usual failure mode for welds of adequate width when properly supported is web tension.

2. Where flange thickness is significantly greater than the web, bending of the flange is unlikely to occur for components of similar steels.

3. For cases of equal web and flange thickness
   a. Flange support span must be less than 130% of thickness even if some weld upset must be removed to get the supports close enough to the web.

b. Failure may occur by double shear in the flange
      even when it is properly supported as seen in
      the case of Run 34.

4. With one exception impact testing produced no evidence of low temperature shock load sensitivity of this sort of joint in these materials. In Run 105 short transverse (Z) tensile failure occurred in the flange with a reduction of fracture energy.

5. The unexpected and unusual tensile failures seen in the flange of Run 105 at low temperatures suggest the need of low temperature Z direction ductility and impact study of certain HSLA steels. Ref. 7.

6. The flange shear failures of Run 34 raise some questions about the relative shear and tensile yield strengths of this material.

7. Within the temperature limits of these tests, 19C and -45C, and the extent to which this geometry represents a conservative simulation of service loading, it is clear that H.F. welded HSLA structural Tee welds are satisfactory for low temperature shock load service if weld width and centering are adequate and if selected materials are used.

8. It must be remembered that the test welds were made from small sample lots taken from a single heat of each of the steels used. The test samples were all cut from a short length of each run of welded Tee or beam. They, therefore, do not represent a statistical sample of the heat concerned.

FUTURE WORK

Future work will be needed to examine the effect of testing a greater range of materials over a wider range of temperatures. Fatigue testing of the welded assembly is obviously of great interest and a study of this nature is now under way.

ACKNOWLEDGEMENTS

The authors wish to acknowledge the support that they have received during the course of this work from many members of THERMATOOL Corp. Staff. In particular we would like to mention Mrs. B. Larson for her patience and cooperation during the course of numerous drafts and revisions of the paper, Mr. A. Zausner who produced all of the test welds made in the THERMATOOL Research and Development Laboratory and Mr. W. Schedlbauer who performed all of the metallographic and hardness tests and also assisted with the Tee Tensile Impact tests. We are also grateful to the various organizations who supplied us with steel for our test work and to Dr. J. Malcolm Gray of Molycorp for his encouragement and advice.

REFERENCES

1. K. Norring in Proc. of Micro-alloying '75 Conference Washington, DC   Oct. 1975 p.p. 30 - 37 Applications Volume

2. I. G. Hamilton   ibid p.p. 52 - 57

3. A. Z. Lubuska   ibid p.p. 65 - 76

4. J. M. Sawhill and T. Wada, Wldg. Res. Suppt. (54)   January 1975 p.p. 1s - 11s

5. T. Greday and M. Lamberigts in Proc. of Micro-alloying '75 Conference, Washington, DC Oct. 1975 p.p. 145 - 158 History and Theory Volume

6. N. E. Hannerz, Wldg. Res. Suppt. (54)   May 1975 p.p. 162s - 168s

7. A Technique For Investigating The Short Transverse Properties Of Sheet And Plate Materials Exhibiting Texture And Lamellar Discontinuities - R. G. Kumble, J. T. Berry, E. D. Oppenheimer - Proceedings of 4th European Conf. on Texture, Cambridge 1975

HIGH-FREQUENCY WELDING
OF PIPE AND TUBING

Dr. H. B. Osborn, Jr.

Marketing Manager, TOCCO DIVISION
The Ohio Crankshaft Company

INTRODUCTION

The art of high-frequency induction welding of tubing would appear to be quite old. A patent filed in 1928, and subsequently issued, carries claims which read in part, "The method of continuously welding open-seam metal tubes by induction . . . progressively advancing an open-seam metal tube, bringing the seam edges together into closer and closer proximity during the longitudinal movement thereof until said edges are sufficiently close . . . inducing in the metal a current flow for heating the seam . . . bringing the edges completely together to cause concentration of flow of induced current and impressing the edges together to weld the pipe. . . ."

It sounds very simple, but it took many years of research and development and refinement of technique to put the process on a production basis.

There are two basic designs used for the inductors. The first, which we shall call "nonencircling" generally consists of an elongated construction held above and parallel to the seam. The second, which we shall call "encircling inductor," is nothing more than one or more turns surrounding the tube, generally in advance of the seam junction point. In either case the copper tubing is water cooled.

Nonencircling Inductor

The open-seamed tube is fed progressively under the inductor and is gradually closed to produce a weld from the heat which has been developed along the seam edges. The point at which the seam is closed may be under the inductor or after the tube has left the inductor, just so long as the temperature is high enough when the pressure rolls "forge" the edges together.

With this design, a seam guide is advantageous to provide an open seam as the strip goes under the inductor and to control the position of the seam in the pressure rolls. However, it is absolutely necessary to have this guide to accurately position the seam with respect to the inductor so that the edges to be heated are kept beneath the center leg of the inductor which carries a flux concentrating device. This heats a minimum depth of metal back from the seam edge which results in a small amount of upset.

This type of inductor usually consists of three parallel conductors with the center conductor opposite the edges of the seam to be heated and provided with magnetically permeable material so as to induce high frequency current in the edges of maximum concentration or density. The other two conductors are spaced away from the edges but more or less adjacent to the surface of the tube so as to create electrical coupling and induce current in that area, but at low density. (In some installations these side arms are remote to the tube surface even above the level of the central conductor in order to provide easier clearance for the pressure rolls.) An air gap of approximately 0.080/0.100 in. is maintained between the tube and the center leg (Fig. 1).

A set of data for E.M.T. conduit welded at 10,000 cycles using 250 KW is as follows:

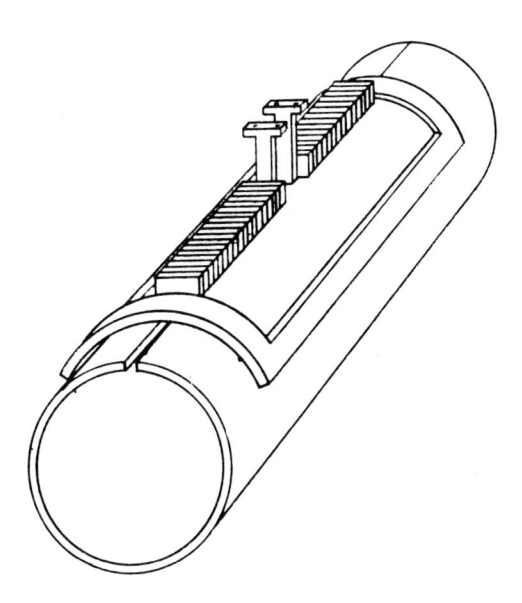

Fig. 1. Non-encircling inductor for high frequency induction welding.

Tube Size, In.	OD, In.	Wall, In.	Welding Speed, fpm
1/2	0.706	0.042	210
3/4	0.922	0.049	200
1	1.165	0.058	180
1-1/4	1.510	0.065	150
1-1/2	1.740	0.065	150
2	2.197	0.065	150

The mill processing conduit per the above data has been in operation since 1950 and averages approximately **50,000,000** ft. per yr. The strip-to-conduit yield is 96%. This figure includes trimming losses for the wide coils, material consumed for test set up on size changes and end weld losses.

The air gap between the inductor and the seam affects the efficiency of welding. Pipe has been run at 10,000 cycles with an air gap considerably less than that applicable to the above data. Compare, for example:

Source: *Mechanical Working & Steel Processing 4*, Gordon & Breach, 1969

Diameter, In.	Wall In.	Power, KW	Speed fpm	Air gap, In.
4	0.187	100	30	0.040/0.050
8-5/8	0.141	130	25	0.080/0.100

With the smaller gap, more metal has been welded with less power and at greater speed. The improvement has been about 100%.

Unfortunately, it has been found impractical to maintain, on a continuous production basis, the small air gap indicated above. Further, under such conditions momentary changes in gap corresponding to deviations of the seam from parallelism with the inductor result in undesirable temperature variations. Additionally, scale from hot-rolled strip will introduce serious problems by often filling up the gap despite wiping and blowing devices. Although at considerable loss in efficiency, these difficulties are not encountered at the 0.080/0.100 in. air gap which is recommended for production.

The use of this type of inductor has been limited to motor generator frequencies of 1, 3, and 10 KC.

Encircling Inductor

The open-seamed tube is fed progressively through the inductor and is gradually closed to produce a weld from the heat which has developed along the edge of the seams and has concentrated at the seam junction point. It is necessary to maintain the seam open as it goes into the coil, and a nonconductive guide is generally used. This also serves the additional purpose of maintaining the position of the seam through the pressure rolls—a very essential control.

The encircling inductor induces, by the usual process of electromagnetic induction, a circumferential flow of current. This current, however, cannot flow across the open seam and, therefore, flows down the tube to the seam junction point. There is an increasing concentration of current density as the weld point is approached (Fig. 2).

We must also take into consideration the flow of current in the cross section of the strip itself (Fig. 3). The depth of this current flow depends on the frequency of the current and both the magnetic permeability and resistivity of the material which are affected by temperature. For steel (magnetic) at

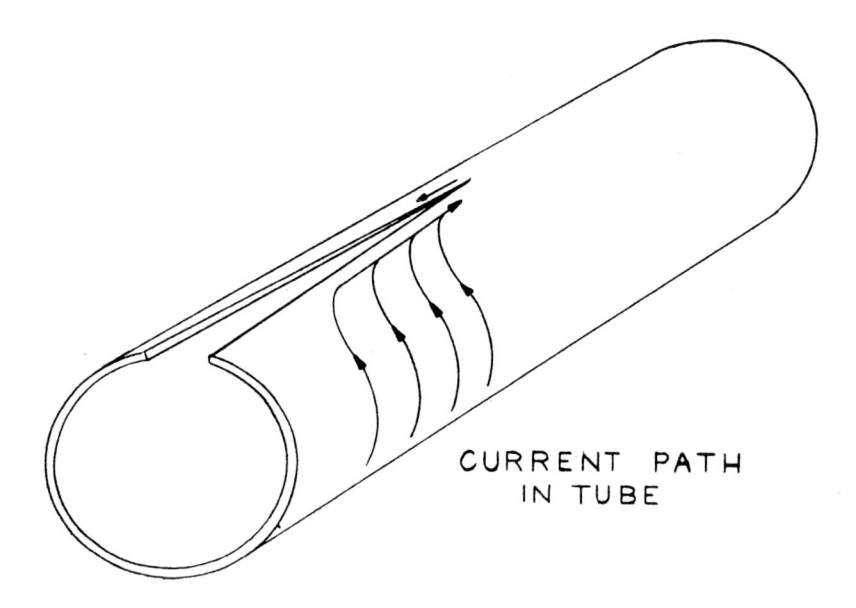

Fig. 2. Current path in tube produced by encircling inductor.

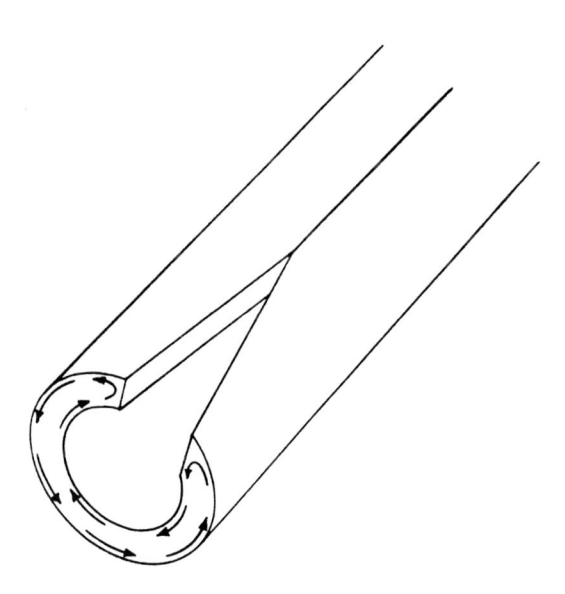

Fig. 3. Current path in strip cross section produced by encircling inductor.

room temperature

$$D = \frac{2.3}{\sqrt{F}}$$

D - Depth in inches
F - Frequency in c.p.s.

At welding temperature this becomes $D = \frac{23}{\sqrt{F}}$

Radio frequency oscillators are generally used with such inductors; however, 10 KC from motor generator sets can be used provided the depth of current is less than half the strip thickness.

Further details of the encircling type of inductor are covered in a later portion of this article.

HIGH FREQUENCY INDUCTION RESISTANCE WELDING

Chambers' Technical Dictionary, printed in 1961, defines resistance welding as: "Pressure welding, in which the heat to cause fusion of the metals is produced by the welding current flowing through the contact resistance between the two surfaces to be welded, these being held together under mechanical pressure."

This is exactly what happens on a high frequency induction resistance weld mill using an encircling type of inductor.

The current which is induced to flow by the inductor (Fig. 2) finds a path across the point at which the strip edges join and the resistance to this current flow results in intense I^2R losses responsible for the heat necessary to produce a weld. It is because there is a flow of current across the point at which the strip edges come together that the method of welding discussed in this article is "resistance." Further, when the current has been caused to flow by virtue of the tube being placed within the confines of an encircling type inductor energized from a high frequency power source, the method of welding is called "High Frequency Induction Resistance" as contrasted to "Contact Resistance" where the high frequency is introduced through electrical contacts.

ADVANTAGES OF HIGH FREQUENCY INDUCTION RESISTANCE WELDING

Many advantages are obtainable with high frequency induction resistance welding. These include:
1. Increased welding speed.
2. Hot rolled strip can be welded without cleaning.
3. All kinds of metal can be welded—both ferrous and non-ferrous.
4. Thin walled material can be processed.
5. Extreme flexibility and low maintenance.
6. Improved quality of product.
7. Greater efficiency.

Increased Weld Speed

When a frequency of 60 cycles, or even 180 cycles for that matter, is used with the conventional resistance method, the speed of welding is limited because of the well-known "stitching" effect which appears and which is determined by the frequency.

With high frequency welding, this is eliminated and tube mills are in production processing tubing at the rate of 1000 fpm. Before this was an accomplished fact, a considerable amount of mill redesign was necessary for operation at these higher speeds and the present speed is limited only because of the mill—primarily the cut-off.

Surface Condition

The low-frequency contact-resistance method requires a clean surface. Thus, it is necessary to use strip which fulfills this specification; that is, a hot-rolled strip which has been pickled or one that has been cold rolled. At radio frequencies, however, this is not always a factor and usually no surface preparation is required. With the induction resistance method there is no contact made by the "inductor" with the material; therefore, it is practical to weld regardless of surface condition, e.g., even hot-rolled skelp. Edge preparation from a splitting operation is recommended in all cases so that the abutting surfaces are clean.

The cost savings to be gained because of this advantage are not applicable if the tubing produced must eventually be cleaned before subsequent use; that is, if plating or other operations follow.

Material

Conventional resistance mills have been in successful operation for years on ferrous materials, generally low and medium-carbon steels. Certain other materials such as alloy steel, stainless steel, aluminum, bronze, and brass had never been satisfactorily processed. Radio frequency induction resistance mills are now a reality on all these materials.

Thin-Walled Tubing

With rotary electrodes, the heat pattern is generally broad. Thus, it had been practically impossible to weld very thin-walled (less than approximately 0.030") tubing, especially since it cannot withstand the extreme pressure applied. This is particularly true with those materials having a narrow plastic range such as aluminum.

The high frequency technique has made it possible to weld tubing but .005" thick.

Flexibility and Maintenance

The same high frequency induction resistance mill can be used for a wide variety of materials so that a given manufacturer can produce a wider product range. Since there is no contact made with the tube by the inductor as in the case of conventional electrodes, there is no refinishing or replacement cost involved.

Product Quality

The "no contact" technique (induction resistance) makes it possible to
1. Produce highly polished tubing with no scratches.
2. Produce tubing from coated or galvanized strip.
3. Produce tubing from "patterned" surface strip.
4. Eliminate risk of weakness due to surface marking. This becomes most important with the growing trend to large diameter thin-walled pipel.
5. Produce tubing more suitable for subsequent processing.

6. The narrower heat-affected zone lowers percentage of hardened metal in the weld zone. This results in less distortion in the postweld swaging or bull-dozing types of fabricating operations.
7. Improved weld reliability—since the strip edges are heated to fusion temperature while in their vee pattern prior to entering the pressure rolls, edges have time to reach a temperature equilibrium before being joined.
8. Complete absence of copper contamination—such contamination from low-frequency rotary-electrodes can prevent proper case hardening of parts made from tubing so fabricated. It can also cause intergranular cracking in certain high-temperature applications such as welded tubing used in heat exchangers and condenser tubes.

Efficiency and Frequency

The high frequency current tends to flow on the strip edge and the higher the frequency the more pronounced is this effect. Even without this advantage, the conventional electrode method contacts a greater area. In any case for a given wall thickness and weld speed, a much smaller amount of metal is heated with the high frequency method. As a result, power requirements are reduced representing a substantial savings in operating cost and capital outlay.

Further, with radio frequency oscillators, the current for even equivalent power is substantially reduced due to the high voltages inherent with the radio frequency. At 500 KW for example, the reduction in current can be as much as 100:1. With the lower current levels, cooling problems are substantially reduced.

High frequency induction resistance welding has been successfully accomplished at both 10 KC and 400 KC. As stated earlier in this article, 10 KC can be used on heavy walled pipe. However, the greater amount of metal heated because of the greater depth of current flow at the lower frequency results in the use of more power than at 400 KC for the same weld speed on the same pipe or tube. This more than offsets the higher initial cost per KW and lower operating efficiency of the radio frequency equipment. As a result, the high frequency induction resistance method has been almost exclusively R.F. and the balance of this article will deal only with the Radio Frequency Induction Resistance Method. Small oscillators (below 100 KW) are operated at approximately 400 KC. Larger units (several hundred KW) are operated at 300 KC.

EQUIPMENT

An installation for radio frequency induction resistance longitudinal seam welding of tubing or pipe will consist of a device for converting strip into a C-shaped form, bringing the edges together in proper alignment, providing heat and pressure at the point of convergence of the strip edges, scarfing of excess weld upset, sizing, maintaining straightness and cutting to length—all under uniform conditions of speed and weld power input (Fig. 4).

This, then, may consist of a conventional low-frequency mill which has been converted to high frequency, or a mill installed initially with high frequency as the weld power source. (Mill operating practice, product requirements, and customer preference introduce variables with which it is sometimes difficult to cope.)

The power source is a radio frequency oscillator which takes line frequency power, converts it to a high voltage, rectifies it, applies the D.C. to an oscillator tube and generates a high frequency output of around 300 to 400 KC. It is these same radio frequency oscillators that have been so successfully used for many years for a wide variety of metal-heating operations such as brazing, surface hardening, and heating for forging. Radio frequency oscillators are now available for welding with power outputs up to 600 KW due to the continual development of new electronic components having higher electrical capabilities.

Much of the oscillator component development can be attributed to the progress made in other fields such as radar and high power broadcasting. For example, the large oscillator tubes used today were originally used for radar pulse applications in the megawatt (one million watts) power region. The advances made in the past few years with high current semi-conductors, which now also have higher voltage ratings, make them more feasible for use in the plate power supplies

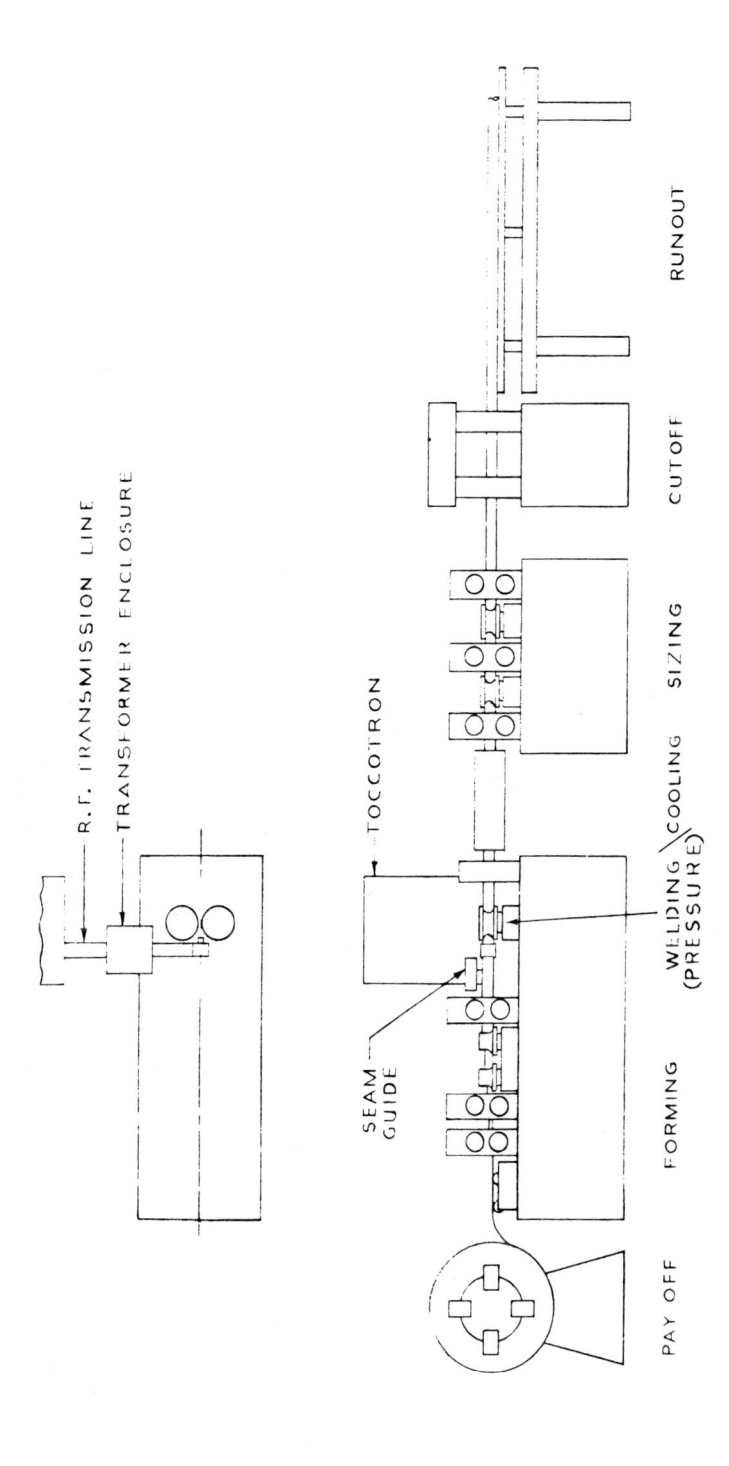

Fig. 4. Typical radio frequency induction resistance longitudinal seam welding installation.

necessary to operate these oscillators. Advancements in transformer and capacitor designs have led to higher power capabilities at lower costs and physical size.

The oscillator can put out its maximum power only if it is properly "matched" to the load presented by the tube or pipe. This is accomplished with series tuning coils, parallel tuning capacitors and an output transformer. The current design and adjustment of these components is vital to the successful operation of any installation.

The output of the oscillator must be controllable to allow for different power requirements depending on material, speed, and wall thickness. Further, steps must be taken to prevent changes in this power output once the required level has set.

The equipment details are covered under the following headings:
1. Basic Oscillator
2. Protective Features
3. Power Control
4. Line Voltage Regulators
5. Transmission Line
6. Output Transformer
7. Inductors
8. Impedors

Basic Oscillator

Although an industrial oscillator operates on the same electronic principles as other types, it has some physical differences because of the ultimate application.

Broadcast oscillators, for example, are (1) situated in clean, cool atmospheres that are fairly dry, (2) operated by specially trained personnel, (3) receive the best of maintenance by skilled hands, and (4) operate steadily at conservative power levels.

Industrial oscillators, on the other hand, are generally situated in an industrial atmosphere, subject to heat, smoke, oil fumes, dirt, and physical abuse. Ambient temperatures may in some cases reach 110°F. They are operated by ordinary factor labor and generally are serviced by semi-skilled help when a breakdown occurs.

Sometimes the time allocated for preventative maintenance is skimpy, especially if the equipment is used on a high production basis. Also, it is common practice for the people involved to have little or no knowledge of the operating theory of these oscillators.

To compensate for the physical and electrical abuse that industrial oscillators receive, they must be built with a greater component safety factor. Component ratings must be more conservative to withstand overloads and transients. The mechanical durability of external components must be capable of withstanding unusual physical stresses as well as severe surface contamination in order to cope with the worst conditions that might be encountered. Totally enclosed heavy gage and pressurized cabinets are a prime requisite.

From an economical standpoint, the increased cost of a more durable oscillator for industrial use is more than offset by the reduction of unnecessary and unscheduled equipment downtime. This is especially true when the total cost of such unproductive time is considered. In some instances, downtime can amount to $4,000.00 per hour.

Because of the increased exposure of personnel to the hazards of high voltage, certain safety precautions have to be observed with industrial oscillators. An oscillator design must have some flexibility to accommodate all of the plant situations that it might encounter, since various versions of "safety" exist throughout industry today. While one plant is very rigid in what is acceptable equipment, other plants are lax and accept a more simplified equipment design to minimize the cost. The minimum safety standards, however, cannot be ignored in any case.

Protective Features

All critical points in the oscillator are either water or air-colled. Protective switches guard against high temperatures or loss of coolant flow through key components. Air and water flows are monitored where necessary with protective limit switches to insure that adequate heat removal capacity is available for all thermally sensitive components during normal operation.

To insure long tube life and achieve proper operation, the voltage applied to the tube filaments must remain constant

when changes in line voltage occur. A regulation system not only provides this stability, but also insures against damaging high current surges which can occur when filament power is first applied. The following specific protective devices are used:
1. Plate overload relay.
2. Grid overload relay.
3. Flow switch in oscillator tube cooling drain.
4. Flow switch in R.F. tank and output circuit cooling drain.
5. Thermal switch in oscillator tube cooling drain.
6. Water pressure switch.
7. Time delay relay to prevent the application of high voltage power to the oscillator tube until the filaments have reached their proper operating temperature.
8. Door interlocks remove high voltage plate power when access doors are opened.
9. Control fuses.
10. Filament circuit fuses.
11. Shorting stick for use by maintenance personnel.
12. On high powered units an additional safeguard is provided—either a vacuum switch or a "crowbar." Since this latter type of device may be new to the readers of this article, the following information is included.

Power vacuum tubes are subject to internal arcing if subjected to high dv/dt transients, which may occur if the r.f. oscillations are suddenly interrupted by a fault, or to corona discharge occasioned by physical contamination left within the tube at the time of manufacture. If the tube does arc internally, excessive current can flow within the tube. These currents can be of a magnitude that will make the tube inoperative or they can simply reduce its emission characteristics and impair operating efficiency. In low power devices where the impedance of primary source is relatively high, primary circuit breakers and/or over current trip line contactors are employed. These devices remove the energy source from the system that feeds the tube. In other instances current limiting resistors are employed to prevent damaging currents from flowing within the tube in the event of an internal arc.

The above protective devices are employable with reasonably good assurance of safe operation provided the output power of the oscillator tube is 150 KW or less. On oscillators of powers exceeding 150 KW, vacuum switch circuit breakers or crowbars should be employed. The normal operation time of a vacuum switch is the order of 8 to 9 milliseconds. Crowbars are available and operable in the same general time area. The difference between the two approaches is the vacuum switch breaker removes the equipment from the primary power source by opening incoming power lines. The crowbar physically short circuits the power supply, removing the anode potential from the oscillator tube and at the same time causes the line contactor to interrupt primary power to the supply. The crowbar is considered **the** ultimate in assuring reliability of tube operation and protection and is particularly recommended in high power installations. The crowbar is valuable in smaller powered oscillators if considerable system interruptions are anticipated which will generate high transient voltages that might lead to internal tube arcing.

Power Control

Power control of the oscillator can be obtained in a number of ways. Low power machines often use control of the D.C. rectifiers. In high power machines, this is not desirable because of the voltage transients that may take place. For these, either saturable reactors or regulating transformers are placed in the line ahead of the plate transformer and weld power control from 100% down to a low value (essentially zero) is obtained.

Line Voltage Regulators

Since plant line voltage may vary from hour to hour or even minute to minute, it is usually desirable to supply a line voltage regulator. This device will maintain a stable output voltage with a variation of ±10% of the input voltage. If power control is also desired, it is possible to combine these two features into one unit and have the ability to control weld power from 16 to 100% and regulate within ±1% at any point within this range. When a wide range of products is to be welded on a single mill, this arrangement is preferred.

An instantaneous voltage drop can occur when a large current drawing piece of equipment is thrown across the line. No voltage regulating system has a quick enough response to maintain constant voltage to the oscillator during such periods. While the effect on weld quality will not be evident in most applications, it could not be tolerated for high pressure stainless steel tubing or high quality brass, copper, or even aluminum tubing. In such cases, the only answer is the use of a 60 cycle to 60 cycle voltage regulating motor generator set which completely isolates the oscillator from the line. The MG set has enough inertia so that such input voltage changes have absolutely no effect on the output.

Transmission Line

It is generally desirable to place the oscillator as close to the point of weld as possible to reduce any losses that might occur in the transmission line. However, it is practical to remotely locate the rectifier cabinet by a distance of several hundred feet and transmit the D.C. power to the oscillator by means of regular high voltage X-Ray cable. Such cable should be selected so that if any physical damage does occur, the voltage is completely contained within the cable and no injury to personnel can result. It is possible to separate the oscillator from the mill by a distance up to 20 feet. The circuit between the oscillator and the output transformer sometimes carries currents as high as 800 to 1000 amperes and the line must be carefully designed to keep losses to a minimum.

For a distance of 4 feet or so, two parallel lines consisting of regular water-cooled copper tubing can be used. Distances greater than 10 feet, however, require a co-axial line and the design of this should be done by someone skilled in the field. Such co-axial lines are available from equipment manufacturers. If a parallel line is constructed, it is generally shielded for two reasons. The first is safety as any inadvertent contact with these lines while the power is on could result in a serious burn. Additionally, the Federal Communications Commission has Rules and Regulations regarding radiation that place a limit on allowable radiation. In order to keep the radiation from the parallel lines within the limits, shielding is generally required.

Although the FCC Certification is solely the responsibility of the user, the manufacturer provides all necessary forms and technical advice and, on package units, type certification.

Output Transformer

The oscillator operates best when it is matched to the load. A transformer performs its part of this matching function by transforming the high voltage at relatively low current from the tank circuit to the low voltage and high current required in the welding inductor. Depending upon the size of the oscillator, different types of transformers are used. However, they are all air core, water cooled, isolating transformers. To obtain optimum results, their design will vary as a function of the diameter of the work to be welded and the material of the welded product. The reason for this is that the resistance presented by the path the current takes in the tube or pipe will vary considerably from, say, magnetic steel to aluminum although it is possible to weld either material with the same inductor and transformer. To gain maximum efficiency, it would be desirable to use transformers of different ratios. Generally, the range of sizes of tubing to be welded on a mill is within the capability of a single transformer. If, however, two similarly sized oscillators are used to weld, say, from 1 to 3" on one mill and 4 to 8" on another mill, transformers of slightly different design would be provided for each of the mills.

The solid potted type transformers used represent a considerable improvement over earlier designs. The insulation is poured as a liquid into the transformer as it is held in a form. It is then cured until solid and maintains the physical relationship between the primary and secondary. After removal of the form, we have a superior product. No space between the primary winding and secondary winding is wasted by extraneous insulation or supporting structures. The net result is that we have a "tight" or closely coupled unit with low losses. Since the unit is solid, with no loosely held parts, it can be mounted in any position at the user's convenience.

Proper provisions must be made for housing and mounting the radio frequency output transformer on the mill. The inductor is attached to the transformer and accurate positioning with respect to the open seam and weld point is essential.

Therefore, as diameters and inductor are changed, three-way motion is essential.

1. ± 1" Vertically — to maintain proper air gap between the top of the tube or pipe and the I.D. of inductor and nose.
2. ± 1" Horizontally — to keep split in inductor and nose positioned directly over seam.
3. ± 1/2 Pipe Diameter Longitudinally — to properly position inductor with respect to weld point and weld pressure rolls.

Inductor

The purpose of the inductor is to transmit the energy from the transformer to the tube or pipe and to do this with the least loss of energy. It is designed to produce maximum current in the work and to concentrate this current in the edges of the strip as they approach the weld point. See Fig. 5.

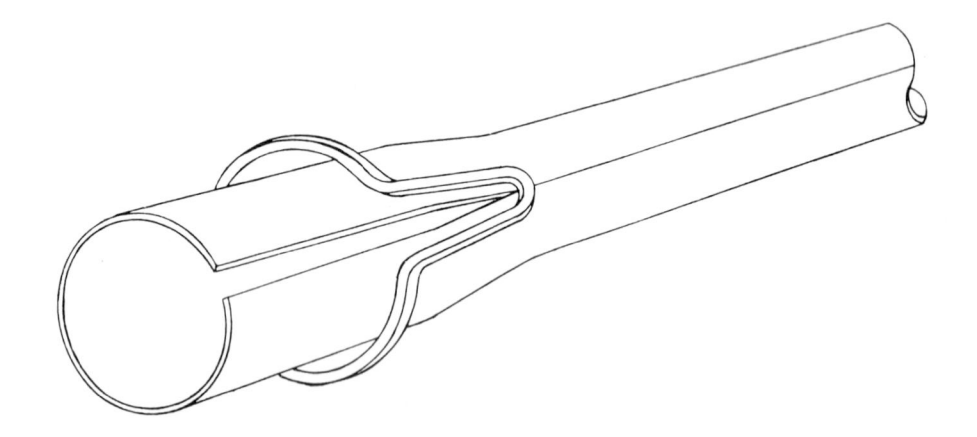

Fig. 5. Basic design of encircling inductor.

The inductor (see Fig. 6) is made wide as it encircles the tube. This portion is often referred to as the barrel. The reason for the width is to distribute the current in a wide path so that a minimum loss is incurred around the back of the tube or pipe. As the current reaches the open gap, it must flow along the edges toward the weld point and preheat

Fig. 6. Actual 6-5/8" diameter encircling inductor for R. F. induction resistance welding.

effect is obtained. The nose or hairpin extension may extend considerably as shown in Fig. 5 and covers the open seam just before the weld. It tends to aid in concentrating the current along the edges as they approach the weld point by virtue of the "Proximity Effect." This design is particularly effective below 2" diameter. As the diameter is increased and/or power goes up, the nose is shortened as its concentrating effect on the weld current is overshadowed by the dimensions of the pipe and as power increased, its size must be increased in order to keep it cool.

The inductor must be water cooled. Cooling consists of two or three independent paths of cooling water and it is most important that there is a high pressure supply to the nose;

80 to 100 psi should be available with unrestricted drains from the two tubes at the outlet of the nose. The inductor bus and barrel are cooled with water at normal pressures, 30 to 60 psi. This can be done by connecting to the plant water system where the system provides proper characteristics of pressure, temperature and hardness. In the event the existing plant system does not offer the proper requirements, a suitable system can be provided by the equipment manufacturer. (See Fig. 6.)

The inductor by itself is rugged; however, it can be damaged if it is struck by the leading edge of the tube as it is being threaded, or by malformed tubing entering the inductor and bumping it. It is generally coated completely with a thermo-setting resin to prevent entrance of mill coolants between the bus from the transformer to the inductor barrel. Accumulation of carbonaceous material in this area can lead to voltage breakdown and failure of the inductor. As supplied, the inductor is coated with an Epoxy resin. Various users may have other materials that more adequately meet their particular needs.

The single turn inductor is designed to provide the optimum coupling to the skelp and allow the user to realize maximum efficiency in the operation of his pipe welding installation. The single turn inductor geometry provides a maximum of adjacent coupling surfaces between the inductor and the skelp. This characteristic lowers the effective resistance of the inductor, minimizing I^2R losses and maximizing induced currents in the skelp. Because of the aforementioned geometry and the resulting electrical characteristics, the voltages appearing on single turn inductors are substantially lower than those apparent on other designs. This results in less potential arcing between inductor and pipe and produces pipe unmarked by spark or arc burns. Multiturn inductors made from copper tubing are often used but they are not always as effective in the tube welding operation as the single turn inductors with the nose or hair-pin extension. While inductors made from tubing are less expensive, they are at the same time somewhat "fragile." The rugged single-turn inductor can absorb a substantial blow from the leading edge of the tube as it is being threaded or if a strip butt weld breaks. Under such circumstances, the multiturn inductor would be completely destroyed.

Impeder

When welding non-ferrous material under 3/4" diameter and for all sizes of ferrous material, it is necessary to use an impeder to maintain optimum weld efficiency. The purpose of the impeder is to increase the welding speed by reducing the current that flows circumferentially on the inside of the tube as shown in Fig. 3. It can do this best when it completely fills the volume within the tube and is positioned as close to the seam as mill practice will allow. The impeder length is such that it extends upstream from the inductor at least one tube diameter and downstream beyond the weld point at least 1/4". As diameter increases it becomes less important to place the impeder material in the bottom half of the tube and from 6" diameter and above, only the upper third or quarter sector may be required to obtain optimum benefits.

A ferro-magnetic material gives the best compromise between operating characteristics and electrical characteristics. This ferro-magnetic material may be either a powdered iron or one of a family of magnetic oxides called ferrites supplied in rods or shapes which are built into an impeder assembly to suit the specific job. In any event, for the material to function effectively it must be kept cool. Specific designs for accomplishing this are proprietary.

The holder for the impeder must be custom-designed to meet the requirements of each installation. For example, if no further processing is to be done, the impeder can be supported on a wire and left to drag in the tube with cooling supplied by coolant within the tube. It may also be placed in a laminate tube holder and force cooled. If scarfing or seam rolling is required, the impeder must be designed around the existing rods or new rods must be designed to allow proper impeder operation.

FACTORS AFFECTING WELD SPEED

1. Material and Product Characteristics
2. Mill Performance
3. Welder Performance

Material and product characteristics are established by user requirements and these can vary over an almost limitless range of combinations. Therefore, it is essential that the mill and welder design be carefully engineered to tailor them to the specific material and product requirements. Only by properly balancing and effectively controlling the many variables which are available to him can the designer be certain that he will provide a quality weld at the highest speed practical. This, of course, is most convenient to do with new mills where mill and welder designs can be integrated from the outset.

Material and Product

The material being welded plays an important role in determining the power required to obtain specific speeds. Nonferrous materials such as aluminum, for example, can be welded at significantly higher speeds than steel at a given power input because of the lower welding temperature.

Welding speed of the installation is also governed by wall thickness, which directly affects the mass of metal which must be heated. While it might at first seem that the relationship between wall thickness and speed would be linear, this is not always so because at low speeds heavier walls do take slightly higher power per unit thickness than do thin walls. This is due to the greater thermal conductivity of the heavy wall material and is illustrated by the fact that power required for welding does not decrease directly as mill speed is reduced for a given thickness of material since the conduction losses away from the weld zone are greater at the slower welding speeds. On the other hand, at high weld speeds this condition is not effective. There are certain electrical systems loss regardless of speed and at high power level these become smaller percentage-wise. As a result, doubling of weld power will more than double weld speed.

High pressure boiler tube requires a higher weld quality than structural tubing. Weld speeds for the latter can be higher for equivalent stock size and weld power. The same is true for ornamental stainless when related to that used for re-draw operations and pressure tubing.

Diameter of the pipe or tube, in general, has little effect on the weld speed since this is accommodated in a specific design of the inductor and impeder. Likewise, surface condition of the material being welded does not affect the speed since no contact is made with the material, but rather the radio frequency induces the welding power directly into the material regardless of its surface condition. Fig. 7 provides weld speeds for steel pipe, Fig. 8 for aluminum, and Fig. 9 for copper for 1" diameter and larger. For smaller diameters, some decrease in speed will result.

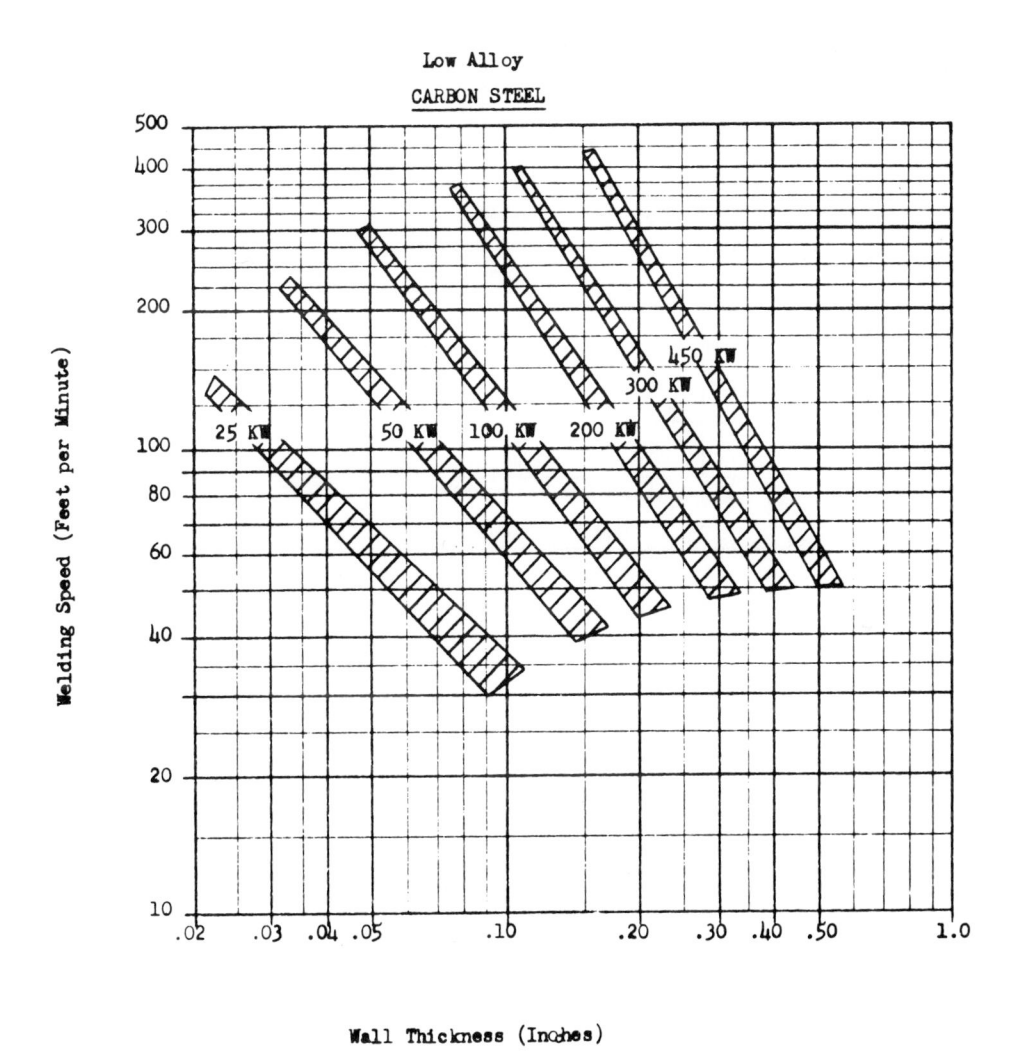

Fig. 7. R. F. induction resistance weld speed vs. wall thickness for low alloy or carbon steel for various power levels.

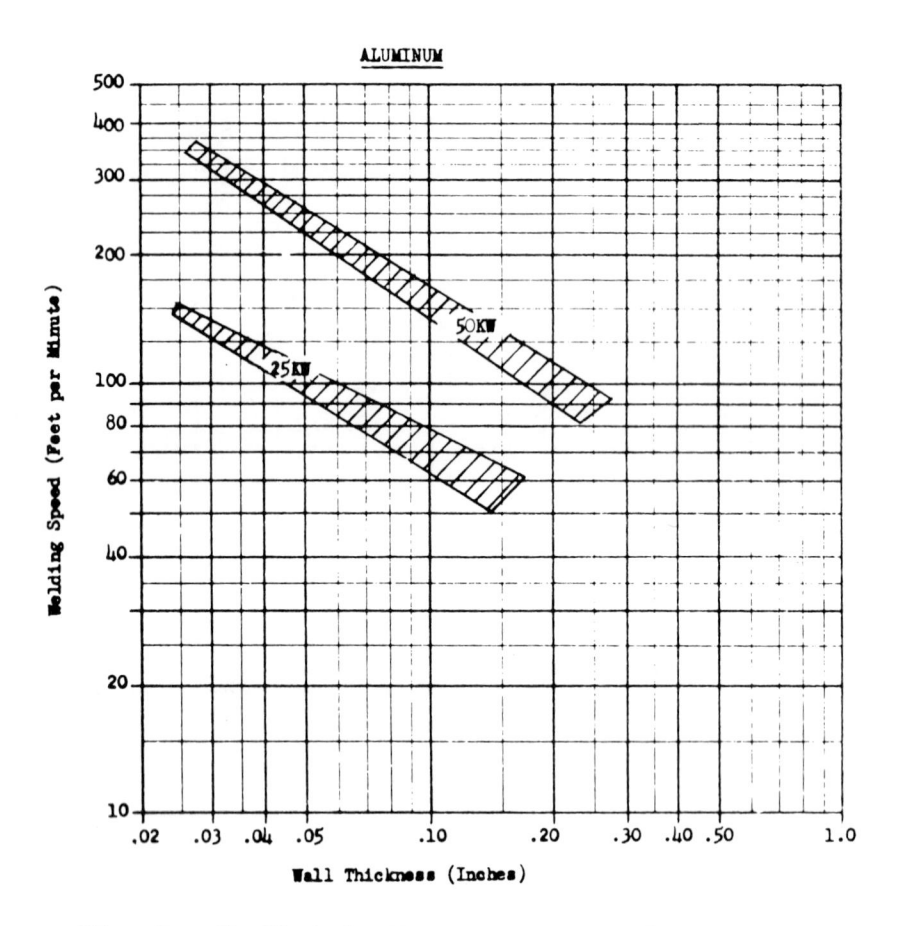

Fig. 8. R. F. induction resistance weld speed vs. wall thickness for aluminum.

Mill Performance

The requirements for mill practice are not significantly different from those of low frequency contact resistance, except as they may be affected by the higher welding speeds and the lack of overhead pressure at the weld. The induction resistance welding system is more tolerant of occasional malformations in the tube than any other process. Laps, ripples, and wandering of the seam that would be disastrous with other methods, although they may affect the weld quality, will pass through the inductor and weld stand without difficulty. The tolerance in the width of the strip is only important in its effect on the pressure at the squeeze rolls. A nominal maximum variation of .005"/inch of width will generally produce a reasonable result.

COPPER

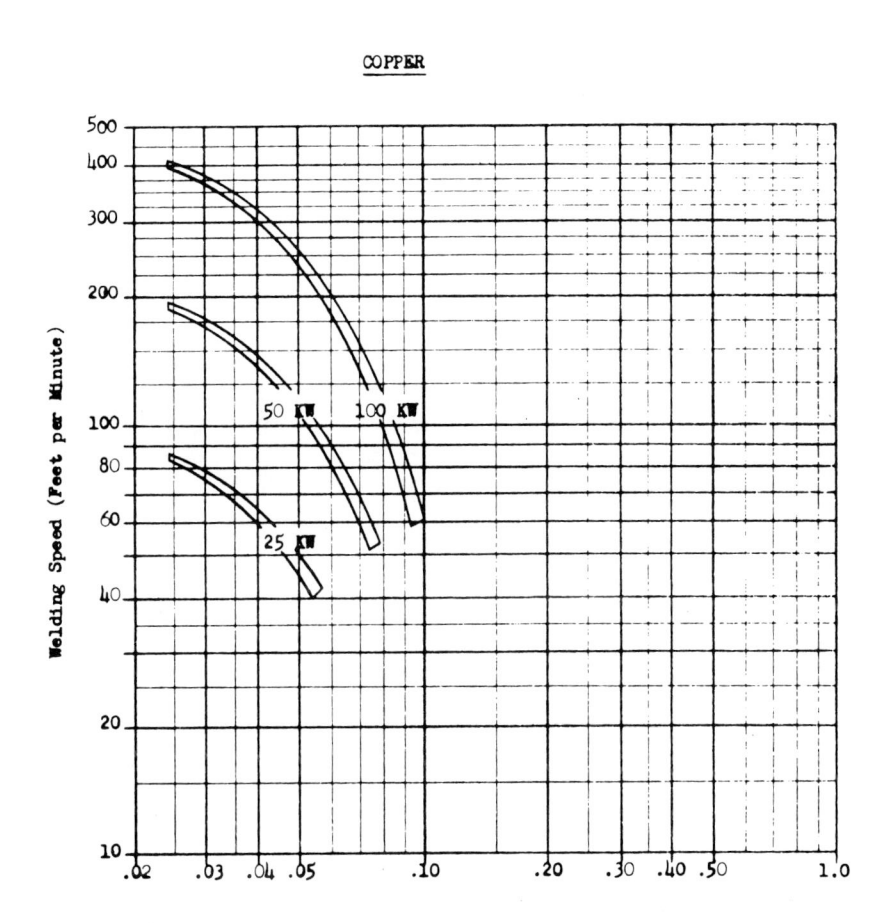

Fig. 9. R. F. induction resistance weld speed vs. wall thickness for copper.

The welding process requires more than heat alone; it is also dependent upon the manner in which the metal is brought together, the pressure applied, the consistency of the mill itself, and the control which can be maintained over these mechanical factors. The tube mill must be capable of forming the tube or pipe so that the metal flows smoothly without instantaneous variations in speed from the flat skelp to the welded tube.

Source: *Mechanical Working & Steel Processing 4*, Gordon & Breach, 1969

In forming the strip, it is essential that the edges come out of the final fin pass fully formed so that a minimum amount of pressure is required at the weld rolls to bring the edges together for the weld. The edges should be sufficiently clean to prevent inclusions in the weld and should be parallel to each other at the weld point. The forming should be adjusted so that there is no tendency of the weld line to drift radially as the welding progresses, although the induction resistance process is inherently tolerant of some drift.

In accomplishing the forming operation the axial movement of the weld juncture or point must be held to -0.000/+.030" for carbon steels and to -0.000/+.015" for alloys and non-ferrous materials. The seam must be kept straight and circumferentially fixed within ±4° of the 12 o'clock position. Vertical and lateral skelp stability must be maintained within ±.005" from initial forming to weld juncture.

Between the fin pass and the weld point, it is sometimes desirable to use a seam guide for the purpose of additional control of seam stability, both in a radial direction as well as controlling the axial location of the weld point. This seam guide is preferably a blade with insulated carbide inserts that rub on the edges of the open seamed tube or a rotating roll similar to a fin pass; however, the guide used must not short circuit the edges. Its mounting, to be most useful, should provide for adjustment for positioning it perpendicular to the tube axis, axially along the tube length, as well as horizontally. The convergence angle formed by the edges of the tube as they approach the weld point should be between 4° and 7°

The sizing mill should supply a tension on the tube by operating at a higher speed than the forming mill to assure maintenance of the limits set forth above.

It is preferable to have the tube welded in the round shape and approaching skelp edges should be parallel outside to inside from the last forming fin pass to the weld juncture.

Various combinations of weld rolls, material and configurations can be used successfully with induction resistance welding.

Although no hard and fast rules can be given, the most practical arrangements as a function of tube diameter are listed below. In all cases, it is recommended that the rolls be insulated from the mill bed and each other to prevent passage of stray currents through the bearings. Recommended maximum diameters:

    Tube 1" Roll Diameter, 3"
    Tube 4" Roll Diameter, 7"
    Tube 8" Roll Diameter, 14"

As a general rule, the diameter of the rolls should be as small as possible consistent with bearing diameters and scuffing of the tube. This will allow the inductor to be placed nearest to the weld point and provide the greatest productivity. For tube diameters below 2", the weld-rolls, to obtain maximum speed, should be made from non-magnetic materials. The hard bronze materials have found wide acceptance. Above 2" diameter, the rolls may be made of either magnetic or non-magnetic materials; however, slightly reduced speeds and heating of the rolls may be experienced if magnetic large diameter rolls are used. This is due to the induction heating of the roll edges in proximity to the inductor nose and barrel.

The weld pressure rolls should be designed to do as little forming of the tube as possible and should be used only to position the tube and apply the pressure for welding. The weld pressure rolls should not be used to correct for poor forming or to augment the forming section of the mill.

The basic design of weld roll assemblies varies considerably between different manufacturers and users; however, as previously recommended, the weld rolls should be insulated from the shaft and stand to minimize the heating of the rolls from stray currents.

The amount of weld pressure or squeeze is that amount required to bring the two approaching edges of the strip in full contact with each other to provide for a uniform forging effect. An increase or decrease in this pressure will affect this forging effect. An increase or decrease in this pressure will affect this forging action which will directly affect the quality of the weld. Strip width and weld roll concentricity will have a direct effect on the weld pressure.

The seam guide or seam separator can also be used very effectively to provide and maintain the proper convergence angle in practically all cases of tube welding. The seam guide support should be made adjustable, so as to provide the three-way adjustment. In the case of some pipe installations, the use of a seam guide is not practical. The last fin pass of the

Source: *Mechanical Working & Steel Processing 4*, Gordon & Breach, 1969

forming section followed with a pair of horizontal rolls is used to control the convergence angle.

There are generally two types of weld roll stands -two-roll and multiple-roll. Regardless of the type, the supporting structure should be insulated from the mill bed and the rolls from the stand to prevent flow of stray current. It should also be kept as high as possible to reduce the amount of metal which may pick up flux from the inductor. Both factors, if disregarded, can waste useful power.

The two-roll stand consists of two horizontal acting squeeze rolls. The basic design of the weld stand used is quite similar to that employed in low frequency contact welding. The roll must be a full containing type roll. In other words, the roll should have a full top flange with the minimum of clearance between flanges to allow for the outside upset. The top flange of this type of roll should not be used to form the strip, but should be used to support and maintain alignment of the top edges during the welding operation.

The multiple roll stand consists of a group or cluster of three or more rolls. Three rolls at 120° or four or five rolls so positioned so as to have two horizontal squeeze, bottom support and two independently adjustable rolls to support the top edge of the tube are examples of this type of setup. The advantages of this type of stand are more control of the tube at the weld point and smaller diameter rolls, along with the fact that the possibilities of marking or scuffing are reduced. Disadvantages are complexity of adjustment and bulk of the support housing.

Scarfing of the I.D. and O.D. can be performed in a manner similar to standard practice. The only variation might be in the design of the bar for the I.D. scarfing as mentioned in the discussion of Impeders.

Sizing by the rolls downstream should follow standard practice. However, as stated previously, in the induction resistance method of welding it is particularly important that sizing mill always attempts to run slightly faster than the forming mill so that tension is maintained in the weld area. In this way, the desired stability of the weld will be more easily obtained.

Welder Performance

The smaller the spacing between the pipe and the inductor, which is called "coupling," the greater will be the power which is induced into the pipe. Nominal inductor I.D. will be 10-20% (Maximum) greater than the O.D. of the tube at the weld point. However, operating practice may require that this coupling be increased and a reduction in welding speed may result. The operator of the mill must determine for himself the best balance between these two conflicting factors Generally, up to 1" diameter, a coupling of dimensions of 1/16" is reasonable from 1 - 2; 1/8" from 2 - 3; 5/16" from 4 - 8; and 1/2" or more above 8". It is important that this spacing, particularly at the "nose," be retained at the minimum possible in order to maximize speed. Thus, an inductor is generally designed to weld a particular size of pipe. The inductor can be used to weld smaller diameters than that for which it was designed, but welding speed will be reduced —all other things being equal. The extent to which an inductor can be used on smaller sizes is dependent upon the speed required for the smaller diameters. Further, the "nose" should be close to the weld point—usually a minimum of one and a maximum of two tube diameters.

To maintain weld speeds, impeder specifications must be adhered to as described previously.

Fig. 10 shows the weld section of a 300 KW unit on 6-5/8" x .238" wall pipe.

METALLURGICAL CONSIDERATIONS

The radio frequency welding method (either contact resistance or induction resistance) has made it possible to produce welded tubing from copper, brass, aluminum, cupro nickel and titanium. This was not possible with low frequency.

Low frequency welding has a limited speed. Radio frequency welding is limited only by the ability of the mill to properly form and cut off.

Heliarc (or similar) welding of stainless steel is a slow process—speeds expressed as inches per minute. Radio frequency welding of stainless steel is limited only by the ability of the mill to operate at a constant speed.

Fig. 10. Weld section of 300 KW R. F. induction resistance unit on 6-5/8" diameter x 0.238" wall steel pipe.

High frequency induction post weld heating using the same kind of inductor as described in the beginning of this article as a "non-encircling" welding inductor has been common on many low frequency mills producing line pipe. This operation was performed to normalize the weld and eliminate the martensite which tended to form as the weld cooled. It was never clearly established whether it was the greater tendency of the martensite to form in the higher strength materials demanded for the pipe or a merchandizing angle which forced the installation of post weld heaters on practically all of the line pipe mills. However, when these mills were converted to high frequency or a new high frequency mill was installed, the considerations of post weld heat took on another aspect.

As suggested, it was sometimes doubtful that the post weld heating actually produced an improved product from the low frequency mill. With the high frequency mill this doubt was completely eliminated. Because of the amount of metal heated for welding being so small, the chilling effect of the adjacent pipe material was always enough to take the weld down through the critical at a speed sufficiently fast to form appreciable quantities of martensite. Thus, with high frequency welding, post weld heating is a must for those applications where martensite can form in the weld and is objectionable or not removed in some other subsequent process such as hot stretch reducing.

Fig. 11 shows a typical post weld heating unit operating from two 500 KW, 1 KC generators. Fig. 12 gives post weld heating power requirements for various wall thicknesses.

Fig. 11. Dual head post weld heating package for installation on pipe mill. Each head powered by a 500 KW 1 KC generator.

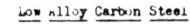

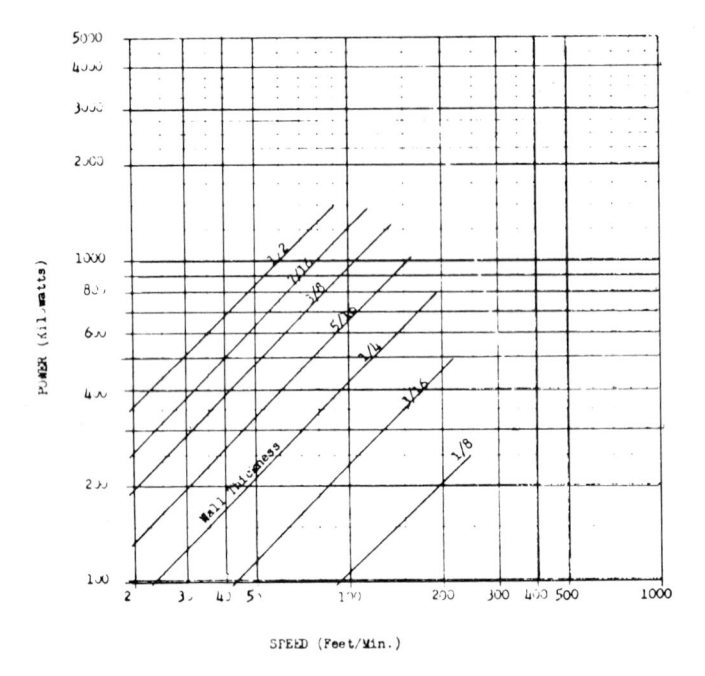

Fig. 12a. Post weld heating power requirements for various wall thicknesses for heating weld area from 750°F. to 1,750°F.

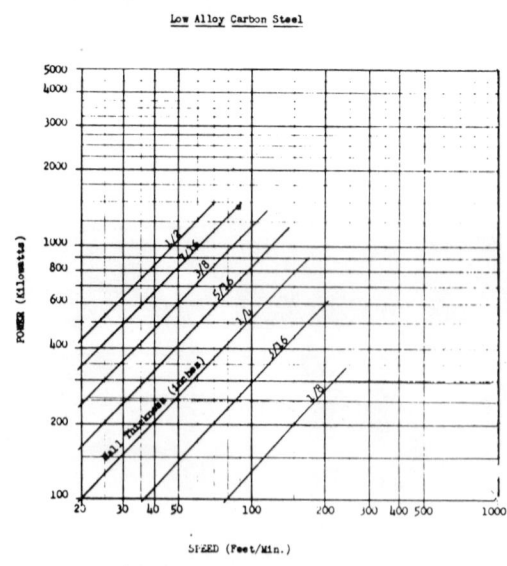

Fig. 12b. Post weld heating power requirements for various wall thicknesses for heating weld area from 200°F. to 1,750°F.

WELDING ON PIPE LINE RIGHT-OF-WAY

Although this has been reported previously, this article would not be complete without mention of the radio frequency induction Resistance Welder that produces the pipe on the right-of-way.

Conventional pipe lines are constructed from 40 feet lengths shipped to the field and welded together. A mobile unit was developed which "on site" takes 1,400 ft. coils, forms it into pipe, welds it into 8-5/8" diameter x 0.156" thick X-52 Grade, API line pipe with a radio frequency induction resistance welder, tests it ultrasonically and hydraulically and places it in the right-of-way. The equipment can be tooled to handle 10-3/4" diameter with 1/4" wall. A larger unit will handle up to 16" diameter.

The equipment was developed by the Mobile Pipe Corporation and uses a 150 KW TOCCO radio frequency welder (Fig. 13) The potential contribution to pipe line construction is enormous—it cuts freight and pipe laying costs and can eliminate stringing and line-up gangs. For example, it reduces end butt welds from over 100 per mile to 6 to 7. The greatest advantage of the mobile feature will be in areas far removed from standard mill facilities—shipping one compact coil instead of 25 fully formed 40-foot lengths of pipe.

Fig. 13. Mobile unit for producing 8-5/8" diameter line pipe on the right-of-way.

Source: *Mechanical Working & Steel Processing 4*, Gordon & Breach, 1969

    Equipped with a bending device, it can make horizontal or vertical bends as it moves to form the line to the contour and route of the right-of-way. As the steel moves through the mill, the mill moves on caterpillar tracks in the opposite direction at the same rate of speed.

    The radio frequency induction resistance (no-contact) method is especially well suited for use with a mill "on the move." Since it is tolerant of motion and vibration, a good weld is obtained despite movement of the mill over rough terrain.

HF contact or induction tube welding: which is better?

Each has advantages, but contact welding can use half as much power

by C. A. TUDBURY, *technical advisor, Thermatool Corporation*

Manufacturers of longitudinal butt seam welded pipe and tube use high frequency (400 KHz) welding. Some use induction welding, in which induction coils that do not touch the pipe generate the high frequency current required to heat the work to welding temperature. Other makers prefer contact welding. It uses small metal contact shoes that ride along the seam to introduce the HF current directly into the metal. These methods give significantly different current and heating paths.

Electrical "skin" and "proximity" effects, phenomena peculiar to high frequency current, make high frequency welding possible.

Skin effect is the tendency of high frequency current to concentrate in a thin layer of metal at the surface of the workpiece.

Proximity effect is the tendency of high frequency return current to concentrate nearest a return conductor in a path that mirrors the path of the entering current.

With the right geometry, the combination of the two effects gives a heat path concentrated in the surfaces to be joined. The greater the concentration of heat and the greater the accuracy of the path position, the greater the energy and production efficiency of the welding operation.

Contact welding

Contact tube welding is a practical use of the combined skin and proximity effects. The go-and-return conductor is the tube edges themselves. Tube diameter has no effect on the

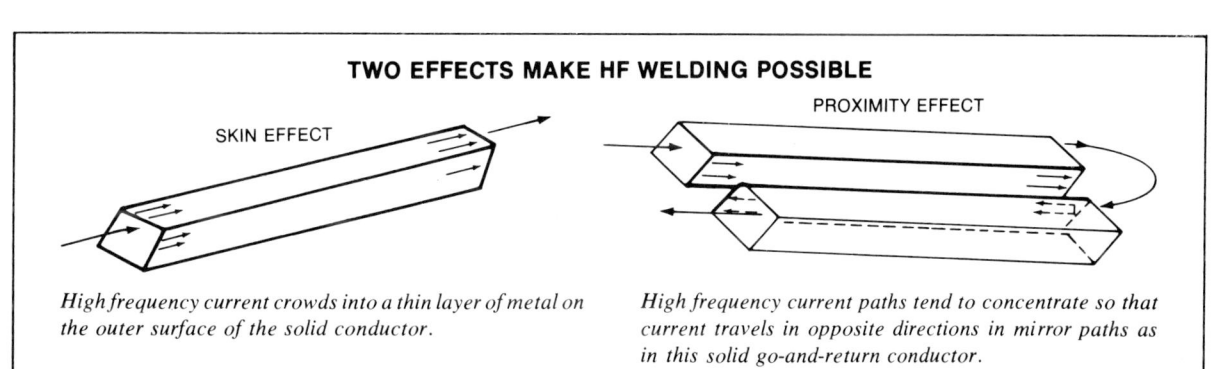

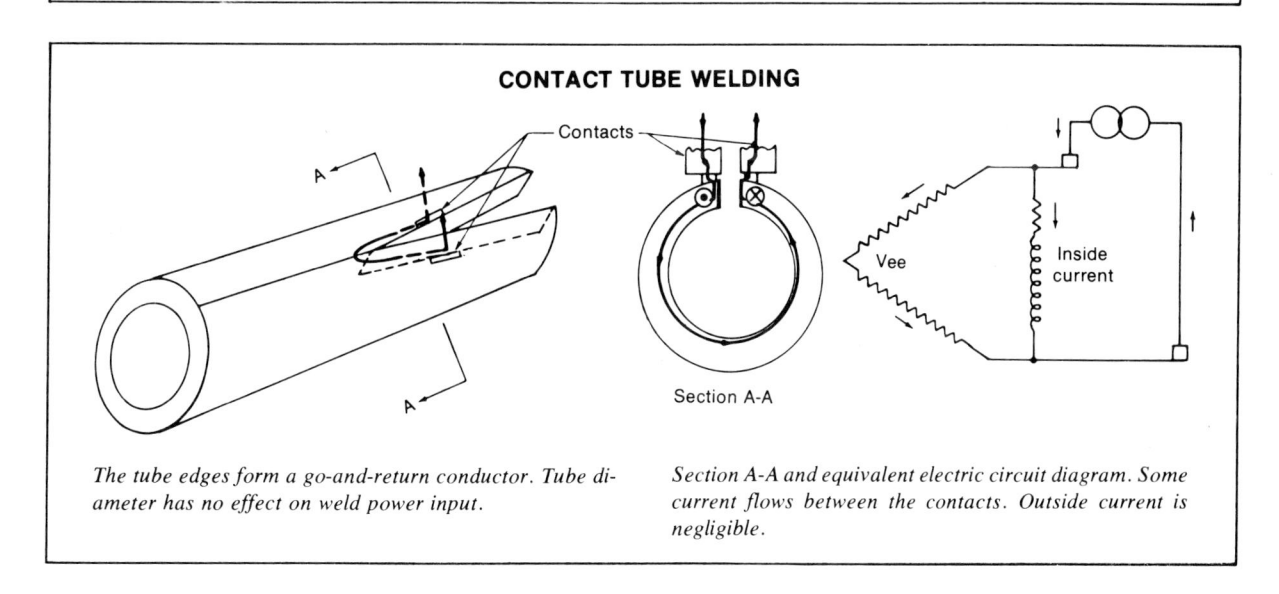

TWO EFFECTS MAKE HF WELDING POSSIBLE

High frequency current crowds into a thin layer of metal on the outer surface of the solid conductor.

High frequency current paths tend to concentrate so that current travels in opposite directions in mirror paths as in this solid go-and-return conductor.

CONTACT TUBE WELDING

The tube edges form a go-and-return conductor. Tube diameter has no effect on weld power input.

Section A-A and equivalent electric circuit diagram. Some current flows between the contacts. Outside current is negligible.

RESISTANCE WELDING

welding process. The same power input welds the same wall thickness of tube at the same speed, no matter what the tube diameter.

Some high frequency current can flow from one contact to the other around the inside of the tube. This current contributes nothing to the welding. The path between contacts around the outside of the tube carries negligible current, so power loss is small.

Induction welding

For induction heating an induction coil acts like the primary of a transformer, inducing a current that flows in the opposite direction all around the surface of the work. In induction tube welding the current flows along the inner surface of the tube and along the joint surfaces.

All of the current that travels through the vee for welding must also flow all the way around the tube in a continuous circuit. The path around the outside of the tube cannot be neglected. As the drawing of the equivalent electrical circuit shows, the outside current is in series with the vee (weld) current. The induced current flowing in parts of the pipe other than the vee edges in the tube is worse than useless, because it heats the tube and is a power loss. It causes a higher voltage requirement at the coil terminals.

What is an impeder?

When a bar of soft iron is placed inside a coil carrying high-frequency current, the coil induces a secondary current on the surface of the iron. This secondary current flows opposite to the coil current. Its effect is to cancel the induced magnetization of iron beneath the surface of the bar. In effect, this opposing surface current shields the iron center from induced magnetization by the coil. Even though the cross section of the iron has good magnetic permeability (it magnetizes easily), it cannot affect current in the coil. The iron has a low resistivity. Current will flow through it easily.

Unlike soft iron, ferrite has high magnetic permeability and high electrical resistivity. Resistivity is so high, in fact, that large currents cannot flow.

Impeders are ferrite cores covered by insulating tubes. When an impeder is placed inside a coil carrying high frequency current, the high resistivity of the ferrite prevents current induction, and the cross section is not shielded from the coil. The magnetic effect of the coil penetrates the entire cross section, letting the magnetic permeability of the ferrite core take effect, and increasing the inductance of the coil. Coil current with a ferrite core inside is lower than when the coil is empty.

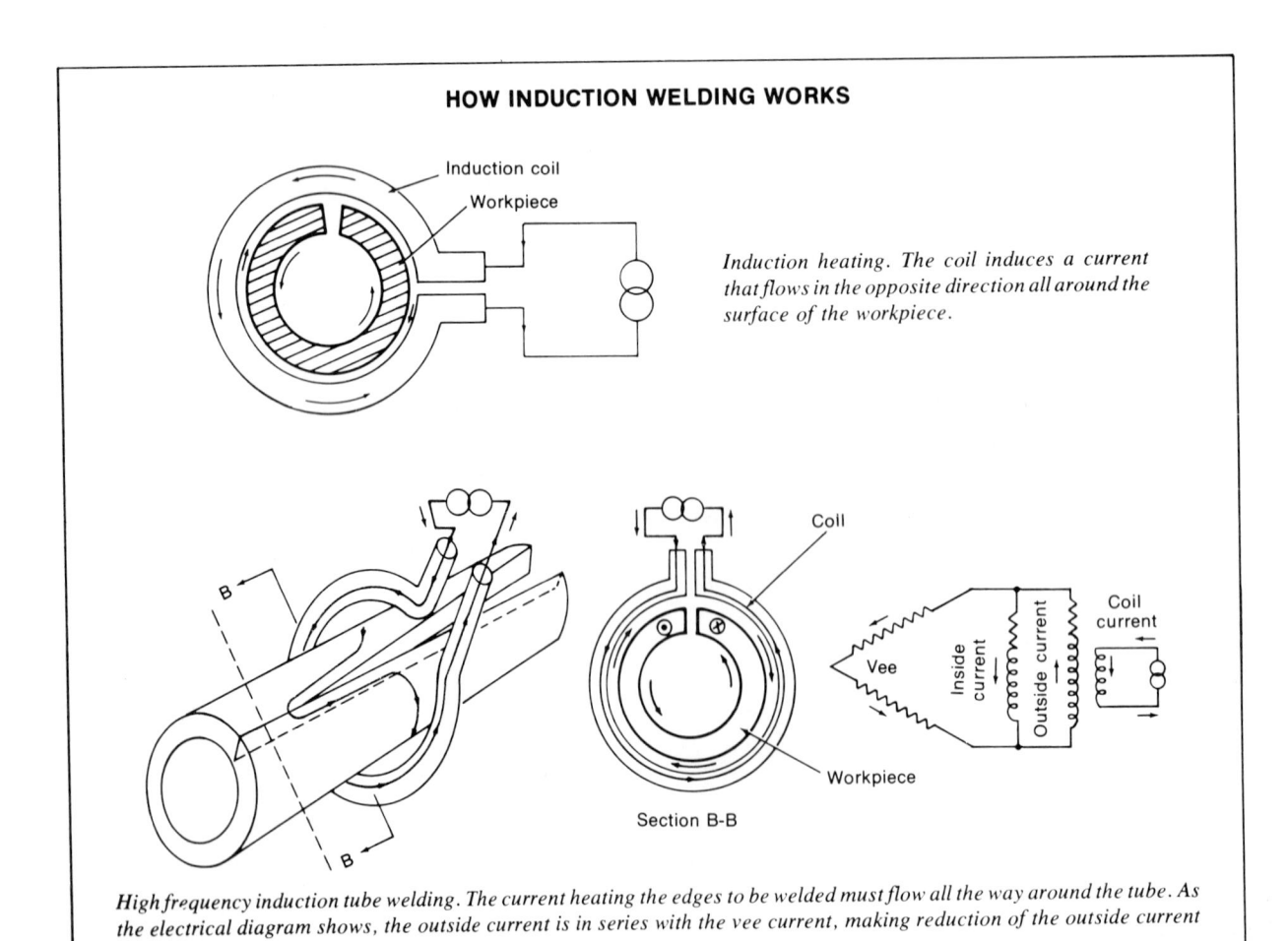

HOW INDUCTION WELDING WORKS

Induction heating. The coil induces a current that flows in the opposite direction all around the surface of the workpiece.

High frequency induction tube welding. The current heating the edges to be welded must flow all the way around the tube. As the electrical diagram shows, the outside current is in series with the vee current, making reduction of the outside current impossible.

Effect of impeders on tube welding

A ferrite impeder placed inside a contact welded tube can increase the inductance of the path around the inside of the tube so that current flow practically ceases. The remaining path between the contacts is the vee to be welded. Little current is lost.

An impeder placed inside an induction welded tube has the same effect: it eliminates current flow inside the tube. But it has no effect on the current around the outside of the tube, nor should it, since the same current goes where it's wanted, to the weld vee. The power loss and the voltage drop due to current flow around the tube exterior are not eliminated. Induction tube welding is less efficient than contact tube welding, and an impeder does little to improve its efficiency. For an equal welding rate, induction takes more power from the high frequency generator and will demand a larger generator. In other words, for a generator of given size, weld rate will be higher for contact than for induction welding.

Effect of tube diameter

At the same KW input, induction welding rate for 1-inch (25 mm) o.d. tube is about 85 percent of the rate with contact welding. As tube diameter increases, the rate becomes even more favorable to the contact method. The reason: with induction, as tube diameter increases, the induced current path around the tube becomes longer relative to the useful path in the vee. The bigger the tube diameter, the higher the proportion of the current wasted in heating the outside surface of the tube and the smaller the proportion flowing through the vee. For each 1-inch increase of tube diameter, welding rate drops about 15 percent.

In contact welding, efficiency rises as tube diameter increases because impedance of the current path around the back of the tube increases with length of the path. Because HF current seeks the path of least impedance, more of it will concentrate in the shorter vee path in the tube edges.

What about nonferrous tube?

Low resistance materials, like copper, are harder to heat by induction than is steel. The best electrical efficiency for induction heating copper is 50 percent. In other words, less than half the KW input enters the copper, and even less than that goes to heat the joint. The rest of the power goes down the drain with the cooling water. The power efficiency difference between contact and induction welding is even greater for welding copper.

Scarfing mandrels

Metal scarfing mandrels passing through the tube decrease the efficiency of both high frequency welding processes. High frequency current flowing around the inside of the tube acts like an induction heating coil. The mandrel is induction heated, reducing its mechanical properties and causing a power loss. What is needed is a strong material that has the electromagnetic properties of ferrite. As yet, no such material is available. Most mandrels now in use have a non-magnetic stainless steel section in the weld area.

Why consider induction at all?

In certain cases there are good reasons for using induction. When welding aluminized or galvanized steel, coating pickup by the contacts can adversely affect production. Also, if the metal being welded has an insulating coating on its surface (not on the edges), then induction has an advantage. Finally, if a tube maker wants perfectly mark-free tubing and doesn't want to take the care needed to achieve it with contacts, then induction, even with its reduced welding rates, is an easy way out. ∎

HOW IMPEDERS AFFECT TUBE WELDING

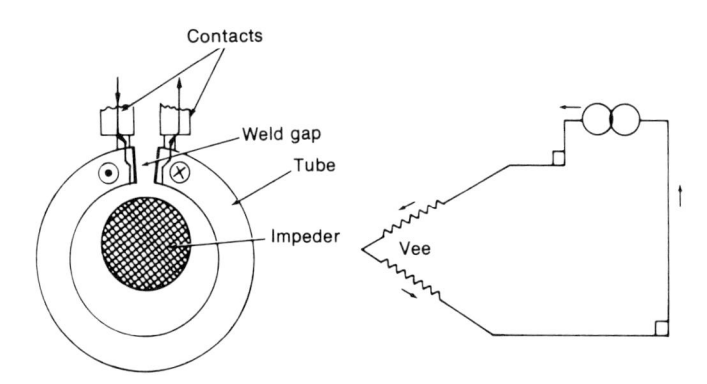

With an impeder in place, current flow along the inside of the tube is eliminated. As the equivalent electrical circuit shows, all current flows through the vee.

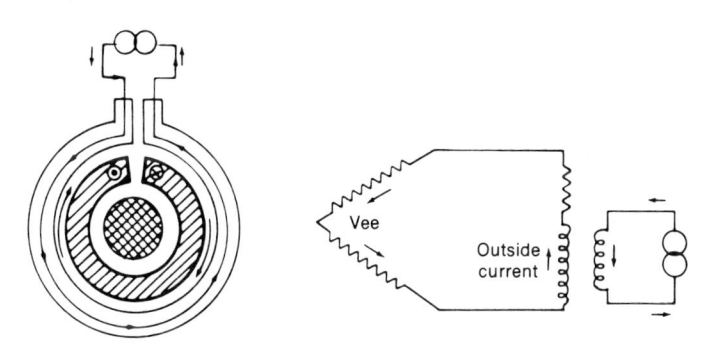

Effect of an impeder on induction welding setup. As with contact welding, the impeder neutralizes current flow on the inside surface of the tube. The induced current on the tube exterior must remain because it is the necessary link to the power generator.

High-frequency 'bar-butt' welding

By WALLACE C. RUDD (AMF Thermatool Inc., USA)
(Presented by A. ZARINS, AMF International Ltd.)

In this paper flash welding which is used for the rapid joining of strip, plate, bars, rods and shapes is compared to a new process called 'High-Frequency Bar-Butt Current-Penetration Welding' which is most suitable for joining flat strips of steel and stainless steel. Results are given. This paper was presented at a conference on 'Steel Sheet and Strip Welding' organized by the Institute of Sheet Metal Engineering, the Welding Institute and 'Sheet Metal Industries', March, 1972.

FLASH WELDING has been used for over seventy-five years for the rapid joining of strip, plate, bars, rods and shapes. It is a simple process and has much to recommend it. Recently, a new process, which is expected to compete with flash welding in several areas, has been developed and is being used commercially in several installations. This new process, called 'high-frequency bar-butt current-penetration welding', is most suitable for joining flat strips of steel and stainless steel very rapidly and with very little fuss.

Flash welding

Conventional flash welders consist of two pairs of clamping electrodes as shown in Fig 1. These copper electrodes are connected to a single-phase power line frequency transformer. A low voltage is impressed on the two clamping systems. The work is clamped in the copper jaws and one of the work-pieces is moved very slowly toward the butt edge of the other. When they touch lightly, arcing and sparking, called 'flashing', takes place. This occurs across the length of the work-pieces. The flashing period is actually the heating cycle. The electrode system that is gradually moving in, moves at a fixed rate until by previous experience it is indicated that there is sufficient heat in the edges of the work to produce a weld. At this point, forging pressure is applied, and an upset in the edges of the two pieces being joined is created. At this point, the weld is complete, is allowed to cool, and is then removed, or if the weld is in a material in which hardness is produced by the rapid self-quenching after the weld, then current from the power source may be permitted to flow, reheating the joint and normalizing or annealing it in smaller sizes.

In flash welding the current flow is at right-angles to the joint and is distributed over the entire width of the seam by the resistance of the circuit consisting of the jaws or electrodes, the strip, the contact resistance, and the arcs taking place during flashing.

A large variety of metals, lengths, thicknesses, and shapes are capable of being welded by this process. It has some disadvantages and many advantages as compared to the various forms of arc welding.

After the weld is completed, an upset bead, or flash, exists on the outer surfaces of the weld. Normally, this has to be removed and can be cut off by a travelling plough-type cutting tool or in some cases may be removed by grinding or belt sanding.

High-frequency bar-butt welding

Until recently, flash welding has had no competition in the area of butt welding flat strips of metal together. Its only real competition in this area was arc welding, which is slow and produces a cast weld. Recently a new high-frequency process developed by AMF Thermatool Inc., has proved itself in the area of welding flat strips of steel and stainless steel very rapidly and without most of the disadvantages of flash welding. This new process, instead of operating at power line frequency, uses a motor-generator power source at 10kHz. In flash welding, as indicated above, the current that performs the heating of the joint crosses the joint at right-angles. In bar-butt welding the 10kHz current travels longitudinally in the joint the whole length of the joint.

There are two physical arrangements for performing bar-butt welding, one of which is more suitable for commercial operation. The simplest form is shown in Fig 2. In this case, two flat strips of metal to be butt welded are clamped in metallic or non-metallic jaws with the butted ends pressed together with full forging pressure at the beginning of the heating cycle. The butt ends rest on top of a copper anvil which is in the form of a bar underneath the weld area. A proximity conductor, actually a piece of 4.8mm diameter copper tube, is placed over the top of

Fig 1 Diagram of flash welder.

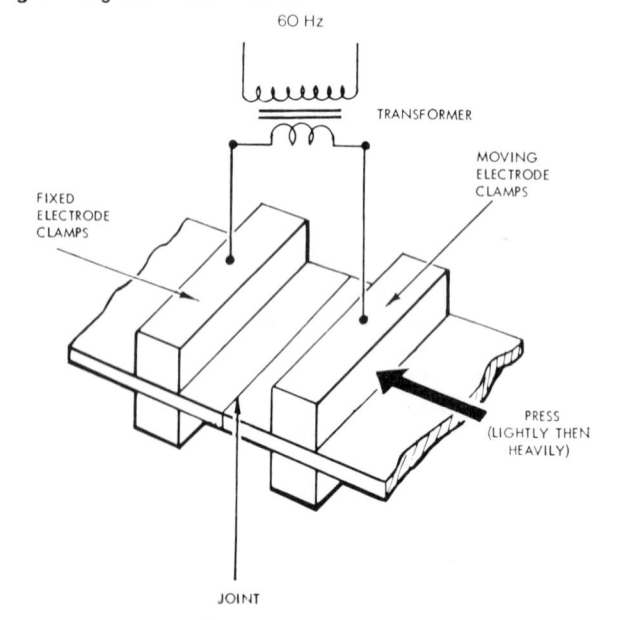

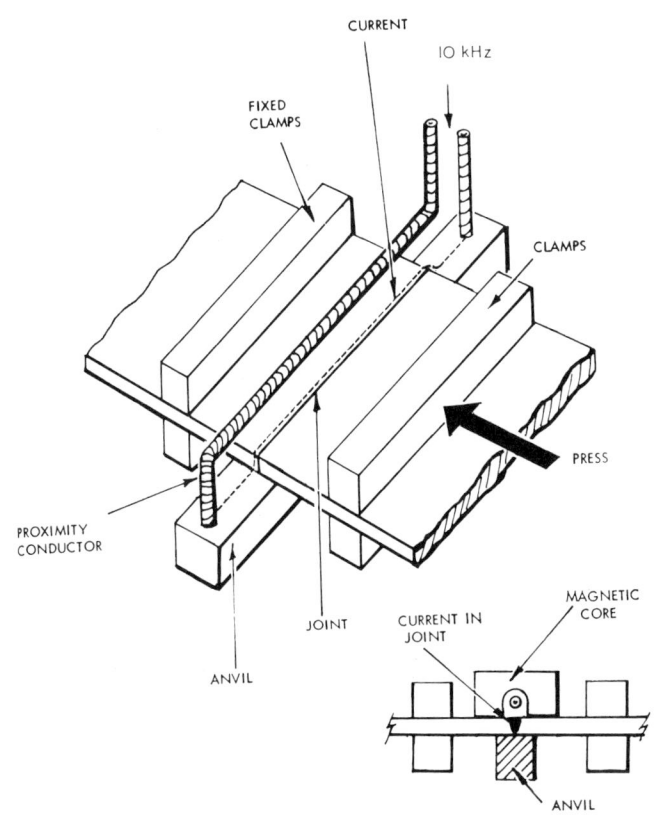

Fig 2 Single-anvil bar-butt weld. **(Bottom)** Section through weld area.

the seam. One end of this proximity conductor is connected to the copper anvil as shown. The other end of the conductor and the other end of the anvil are connected to a 10kHz low-voltage transformer fed from the generator.

High frequency is used because it has three attributes which make it particularly advantageous for joining strips of metal together. The first attribute is that high-frequency current tends to flow on the surface of a material, called 'skin effect', and in the case of red-hot steel, 10kHz would flow to a depth slightly less than 6.35mm; whereas, 60Hz current would flow to a depth of nearly 72.2mm.

The second attribute is that high-frequency current flows in the low-inductive-reactance path rather than the low-resistance path. The low-inductive-reactance path is the confined path closest to the return current or conductor. This is called the 'proximity effect'. The third effect is that current-carrying ability between two lightly contacting surfaces is significantly better at high frequency than at low frequency; therefore, simple contacts can be used to carry large high-frequency currents into a work-piece.

In Fig 2 the current enters the proximity conductor from the high-frequency transformer terminal, flows across the proximity conductor, and into the copper anvil. It then flows across the upper surface of the anvil until it reaches the edge of the seam to be welded. It then rises out of the copper anvil and preferentially flows in the seam (proximity effect) and then at the other end of the seam returns to the anvil travelling in the upper surface again and thence to the other terminal of the high-frequency transformer.

The current flowing in the work flows to a depth in steel as previously indicated but also flows in a very confined line in the seam, also due to the proximity effect. A cross-section through the weld area is shown in Fig 2 (bottom). A cross-section of the current is shown in the seam indicating that the width of the current flow might only be a little more than the diameter of the proximity conductor. In effect, they are a mirror image of each other. The net result is that a very narrow shallow line of current flows in the seam. The seam very rapidly heats, typically in one

to two seconds, and since the butted seam is under forge welding pressure, the right-hand clamp system pushes the butted ends of the strip into a forge weld with a total time of one to two seconds.

Fig 2 (bottom) shows a laminated magnetic core surrounding the upper portion of the proximity conductor. This is omitted in the main figure in order to show where the current flows. The magnetic core performs two functions: First, it electrically narrows the width of the current flow in the work; second, it holds the two edges of the strips in alignment against the anvil so that when a weld occurs there is no offset of the edges.

The system shown in Fig 2 has one disadvantage and that is if relatively thin work is being welded, for example 1.59mm steel, and a frequency of 10kHz is used with a depth of penetration of nearly 6.35mm; then the current flowing in the seam will extend downward into the copper, and some of this current will be heating the copper anvil, not the work. From a purely theoretical point of view, with the system shown in Fig 2, the frequency should be selected so that the depth of penetration is equal to or a trifle less than the thickness of the work-piece being welded. Changing the frequency of the power source is not practical; therefore, another means must be resorted to in order to prevent some of the current flowing in the copper anvil.

The solution to this problem is shown in Fig 3. This is the version which is used commercially. In Fig 3, there are two copper bars representing anvils, one on each side of the weld area. In this case, the proximity conductor is placed between the bars and underneath the seam (it does not make any difference whether it is above or below the seam). The strips are placed on top of the copper bars and are clamped against them. One clamp and its associated bar press the butted ends together. High-frequency current always tries to flow as close to the proximity conductor or the return conductor as it can (proximity effect) Therefore, when one terminal of the high-frequency transformer is connected to the proximity conductor, the current will flow

Fig 3 Double-anvil bar-butt weld. **(Bottom)** Section through joint.

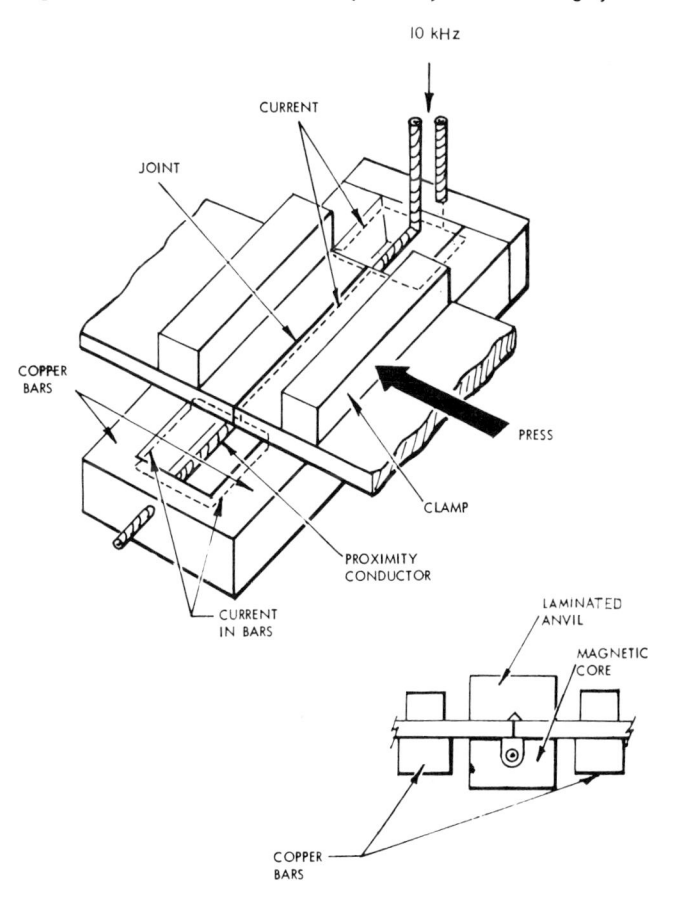

Source: *Sheet Metal Industries*, Aug 1972

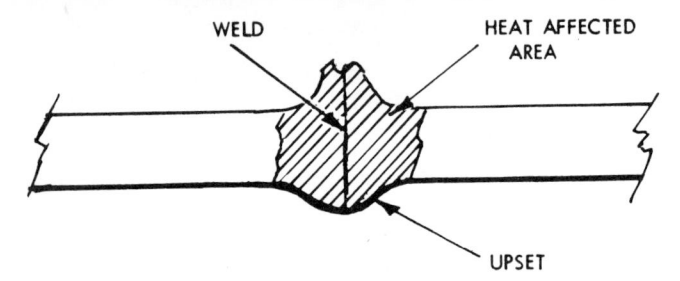

Fig 4 Bar-butt-weld cross-section.

through the proximity conductor, underneath the seam, and into the tie bar at the near end. The current will then divide in half, each half travelling back on the inner face of the side bars. When the current gets to the spot where the work is clamped on the side bars, the current will prefer to travel as close to the proximity conductor as it can, and each half will leave the copper bar and travel in the edge of the work as shown and combine as a single current in the seam exactly as occurs in Fig 2. When the current reaches the other end of the seam, it will again divide in half and enter the two side bars as half currents on each side, thence returning to the far-end tie bar and to the other terminal of the transformer.

In this case, there is no electrical conductor such as an anvil immediately below the joint; therefore, regardless of the thickness of the joint, a high-frequency current even though its depth of penetration is greater than the thickness, must flow only in the joint, thus all of the current is contributing heat in the joint, not partly in the anvil.

Fig 3 (bottom) is a cross-section through the joint in this case. In order to show current flows, Fig 3 has omitted the laminated core around the proximity conductor and also the laminated core pressed down on the top of the joint. These cores have exactly the same effect as was stated for Fig 2, that is, the one around the proximity conductor confines the current to a narrower band, and both of them act to hold perfect alignment between the butted ends of the work.

In both of these bar-butt techniques, the current is flowing in the work along the seam, not at right-angles. The current is only heating the seam, not the strips of metal between the clamps and the seam as it does in flash welding. In bar-butt welding a relatively low current is flowing in a long high-resistance path; whereas, in flash welding a very large current is flowing in a short, low-resistance path.

Fig 4 shows a cross-section through the weld produced. In effect, it is very similar to that produced by a flash weld, that is, there is an upset on the top and an upset on the bottom. Typically, the upset will be a trifle larger on the side toward the proximity conductor. This is due to the fact that the side near the proximity conductor is a trifle hotter than the other side. The heat-affected zone is very narrow with weld quality being extremely good.

Generally, there is very little, if any, molten metal present at any time during the welding cycle; therefore, it may be said to be a solid-phase weld. Welds can be made by timing the current or heat cycle, but preferably a self-controlling device without a timer is used. In this case, a micrometer adjustable switch is arranged in conjunction with the moving platen of the bar-butt welder. Power is turned on, the seam heats, and when it is soft enough, the forge pressure causes an upset to occur which results in one platen moving. Movement of this platen is used to turn off the weld power automatically. Therefore, by pre-setting the micro-switch the amount of upset is used to shut off the power. Typically, the amount of upset is equal to the movement of one-half to one thickness of material. This system eliminates problems from certain variables such as differences in material, slight differences in thickness, etc.

If steel is welded with this technique in which hardness is produced in the weld area due to self-quenching, normalizing or annealing can take place without any change in the piece of equipment other than reducing the power level and re-cycling the heat for a longer time. In effect, this is simply a slower re-heating of the weld area.

The upset, bead, or flash which is produced on the weld is quite similar to that produced in flash welding and is removed the same way.

Comparison with flash welding

1. Typically, bar-butt welding has weld times lying from 0.75 second to 2.5 seconds with thickness ranges in steel and stainless steel of from 0.6 to 4.8mm and lengths up to 914.4mm. Flash welding has a flashing cycle and then a subsequent forging cycle with times usually in the order of double or more that achievable with bar-butt welding.

2. In order to produce the heat in the seam with flash welding, a period of flashing and arcing precedes the forging cycle. During this period, metal is being burned away and thrown off into the air in the form of molten droplets. The metal being burned away can be from 2.3mm to 19.05mm of length depending on thickness and material. This burned-off metal is wasted and must be compensated for in the original dimensions of the pieces being welded. In bar-butt welding there is no burn-off. The only loss of metal is that produced in the upset, typically from one-half to one thickness.

3. The burn-off which occurs in flash welding has some severe side effects. The burn-off is actually droplets of molten metal. They are blown into space and stick to machine surfaces causing all kinds of problems in cleaning and maintenance. In addition, they cause considerable fire and personnel hazards in operation of the machine. Bar-butt welding does not throw any spatter whatsoever and one of its attributes is the lack of spectacularity or fireworks.

4. Bar-butt welds may be normalized in place as indicated previously regardless of the length of the seam since the high-frequency current flows only in the seam and along the whole length of the seam; whereas, with a flash welder, heat treating after the weld in wider materials is difficult since the distribution of current is set by the resistance of the paths and the current flows across the seam, not along it. Hot spots result. In flash welding, in-place normalizing after welding is sometimes done in wider widths by re-gripping the sides of the weld with the welding jaws so that more length of metal exists between the jaws, then reapplying power to the system so that the weld area heats by resistance heating. However, here too, hot spots can result regardless of length. Hot spots cannot occur with the bar-butt normalizing system. Heat treating flash welds in short lengths in place is quite practical.

5. Bar-butt welding does not require any controlled movement of the platen during the heating cycle. If desired, the full force of forge welding may be applied prior to heating and then no control of the platen is required since it moves in of its own volition due to the softening of the metal in the weld area due to heating. Flash welding requires a very careful control of the rate of movement of the platen during the heating portion of the cycle, otherwise improper heating results.

6. If different thicknesses of metal or dissimilar metals are being welded, the differences in rate of burn-away in flash welding must be compensated for. This makes for complication in adjusting a machine for operation. With bar-butt welding this is not the case. The difference of heating requirements on the two sides of the seam is simple to compensate for by arranging the proximity conductor to be offset slightly from the centre-line of the seam. The proximity conductor would be adjusted so that it was nearer to the side requiring more heat, for example, the thicker side. This off-centring of the proximity conductor off-centres the current slightly in the work and thus more current would flow in the piece requiring more heating.

7. In welding circular rings such as wheel rims, bands, etc., with a flash welder in which the current flows at right-angles to the seam, there is a problem with current travelling around the back side of the ring in addition to that going across the joint.

Fig 5A (×14) 1.5mm aluminized steel HF CP butt weld.

Fig 5B (×14) 1.6mm 302 stainless-steel HF CP butt weld.

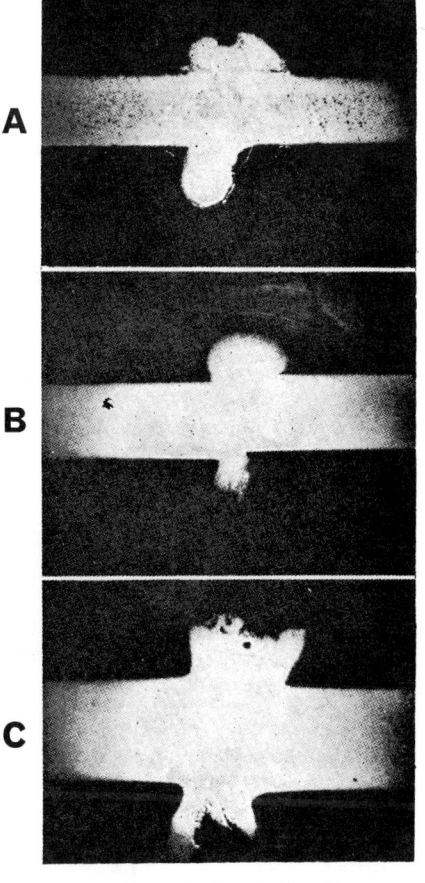

Fig 5C (×14) Low-carbon steel HF CP butt weld.

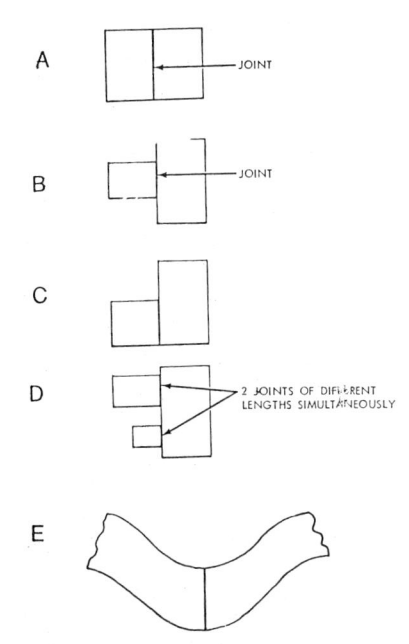

Fig 6 (above) Joint configurations.

Fig 5 Three high-frequency current penetration butt welds.

This is a bypass current and is wasted. It occurs simply because the two sides of the ring are electrically in parallel. With bar-butt welding this does not occur at all since the current is running longitudinally in the seam, and there is no voltage difference that would create a current around the back side of the ring. In smaller diameters this is a major problem with flash welding.

8. With flash welding the edges of the pieces of strip being welded protrude from the electrical and forging jaws. They are unsupported in space; therefore, alignment of the two edges with this overhang is difficult. With bar-butt welding the two edges being welded are contained between an anvil and a laminating steel structure during the heating and forging cycle. Therefore, they are held in complete alignment with no opportunity to deviate.

9. The structure used electrically and mechanically in bar-butt welding permits multiple- or variable-length welds to be made simultaneously in the same fixture. For example, if the fixture were capable of welding a 254mm wide piece, there would be no difficulty in welding two 76.2mm wide pieces without changing mechanically, or even a 101.6mm and a 76.2mm piece. Weld power level would have to be changed. Multiple welds are difficult to do in flash welding because the rate of burn-off may be different in the two pieces. It is also possible to bar-butt two pieces on one edge of another larger piece simultaneously. This is shown in Fig 6. This, too, is difficult with flash welding.

10. Flash welding requires controls to set the rate of movement of the platen, time of flashing, and time to apply forging pressure. Bar-butt welding requires none of these controls and simply uses an upset-type limit switch described above.

11. In welding some metals, it is desirable to use a gas atmosphere in the weld area to prevent oxidation. This is very easy to do with bar-butt welding and difficult with flash welding.

Suitable for flat strips and stainless steel

Any new process cannot be expected to make an overall improvement in an older process, and here, too, this is the case. Bar-butt welding at the present time is only suitable for butt-welding flat strips of steel and stainless steel up to 914.4mm long and in the thickness range of 0.6mm to 4.8mm. It is expected that both length and thickness limits will be increased in the near future. In addition, it is expected that materials other than steel and stainless steel will be welded. Flash welding can make very long welds in very thick material very satisfactorily. For example 9.14m pipe, 12.7mm thick has its side seam welded by flash welding in one installation. Flash welding can do almost any metal. Flash welding can do bar ends, rod ends, tubes and certain odd shapes. At the present time it does not seem that the bar-butt-welding technique would lend itself to work in this area. It is expected, however, that certain curved and C-shapes will be done in the future.

High-quality welds

The welds produced by bar butt welding are of extremely high quality. Steel which is coated with zinc, aluminium, or lead is easily butt welded, and within limits it is practical to weld different materials together, for example, M-2 high-speed steel to 6135 carbon steel for hacksaw blades. Fig 5 shows the etched cross-section through several different types of joints.

Fig 6 shows typical joints which can be performed with bar-butt welding within the limitations specified above. At present, 1971, there are two commercial installations welding automotive chassis sections together in mild steel. A third is under construction. In this case, a weld is made in one second between pieces 190.5mm long and 2.3mm thick.

All indications are that the new patented high-frequency bar-butt welding process will be used in many of the strip joining areas of flash welding. It has numerous advantages particularly in less consumption of metal and the lack of spatter produced as well as its high speed.

Source: *Sheet Metal Industries*, Aug 1972

Cut rim blanks moving along the center mandrel in successive stages of fabrication (left to right): roll forming, welding, trim, planish, end trim.

How to make a wheel rim — in 4 seconds

An automobile manufacturer uses a super fast machine to turn coiled strip into finished car wheel rim blanks at 1,000 pieces per hour. High frequency resistance welding makes it possible.

by R. BROSILOW, *associate editor*

Chevrolet Motor Division of General Motors Corporation wanted a machine capable of turning out 1,500 automobile wheel rims an hour, Grotnes Machine Works Inc., Chicago, makers of metal expanding, shrinking, and rotary roll forming machinery, set out to meet that high production challenge.

The old way

In the standard production method, rim makers start out with coiled strip. The material passes through many operations:
- Unroll and flatten the coil, then shear it to length for the rim.
- Feed the cut strip through one or more roll formers to turn it into a cylinder with the ends butted for welding.
- Clamp the rolled strip into a fixture for a resistance weld.
- Cut or grind off the weld upset material inside and outside the rim.
- Trim off the material at the ends of the weld.

Each of these steps requires a worker to fit the rim into the machine with the weld properly oriented for the operation. A worker usually removes the work from the machine after the operation. And the work has to be transported, either by machine or hand, from one operation to the next.

The new way: a system approach

To meet Chevrolet's requirements, designers looked at the whole rim blank production line. They decided to design one machine which would take the raw material, coil strip, and turn out a finished rim blank. The key to automatic rim blank preparation was orientation of the weld. In the standard method, after the strip is rolled into a cylinder the weld area gets all the attention. Every time an operator moves the work to the next machine he orients the weld to the tool. A continuous process would hold the rim in the same position as it moves through succeeding operations. On Grotnes' machine the cut sheet, typically SAE 1008 steel, is rolled around a mandrel. Hydraulically powered dog clamps reach up to move the rim along the mandrel from one operation to the next:

- End conditioning press — a small hydraulic press forms the cut ends to

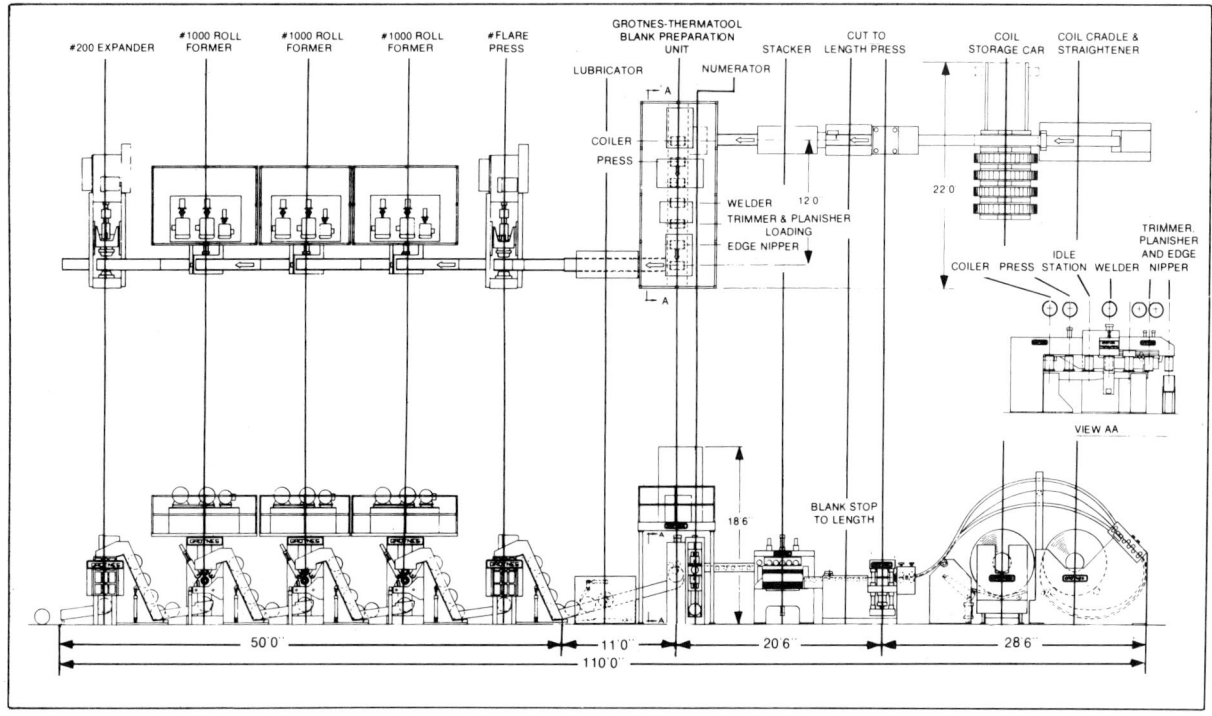

The automobile wheel rim line starts with a coil storage rack. The coil feeds into a coil cradle and straightener, then to a press which shears it to length and stacks it. From these stacks the material feeds automatically to the rim blank preparation machine. The finished blanks feed to forming machines which roll and expand them into finished rims.

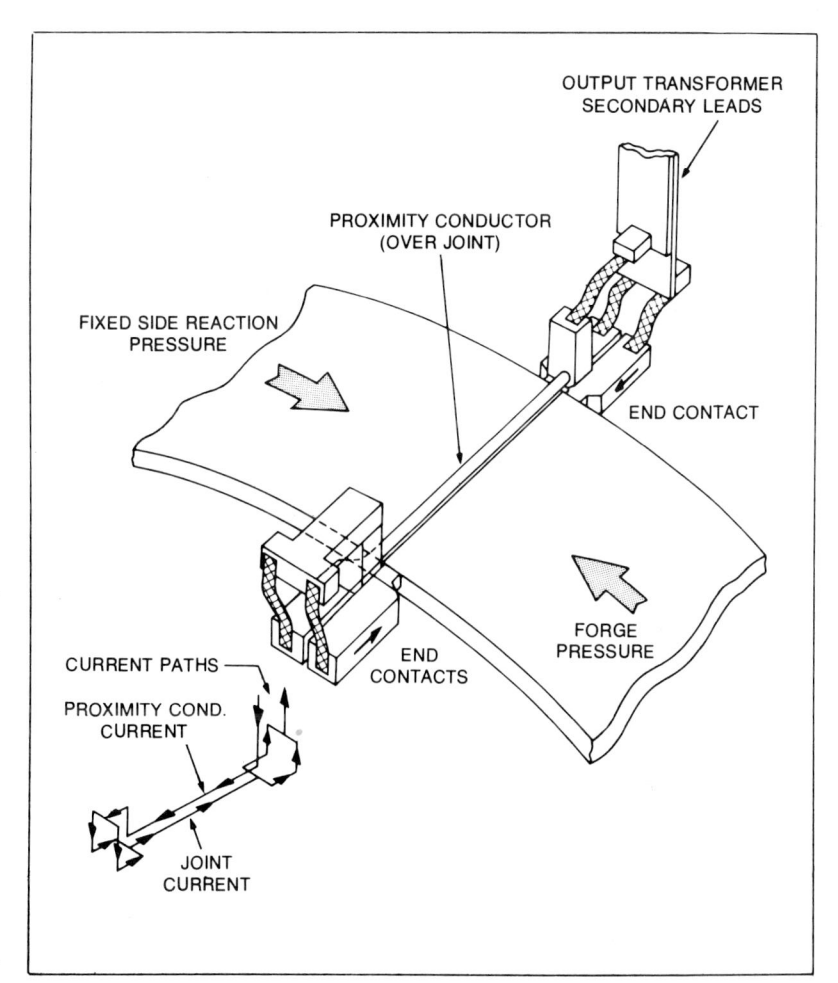

the required fit for welding, a 0.010-inch gap.

- High frequency current penetration bar butt welder — a spreader bar aligns the ends of the coil, then current penetrates the edges, heating them to welding temperature. A clamp moves in to apply forging pressure, 10 to 15 Kpsi, to the heated edges.
- Weld upset trimmer — trims the weld while still hot (above 600°F.) inside and outside the rim blank.
- Planisher — flattens the weld and refines the microstructure.
- Weld end trimmer — cutting tools trim off the ends of the weld.

As each operation is completed, the dog clamps open flat and move back to bring up another rim blank.

HF makes it possible

The high speed blank preparation machine needed a high speed welder. High frequency welding made the job possible.

At low frequencies, like 60 Hz, current flows through the whole

Thermatool designed this butt penetration welding fixture for Grotnes' wheel rim blank machine. Sketch at lower left shows current flow pattern from transformer through proximity conductor, end contact, through the weld joint and back to the current source.

Source: *Welding Design & Fabrication*, Feb 1976

cross section of a joint to be resistance welded. But at high frequencies, like 10 KHz, changes in current flow patterns turn resistance welding into a different ballgame. The current bunches up to give a so-called proximity effect. It flows as close as possible to the current conductor, called a proximity conductor. The depth of the current penetration in the metal, and the depth of the weld, depends on the frequency of the current, resistivity of the material to be welded, and its magnetic properties. When a magnetic core surrounds the conductor, the current in the return path concentrates in a very narrow line. This effect means that very high power levels can be reached quickly in very narrow areas: high speed welding becomes possible.

When the rim blanks enter the welding fixture, the proximity conductor comes down on the joint. High frequency current flows through the conductor, the end contacts, and the joint, to the other end contact where it returns to the source. When the joint is up to temperature, a clamp applies pressure, and upset occurs, producing a forge weld. The edges are red hot at forging temperature, but are not molten. The amount of upset material is small. A ceramic block under the seam supports the weld. The block, grooved to allow room for the upset material, assures uniform weld upset going into the upset trimmer. Sparking and spatter are non-existent, unlike the more common resistance flash welding process.

Size changes

Chevrolet can use the machine to make a family of rims. The table shows the maximum and minimum capacities:

BLANK SPECIFICATIONS

	Min.	Max.
Width	5.5"	12.5"
Thickness	.100"	.186"
Length	37."	54."
Blank Dia.	11.5"	17."

Tooling change to modify rim diameter and width takes 4 hours.

Each blank preparation machine can turn out 1,000 rim blanks per hour. Chevrolet will use two machines in parallel, both working off a common stack feeder. ■

The steel blank after each step in the continuous welding operation (left to right, top to bottom): coiling, end conditioning, welding, weld trimming, weld planishing, edge trimming. At left, formed wheel rim with hub inserted.

MELT WELDING -- A NEW HIGH FREQUENCY PROCESS

By

W. C. Rudd and
H. N. Udall
Thermatool Corporation

ABSTRACT

A new high frequency welding technique has been devised and patented that will compete advantageously in certain areas with spot and edge seam welding. Called high frequency melt welding, the new technique involves applying high frequency electrodes to several pieces of metal that are to be joined. Current flowing between the two contacts heats and melts the area between them. Since the melt is composed of two or more pieces of metal immediately adjacent to one another, the metal from both pieces will run together and produce a small cast weld.

INTRODUCTION

Melt welding can be done at a frequency of 400 KHz for light members and welds, and at 10 KHz for heavier, deeper welds. The high frequency current is produced by vacuum tube oscillators for 400 KHz and by solid state SCR inverters for 10 KHz. The basic principle is shown by Nos. 1A and B in Figure 1. In this case, a single piece of metal has high frequency contacts applied to one edge as shown. The pressure on the contacts is very light being only a few pounds. When the power is turned on, the edge will heat and melt, and molten metal will fall away. Typically, the time might be in the range of 1 second. The result is a melted-out section as shown by No. 1B in Figure 1.

APPLICATION FOR SPECIFIC TYPES OF JOINTS AND WELDS

If the technique is applied to a joint in a lap weld as shown in No. 2A of Figure 1, the current will melt both pieces provided they are not too thick, i.e., 0.1 inch maximum (2.54 mm) and cause the liquid metal to flow together as shown in Figure 1 by No. 2B and in cross section by No. 2C.

The lap weld produced would be similar to a spot weld but not have as good an appearance. It, however, does have the advantage that no welding electrode is required on the bottom side and no forge welding pressure need be applied.

The same high frequency melt welding technique can be applied to a butt weld as shown in Figure 1 by No. 3A, with the result shown by No. 3B. A lip weld can also be done as shown in No. 4A of Figure 1. In this case, the high frequency contacts are applied either on the top edge of the lip or on opposite sides as shown in No. 4A. When power is turned on, the lip melts and produces a melted notch with the corresponding weld as shown in No. 4B.

It is even possible to make a T-joint as shown in Nos. 5A and 5B of Figure 1. Heavy pieces can be welded to light pieces as shown in No. 7. For example, this weld could be used for welding the starter ring gear on a car to the stamped center spider. A photograph of an actual weld of this type is shown in Figure 2.

Another advantage of this system is that several melt welds can be made simultaneously with the electrodes electrically in series as shown by No. 6 in Figure 1. These several welds may be on a single object. With high frequency, currents can be caused to flow in various sections of the same piece of metal without having short-circuit problems.

Another asset of the process applies in places where electrodes for normal spot welding make a heavy mark on a surface to be painted such as the edge of a car door. Here the melt welding technique, by contacting the inside surface only, makes a weld without marking the outside significantly as would be the case with spot welding. The weld itself, however, if made using present techniques, has a considerably poorer appearance than a spot weld.

Typical times for melt spot welds can run from 1/2 to 2 seconds. At present, it is only available for steel and stainless steel. It is also possible to spot weld steel to selected aluminum alloys in certain configurations. Typical weld strength for the melt welding of a spot on two 1/32 inch (0.8 mm) thick sheets in a lap as shown in No. 2B of Figure 1 would be 300 to 400 pounds tensile strength (136 kg to 181 kg). Of course, in none of these melt welds is it required to add metal. It is simply a case of quickly melting two pieces together and letting them solidify.

LINE MELT WELD FOR LAMINATIONS

Another area that appears very promising is to use the same technique of high frequency melt welding for welding laminations together in electrical motors or transformers. At the present time, stacks of laminations are held together with bolts, rivets, clamps, and sometimes a gas tungsten arc weld (GTAW) across the outer edges of the laminations.

The melt welding process lends itself to producing the same type of weld as the GTAW but is faster, cheaper, may be automatic, and can be done in multiples. In effect, a line melt weld is produced by an extension of the principles indicated above. No. 1 in Figure 3 shows a stack of circular laminations that are to be fastened together by producing a very shallow weld across the outer periphery in about 1/2 second.

With the high frequency melt welding process, contacts press the stack of laminations together and a proximity conductor is placed over the area where the weld is to occur. This proximity conductor will serve reactively to confine the high frequency current to a narrow line. High frequency current at about 400 KHz is applied to the contact/proximity conductor system, and a fine line of current flows across the laminations underneath the proximity conductor. With the right power level and timing, a small line melt is produced across the laminations and as soon as the power is turned off, the melt solidifies, joining all of the stack together.

The same operation is repeated at several places around the periphery in order to hold the stack together and in alignment. No. 2 of Figure 3 is a side view of No. 1 showing a cross section through the laminations in the proximity conductor area; No. 3 shows the resulting shallow melt weld. Figure 4 is an actual photograph of such a weld. Figure 5 is a (X50) micro section.

With lower grade laminations, there is sufficient electrical contact between laminations along the path underneath the proximity conductor to allow the high frequency current to flow. Thus, the melt will occur immediately underneath the proximity conductor, where maximum current is concentrated. Higher grade laminations may have thick insulation in the form of oxide or even possibly lacquer; here it is necessary to have a light scribed line or light indent made by a sharp chisel underneath the proximity conductor. This scribed line, or indent, produces burrs between the laminations allowing initial current to flow in a line underneath the proximity conductor. The indent device can easily be incorporated in the welding equipment.

With the high frequency process, it is possible to do a number of lamination stacks in series as shown by No. 1 in Figure 6. In this case, melts are being made across three identical groups of laminations. It is also possible to electrically connect a number of proximity conductors on a single piece so that they are electrically in series. Thus, two, three, or four welds could be produced simultaneously on a single stack of laminations from a single power source.

Source: SME Technical Paper No. AD78-745, 1978

In certain cases, it is desirable to make a melt weld inside a hole. Nos. 2A and B in Figure 6 show how this can be done by placing the proximity conductor through the hole with contacts on the outboard edges of the hole. No. 2A shows that with an oval proximity conductor two welds can be made simultaneously in the same hole.

If it is desired that the melt weld not project above the outboard surface of the laminations, then the laminations could be stamped so that there is a notch and a bump as shown in No. 3 of Figure 6. In this case, the bump or projection in the center of the notch would be a preferential spot for the high frequency current to flow and for the melt to occur. The resulting weld would be below the surface of the outboard edges of the laminations and would not interfere with any mounting devices that may be used. This technique may also be used in a hole as shown by No. 1 of Figure 7.

A 1-inch (25.4 mm) long stack of laminations would require on the order of 25 KW of high frequency power for the production of a weld in a time between 0.5 and 1.0 second. Greater length of path would increase the power. Preferably the time should be kept short so that not too much metal is melted. A dial-type feeding arrangement could be used so that the laminations were stacked automatically and fed into the high frequency welding system without human assistance.

It is preferable that the molten path be in a horizontal position rather than vertical to avoid the effects of gravity. In other words, it is preferred that the weld be on the top, side, or bottom of a stack of laminations with the axis of the system parallel to the earth's surface. However, for many applications the weld can be vertical.

In some line type welds it is quite practical to supply added metal to the melt in the form of wire or particles. The added metal is put in position before heating, and when the current flows, it melts the base and the added metal together, producing a stronger weld.

CONTINUOUS EDGE MELT WELD

A continuous form of the high frequency melt welding process is shown by No. 2 in Figure 7. In this case, two or more strips with matched edges are traveling at a relatively high rate of speed, i.e., 10 to 100 ft/min (3.1 to 31 m/min) with the edges aligned that are to be welded together. The two or more strips held with guides pass through the current area, their edge temperatures are raised to melting, and just as they leave the downstream contact, the edges flow together producing a rounded, melted weld - No. 3, Figure 7.

This continuous process would be suitable for steel, stainless steel, Inconel, Incaloy, and possibly other materials. Total thickness of the combined edges of strips at the present time should not exceed 3/16 inch (4.8 mm). Minimum thickness would be about 0.020 inch (0.5 mm). This type of weld could be used for sealing and finishing the edges of several strips of metal for a variety of purposes.

This new melt welding process is extremely fast and lends itself very readily to completely automatic equipment and therefore is suitable for high-production work. Argon gas may be used in certain special cases to improve the quality of the welds produced.

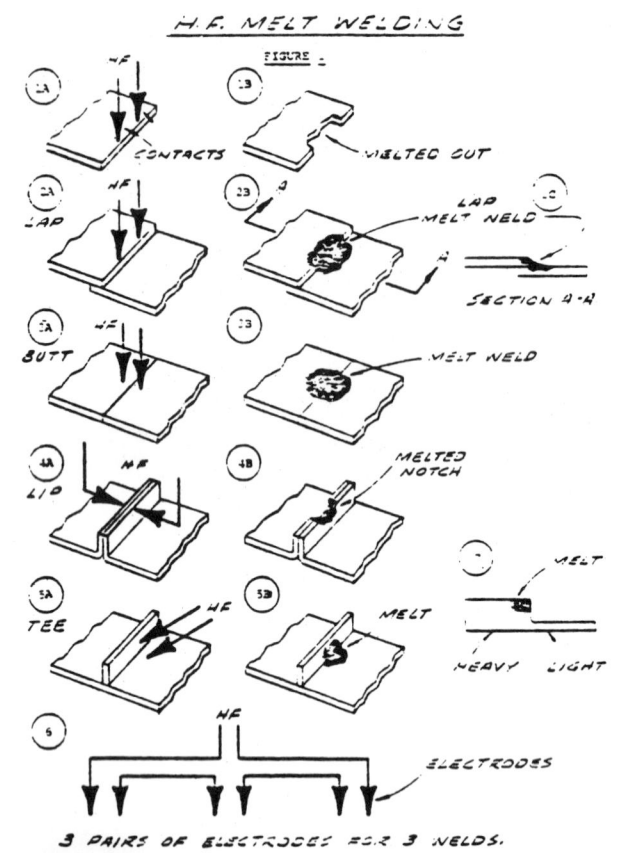

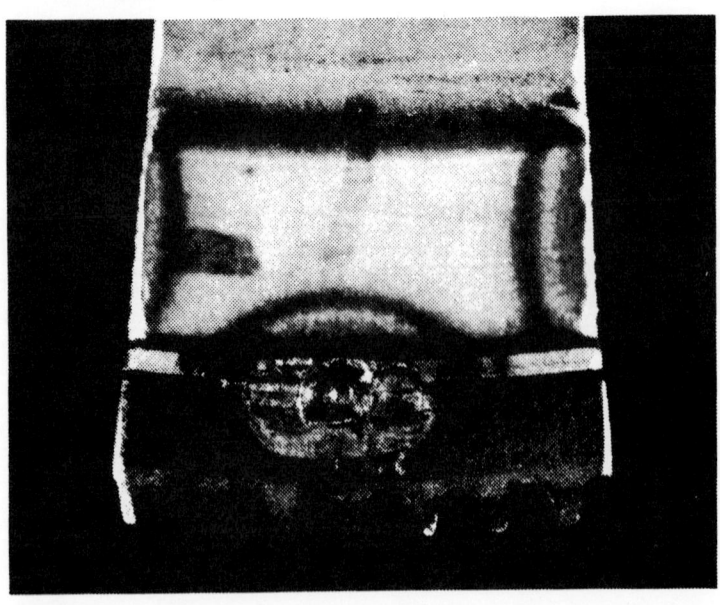

Figure 2: Weld Between Automotive Starter Gear and Center Spider

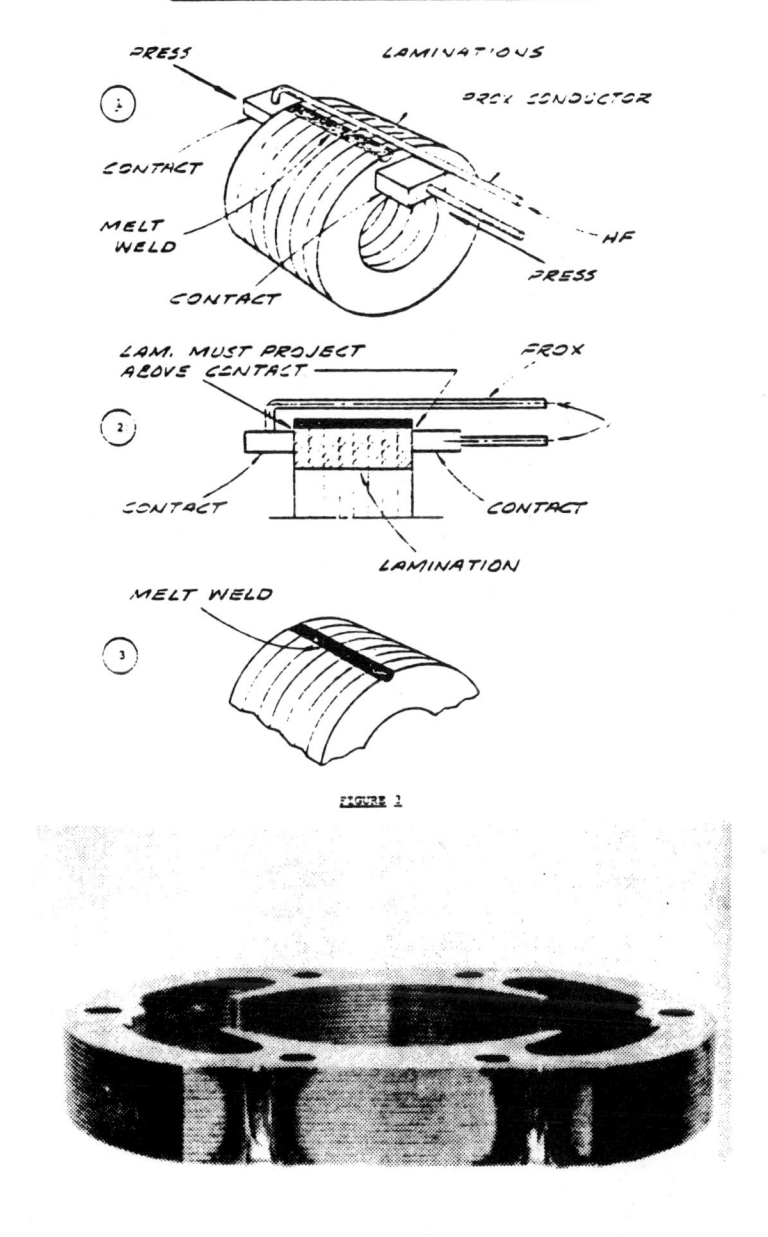

Figure 4: Melt Weld on Motor Lamination Stack

Figure 5: Micro Photograph (50X) of Lamination Stack

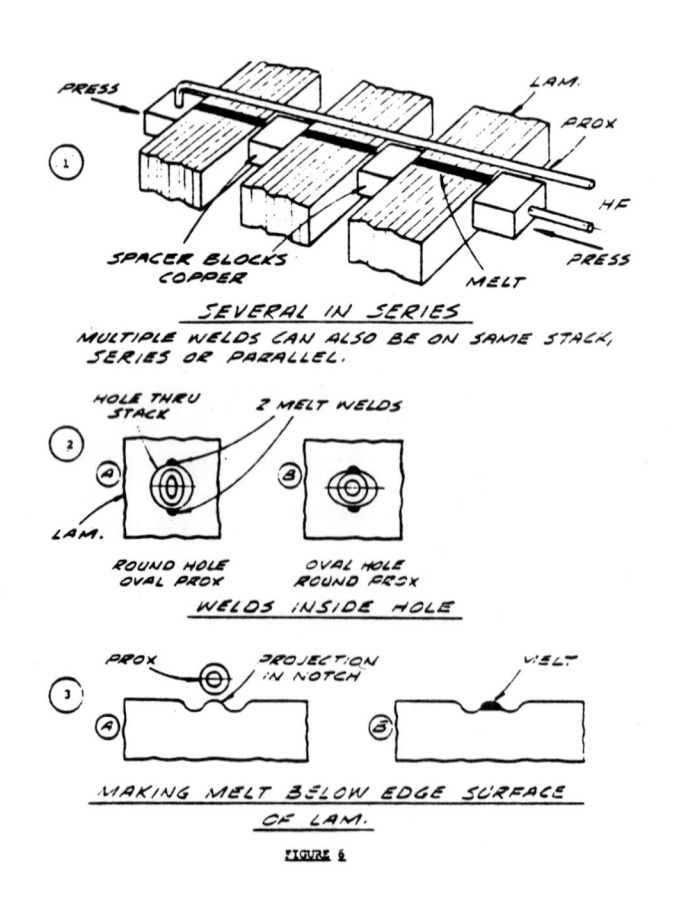

FIGURE 6

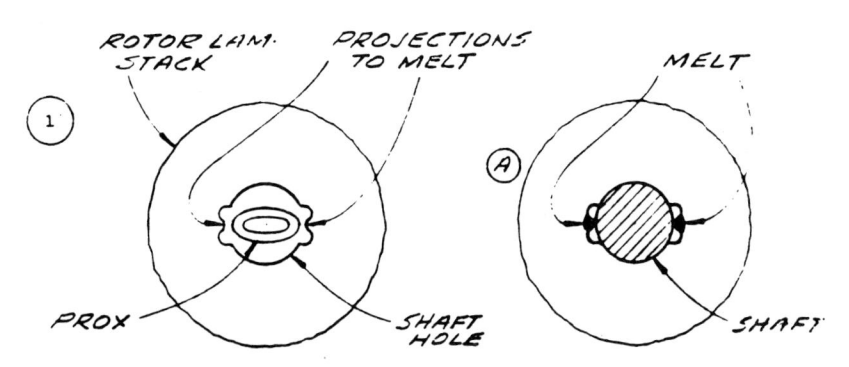

① ROTOR LAM. STACK / PROJECTIONS TO MELT / PROX / SHAFT HOLE

Ⓐ MELT / SHAFT

BALANCED WELD INSIDE HOLE FOR SHAFT.
MELT WILL NOT INTERFERE WITH SHAFT.

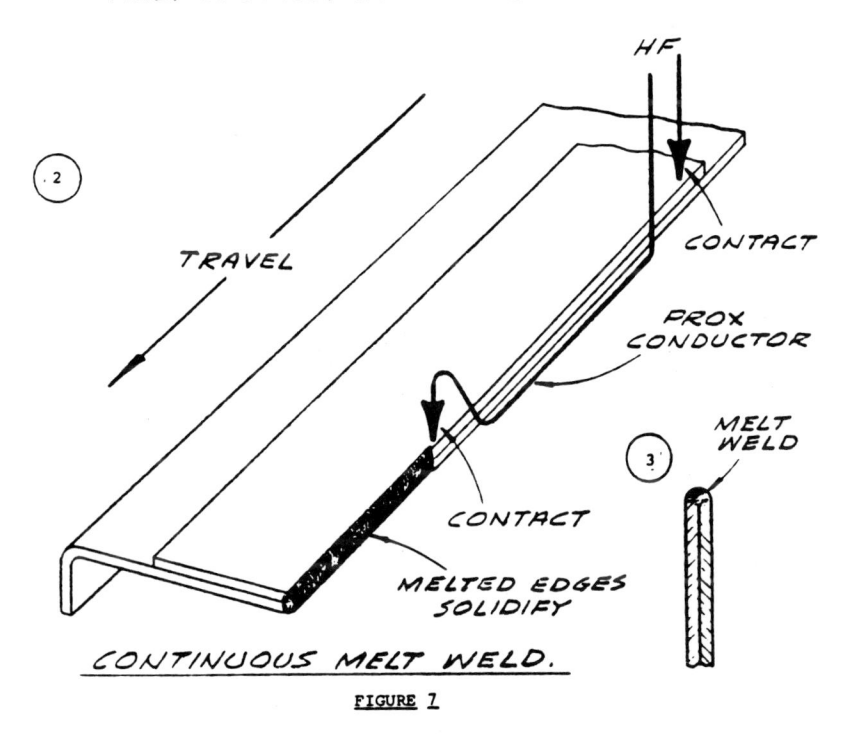

② TRAVEL / HF / CONTACT / PROX CONDUCTOR / CONTACT / MELTED EDGES SOLIDIFY

③ MELT WELD

CONTINUOUS MELT WELD.

FIGURE 7

SECTION IV:
Diffusion Welding

Introduction to Diffusion Bonding 255
Diffusion Bonding: No Longer a
   Mysterious Process 259
Pressure Welding by Rolling 263
Effect of Heat Treatment on Cold
   Pressure Welds 279
Diffusion Welding of Molybdenum 289
Diffusion Bonded Columbium Panels for
   the Shuttle Heat Shield 298
Advanced Diffusion-Welding Processes . . . 309
Roll Diffusion Bonding of Boron
   Aluminum Composites 338
Fabricating Titanium Parts With
   SPF/DB Process 353
Mechanical Properties of Diffusion-
   Bonded Beryllium Ingot Sheet 358

Introduction to diffusion bonding

by P. M. Bartle, LIM, AWeldI

The question 'what is diffusion bonding?' can be answered simply, but there are always subsidiary questions associated with any joining process that are more difficult to answer specifically. The objective of this article is to stimulate thought on these latter questions, particularly those concerned with the 'How, Why and When' of the application of diffusion bonding.

In the simplest case two flat clean surfaces can be joined by diffusion bonding, if they are heated while they are held together by an applied pressure. A metallic bond forms as a result of microdeformation of surface asperities and diffusion across the interface. The temperatures used are well below the melting points of the parent materials, and the applied pressure should be sufficiently low to avoid bulk deformation of the components. With like materials the microstructure of the joint should be indistinguishable from the surrounding material, and joint properties should equal those of the parent materials.

Since such an 'ideal' process is not commonly used it would not be unreasonable to assume that either it was applicable to very few materials or that it was a difficult process to apply. Neither assumption would be valid—the process can be applied to most materials and most combinations of materials; application is basically easy. The reason that it is not widely used is lack of knowledge of the process, and because for many run of the mill joints there are alternative established procedures which are economically more viable. The economics of a process are always a major concern, and therefore at present diffusion bonding is usually only applied where it has a clear technological advantage over other possible processes.

In many cases its advantages lie in the relatively low temperatures employed, coupled with the lack of macro-deformation of the components being joined. Once a condition has been established for diffusion bonding a particular material, a requirement often arises for the materials to be joined at a lower temperature. Techniques are available for modifying bonding conditions: these techniques are extremely important and I shall try to put them into perspective.

Many materials, metals and non metals, can be diffusion bonded directly to themselves and to other materials (Fig. 1). The range of materials that can be joined by the process can be greatly increased by using interlayers, the use and application of which are discussed. Equipment is often simple and is usually custom made. The process is inherently slow, but steps can be taken to reduce non bonding time to a minimum. At least for the time being diffusion bonding is a 'shop floor' rather than an 'on site' technique, and this limits its range of application. Even so, there is a wide range of possible applications, and a few basic types are discussed.

BASIS OF THE PROCESS

There are two variations of the process, solid state diffusion bonding, and liquid phase diffusion bonding. With the former all reactions occur in the solid state, while with the latter inter-diffusion occurs between two dissimilar metal parts such that the change of composition at the interface results in the formation of a liquid phase at the bonding temperature. Until the liquid is formed the liquid phase process is identical with the solid state process, but once the liquid has formed the process becomes virtually a brazing operation.

With diffusion bonding the objective is to produce a metallic bond such that the metallic structure extends without interruption from one component into the other across the whole area of the original interface. This means that the metal surface atoms of one component have to be brought to within interatomic distances of the metal surface atoms of the second component, and interatomic distances are usually less than 5 Å.

When two metal surfaces are brought together, true contact is only made at isolated points due to surface roughness (Fig. 2). Even at these points the metal surface atoms are separated by layers of oxide and contamination, which are thick by comparison with the interatomic distance. Before a diffusion bond can be completed the oxide and contamination layers have to be dispersed, and the surfaces have to be made to conform perfectly with each other. The procedures and requirements of the process are usually aimed at speeding these operations.

Mr Bartle is a Senior Scientific Officer in the Process Applications Department of the Institute.

THE PROCESS

It is perhaps convenient to consider diffusion bonding as a two reaction process, and review these reactions separately although they both proceed simultaneously. The two reactions are—the displacement of the two surfaces until they contact and conform perfectly with each other on an atomic scale; and the forming of the bond.

To reduce the difficulties involved in bringing two surfaces into perfect contact a high quality surface finish is normally specified. There has been little or no work in this country on surface finish requirements, but the Russian literature suggests that an 8μ in finish is a typical requirement. This is expensive to achieve and is one of the reasons why diffusion bonding is usually used only when it shows technological advantages over other processes.

When parts to be joined have been assembled the bonding load is applied, and this results in deformation of contacting surface asperities with consequent increase in the area of true contact. As the yield stress of the material decreases during heating to the bonding temperature, further deformation of the contacting asperities and increase in the area of true contact occurs. Once the bonding temperature has been reached yielding ceases, but deformation of the surfaces continues by creep mechanisms until the only interface regions not in contact are in the form of lenticular pores. Finally these are eliminated by a vacancy diffusion mechanism. Each of these stages is slower than its predecessor so that the final stage is reached very early in the process.

The need to apply pressure imposes a limit on joint design, and unless specialised techniques are used to apply the bonding pressure, joint configurations are restricted to those shown in Fig. 3. A butt and lap joint can be made between flat surfaces, or conical mating surfaces can be used. In some cases cylindrical mating surfaces can be bonded, but here the pressure has to be developed as a consequence of differential expansion either of the parts themselves or by using a low expansion collar and a high expansion mandrel.

Cleanliness is a pre-requisite to bonding, and steps have to be taken to ensure that surfaces are cleaned of oxide, grease and grit before assembly. Although sophisticated techniques such as ionic bombardment are used in some laboratory experiments, in commercial practice freshly machined surfaces, degreased immediately prior to assembly, would normally be used.

In addition to starting with clean surfaces the interface must be protected from contamination throughout the thermal cycle, and thus some form of protective atmosphere or a vacuum is required with most materials. Vacuum is commonly used since it is a suitable environment for a wide range of materials, but argon, nitrogen and a range of 'controlled atmospheres' have been used with some materials. With some of the noble metals, such as silver and gold, protective atmospheres are not required, and coatings of these metals can be used to obviate the use of protective atmospheres with lesser metals.

The mechanism is not understood whereby the layers of oxide and contaminent are dispersed so that the metallic bond can form. They are dispersed however and diffusion across the interface occurs, establishing the metallic bond and aiding elimination of the lenticular pores left after the early stages of the reaction which allow the surfaces to conform with each other. It is doubtful whether the atmosphere used to protect the joint plays a large part in the dispersal of the oxide and contamination, since the surfaces to be bonded are normally clamped together,* and once a bond has started to form large volumes of oxide and contamination will be sealed within the joint interface. It appears likely that much of the oxide and contamination is dissolved in the parent materials.

---

*Some equipments allow parts to be heated in a protective or reducing atmosphere before they are brought together for bonding, but these are the exception rather than the rule.

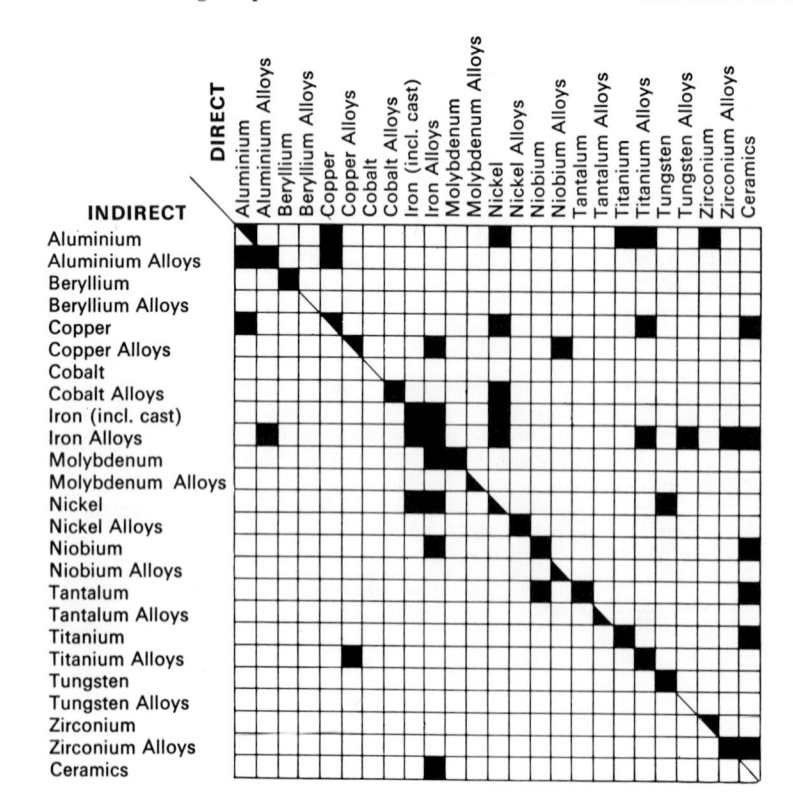

1 Some of the combinations of materials joined (directly or indirectly) by diffusion bonding

PROCESS PARAMETERS

In theory there are four parameters which can be controlled—time, temperature, pressure and surface condition. In practice temperature is the only one that can be effectively controlled. Usually the cost of improving surface finish is prohibitive and similarly time is kept as short as possible consistent with achieving reproducibility and producing high integrity joints. Pressure is usually kept below the bulk yield stress of the material, but close to it, leaving little margin for variation. An increase in temperature can, in some cases, counteract the adverse effect of a rough or inefficiently cleaned surface.

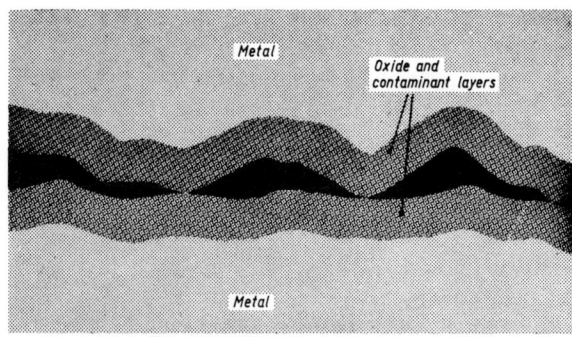

2 Schematic representation of two metal surfaces in contact

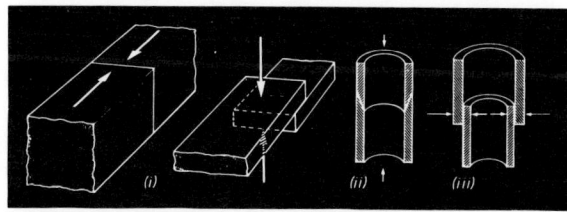

3 Basic joint configurations (arrows indicate direction of applied pressure). (i) Flat surfaces (butt or lap). (ii) Conical surfaces. (iii) Cylindrical (pressure developed as result of differential expansion)

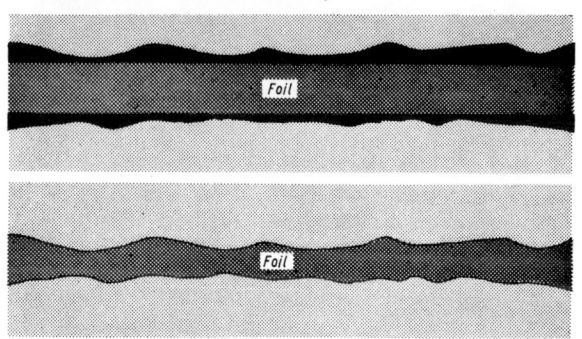

4 Use of a soft foil interlayer to aid the process of making mating surfaces conform perfectly with each other. (*top*) Before load is applied. (*bottom*) After load is applied

5 Example of a simple differential expansion vice, holding a cutting tool and tip ready for bonding

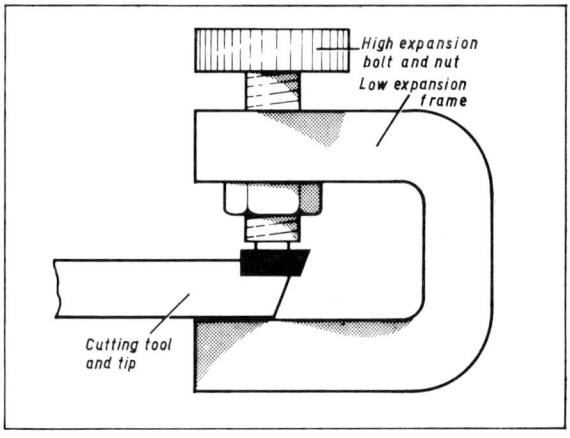

Typically, under commercial conditions, the temperature used for joining two parts with the same composition would be about $0.7\,T_m$, where T_m is the melting point of the material in degrees kelvin. With dissimilar materials the bonding temperature is probably near that for the material with the lower melting point.

Diffusion bonding is considered when a joint has to be effected without heating the parts above a critical temperature. If this temperature is below that at which bonding is normally carried out steps can be taken to reduce the bonding temperature. Substantial reduction is possible by using special surface preparations and cleaning procedures, but for normal commercial production these techniques are economically, if not practically, non-viable, although interlayers may be used to reduce diffusion bonding temperatures.

INTERMEDIARY MATERIALS

Similar and dissimilar materials can be used as intermediaries to promote diffusion bonding. They are normally used in the form of foils or coatings. Interlayers of the same composition as the materials being bonded act by aiding the mating of the surfaces to be bonded. In this case the intermediary is usually in the form of a soft foil into which both parts are impressed, so that the foil takes up the shape of the surface on either side of it (Fig. 4). Foils of a dissimilar material may act in the same way, but may be of a material that bonds readily to the parent material giving an alternative or added aid to bonding.

Surfaces can be plated with materials that bond readily to each other, but if the bonding temperature is to be substantially reduced by this technique the surface preparation before plating has to be to a very high standard. An intermediary material of one composition can be used with a material of another composition to give a suitable combination of dissimilar materials to allow liquid phase diffusion bonding to occur.

Interlayers can be used for reasons other than to reduce the bonding temperature. With dissimilar metal joints they can be used to act as a barrier between incompatible materials. With dissimilar material joints difficulty is often encountered because of the differing expansion coefficients of the materials being bonded. Joint failure often occurs from this cause, either during the bonding thermal cycle or thermal cycling in service. Interlayers have often been used to counteract the detrimental effects of dissimilar expansion with great success.

EQUIPMENT

Essentially diffusion bonding equipment consists of a press or similar device for applying pressure, a suitable heat source and usually means of protecting the joint from atmospheric attack while hot. There is considerable scope for variation in equipment design. Although in Russia standard models of diffusion bonding equipment appear to be available, this is not the case in the western world, where all equipment is custom made. To some degree this reflects the nature of the process, in that at first sight it is not a mass production technique, but one for application to specialised production. Lack of standard equipment is a drawback and undoubtedly accounts in part for the lack of utilisation of the process. The only advantage of custom made equipment is that, if properly designed, it can work better in some instances than standard equipment, which is often designed on a compromise basis to give a wide range of application.

Source: *Metal Construction and British Welding Journal*, May 1969

Probably the largest difference between various types of equipment results from the choice of the method of heating, in particular whether heating is to be localised to the joint area or the whole assembly is to be heated. This in turn effects the mode of application of pressure. With localised heating elaborate presses can be used, but if the whole assembly is to be heated then some system of rams or levers must pass through the furnace walls or into a cold zone of the furnace, or a clamping fixture which can be heated must be used. Differential expansion vices are examples of the latter (Fig. 5).

Diffusion bonding often involves the use of a vacuum chamber, and this greatly increases the total time required to produce a joint. Regarding the thermal cycle, the time at temperature is fixed, but the heating and cooling periods will depend on the equipment. With induction heating the time taken to reach bonding temperature can be reduced to a minimum, and even with vacuum equipment, back-filling techniques can be used to speed cooling. Where small units are involved multi-chamber systems can be used to increase the duty cycle on a high frequency power source, if this method of heating is used.

Assembly and unloading times form an appreciable proportion of the total time cycle, and this detracts from the economic viability of the process. This difficulty has been encountered in both electron beam welding and vacuum brazing, where pre-assembly and other techniques have been developed to reduce non productive time to a minimum. In many cases the total time cycle for diffusion bonding should be similar to that for vacuum brazing.

APPLICATIONS

So far diffusion bonding has found only limited application, but it is envisaged that eventually this will increase. It is interesting to note that in Russia diffusion bonding has superceded brazing for the tipping of heavy duty cutting tools. Tipping is being carried out on a semi mass production basis with 4–500 tools being tipped per shift.[1] Economies are claimed from using diffusion bonding instead of TIG welding on a 0·2% C:13% Cr steel (where cracking has been a problem), and for putting splines into tubes instead of machining from the solid.[2] Applications mentioned in the literature include electronic components, and parts for use in nuclear reactors and missiles. The joining of materials that are difficult to weld by conventional processes may be achieved by diffusion bonding, for example joining high speed steel wires,[3] and in the fabrication of dies.[1]

More generally there are two fields where the process is expected to find wide application, viz., joining dissimilar materials, and the production of complex components at present produced as castings or forgings. Regarding joining dissimilar materials, in addition to the fabrication of components from parts of differing composition, the process will also be used for the production of transition pieces. These will be of enormous benefit to many designers and fabricators, especially since they can be made under closely controlled conditions and joints between like materials can be used to fix them into the structures where they are to be used. Hardfacing may well prove to be an important future application for the process (Fig. 6) which is, for example, well suited to bonding seats into flow valve bodies.

The properties of complex forgings and castings are often below those desired because of the inherent limitations of the production processes. They are also very expensive. There is the possibility of using diffusion bonding to produce the same components from simple castings, forgings and sections cut from rolled plate or extrusions. Components produced in this manner may be both cheaper and of higher quality.

CONCLUSIONS

Diffusion bonding is a simple and versatile process that has great potential, but has yet to be widely applied. It can be used to join a wide range of similar and dissimilar materials (including non-metals), and techniques exist for modifying bonding conditions so that critical temperatures are not exceeded during bonding. These are the features of the process which attract interest at the moment. In many instances economic considerations are such that diffusion bonding will only be considered at present if it shows a technological advantage over alternative processes. The Russians, however, claim economic advantages from using the process, and therefore future developments may alter the criteria on which decisions to apply the process depend.

REFERENCES

1 KAZAKOV, N. F., *Autom. Weld.* 1960, **13,** 2, pp 32–37.
2 USHAKOVA, S. E., *Weld. Prod.*, 1963, **10,** 5, pp 34–36.
3 MASHKOVA, N. A. *and* KAZAKOV, N. F., *Weld. Prod.*, 1963, **10,** 12, pp 40–42.

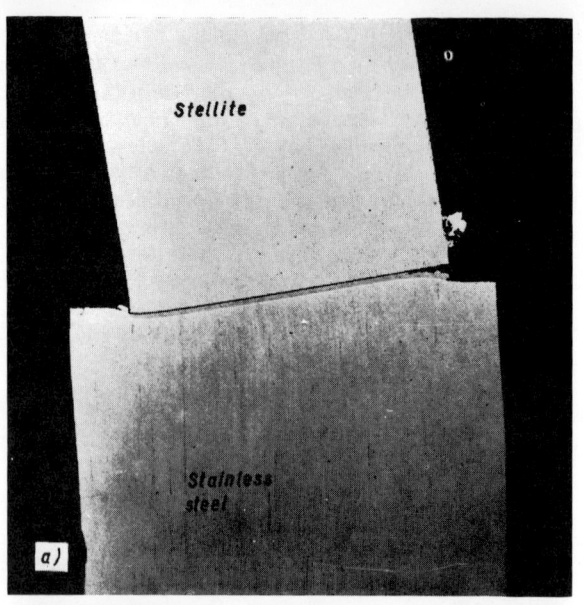

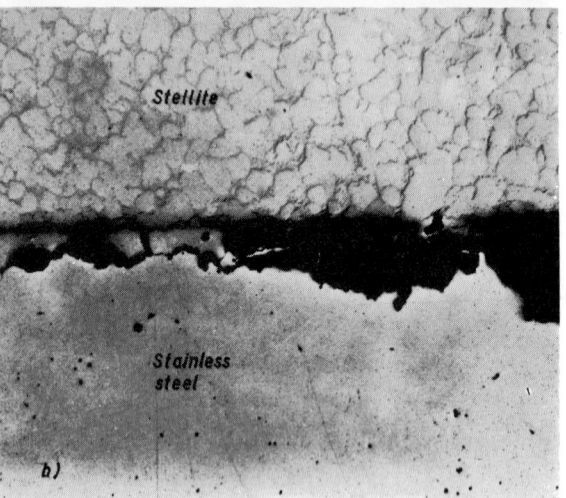

6 Stellite/stainless steel specimen after hammer testing. Specimen produced using pure nickel interlayer. (a) Macrosection, note deformation in stainless steel. (b) Detail of failure through nickel interlayer

Diffusion bonding: no longer a mysterious process

by SCHOLER BANGS, *western editor*

The massive titanium structure on the cover of this issue makes history. Suddenly titanium is cost-weight-strength competitive with high-nickel steels. Look again. It leapfrogs, in the way it was made, all conventional fabricating methods. Each bolted segment could have been machined from a forged titanium billet, but wasn't. Precision casting? No. Weldment? Yes, but then again, no.

Consider 533 separate pieces of titanium plate positioned to form the structure on the cover. Heat, pressure, and time combined to fuse all the bits to substructures. The final result: a single piece of parent metal. Take a macrograph of a cross section at any bond and no weld line will be visible. All you'll see is homogeneous parent metal, yet composed of pieces joined by diffusion bonding.

Diffusion bonding isn't anything new. It has been kicking around and kicked around for 20 years as a costly way to put together aircraft and engine components. Diffusion bonding is generally defined as a solid-state process using pressure and heat to join either like or dissimilar metals with or without the use of another material between.

Two controversial aerospace vehicles have brought diffusion bonding into the light—the B-1 bomber and the still-under-construction space shuttle. The imagination of the Rockwell International designers has helped to launch diffusion bonding as an actual production method. They succeeded in diffusion bonding the assembly on the cover which will be joined to the movable swept wings of the B-1 bomber eventually.

Figure A. *Just inside a retort which will hold titanium beam parts is an end rail of 22-4-9 stainless steel. Machined grooves in this rail allow air to escape when the vacuum pump is turned on.*

The B-1 program is scheduled to move into full production before September 30, 1976, if the money is available. For the first time, production quantities of diffusion-bonded parts will be made with ordinary hydraulic presses. Parts will come from the press at nearly finished size, saving prodigious amounts of machining. Fillets will no longer have to be machined in after bonding. A fleet of 244 bombers, to cost $21.5 billion is proposed.

The technological advances at the B-1 division of Rockwell International, Los Angeles, are already making diffusion bonding more attractive for other applications.

In mid-1979, the first shuttle ship of the U.S. Space Shuttle program should begin round trips to and from an earth-orbiting space station. The first two of five shuttles are being assembled at Downey, Cal., at Rockwell's Space Division. Each shuttle can carry a payload of 65,000 pounds in a cargo hold behind a sealed cabin carrying a crew of four. Rockwell is building the engine mount and tail assemblies of titanium. The engine mount of each space shuttle will have 28 diffusion-bonded titanium parts. They'll range from large frames to interconnecting box tubes. Although the aftersection of a space shuttle's body is light, the engine mount will have to absorb a combined thrust of nearly

1.6 million pounds from three main rocket engines during at least 100 space flights. Allan H. Carlson, Rockwell's supervisor of manufacturing productivity, tells why the diffusion-bonded titanium structure won out for the job:

In November 1972, we were considering aluminum because it would be light enough for an engine mount. But aluminum would offer production drawbacks in machining shapes and fillets from solid stock and in assembling finished parts. We had just about decided on an aluminum-and-titanium open truss structure for accessibility to engine areas, but by April 1973, we had swung toward a structure of hat sections, mechanically assembled.

We could build the hats in a number of ways:
Machined plate—straightforward, but expensive and high risks of rejection owing to distortion during machining.
Hand forgings—production-proved, but too costly.
Closed-die forgings—a lot less machining, but they would be limited in ease of design changes during production. Any adjustment would involve a new set of costly dies and a long wait. Closed-die titanium forgings cost least, but only if no complications arose. We had to assume that complications were inevitable.

The remaining, and best choice (also the one we picked) was a diffusion-bonded titanium assembly using the tooling and process of our division that had produced, successfully, quantities of diffusion-bonded parts for the B-1 bomber.

The scheme might even show a profit in a production run. The process offered ease of change, with little lost time. Rockwell designers could change shapes simply by machining new forming blocks. Long waits for delivery of new forging dies from eastern suppliers would be avoided.

The diffusion-bonded parts would have the best mechanical properties, with no loss of strength at joints. Further, the engine mounts would be lighter than those made by a closed-die forging. As Carlson said, advance design studies showed diffusion bonding to be the only feasible way. Diffusion bonding was also strong for its weight. Bonding laminations of high-strength thin titanium strips into a homogeneous mass stronger than a forged part. Why, then, if it's so good, isn't diffusion bonding more popular with metal fabricators? Rockwell's B-1 diffusion bonding manager answers:

Diffusion bonding has long been considered a mysterious process, has lacked publicity, lacked someone to sell the process. One of the best selling jobs for diffusion bonding took place 5 years ago when under Air Force sponsorship, we diffusion bonded a Sikorsky H-53 helicopter rotor hub; which had been in production as a titanium forging that just about taxed the capacity of the only 50,000-ton forging press in the United States. The hub was 5 feet in diameter, weighed

DIFFUSION BONDING: AN OVERNIGHT SUCCESS?

In a 1967 report, Rockwell engineers Joseph Melill and Julian P. King, Jr., stated that technically one can diffusion bond just about any metal to any other. They listed as diffusion bondable titanium, steel, stainless steel, aluminum, nickel, copper, beryllium, and others, and combinations of two or more different metals. At that time, nobody questioned the industrial potential of the process. As early as 1953, North American Aviation, now a part of Rockwell International, used the diffusion bonding process in production. They bonded uranium into aluminum cylinder containers and then into nuclear reactor fuel rods. This process produced a complete metallurgical bond. Vastly expanded technology was needed to expand the capability of the diffusion bonding process to large volume structures and complex ones.

In the early 1960's, research covered many types of structures. One of the first diffusion bonding production contracts covered the manufacture of 116 T-stiffened titanium sandwich panels, each 3 by 11 by 0.5 feet for the Argonne National Laboratory. A series of research contracts led to the fabrication of complex hollow channeled parts, roll diffusion-bonded structures 36 feet long and massive 5-foot-diameter rotor hubs.

One great advantage of the process is that it maintains the strength of the individual pieces of material combined by the process. If a tough fracture resistant structure is required, the materials that go into the structure have to be tough.

Finding or even defining tough materials in 1970 at the beginning of the B-1 program was a difficult task. Mill suppliers could not guarantee delivery of fracture tough materials. Since the B-1 is to be the toughest aircraft ever built, tough materials were called for: After intensified research, Rockwell produced 39 different specifications for production and processing of fracture tough steel, titanium, and aluminum. Changes were made in mill heat treating processes and blast furnace chemistry along with improvements in rolling practices. By 1973, a few basic mill suppliers were able to guarantee delivery of fracture tough materials. Now fracture tough materials are available to all industry.

Diffusion-bonded structures in all critical areas of the B-1 not only live up to but also surpass the new fracture toughness standards. A 13-foot-high, 1750-pound vertical stabilizer support frame completed fatigue tests equivalent to four vehicle lives (54,000 hours, the longest life of any U.S. bomber ever produced). The structure was then subjected to a static residual strength test equal to 165 percent of limit load. When the part finally did fail, one of the side rails bent slightly.

To date, Rockwell has produced 484 diffusion-bonded structures; only nine have been rejects. Of the nine, five proved to be lemons during the early design verification tests.

1,000 pounds and strained the dimensional limits of the press. We duplicated the forged hub with diffusion bonding. We assembled laminates, cut starting weight from 1,000 pounds to 470 pounds, and reduced by two-thirds the machining needed to get the finished hub to 230 pounds. Even more important, all we used was one-half the capacity of a 4,500-ton press.

Dimensionally, diffusion-bonded parts probably are limited only by the length and breadth of press beds available. Rockwell has evaluated for Boeing a 22-foot-long, 2,000-pound diffusion-bonded substitute for the 747 jetliner main landing gear beam, now being fabricated from extrusions and sheet parts with mechanical fastenings. The largest of Rockwell's space shuttle diffusion-bonded assemblies are produced with two-thirds the capacity of an in-line composite of three 6,000-ton hydraulic presses.

Rockwell's titanium diffusion bonding process

Rockwell's process starts with the designer who breaks down complicated structures into simplest forms to speed production machining. Then the parts are fitted into a container (called a retort). Tooling blocks and spacers of 22-4-9 stainless steel fill the voids between the titanium pieces to hold the structure to dimensions.

The container, or retort, is welded of a cheap grade of 0.063-gage ⅜2 muffler steel, conforming to the shape of the structure. A lid, welded on the retort, seals the container, which is to be evacuated and maintained at no more than 10^3 torr during the entire bonding process. After a part has been bonded only the stainless steel spacers are reused. The retort, which is cheap, is scrapped. Figure A shows a retort holding titanium parts for a beam in position. The steel tube at the end of the retort leads to a vacuum pump. Just inside the retort is a width-spanning end rail of 22-4-9 in which grooves have been machined to allow air to escape when the vacuum pump is turned on. Similar 3-inch, 22-4-9 plates line the bottom, walls, and opposite end of the retort, and one will cover the filled retort before the lid is welded on.

With titanium parts in place, filling the empty spaces with tooling blocks begins. This process takes place in a semi-cleanroom to prevent contamination. The tooling, or detail blocks, come in a vast variety of shapes and sizes. Possibly 600 may be required, tier upon tier, to fill the retort to its brim.

During the loading, the builders insert titanium slip shims to separate the tooling blocks slightly from titanium parts. Afterwards, workers remove these shims to create a vacuum path that allows air to escape during evacuation of the sealed retort. Because titanium expands at a rate different from the steel tooling, the designers size the tooling to compensate.

The volume of the titanium parts is slightly greater (by a measured amount) than the amount of space left in the retort. After bonding, excess titanium appears as flash. Equality in the amount of flash all over the part is one means of confirming the uniformity of pressure and titanium flow, which indicates the quality of the bond.

Once the lid is welded on the retort, evacuation begins. To be certain that the lid is perfectly sealed, the technicians flow helium along the weld bead. Even the smallest amount of helium drawn into the retort will be detected immediately by a mass spectrometer connected to the vacuum pump. Should a leak show up, the technicians locate it and seal it with fresh weld. A high vacuum, below 1 micron, is essential. Any nitrogen and oxygen left behind will contaminate the bonding.

In final preparation for the press, the technicians cloak the retort with reusable ceramic blocks which are covered by steel plates containing thermocouples. The ceramic blocks pressing against the top and bottom of the retort contain heating wires that bring the entire assembly to 170°F. The ceramic blocks on the sides and ends of the retort transmit heat and pressure to the assembly during bonding and at the same time insulate the press from excessive heat.

On the bonding press, Rockwell's largest, side and end hydraulic jacks, together with pressure from the vertical cylinders, exert 2,000 p.s.i. on the retort in all directions. In actual production, the technicians heat the completed assembly packs, (retort,

DIFFUSION BONDING PROCESS

Design — Part details — Tool details — Layup in retort — Press — Evacuate-heat-press — Reuse tooling — Bonded part — Remove part

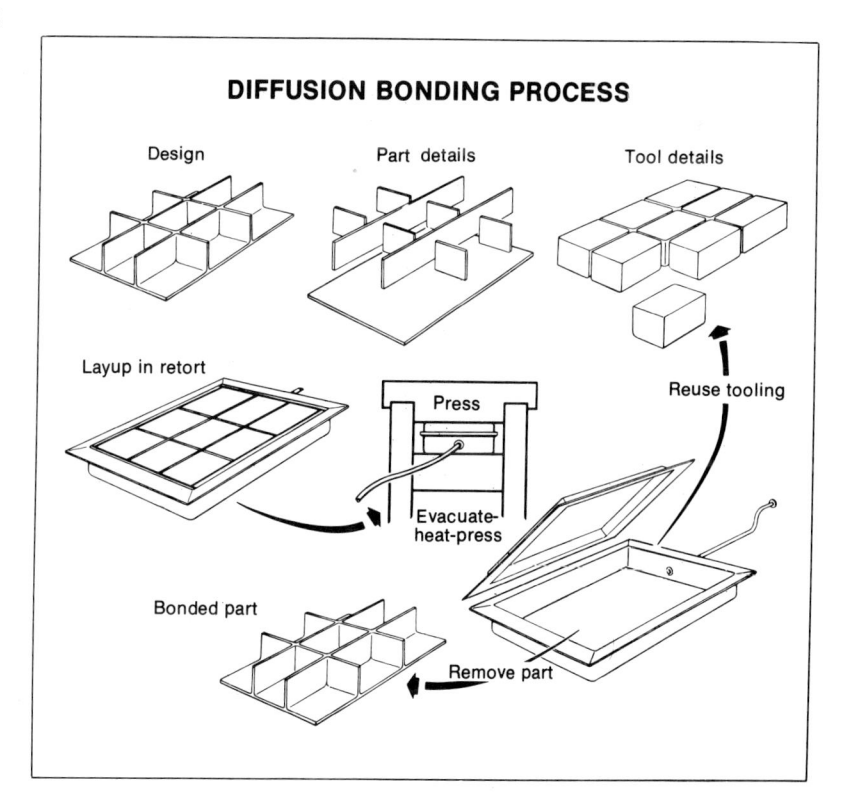

When a retort is positioned in Rockwell's largest press, side and end hydraulic jacks together with the vertical cylinders exert 2,000 p.s.i. in all directions.

Corner fillets require no machining. Reverse fillets machined on 22-4-9 stainless steel tooling blocks accurately mold fillets during bonding.

heating pads, insulation and evacuation equipment) before moving them into the press. Large structures may require a preheat and soak of possibly 40 hours. Several packs may be in assembly and preheat at one time. The actual time in the press at 2,000 p.s.i. will vary from 2 to 12 hours depending on the shape of the structure and the mass of titanium. Four days of cooling follow before the assembly is dismantled and the retort cut open. The pieces of titanium now are bonded.

Machining titanium details for bonding is a low-cost operation because all cuts are straight—and planned to be. Bonding of straight-cut ends to opposing flat surfaces saves, because bonding pressure produces corner fillets which require no machining. Reverse fillets machined on 22-4-9 tooling blocks, where they meet titanium intersections, accurately mold fillets during bonding. The pressure crushes out any irregularities at an intersection.

Diffusion bonding of titanium requires the utmost cleanliness. Rockwell experience dictates the need to assemble the parts in the retort to get them under vacuum within 3 days after cleaning. Otherwise they are stored in argon. To prevent sticking to titanium, all 22-4-9 tooling is surface oxidized by oven baking at 1,400°F for 4 hours and must be used within a week.

Testing procedures

Quality control starts well before the bonding process takes place. Careful preparation of part surfaces, keeping them smooth and free from foreign matter helps assure bonding. Carefully controlled heat application and pressing is also important.

When the part is removed from the retort, after bonding, it is given a thorough dimensional and visual inspection looking particularly to see if fillets have formed properly. No attempt is made to measure radii of fillets. Then extensions of the part are cut off and tested. These extensions, looking rather like metal ears, are excess material, but have the exact same composition as the rest of the part. They are tested for metallurgical content.

Other tests: one ear section is tested for mechanical strength; another undergoes chemical analysis. A metallographic evaluation—in which the part is ground, polished, and acid etched, then viewed under X350 magnification—makes sure no voids or contamination are present and also sees that grains grow clear across the bond interface. Before machining, the part will be ultrasonically tested. After machining the bond undergoes a fluorescent penetrant test.

Rockwell's recent successes in diffusion bonding titanium for vastly more complex structures of the B-1 bomber and space shuttle might not lead to enthusiasm for structural diffusion bonding of all the other metals suggested by Melill and King. Rockwell's John Welms, a program relations engineer, explains by saying that the process pays for complicated titanium structures because a relatively inexpensive steel, 22-4-9, is perfect for detail parts under bonding pressure. At bonding temperature, as titanium becomes super plastic, the 22-4-9 steel used in the detail blocks is still solid, maintaining the dimensional integrity of the structure while the titanium is bending. 22-4-9 steel is also ideal in that its high carbon and low nickel content (Cr 22-Ni 4-Mn 9) does not contaminate the nickel-hungry titanium. In contrast, diffusion bonding of a steel structure would require ceramic blocks, extremely costly to make for detail tooling. Ordinary welding, forging, casting, and machining would cost less than bonding every time. Ceramic blocks would also be necessary with other metals, such as beryllium, already in use in aerospace. ■

262

PRESSURE WELDING BY ROLLING

By L. R. Vaidyanath,

M. G. Nicholas,

and D. R. Milner

The mechanism of pressure welding as it is operative in roll bonding has been investigated. A test has been devised and bond strengths measured for aluminium, copper, lead, tin, and zinc welded at room temperature. In all cases a 60–70% reduction in thickness was required to approach maximum bond strength, although the threshold deformation required for the initiation of bonding varied. A theory is proposed for the maximum strength that can be attained on the assumption that bonding is prevented by the presence of the oxide film, and that at an early stage in the progress of the composite through the roll gap the two oxide films come together and thereafter behave as one. The virgin metal area of potential bonding is then given by the increase in area of the interface, and taking account of the triaxiality of stressing this results in a maximum ultimate shear strength given by: $U.S.S._{weld} = R(2-R) U.S.S._{solid\ metal}$, where R is the reduction in thickness of the composite.

However, there are many conditions under which this maximum strength is not achieved, since the virgin metal areas do not always join. Hence the initiation and development of bonding has been studied and was found to be promoted by increasing the pressure, increasing the temperature, increasing the roll diameter and decreasing the roll speed. Prolonged exposure of the surfaces before rolling led to the formation of adsorbed layers of water vapour which inhibited welding. This information has been related to the possible mechanisms which could control bond formation; it is concluded that a correct model must involve considerations of atomic movements during deformation, local readjustment at the interface by diffusion processes, and the solution of entrapped gases.

Introduction

AMONG the welding processes, pressure welding is regarded as a relatively specialized technique, applicable only to specific types of joining problems. For this reason it has not been extensively investigated, although the important factors and the conditions of application of the process have been fairly well defined empirically. The mechanism of bonding is not clear despite a number of researches into the subject, and various workers have postulated that it is dependent on a variety of different factors. The present investigation, while it may contribute to the relatively limited data available on the practice of roll welding, has been primarily concerned with the mechanism of bonding, and the experimental approach has been chosen with this end in view.

The present state of the practice and theory of pressure welding has been reviewed in the first paper in this series by Professor E. C. Rollason; the factors of temperature, pressure, degree of deformation, oxide break-up, flow of surface asperities, and elastic recovery have all been shown to exert a marked influence on bonding, and some workers have proposed that one factor in particular controls the process. Since the advocates of the importance of a specific parameter can produce considerable supporting evidence for their ideas, it is probable that the controlling factor will be shown to vary with the conditions of welding.

A preliminary analysis of published information led to three conclusions. Firstly, whilst the importance of many factors has been demonstrated, for instance, surface preparation before welding, the inherent basic parameter (which might, in this example, be oxide film thickness, surface roughness, or the presence of adsorbed gas layers), is not always obvious, and cannot be evaluated because information gained from the various welding systems can not be correlated. Therefore, before carrying out an analysis to decide on the model that has to be developed for an understanding of the mechanism of pressure welding, it is necessary to amass a body of relatable data on one simple welding system. Secondly, whilst the purpose of pressure welding investigations is to determine the processes that control bonding, no really satisfactory test method has been set up for the measurement of bond strength,

---

Manuscript received 5th August, 1958.
The authors are at the Department of Industrial Metallurgy, University of Birmingham.

and some of the techniques used, particularly for small-tool welding, give results which are in fact misleading. Thirdly, a major difficulty in the study of the mechanism of pressure welding is that it is linked with the complex problem of metal flow and deformation patterns in metal working. Thus, whereas the formation of a bond between two metal surfaces is likely to be dependent on such factors as the presence of oxide or other films, and the closeness of approach of the two surfaces, the degree of oxide break-up and the deformation system will depend on the shape of the deforming tool. It is therefore important to separate, as far as is possible, the study of bond formation across an interface from the problem of the plastic flow of metals under various tool shapes, as this is a major field of study in its own right.

Pressure welding by rolling was therefore chosen as the most suitable system for investigation, because the deformation pattern is simple and is reasonably well understood. In addition, for this method a test could be envisaged, the results of which would be more amenable to analysis than in other cases. Using the roll bonding technique and this test the mechanism of pressure welding is being investigated, and this paper describes the first phase of this work, which has been concerned with the welding of similar metals at room temperature. Complementary to the determination of bond strengths, a microscopic survey has been made of the characteristics of weld interfaces. The results obtained have led to a model which explains the role of the oxide film and allows the calculation of the maximum bond strength that can be obtained for any given deformation. However, under many circumstances there is no bonding, or the maximum strength attainable is not achieved; no positive contribution can be offered to the solution of this problem, but the evidence is assessed to show the factors which a correct theory would have to take into account.

Measurement of Bond Strength

A satisfactory test should measure the strength of pressure welds by a method that allows comparison with the inherent strength of the metal. This can be done for roll welded material by using the test shown in Fig. 1. A test strip, about ½ in. wide and at least 0·2 in. thick, is cut from the rolled composite, and saw cuts are made so as to isolate an area which is then tested in shear by pulling the test strip in tension. Originally it was thought that, to get a true shear strength result, it would be necessary to work with triple composites so that the stress conditions would be symmetrical and thus eliminate any twisting of the specimen and the consequent introduction of a tensile component. To isolate the test area with the triple composites, a central cut was made by drilling a small hole and then cutting outwards with a jeweller's saw; the alternative of isolating the area under test by three drilled holes was found to be equally satisfactory, although the accurate delineation of the area was more difficult in this case. However, working with triple composites is tedious, and it was found that the same results could be achieved with duplex specimens, provided that the length, W, of the area under test was not more than 0·8 of the total thickness, T, of the composite, so that no twisting or tensile component was introduced. For the testing of relatively weak joints, the length of the area under test can be considerably increased without the introduction of a tensile component, and in fact, to obtain accurate values of low bond strength, this is advisable. To evaluate very low bond strengths, it is essential to increase the length under test, as a small area often provides insufficient strength for the joint to withstand the most careful handling. A final composite thickness of 0·2 in. was

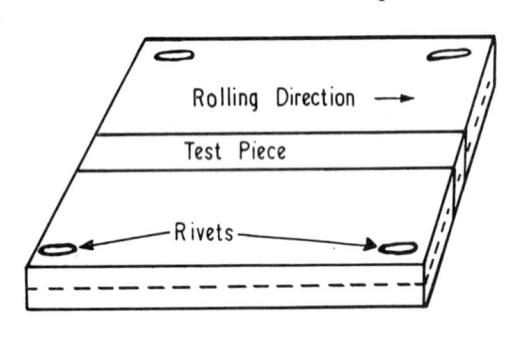

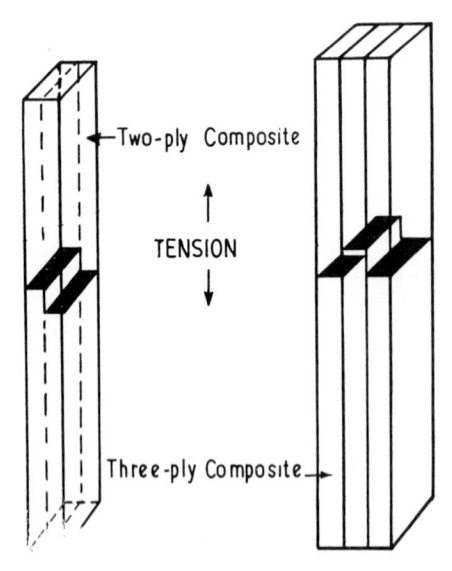

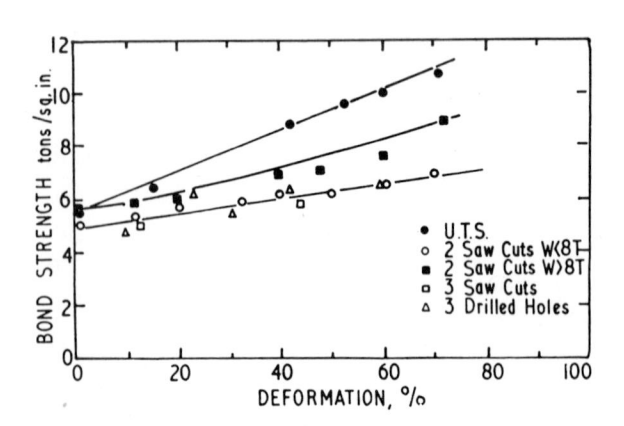

1—*Test to determine bond strengths*

2—*Comparison of the strength of aluminium determined by symmetrical and asymmetrical methods of isolating the area under test, and U.T.S. as determined by normal tensile tests*

found to be advisable, otherwise the location and extent of the saw cuts presented difficulties.

The major advantage of this test is that it permits a comparison with the strength of solid metal which has undergone the same working treatment; throughout this work the general practice has been adopted of rolling a two-ply composite, and a solid piece of metal of the same overall thickness and in the same state of hardness, and then comparing the strengths obtained. Figure 2 shows the strength of solid aluminium as measured by this test for both duplex and triple specimens, and gives also the ultimate tensile strength of the material as determined from conventional tensile tests. The meaning that can be ascribed to the results of a test of this nature cannot be precise because of the nature of plastic flow, but the generalization of Sachs and van Horn,[1] that 'The shearing strength of most wrought metals and alloys ranges from 55% of their tensile strength for severely cold worked metals to 75% for the annealed soft condition', fits very well the results obtained by the present test.

Results of Bond Strength Determinations

Using the test described, bond strengths were measured and compared with solid metal strengths for fully annealed aluminium, copper, tin, lead, and zinc rolled at room temperature. Composites measuring 4 × 4 in. were degreased in trichlorethylene vapour, scratch-brushed, pinned in the corners with rivets to prevent separation in the roll gap, and were rolled, within 2 min of the scratch-brushing treatment, in an 18 in. dia. two-high mill at 200 ft/min. To obtain the necessary 'bite' with high deformations, the rolls were degreased and treated with magnesia powder. The bond strength was then determined at the centre of the composite; the results obtained are shown in Fig. 3.

Attempts to weld magnesium at room temperature showed that no bond strength could be obtained with up to 50% deformation, and the metal disintegrated at higher deformations.

When cutting out the specimens for testing, it was found that there was a difference between the degree of deformation at which the two members of the composite appeared to adhere, and the deformation at which a specimen could be isolated without the joint breaking. Since this difference was not eliminated by the most careful handling and cutting-out techniques, it was concluded that it was due to the readjustment of the residual stress system, leading to a stress in the joint area, as each cut was made in the composite during the process of isolating the test area. Thus, for example, with aluminium the composite would have apparently adhered at a deformation of 35%, whereas no positive bond strength could be obtained for deformations less than 40%.

These bond strength determinations showed that in all cases the solid metal strength was reached at about 60–70% deformation, whereas the threshold deformation for bonding varied from one metal to another. This means that for roll bonding, at least, it is not entirely adequate to state that one metal welds more readily than another, since they all require 60–70% deformation for maximum strength; rather, welding

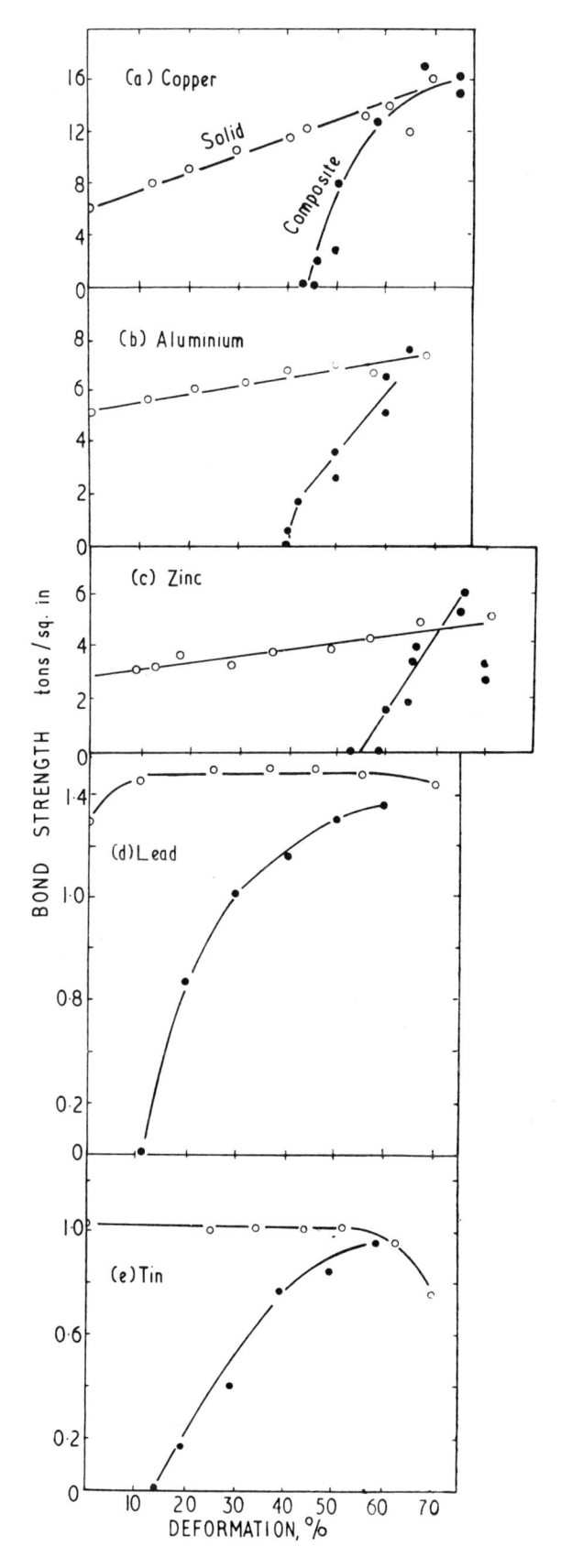

3—*Bond strength as a function of deformation for copper, aluminium, zinc, lead, and tin, and strength/deformation curves, determined by the same test, for solid metal*

Source: *British Welding Journal*, Jan 1959

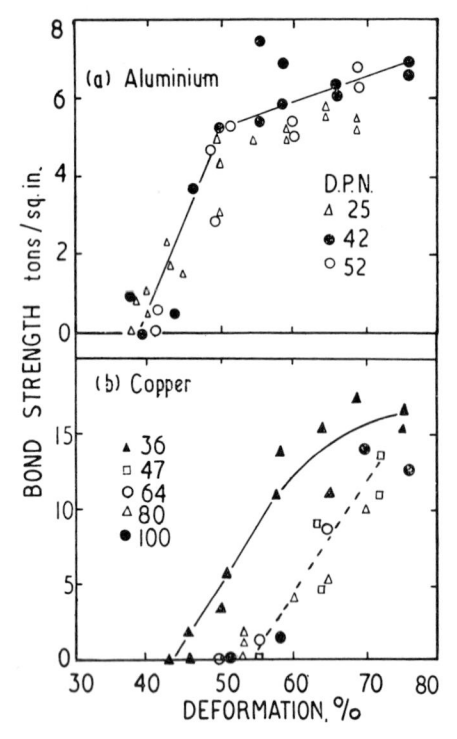

4—*Effect of various degrees of pre-straining on the welding of* (a) *aluminium and* (b) *copper*

can be initiated more readily with some metals than with others.

Since a theory of the mechanism of pressure welding must explain the different deformations required for the initiation of bonding, any trend that can be observed may provide a pointer to the correct model. Two possible causes are immediately obvious for the variation of the threshold deformation: it could increase either with the increasing melting point of the metal, or with the hardness. The latter possibility conflicts with the results of Pugh,[2] and later Tylecote et al.,[3] who showed that increasing the hardness of a metal by pre-straining made small-tool welding easier. For this reason, the effect of pre-straining on the bond strengths in aluminium and copper was investigated; it was found (Fig. 4) that with aluminium increasing the initial hardness had no effect, whereas with copper it made welding more difficult. Since a repeat of the small-tool welding work confirmed the results of the other workers, it was concluded that their results were not fundamental to a mechanism of pressure welding but specific to small-tool welding, and were probably due to an alteration in the deformation pattern and/or the joint area. Unfortunately, it was not possible to extend the roll bonding investigation, with 4 × 4 in. specimens, to include harder metals such as iron and nickel, because of the limiting load-carrying capacity of the rolling mill (200 tons); with smaller specimens (1 in. wide and 0·2 in. thick), it was found that the threshold deformation for Armco iron was 62%, but as will be shown later, welding is not independent of specimen size, so that this result is not strictly comparable with the rest of the series.

Bond strengths at very high deformations were difficult to obtain, because the mill would not bite on the specimen and rejected it. To some extent this could be overcome by chamfering the leading edge, but because of this problem, and the fact that when dealing with harder metals less strain would be imposed on the mill if the deformation could be given in several passes instead of in one pass, the use of a multi-pass rolling technique was investigated, using aluminium and copper. It was found that no bonding could be obtained unless the first pass was sufficient to initiate welding, but thereafter substantially the same bond strength was achieved by multi-pass as by single-pass rolling.

Microscopic Examination of Weld Interfaces

Sections taken from the centre of typical composites (corresponding to the region in which the bond strength determinations were made) were subjected to a microscopic examination and it was found that the weld interfaces could be divided broadly into two types. The first type, which occurred with tin and lead, (*see* Fig. 5), showed a continuous boundary; the second type, characteristic of aluminium, copper, and the narrow Armco iron specimen (*see* Fig. 6), showed a discontinuous interface, having regions of apparently bonded metal interspersed with what have, for reasons which will become clear later, been called primary discontinuities. Zinc appeared to be an intermediate case, in that it showed both types of boundary in different regions of the joint interface. In no case was grain growth across the interface observed, although tin and lead had of course completely recrystallized.

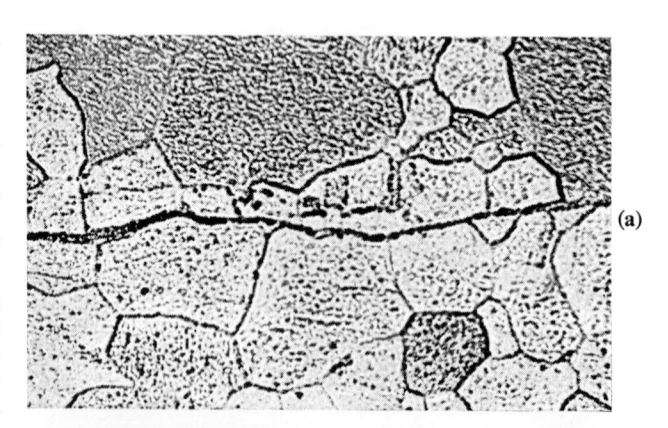

5—*Continuous type weld interfaces*: (a) *tin boundary, 60% deformation, ×250, etch 5% HCl in alcohol*; (b) *lead boundary, 20% deformation, ×90, etch fuming HNO*$_3$

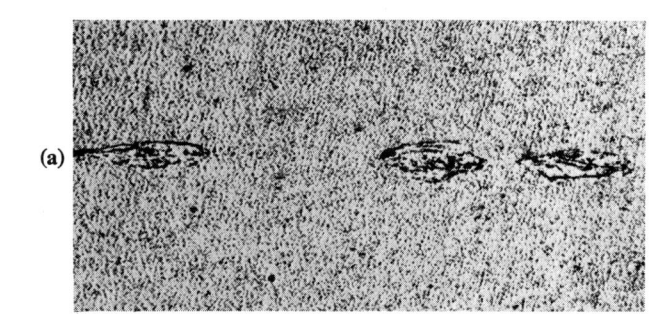

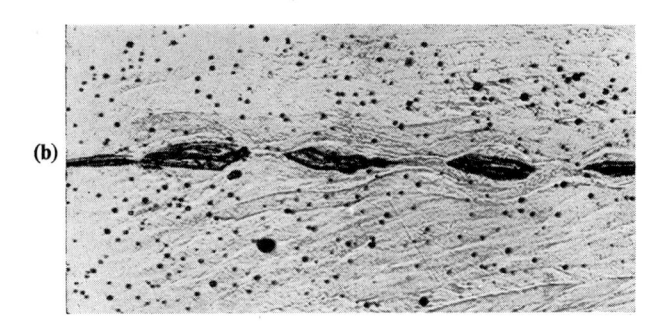

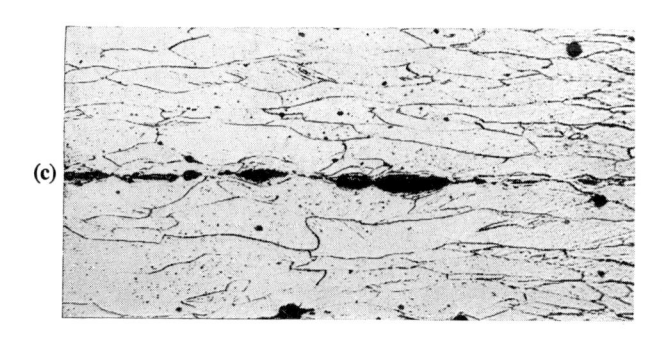

6—*Discontinuous type weld interfaces*: (a) *aluminium boundary, 55% deformation, ×250, etch 5% NaOH;* (b) *copper boundary, 64% deformation, ×450, etch NH$_4$OH/1% H$_2$O$_2$;* (c) *Armco iron boundary, from specimen initially 1 in. wide by 0·2 in. thick, 64% deformation, ×270, etch 2% Nital*

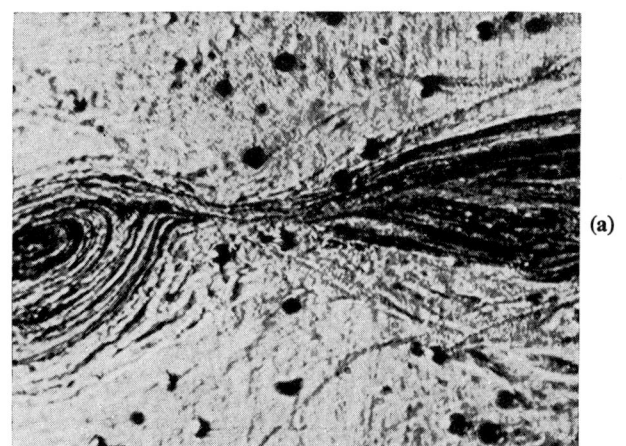

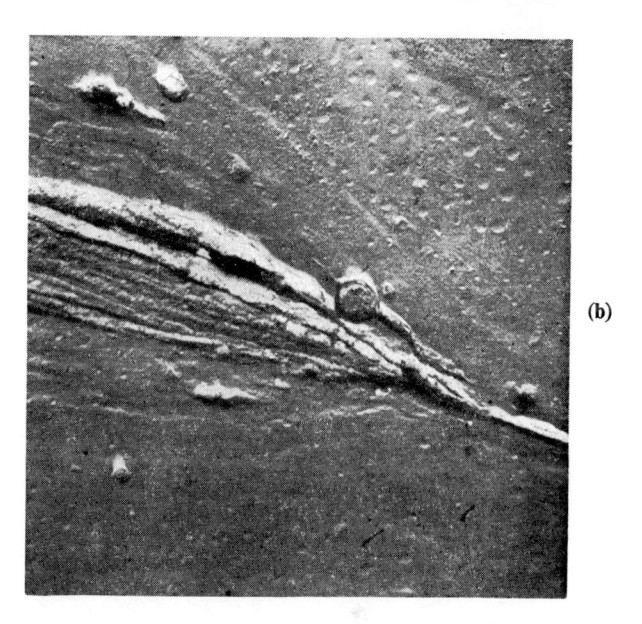

7—*Primary discontinuity in copper showing the heavily worked structure;* (a) *oil immersion micrograph ×2000, etch NH$_4$OH/ H$_2$O$_2$;* (b) *electron micrograph, ×3500; note the unwelded section of the interface*

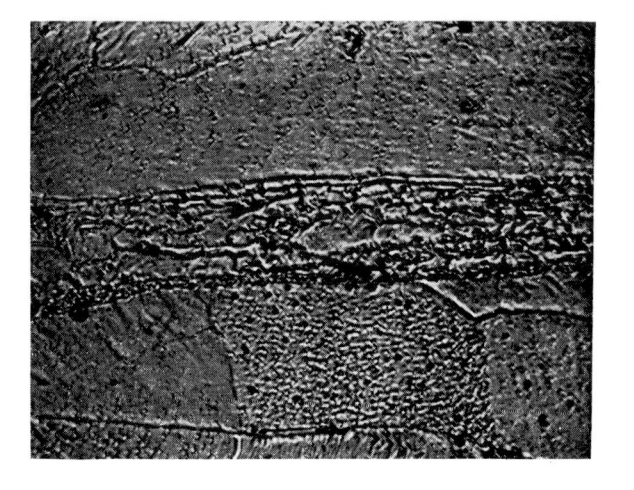

8—*Tin boundary showing fine-grained recrystallized structure, ×2000, etch 5% HCl in alcohol*

In regard to the metallographic technique in this work, conventional emery rough preparation was used, followed by diamond dust polishing, possibly finishing with a light 'Silvo' polish. Electropolishing tended to attack the inclusions very heavily, thus giving the misleading impression that the interfaces consisted of a series of voids. To a lesser extent, prolonged etching or deep-etching reagents lead to similar misleading results.

To learn more about the structure of these boundaries, the investigation was extended to higher magnifications, in one case (copper) using an electron microscope. At these high magnifications the discontinuities in copper and aluminium had the appearance of a heavily worked structure (Fig. 7), whereas tin and lead consisted of a very fine-grain recrystallized structure interspersed with apparent line discontinuities, (see Fig. 8). Since the discontinuous characteristics of the aluminium and copper boundaries lent themselves to a more detailed investigation, a survey was made of the number, size, and distribution of the

Source: *British Welding Journal*, Jan 1959

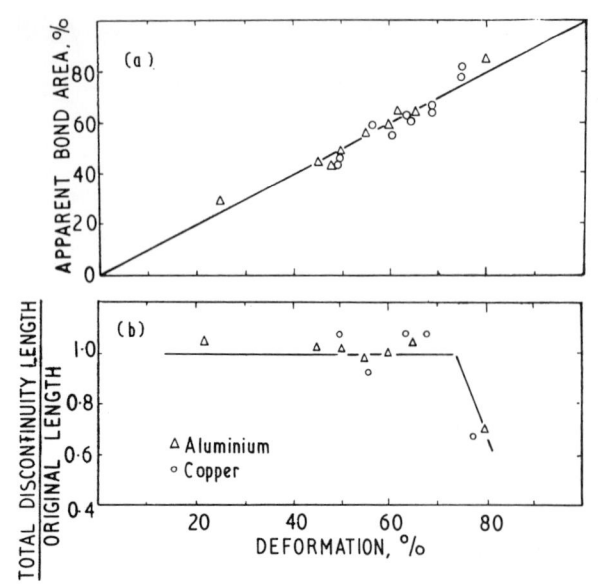

9—(a) *Development of apparently bonded area, and* (b) *total length of the primary discontinuities related to original length of specimen, both as a function of deformation*

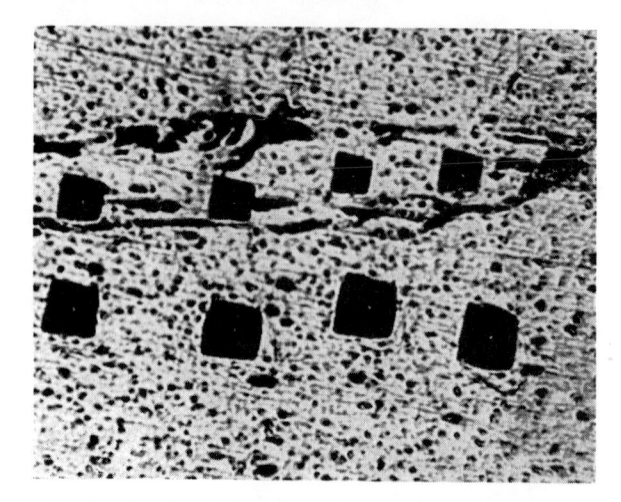

10—*Micro-hardness indentations, showing relative hardness of the primary inclusion and the adjacent metal, ×940, etch 2% HF/27% HNO₃*

side and the primary discontinuities could be clearly distinguished.

Micro-hardness measurements showed the primary discontinuities to be very hard, of the order of two to three times the hardness of the adjacent heavily worked metal; typical comparative hardness indentations for aluminium are shown in Fig. 10. Measurements made on the bonded sections of the interface showed these to have the same hardness (and hence strength) as the bulk of the composite.

The identity of the primary discontinuities was established by examining sections taken through scratch-brushed surfaces of aluminium and copper, which showed that scratch-brushing produced a heavily worked layer (*see* Fig. 11) which was clearly the origin of the primary discontinuities shown in the photomicrographs taken at the same magnification in Fig. 6. The micro-hardness measurements of the worked layer were made, and the hardness was found to correspond with that of the discontinuities. This extreme hardness of the scratch-brushed layer can presumably be ascribed to the inclusion of particles of

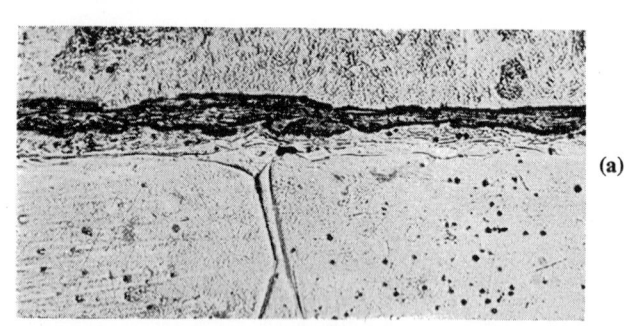

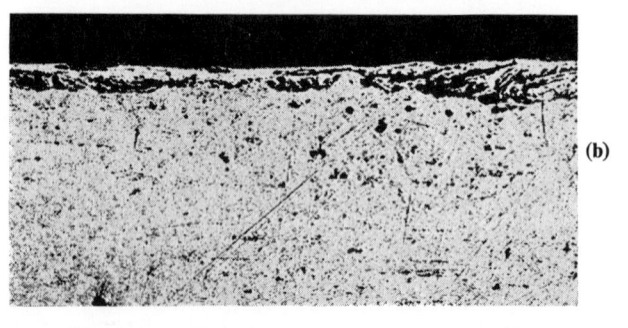

11—*Sections taken through the layer of heavily worked metal produced by scratch-brushing*: (a) *Copper, with surface protected by electro-deposited copper, ×450, etch NH₄OH/ H₂O₂;* (b) *aluminium, ×250, etch 2% HF/27% HNO₃*

primaries for different degrees of deformation. This showed that the length of apparently bonded interface increased linearly with deformation, but that the total length of discontinuity remained equal to the original length of the specimen, except for very high deformations (*see* Fig. 9). The point corresponding to a deformation of 23% given in these curves requires an explanation, since bonding in aluminium, and hence weld interfaces, cannot be obtained for deformations below 40%. In view of the difficulty in deciding the trend of results obtained over a limited range of deformation, one experiment was made in which a ½ in. wide groove was machined along the length of a 4 in. wide specimen, which was then rolled so that, while the overall deformation was 65%, the metal in the groove only underwent a reduction of 23%. Under the microscope, a continuous crack was observed across the region of low deformation, but the specimen was held together by the bonded regions on either

oxide, together with torn-off fragments of scratch-brush wire. There is admittedly some doubt about the accuracy of the measurements made, as the thickness of the layer limited the maximum size of the hardness impressions to within the range 0·6 to 1·6 μ, whereas the handbook[4] recommends a minimum diagonal length of 10 μ. To obtain impressions of less than 2 μ diagonal length, very low loads, of 1 or 2 g., had to be used, and this resulted in hardness values that could not be directly compared with the known macro-hardness of the materials. This difficulty was overcome by deriving a calibration curve, for the same load as used in this work, from a series of metals the

Table I
Micro-hardness determinations

Metal	Scratch-Brushed Layer,* D.P.N.	Scratch-Brushed Layer, D.P.N.	Primary Discontinuities, D.P.N.	Adjacent Metal, D.P.N.
Aluminium	130	90–110	100–120	48
Copper	180	200–230	240–280	125
Iron	600	—	850	212
Brass	210	—	—	—

* Due to Ainbinder and Klokova.

hardnesses of which were determined conventionally at normal loads. It was impossible to carry out micro-hardness determinations for the tin and lead interfaces, since even 1 and 2 g. loads resulted in diagonal lengths greatly in excess of 2 μ. The results of the micro-hardness measurements thus obtained are shown in Table I and these compare well with the only other results of similar measurements known, due to Ainbinder and Klokova;[5] the technique used by these workers was however not specified.

The conclusion reached from the examination of the primary discontinuities is that, with aluminium and copper, scratch-brushing produces a very hard and brittle layer, and that when two members of a composite are rolled together the two scratch-brushed layers come into contact and thereafter behave as one, so that the total length of layer always equals the original length of the composite. Since the oxide film formed after scratch-brushing will be formed on this layer, the position of the primary discontinuities must show the position of the oxide film. In the case of tin and lead, a scratch-brushed layer will recrystallize and deform in a ductile manner on rolling, and thus give the continuous interface observed with these metals. However, if the oxide film is brittle, it will fragment in a similar manner to the work-hardened layer, the total length of the fragments always equalling the original length of the composite. The decrease in the total length of the primary discontinuities at high deformation was due to partial recrystallization in the immediate vicinity of the interface, reducing the volume of work-hardened material, as shown by the micrograph (Fig. 12) of copper which had undergone a 75% reduction.

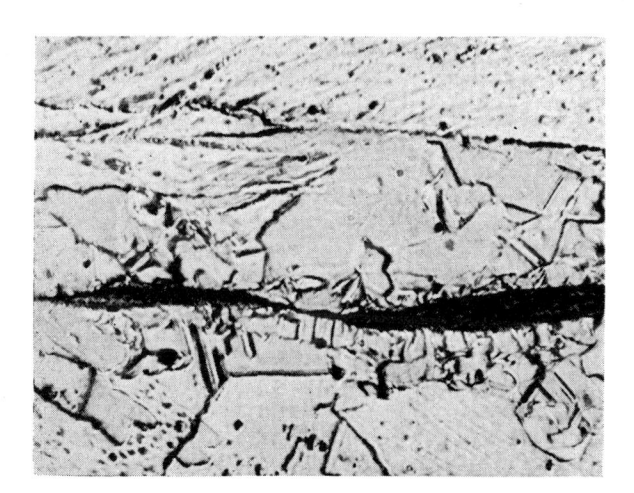

Theoretical Determination of the Maximum Bond Strength attainable

On the assumption that composites which possess an initial oxide film always deform in such a way that the two oxide layers break up as one, and in a completely brittle manner, the area of virgin metal interface which is potentially available for bonding can be calculated. And if account can be taken of the stress conditions which are operative when testing the joint, it should be possible to make a theoretical determination of the bond strength. Thus:

$$\frac{\text{Maximum bond strength}}{\text{per unit area}} = \frac{\text{Metallic area created}}{\text{per unit area}} \times \frac{\text{Strength of}}{\text{bonds}}$$

The amount of metallic area created per unit area can be determined by considering unit width of composite having an original thickness a and length b, which is extended by rolling to length y and thickness x. Then, assuming plane strain, which is certainly true for the centre of the composite, $ab = xy$, and

$$\frac{\text{Metallic area created}}{\text{per unit area}} = \frac{y-b}{y} = 1 - \frac{b}{y} = 1 - \frac{x}{a} = \frac{a-x}{a} = R$$

where R is the reduction which the metal has undergone.

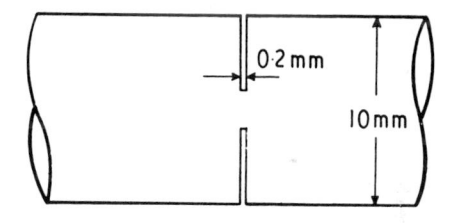

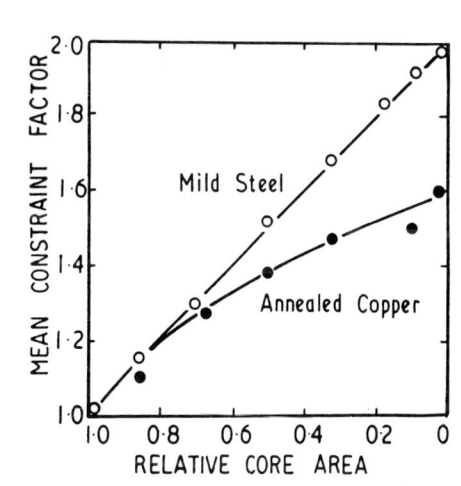

12—*Local recrystallization around a primary discontinuity, resulting from the heat and deformation produced by a 75% reduction in thickness, ×940, heavy etch in NH_4OH/H_2O_2*

13—*Type of test piece used by Orowan, Nye and Cairns,[6] and the relationship obtained by them between the constraint factor and the ratio (area of core/area of bar)*

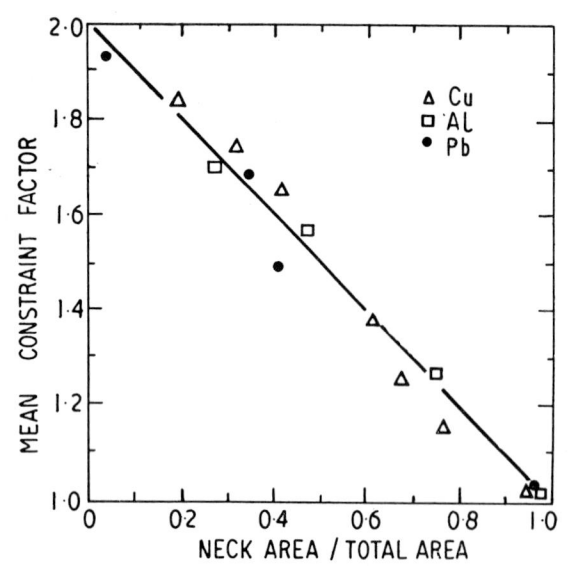

14—*Relation between constraint factor and degree of restraint for lead, work-hardened copper, and work-hardened aluminium*

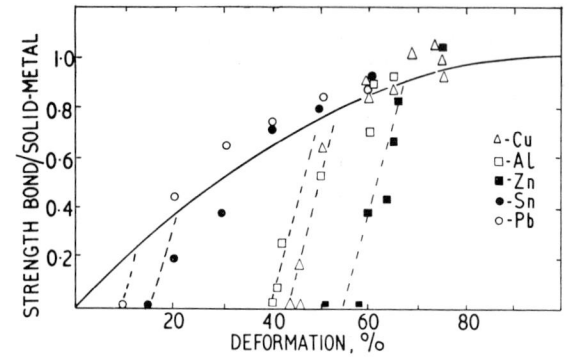

15—*Experimental bond strengths, plotted as a proportion of the strength of the solid metal, compared with the theoretical maximum attainable strength (full line)*

The strength of the bonds is a more difficult factor to assess, and can best be correlated with data available for stressing in tension and then applied to the problem in hand by making an assumption. If the joint is tested in tension the metallic bridges across the interface will not yield at the same stress as they would under pure tension, because of the constraining effect of the bulk metal on either side. The mechanical behaviour of a metal stressed under non-uniaxial conditions has occupied the attention of many workers, and involves the introduction of the concept of a constraint factor C, which in this case would be given by $C = (U.T.S._c)/(U.T.S._t)$, where the subscript c designates the U.T.S. measured under constraint and t the U.T.S. measured under pure tension. Orowan, Nye, and Cairns[6] determined the variation of C with the degree of restraint for steel and annealed copper bars, in which an area was placed under restraint by cutting an annular notch around the bar. The degree of restraint was determined by the ratio of the area of the core to the adjacent area of the bar. These workers found that the constraint factor increased linearly (*see* Fig. 13) to a value of 2 for maximum restraint for steel, and to a value of 1·6 for copper. In the present work, a similar determination of the variation of the constraint factor with the degree of restraint was carried out for lead and for work-hardened copper and aluminium, and it was found that a linear result was obtained (*see* Fig. 14), with the constraint factor increasing to 2, so that

$$C = \frac{U.T.S._c}{U.T.S._t} = \left(2 - \frac{d^2}{D^2}\right)$$

where

d = Diameter of core
D = Diameter of bar

The results of Orowan *et al.*, together with those of the present work, suggest that this equation may apply generally to metals which have a relatively sharp yield point, in this case work-hardened copper and aluminium, and lead and steel, but not to work-hardenable metals such as annealed copper.

Reverting now to the consideration of a weld interface in which bonded and unbonded areas are interspersed, this can be regarded as a series of notched test pieces where the mean degree of restraint is given by the area joined per unit area, and this has already been shown to be equal to the reduction R. Thus

$$C = \frac{U.T.S._c}{U.T.S._t} = (2 - R)$$

or

$$U.T.S._c = (2 - R)\, U.T.S._t$$

so that the maximum strength of the weld (area bonded × bond strength) is given by

$$U.T.S._{weld} = R\,(2 - R)\, U.T.S._{solid}$$

To obtain a comparison with the results of the bond strength measurements, it is necessary to assume that the relationship between the ultimate shear strength and the ultimate tensile strength is the same for both restrained and unrestrained metal. If this is so, then

$$\frac{U.S.S._{weld}}{U.S.S._{solid}} = R\,(2 - R)$$

This theoretical curve is compared with the experimentally determined bond strengths, plotted as a proportion of the solid metal strength, in Fig. 15. It can be seen that the experimental values increase rapidly from the threshold deformation, and then follow the theoretical curve reasonably well.

The theoretical curve derived from this model therefore represents the maximum potential strength for oxidized metal, roll bonded, assuming that no bonding can occur between oxide films and that the oxide remains at the interface. Thus, for example, the curve would not be expected to apply to those cases of high-temperature welding where the oxide film can diffuse into the adjacent metal. It is particularly interesting to note that whilst the theory was derived on the basis of measurements made on the primary discontinuities in aluminium and copper, it agrees well with the bond strength determinations for lead and tin; in these the oxide fragmentation could not be observed, but it must therefore occur in much the same way as with the aluminium and copper.

The problem now remains of why this maximum strength is not attained until some deformation greater than the threshold value has been applied, and this involves a consideration of the reason for the existence of the threshold deformation itself and of the subsequent development of bonding. Unfortunately, no positive contribution can be offered to the solution of this problem, although a number of significant factors have been investigated.

Initiation and Development of Bonding

The approach to this aspect of the work has been to determine the effect on the bond strength of a number of factors generally known to be significant in welding, number of different widths, of aluminium composites which had undergone 60% deformation, and the results are shown in Fig. 16. There is a marked falling off of the bond strength towards the edge of the composite, and as the width of the composite is increased the central strength increases to a maximum, after which it remains substantially constant. However, these results do not show how the bond strength developed with increasing deformation across the width of composites. In particular, it is not clear from them

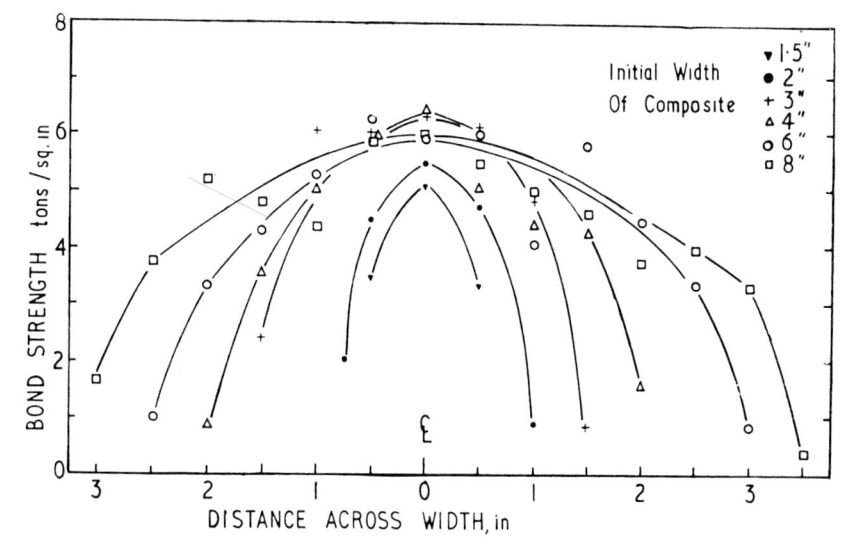

16—*Lateral variation in bond strength for aluminium composites, of various widths, which have undergone a 60% deformation*

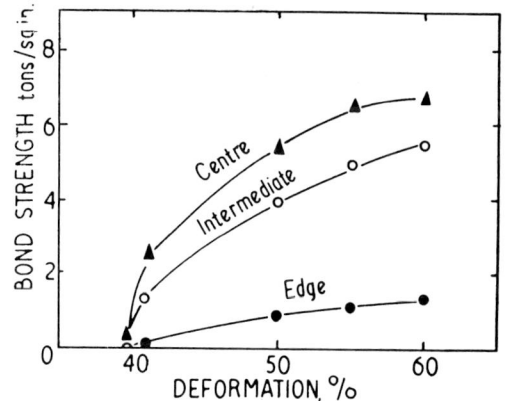

and which might moreover, clarify the mechanism of bonding. As before, bond strength measurements have been supplemented by a metallographic examination.

Rolling geometry

It is well known that the efficiency of pressure welding is dependent on the deformation pattern, and in the present series of papers Donelan[7] and Holmes[8] both stress the importance of the geometry of the deforming tool. In the present investigation, this factor was examined in so far as it arises in deforming metal by rolling; thus, the effect of specimen width, thickness, length, and roll diameter were investigated.

All the results so far reported were obtained from the central area of 4 in. square composites. A survey was now made of the strength across the width, for a

whether the initiation of bonding requires a greater reduction nearer the edge of the specimen, or alternatively, whether the same threshold deformation is necessary throughout but the bond strength develops less readily towards the edge. To examine this point the investigation was extended to further deformations and the results are given in Fig. 17. It can be seen that the threshold deformation is the same throughout, and it is the development of bonding which is pressure sensitive.

A similar investigation of the effect of varying the initial thickness of aluminium composites before rolling showed (Fig. 18) that the bond strength

17—*Initiation and development of bond strength at various positions across 4 in. wide aluminium composites*

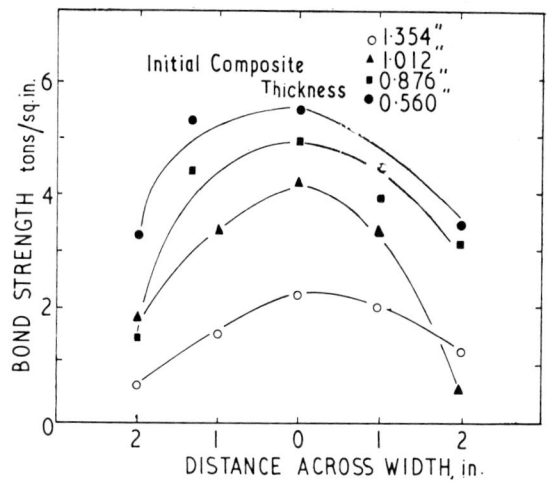

18—*Effect of the initial thickness of aluminium composites on the resulting bond strength (60% deformation)*

Source: *British Welding Journal*, Jan 1959

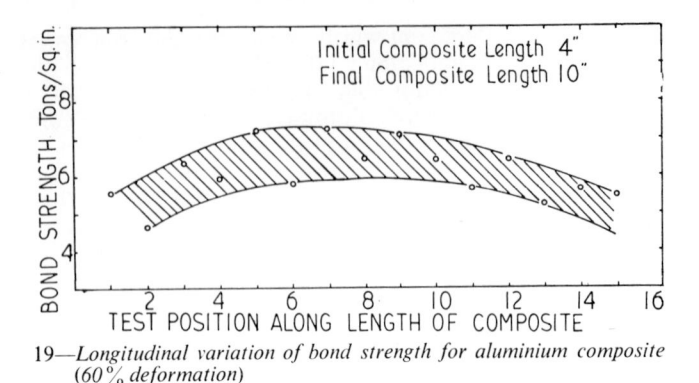

19—*Longitudinal variation of bond strength for aluminium composite (60% deformation)*

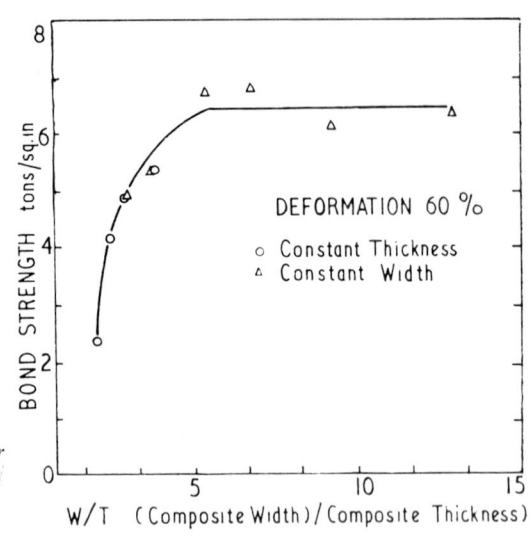

20—*Bond strength at the centre of aluminium composites, after 60% reduction, as a function of the ratio (composite width/composite thickness)*

decreased with increasing initial thickness. An examination made along the length of composites showed that, whilst some decrease in strength took place near the ends, this effect was not as marked as the effect of width and thickness (see Fig. 19).

These variations showed that the weld strength changed appreciably independently of the deformation, and that another factor was therefore playing an important part. From a qualitative comparison of the parameters which were known to be operative in rolling it appeared that the bond strength varied with the pressure, but a quantitative comparison was difficult to make without embarking on a series of determinations of the pressure distribution in the roll gap, which is in itself a major problem. However, the

comparison was pursued further in two respects. Firstly, when the strength at the centre of the aluminium composites was plotted as a function of the composite width/thickness ratio (using the results on the effect of both width and thickness, Figs. 16 and 18) it was found that the strength increased until the width/thickness ratio attained a value of about 6 (see Fig. 20) and thereafter the strength remained constant. It is well known that the mean specific roll pressure can behave in a similar manner, since the load required to reduce the metal has to be increased to overcome the

21—*Comparison between the pressure distribution across the roll gap as determined by Siebel and Leug[9] and the bond strengths obtained in the present work; reduction in both cases 47·5%*

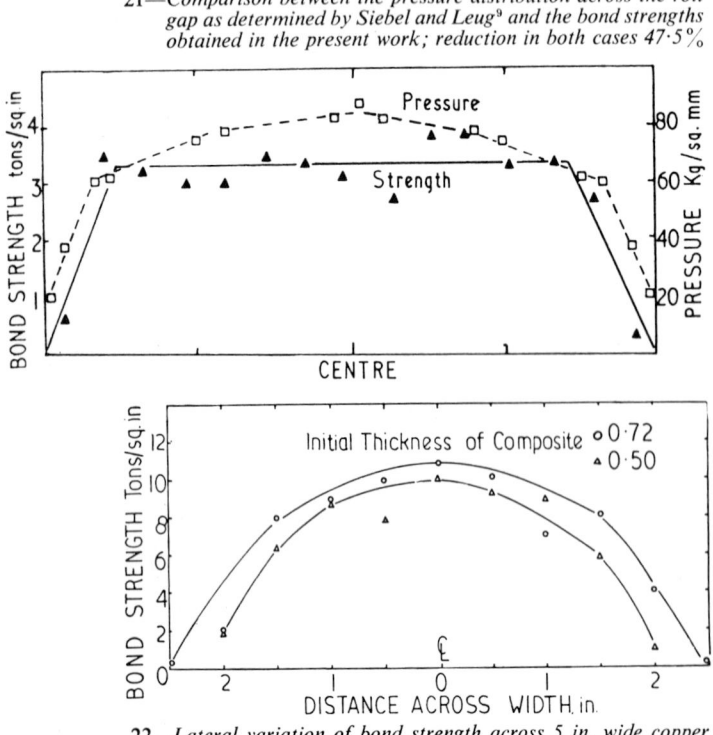

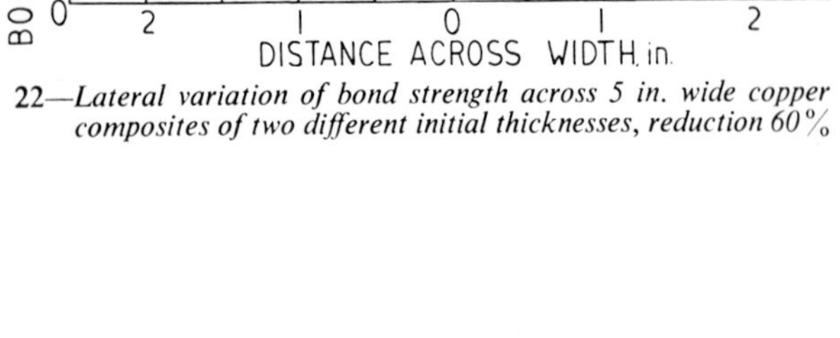

22—*Lateral variation of bond strength across 5 in. wide copper composites of two different initial thicknesses, reduction 60%*

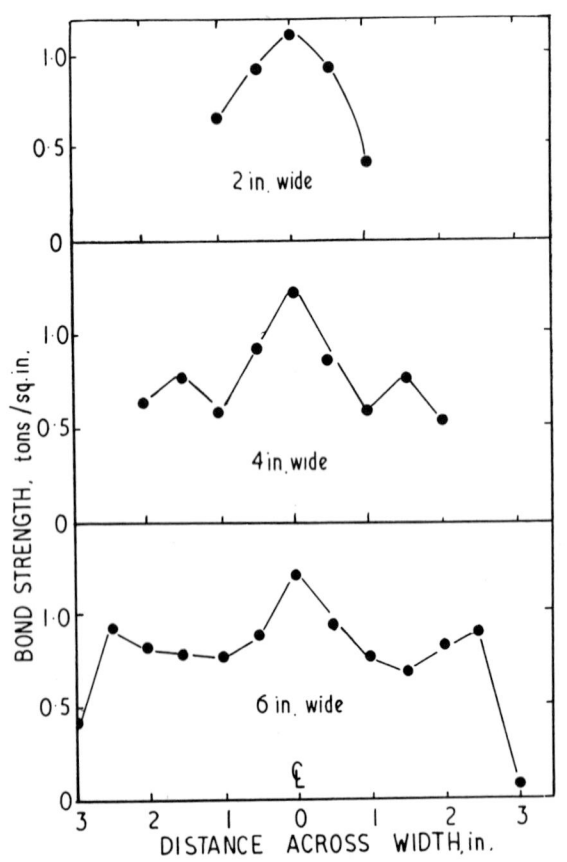

23—*Lateral variation of bond strength across tin composites of different widths, reduction 50%*

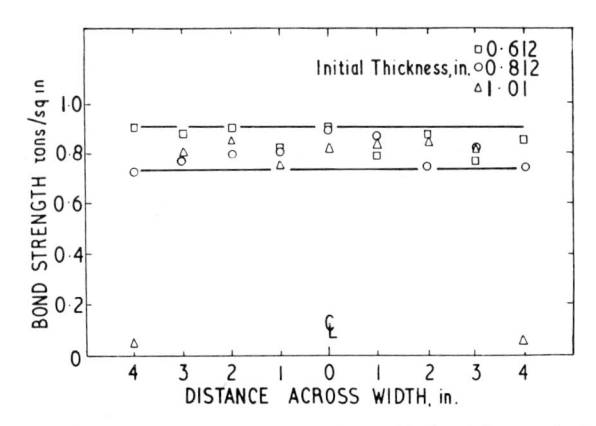

24—*Bond strength surveys across the width for different thicknesses of 8 in. wide tin composites, reduction 50%*

frictional restraint exerted by the roll surfaces on the deforming metal. Secondly, the pressure distribution across the roll gap has been measured by Siebel and Leug[9] for aluminium strip measuring 30 mm. by 2 mm. and reduced by 47·5%, and whilst specimens of this size would not be satisfactory for bond strength determinations, it was possible to use a thicker composite having the same width/thickness ratio and compare the bond strengths with Siebel and Leug's pressures for the same deformation (*see* Fig. 21). It was found that increasing pressure up to 70 kg./sq.mm. increased the bond strength, but further increases in the pressure had no effect.

To determine whether these results were of a general nature or specific to the case of aluminium, lateral variations in bond strength were determined for several thicknesses of copper and tin. In the case of copper results similar to those for aluminium were obtained (Fig. 22). With tin, however, the situation was more complex; for a width of 2 in. the strength varied in a manner similar as with aluminium and copper, but as the width was increased to 4 in. and 6 in., subsidiary strength peaks appeared towards the edges of the composites (*see* Fig. 23). With a width of 8 in., the fluctuation in the bond strength had diminished to give an approximately uniform strength across the width, while changing the initial thickness had little effect (Fig. 24).

Thus, for aluminium and copper, pressure has a clear effect on the bond strength, but with tin, the complexity of the strength variation and the lack of data on pressure distribution in the roll gap make it impossible to arrive at a definite correlation.

Owing to limitations in the control of speed of the rolling mills available, the effect of changing the geometry of rolling by changing the roll size could not be investigated without at the same time changing the rolling speed, and discussion of this aspect must therefore be left until the results on the effect of roll speed have been presented.

Effect of roll speed

All results reported so far were obtained with an 18 in. rolling mill which operated at a fixed speed of 200 ft./min. But a knowledge of the effect of rolling speed is important, since if this parameter had any influence on bonding it would tend to support theories in which time-dependent processes such as diffusion played a part, and would militate against elastic-recovery theories. The only means available for examining the effect of time, while confining the investigation to roll bonding, has been to use a 10 in. two-high mill with a nominal speed variation of 0–145 ft./min. However, the lowest speed at which tin and lead could be fed into this mill was 30 ft./min., whilst with aluminium and copper such a high speed was necessary to obtain the necessary 'bite' that an insufficient working range of rolling speed was left for any investigation to be made. For tin and lead the bond strength deformation curves were obtained for the upper and lower limits of operating speed, and the results (Fig. 25) show that lowering the rolling speed from 145 to 30 ft./min. approximately halved the threshold deformation. No difference was observed in the shape of the strength/deformation curve once the maximum bond strength was attained, in keeping with the theory based on the oxide break-up model outlined earlier in this paper.

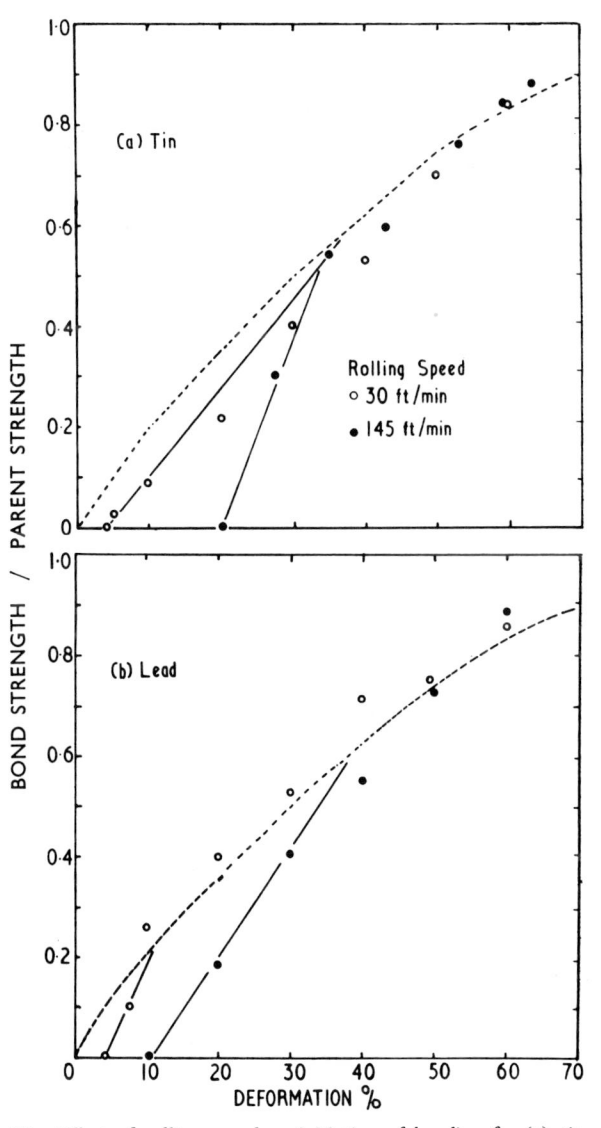

25—*Effect of rolling speed on initiation of bonding for* (a) *tin, and* (b) *lead. Broken line represents theoretical curve*

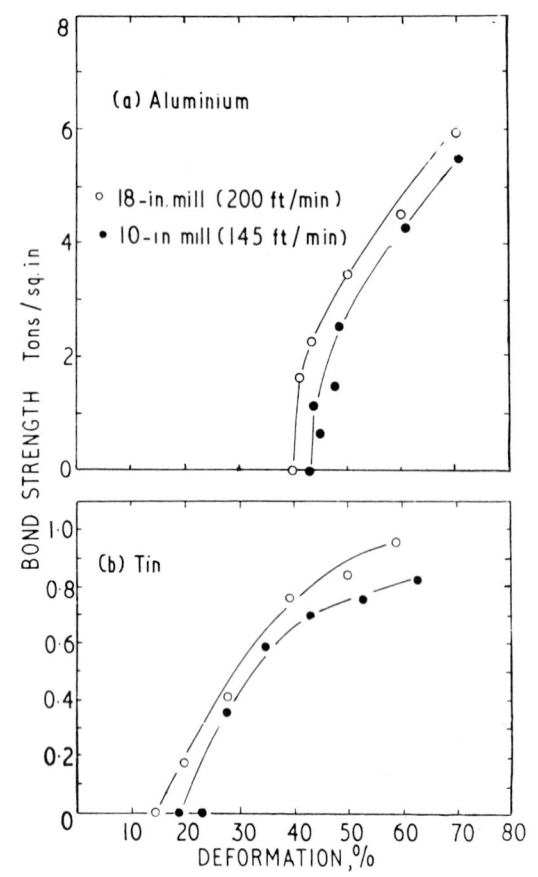

26—*Effect of changing rolling mill diameter on the development of bonding in aluminium and tin*

Effect of roll diameter

In regard to the effect of rolling geometry in so far as the roll diameter has an influence, the difficulty here was that only two mills were available, which, while having different roll sizes, also operated at different speeds; the 18 in. mill had a fixed speed of 200 ft./min., whereas the maximum speed of the smaller mill was 145 ft./min. However, it has already been shown that the effect of decreasing the rolling speed was to initiate bonding at lower deformation; and when the bond strengths obtained with the two mills were compared it was found (Fig. 26) that the threshold deformation was decreased by using the higher speed, larger diameter, mill. Thus increasing the roll diameter promotes the initiation of bonding, and to a greater extent than shown by Fig. 26, since this should be modified to take account of the speed effect by an amount which could not be determined with the available equipment.

Influence of surface contamination

Of the many factors operative in pressure welding, the role of surface preparation and other surface parameters is the most obscure. Scratch-brushing combined with a degreasing treatment invariably gives the best bond strengths, and after some initial experience with machined, electro-polished, ground, emeried, scraped, and 'as-rolled' surfaces, the standard treatment of degreasing in trichlorethylene followed by scratch-brushing was adopted throughout. Even reversing the procedure, and degreasing after scratch-

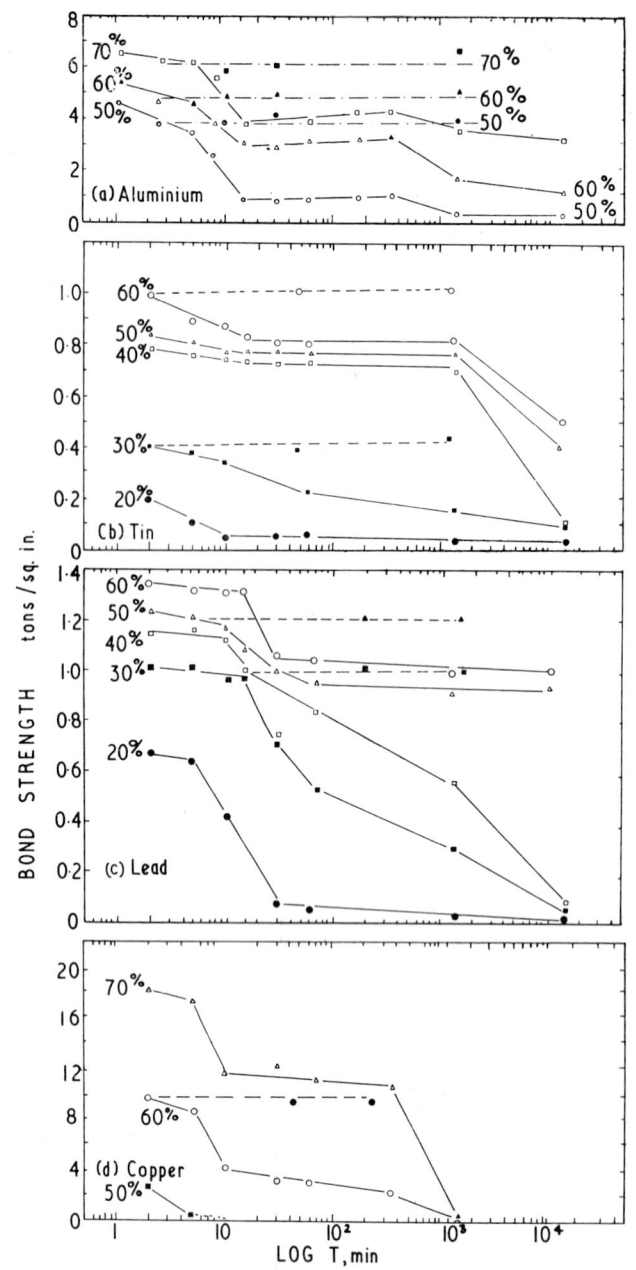

27—*Effect of exposing scratch-brushed surfaces to the atmosphere before rolling. Full line exposed to the atmosphere, broken line retained in a desiccator over P_2O_5*

brushing, was found markedly to decrease the bond strengths obtained.

It is very difficult to make experiments on surface parameters in which a single factor is definitely isolated so that the results can be interpreted unambiguously. Researches on this aspect of pressure welding are now proceeding, but the results of a series of tests on the effect of atmospheric contamination, which appear to lead to a positive conclusion, will be presented here.

Composites of aluminium, copper, lead, and tin were rolled, following routine preparation except that after scratch-brushing the surfaces were exposed to the atmosphere for periods of from 2 min. to 10 days. The resulting bond strengths were found to fall into a regular pattern (Fig. 27), declining markedly with exposure for the first 15 min., thereafter remaining constant for about a day; further exposure resulted in a steady decrease in strength, sometimes to zero, during the next 10 days. The reason for this was not immediately obvious. The thickness of the oxide film would not be expected to increase very much, for instance, in aluminium, where on scratch-brushing sufficient surface heating probably occurs for the very rapid formation of a film which would not be appreciably affected by further exposure at room temperature. Hence the alternative possibility, of the growth of layers of adsorbed material from the atmosphere, was investigated by repeating the experiments with the specimens held in a desiccator above phosphorus pentoxide. In every case the bond strength was found to remain constant. Thus, bonding is very much dependent on the presence of adsorbed layers, the main contaminant probably being water vapour.

Effect of temperature

It is well known that increasing the temperature can result in a very marked lowering of the deformation required for welding. This in turn must be pertinent to the mechanism of bonding, but so many factors and possibilities arise that again this is being treated as a subject for investigation on its own; only one result (Fig. 28) is presented here to show the magnitude of the effect of temperature on the threshold deformation for aluminium. At temperatures above 350°C. bonding was initiated with aluminium at a lower deformation than with lead at room temperature.

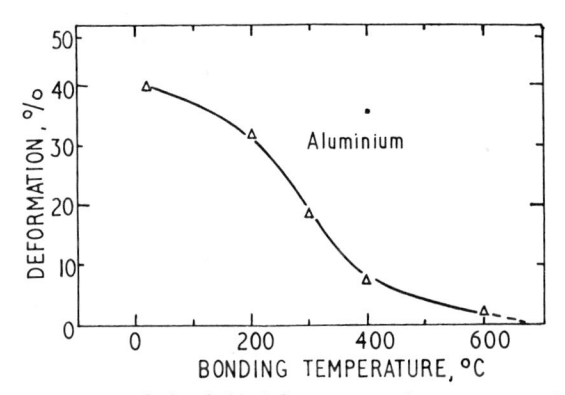

28—*Variation of threshold deformation with temperature for aluminium*

Correlation of the Initiation and Development of Bonding with Interfacial Characteristics

Earlier in this paper it was shown that the macroscopic measurements of bond strength could be correlated with the presence at the interface of primary discontinuities, and that the total length of these discontinuities was determined by the original length of the specimen. The theory was then proposed that the maximum strength was determined by the area of virgin metal which was available for bonding. Since this maximum strength is often not obtained, it might be expected that this would be reflected in the characteristics of the interface. Hence weld interfaces in

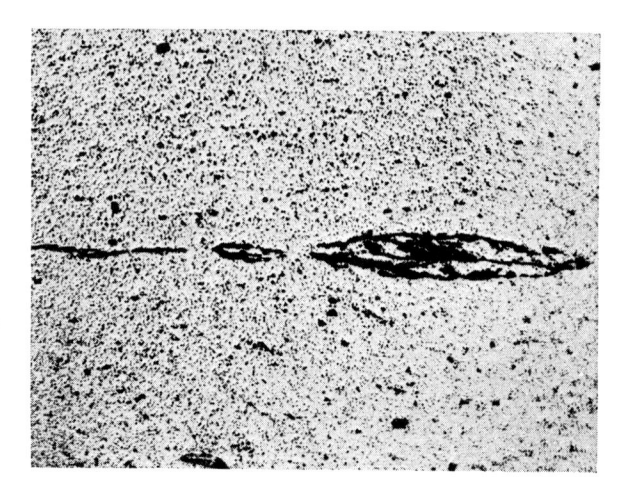

29—*Area of interface in aluminium with very little bonding, taken ⅛ in. from the edge of an 8 in. wide composite. The secondary line discontinuity can be readily distinguished. Deformation 47%, ×250, etch 2% HF/27% HNO₃*

aluminium were examined for those conditions under which the maximum strength is not reached; composites were examined which had been subjected to deformations lying between the threshold value and that at which maximum strength was attained, and specimens were taken of interfaces away from the centre of the composite, and where surfaces had been exposed to the atmosphere before rolling. In all cases a further type of discontinuity, in addition to the normal primary types was observed, having the form of a hair-line crack; a typical example is shown in Fig. 29.

To correlate the presence of these secondary discontinuities with the observed bond strength, a survey was made of their distribution for various cases, and the strength was then calculated by taking account of the triaxial stress conditions as described on p. 20. It will be appreciated, from a comparison of Fig. 29 with Fig. 6a, taken at the centre of a composite and showing only primary discontinuities, that the length of a secondary discontinuity was much more difficult to define accurately, so that there was more scatter with these results. Nevertheless, the correlation was clear enough; Fig. 30 shows the result of such a survey, and the corresponding calculation for the lateral variation in bond strength of aluminium reduced by 47% is given in Fig. 31, which also shows the experimental bond strengths. The distribution of the primary discontinuities always conformed to that expected from the concept of two scratch-brushed layers breaking up as one, as had previously been found for the centre of the standard composites.

A similar examination, although not so extensive, was also made for copper, but in this case the metallographic technique appeared to play a more important part, as a series of discontinuity types could be seen. These varied from the hair-line crack representative of aluminium, through degrees of width, to the relatively thick primary type, so that no clear distinctions could be drawn. With tin and lead, no technique has been found which can resolve the fine structure of the continuous boundaries to show how the interfacial characteristics vary with the joint strength.

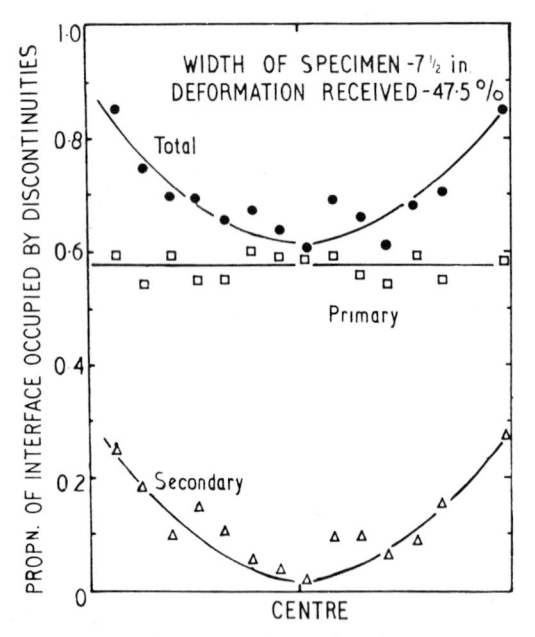

30—*Distribution of primary and secondary discontinuities across the width of an aluminium composite*

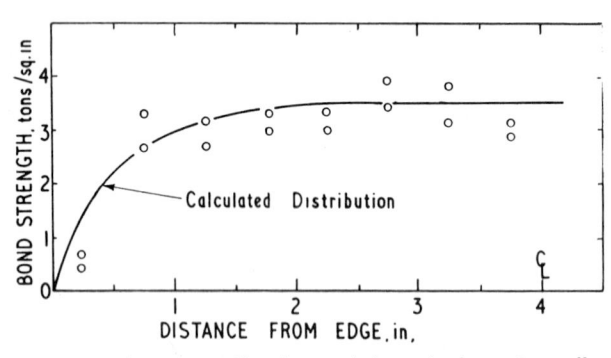

31—*Lateral variation of bond strength determined experimentally, compared with that calculated from the distribution of primary and secondary discontinuities (full line), for the aluminium specimen corresponding to Fig. 30*

Discussion

The purpose of these researches has been to investigate the mechanism of pressure welding in the roll bonding process, and while a body of data has been obtained which provides a basis for analysis, shows the relationship between a number of variables, and allows partial conclusions to be drawn, no overall theory can be offered. However, the research is continuing into some of the more critical aspects, and it is hoped that the information gained will lead to further developments. In the meantime it is possible to arrive at some conclusions regarding the correct model which has to be pursued.

Taking first a broad view, two possibilities exist: either welds do not readily form, or alternatively they do form, but are broken apart by elastic recovery forces, as was suggested by McFarlane and Tabor.[10] There are two arguments against elastic recovery playing a major part. Firstly, it is difficult to visualize how such forces could be effective when two virtually flat blocks are rolled together, as opposed, for example, to the case of a hard spherical indentor pressing into a flat surface, for which this model was originally developed. Secondly, there is the fact that the initiation of bonding shows a marked time dependence, which would be difficult to explain on an elastic recovery model. Thus, it would appear that it is the factors that are operative in causing the surfaces to come into intimate contact which are important, and if this problem is considered at a macroscopic level, these are the presence of an oxide film and the difficulty of bringing two rough surfaces into contact.

On the basis of the results presented in this paper, it has been possible to define the role of the oxide film which is initially present on the metal surface, as preventing bonding over the area which it occupies, this area being equal to the initial area of the composite. This model differs from that previously postulated for the behaviour of the oxide film by Tylecote,[11] who suggested that it was the ratio of the hardness of the oxide film to that of the underlying metal which determined the degree of fragmentation of the oxide film, and that this in turn controlled the initiation of bonding. By contrast, the present work leads to the conclusion that the break-up of the oxide film determines the maximum bond strength attainable, but that other factors control the initiation of bonding.

The problem now remains of why the virgin metal areas, created by the extension in length of the composite during rolling, do not always weld. Pugh and Holmes[3,8] have recognized that contact across the interface may be difficult to achieve, because the restraint imposed by the adjacent bulk metal may make asperities on the surface difficult to deform. In order to determine how much this factor contributes to the difficulty of bringing two surfaces together, the extreme case, of the deformation of a surface into which a series of deep grooves had been machined, was investigated in the present work. Lateral **V** grooves 0·05 in. deep and separated by distances of 0·25 in. were milled out of the surfaces of tin and aluminium plates, 4 × 4 × 0·3 in., which were then clamped onto polished plates of the same material. A series of these composites was rolled, using the 18 in. dia. mill, and the deformation of the grooves was compared with the overall deformation. The immediate observation was that at reductions as low as 5%, when the grooved surface had apparently deformed very little, the surface that had originally been flat had become a complete replica of the grooved surface, even to the extent of

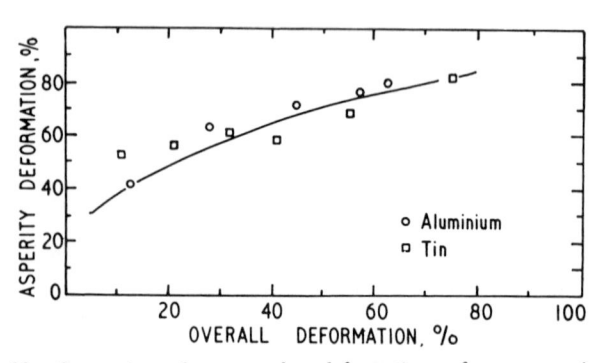

32—*Comparison between the deformation of macroscopic asperities and the overall deformation, of grooved aluminium and tin plates rolled against flat plates of the same material*

reproducing the marks of the machine tool. However, when detailed measurements were made (*see* Fig. 32) it was found that at low deformation the height of the surface asperities had undergone more than the mean reduction. Even so, no flattening of the tips of the asperities could be observed. Thus it is clear that there should be no difficulty in bringing two surfaces into macroscopic contact, even when they are both rough, as at any position across the interface the degree of roughness, and hence the flow stress, will be less at one surface than the other, so the rougher surface will penetrate the smoother to cause overall intimate contact.

Macroscopic contact of the virgin metal areas should therefore be readily obtained, but this is not sufficient for bonding, which will occur only if the surfaces are brought into contact within atomic dimensions, so that the attractive forces between atoms can operate. It is the analysis of how this occurs, at a level in which atomic movements are considered, that constitutes the problem of the mechanism of bonding; this problem of the formation of welds in the solid phase is not specific to pressure welding, but is also relevant to sintering and frictional phenomena. Various considerations are involved; first, there is the flow of the metal, which takes place by slip and other more complex movements of dislocations and vacancies along favoured atomic planes. But it is not possible to bring two surfaces into intimate contact by processes which occur along crystallographic planes which are even only a few atomic distances apart, so that, as well as mechanical flow, local readjustments by diffusion processes in the vicinity of the interface must be essential. Also, unless the welding operation is carried out under a very high vacuum, there is the problem of the solution of the entrapped gases to be considered. Fine, Maak, and Ozanich[12] have shown by special etching techniques that with steels the layer of metal adjacent to the interface is oxygen-rich when welds are made in air, and the present work has shown that adsorbed water vapour can have a particularly deleterious effect. Thus, at a deformation of 60% all the virgin metal area was found to weld if the composite was rolled within two minutes of scratch-brushing, but if water vapour became adsorbed on the surface it did not remain attached to the oxide, but was sufficiently 'mobile' to contaminate some of the virgin metal areas as they formed and thus reduce the bond strength. Held and Hendus,[13] by carrying out their experiments in a 'high' vacuum (using a diffusion pump), eliminated as far as possible the influence of these adsorbed and entrapped gases (although possibly also removing the oxide film), and were able to initiate the bonding of Armco iron with negligible deformation by holding for one hour at 400°C., most unusually favourable conditions for iron.

Since mechanical deformation, atomic readjustment at the interface, and the solution of entrapped gases are all to some extent dependent upon diffusion phenomena, the influence of time and temperature is obvious. Conversely, varying the time or temperature could change any or all of these processes so that, without further control, experiments investigating these parameters yield ambiguous results. It is difficult to isolate the individual processes which can operate in bond formation. Thus, for example, Erdmann-Jesnitzer and May[14] virtually eliminated deformation and drastically reduced the applied load in a series of experiments to determine the temperature at which welding was initiated between copper rods abutted under a light load and heated in a protective atmosphere, and found that bonding occurred at 740°C.; a similar series of experiments was carried out by Parks[15] for a number of metals, but although the conclusion was that welding begins in the region of the recrystallization temperature, it is difficult to relate these results to the mechanism of bond formation in welding as it is normally practised. Cook and Davis,[16] approaching the problem from the purely practical viewpoint of determining the optimum conditions for the welding of high-conductivity copper, found that the U.T.S. of a butt-welded joint made in 10 min. at a temperature of 500°C. and pressure of 1·5 tons/sq.in. was 8 tons/sq.in., but that if the pressure and temperature were maintained the strength increased, until after 30 hr. it had almost doubled, to 15·5 tons/sq.in., thus demonstrating most strikingly the general influence of these factors.

The remaining parameter which this work has shown to be of significance is pressure. Other workers have also found pressure to have an effect; Hoffman and Ruge[17] in particular related welding to pressure, but these workers, having found that there was a critical pressure at which marked flow of the metal readily took place, concluded that it was in fact the flow of the metal that was significant, and the pressure only of importance in that it brought this flow about. In the present work the influence of pressure was demonstrated *via* changes in the geometry of rolling—in particular, the bond strength varied markedly across the width; but since the metal flow had been substantially the same throughout the composite, there is no correlation with the work of Hoffman and Ruge. If in fact the enhanced metal flow were the critical factor, the stronger bonds would be expected at the edge of the specimen where some lateral spread occurred, whereas the bond strength in this region was very low. Whether increasing the applied pressure affects the atomic processes of mechanical deformation is not known, and since it could have an effect on the diffusion of metal atoms, of vacancies, and of entrapped gas in the metal, the exact role of this variable is particularly difficult to evaluate.

The most positive aspect of this work is the role it assigns to the oxide film, and it would be very useful if the concepts involved could be applied to the other pressure welding techniques. This, however, would require a knowledge of the relation between the deformation of the metal and the increase in area of the interface, and this is not known either in small-tool or in butt welding. However, one piece of work does have some relevance; Austin and Jeffries[18] in 1932 carried out butt welding of various types of steel at high temperatures, and measured bond strengths and examined interfaces. As in the present work, they observed discontinuities at the interface, which they regarded as patches of oxide. By some unspecified means these workers determined the apparently bonded area, tried to correlate it with the bond strength, and found that higher strengths were obtained than would be expected on a basis of the weld strength being equal to the apparently bonded area multiplied by the strength of the metal as determined by normal tensile tests. For this reason, Austin and Jeffries concluded that there must be some bonding between oxide and metal. However, when the concept of restraint, and the relation between the degree of restraint and constraint factor, is applied to these results as described on p. 20, it is found (Table II) that on recalculation there is very good agreement between the measured bond strengths and those calculated from the weld area measurements.

Table II
Relation between bond strengths of butt-welded steels and values calculated from weld areas
Experimental Results of Austin and Jeffries[18]

Type of steel	Weld area, %	Metal strength attained by welding, % Experimentally Determined	Calculated
Open Hearth	55	80	80
Open Hearth	50	75	75
Open Hearth	80	105	96
Open Hearth	60	85	84
Bessemer	70	97	91
Bessemer	50	80	75
Bessemer	60	85	84
Bessemer	60	80	84

In summary, an assessment of the available information on pressure welding leads to the conclusion that the initiation and development of bonding are complex phenomena, depending on a variety of factors, but that the maximum bond strength attainable, at room temperature at least, is determined by simple geometrical considerations.

Conclusions

Practical application of cold roll bonding

(1) For room temperature welding, a deformation of 60–70% is required with aluminium, copper, lead, tin, and zinc to obtain weld strengths equivalent to those of solid metal.

(2) The first roll pass should be sufficient to cause the composite to adhere; thereafter the weld strength can be built up by successive small reductions.

(3) Weld strengths deteriorate rapidly as water vapour is adsorbed on the faying surfaces.

(4) Bonding is promoted by the use of low rolling speeds and large-diameter rolls.

(5) With the harder metals investigated (aluminium and copper), the bond strength is low for a region of about 1 in. from the edge of the composite.

The fundamentals of pressure welding relevant to roll bonding

(1) The maximum bond strength that can be achieved at room temperature can be calculated on the assumption that there are present on the surface of the composite brittle oxide films which come together at an early stage of their progress through the roll gap and thereafter behave as one. These films are completely brittle, so that the virgin metal area which is potentially available is the increase in area of the interface. Taking account of the triaxial stress conditions existent when a load is applied, the maximum ultimate shear strength that the weld can have, if the whole of the virgin metal area bonds is given by:

$$\text{U.S.S.}_{\text{weld}} = R(2-R) \, \text{U.S.S.}_{\text{solid metal}}$$

where R is the reduction. (This result would not apply to high-temperature welding where the oxide film can diffuse into the adjacent metal.)

(2) In many cases, no bonding, or less than the maximum strength, is obtained; this is due to the difficulty of bringing two surfaces into atomic contact, and the correct model for the mechanism of the process will have to take into consideration mechanical flow of metal in terms of atomic movements, local readjustment at the interface by diffusion, and the solution of entrapped gases.

Acknowledgments

This work is part of a programme of research into some aspects of the joining of metals, which is being carried out at the Department of Industrial Metallurgy at the University of Birmingham. The authors wish to thank Professor E. C. Rollason, Head of the Department, for his interest and encouragement throughout the course of this work, and the Director of the United Kingdom Atomic Energy Authority at Harwell for financial support for two of them.

REFERENCES

1. G. Sachs and K. R. Van Horn: 'Practical Metallurgy', p. 415, 1951, American Society for Metals.
2. D. Pugh: Ph.D. Thesis, 1954, University of Birmingham.
3. R. F. Tylecote, J. E. Furmidge, and D. Howd: Brit. Welding J., 1958, vol. 5, p. 21.
4. 'Photomicrography with the Vickers Projection Microscope', 5th ed.: 1956, Cooke Troughton and Simms Ltd.
5. S. B. Ainbinder and E. F. Klokova: Latvijas Psr Zinatnu Akademijas Vestis, 1954, vol. 10, p. 113.
6. E. Orowan, J. F. Nye, and W. J. Cairns: Ministry of Supply, Armament Research Department, Theoretical Research Report No. 16/45.
7. J. A. Donelan: Brit. Welding J., 1959, vol. 6, pp. 5–12 (this issue).
8. E. Holmes: Ibid., pp. 29–37.
9. E. Siebel and W. Leug: Mitt. K. W. Inst. Eisenforschung Dusseldorf, 1933, vol. 15, p. 1.
10. J. S. McFarlane and D. Tabor: Proc. Roy. Soc., Series A, 1950, vol. 202, p. 244.
11. R. F. Tylecote: Brit. Welding J., 1954, vol. 1, p. 117.
12. L. Fine, C. H. Maak and A. R. Ozanich: Welding J., 1946, vol. 25, p. 517.
13. H. Held and H. Hendus: Z. Metallkunde, 1954, vol. 45, p. 112.
14. F. Erdmann-Jesnitzer et al.: Ibid., 1955, vol. 46, pp. 756 and 854.
15. J. M. Parks: Welding J., 1953, vol. 32, p. 219 S.
16. M. Cook and E. Davis: Trans. Inst. Welding, 1947, vol. 10, p. 178.
17. W. Hoffman and J. Ruge: Z. Metallkunde, 1952, vol. 43, p. 133.
18. C. R. Austin and W. S. Jeffries: A.I.M.M.E. Technical Publication, 1932, No. 451.

Effect of Heat Treatment on Cold Pressure Welds

By R. F. Tylecote and E. J. Wynne

Cold pressure welds have been made in copper, aluminium, and low-carbon steel strip by the method of small tool welding. Heat treatments have been carried out on such welds at temperatures in the range 200°–600°C. for aluminium, 200°–900°C. for copper and 50°–475°C. for steel. Some heat treatments were carried out on steel under various fractions of the welding load to investigate the effect of elastic recovery.

In all cases it has been shown that short-term heating at low temperatures can produce an increase in weld strength before the onset of the processes of recovery and recrystallisation. The degree of improvement is greater for welds made with small deformations and it is suggested that, in the case of aluminium and copper, thermally activated short-range atomic movements are responsible for producing an improved bond. The apparent activation energies for this process are comparable with those for surface diffusion.

In the case of steel, the application of load during heating is not advantageous until a certain temperature is exceeded, and it is concluded that the improvement which occurs in the absence of load is due to stress relief which can be inhibited at low temperatures by pressure normal to the interface.

Introduction

THE present theory of cold pressure welding suggests that under lateral deformation the surface layer of oxide and metal behaves in most cases in a brittle manner and breaks up into discreet pieces which are here called primary discontinuities. Under the effect of pressure normal to the interface fresh metal is pressed down into the gaps which form between these primary discontinuities and, being clean, it is readily welded. For this reason the strength of the weld may be related to the lateral extension at the interface. According to Milner and his associates[1,2,3] the ultimate shear strength of the weld is a function of the ultimate shear strength of the metal and the deformation, according to the equation:

$$\frac{\text{U.S.S.}_{\text{weld}}}{\text{U.S.S.}_{\text{metal}}} = R(2 - R)$$

where R is the deformation, *i.e.*, the ratio of the reduction in thickness to the initial thickness. This equation applies strictly only under the conditions used by Milner, but is roughly applicable to the type of joint used in this work, where two strips are joined with a circular weld.

The actual shear strength of a joint, however, is influenced by the method of cleaning. Little is known about the factors that influence the 'threshold' deformation, which is the deformation at which welding begins. It has been shown that the threshold deformation is reduced by raising the temperature.[4] It is apparent that this is due, in cases where the oxide film is not soluble in the base metal, to the welding of the hard scratch-brushed and oxide-contaminated regions which were not thought to contribute to the weld strength in the original simple theory.[1]

In metals in which the oxide is soluble, as in iron and copper, raising the temperature may cause oxide to dissolve as well as allowing the surface layer to soften.

The fact that welding at elevated temperatures reduces the threshold deformation has great industrial importance, since one of the factors tending to limit the application of pressure welding is the large deformation necessary at room temperature. But the use of elevated temperatures is itself difficult, and an easier solution is to initiate the weld by deforming at room temperature and to improve it by heating at elevated temperatures without pressure. In this two-stage process, room-temperature deformation breaks up the film and causes bonding between fresh metal surfaces, whilst subsequent heating allows softening, recrystallisation and bonding of the complex mixtures of cold worked metal and oxide which interrupt the interface. Since some of the strength of the joint in the cold worked condition is probably due to the restraint imposed by these pieces of hard material, it might be expected that there would be some loss of strength on heating.

Manuscript received 4th February, 1963.
Dr. Tylecote is in the Department of Metallurgy, University of Newcastle upon Tyne, and Mr. Wynne is in the Tube Investments Research Laboratories, Hinxton Hall, Essex. 801
The work described has been supported by a grant from the British Welding Research Association

The prime object of the work reported here was to investigate the two-stage process, which in many ways may be likened to the compacting and sintering processes of powder metallurgy. Two metals, aluminium and copper, were mainly used as examples of oxide-insoluble and oxide-soluble metals.

One of the reasons why attempts to make joints with small amounts of deformation are not successful is believed to be the phenomenon of elastic recovery. It has been suggested that some of the areas brought into contact by the breaking up of the oxide film are only partially welded and that when the welding pressure is relaxed the elastic forces are sufficient to break the weak bonds.

Some experiments were made on iron to determine the temperature necessary to allow such areas to deform plastically under load and so relieve the stresses.

Previous Information

Most of the previous work pertinent to this investigation has been carried out on metal powders. Although not much is known about the forces which cause individual particles to adhere to each other after compacting, it is fairly certain that they are of a similar nature to those present during the pressure welding of massive material. Since the 'green' compacts are not normally used in their original state, not much interest has been shown in the actual strength of their bonds. Workers in powder metallurgy are almost exclusively interested in the sintering processes that occur on heating compacts in neutral or reducing atmospheres. The processes that assist the improvement of the joint are either viscous flow, evaporation and condensation, or volume and surface diffusion. The shape of a cavity or pore between wires or spherical particles changes in cross section from a triangle to a circle by one or more of these processes. Later, the small pores tend to disappear and the large pores grow in size, since large pores are more stable. If sufficient time is given at a high enough temperature overall shrinkage may occur as the pores diffuse to the surface.

These processes are assisted in powder metallurgy by non-oxidising conditions which may not always be present in pressure welding. In the case of metals which dissolve their oxides, such conditions may be said to exist during the heating of pressure welds at high temperatures.

A small number of tests have already been made on the heat treatment of cold pressure welds.[5] These were sufficient to show that very great increases of strength could be obtained in certain cases. The strength of welds in silver was more than doubled by heating at 700°C. for 4 hr in air. Similar results were found for tough-pitch copper, iron, and sintered aluminium powder. On the other hand, heating welds in aluminium at 400°C. and above, caused a considerable reduction in strength.

A good deal of work involving post-heat treatment has been done on welds made between dissimilar metals.[5,6] This work has shown that considerable interdiffusion takes place, even in the presence of oxide films, and that this may sometimes give rise to the formation of brittle intermetallic compounds. Furthermore, diffusional porosity may cause weakening of the weld.

An investigation by Bruckner and Sayles[7] into the effect of heating on cold butt welds in copper and aluminium showed that the strength of similar metal welds remained constant until recrystallisation began, after which the strengths decreased. The welding deformations were not stated but the micrographs show a considerable dispersion of oxide that suggests a deformation of the order of 70%.

Erdmann-Jesnitzer and Wichmann[8] pressed together aluminium cylinders in a ring at room temperature. The deformations at the interface were small, corresponding to reductions in height of the specimen of 5–20%. These were then heated for long periods without pressure in the temperature range 151°–353°C. Welding occurred with increasing deformation and temperature, and was time dependent. It was found that there was an activation energy associated with the welding process which varied from 31,000 cal/g. atom at 5% deformation to 17,500 cal/g. atom at 20% deformation. Extrapolation to 0% deformation gave an activation energy of 37,000 cal/g. atom. As far as the present authors are aware, no determinations have been made of the activation energy for the volume diffusion of aluminium in aluminium, but it is probably in the region of 30,000–40,000 cal/g. atom. It would seem that at low deformations volume diffusion is the rate-determining process, whilst at higher deformations surface diffusion, with its lower energy requirements, is playing a larger part.

Pugh[9] has shown that the annealing of aluminium cold welded with a low deformation brought about an increased strength owing to the elimination of the interface by recrystallisation and grain growth. In the case of titanium it has been shown[10] that greater ductility and improved impact strength may be obtained by heating welds made at 500°C. for 30 min at 900°C. This had the effect of spheroidising discontinuities which were originally lens-shaped.

Pugh also attempted to determine the role of elastic recovery in cold welds in aluminium. He heated specimens to temperatures between 50° and 100°C. under a portion of the welding load and observed an increase in strength. No such increase was observed with complete removal of the load before heating.

Experimental Methods

Preparation of material and specimens

The copper used was high conductivity electrolytic copper (99·99%) melted and cast under conditions which ensured minimum oxygen contamination, and consequently the oxygen content was less than 0·02%. The aluminium used was of three grades: super purity (99·99%); commercial purity (99·5%); and sintered aluminium powder (S.A.P.) which contains 7–10% Al_2O_3.

The ferrous material consisted of annealed low-carbon mild steel of deep drawing quality containing 0·08% C. This was also used after it had been decarburised in moist hydrogen for 10 hr at 760°C. and vacuum annealed at 850°C. for 2 hr. This results

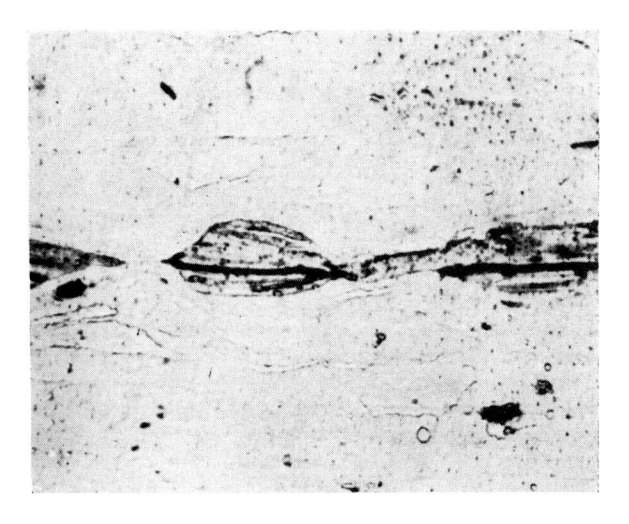

1—Interface of pressure weld made in copper at room temperature with 55% reduction in thickness, showing areas of metallic bonding and primary discontinuities. Lightly etched ×900

in carbon contents in the range 0·003–0·007%, and low nitrogen contents.

The true determination of bond strength requires that welds fail by shearing, as opposed to the tearing-type of failure where the weld pulls out of the sheet. For a tool of given indentor face area, the upper limit of stress for shear failure at any welding deformation is set by the sheet thickness. The strip thickness chosen in this investigation was 0·075 in. as this allowed a full investigation of the heat treatment of welds made with up to 60% reduction in thickness. The welding of super-purity aluminium and S.A.P. was limited by reasons of availability to 0·048 in. thick sheet; whilst weld strengths in sheets of markedly differing thicknesses are not strictly comparable, these results showed the same trend.

Immediately before welding, each individual pair of specimens was scratch-brushed with a coarse wire brush; the time elapsing between brushing and welding was not allowed to exceed one minute. It was found advantageous to degrease the scratch-brush, rinse it in alcohol and thoroughly dry it prior to welding a series of specimens.

The welding tools were of the conventional stepped type[11] with a 0·2 in. dia. indentor face. The compressive welding load was applied in a universal testing machine at a slow rate (approx. 0·5 ton/min) and released at an equally slow rate, which was found to reduce scatter to a marked degree. The application of zinc stearate to the contacting surfaces between strip and tool also assisted in reducing scatter and facilitated the removal of the welded specimen. Welds for micro-examination were made in ¾ in. square pieces of the same material under identical conditions.

Heat treatment of welds

The aluminium welds were made with 40–60% reduction and heat treated in the temperature range 200°–600°C., and the copper welds with 50–60% reduction and heat treated in the range 200°–900°C. Aluminium welds were annealed in an electric muffle furnace, and the copper welds in a vacuum tube furnace at a pressure of 10^{-4} mm of mercury.

To investigate the effect of elastic recovery a deep-drawing quality mild steel (0·08% C), 0·038 in. thick, was used initially, but this was found to give rise to strain-ageing effects during heating after welding. Decarburising in moist hydrogen followed by a vacuum annealing treatment removed such effects, and some of the experiments were repeated using this very soft material. The same standard surface preparation and welding procedures were adopted. Stress relief was carried out under 50, 37·5, 25 and 12·5% of the welding load, or after complete removal of the load, by using a diffuse gas flame. The time to heat a weld to any temperature was found by placing a fine-wire thermocouple at the interface before application of the load and so welding it into the interface. The specimen was merely raised to the required temperature and when this was reached the source of heat was at once withdrawn. The time required did not exceed 5 min. Heating of the weld was uniform, no temperature gradient existing along the interface, and it was found possible to heat to any desired temperature in the range 100°–450°C.

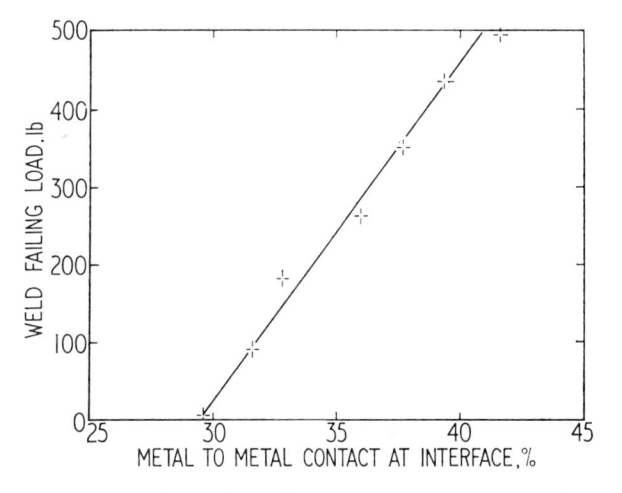

2—Failing load of cold welds in copper as a function of the fraction of metallic contact at the weld interface. (No heat treatment; variable deformation)

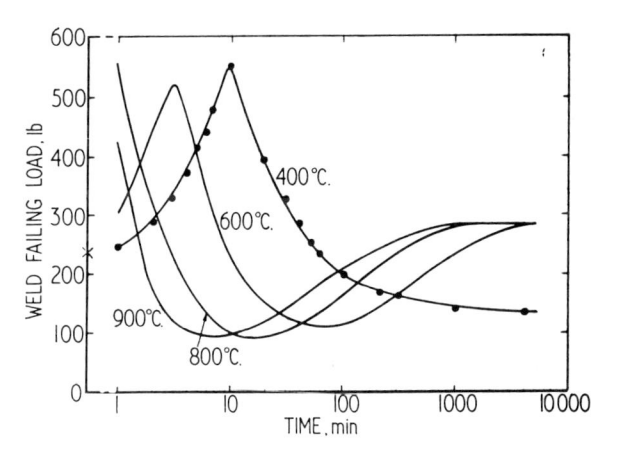

3—Changes in strength of cold pressure welds in copper brought about by heating in vacuo for various times in the range 400°–900°C. (Welding deformation 55%)

Source: *British Welding Journal*, Aug 1963

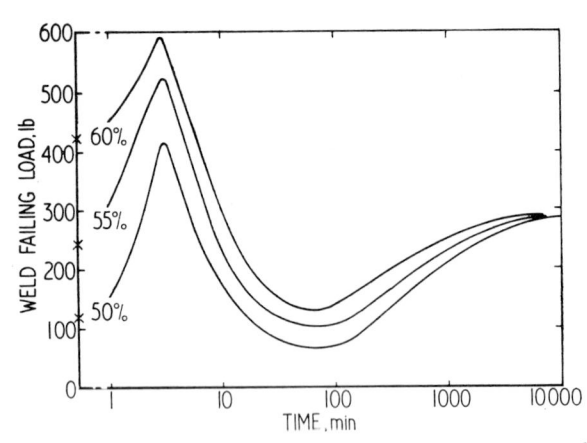

4—Effect of heating time at 600°C. on cold welds in copper with various deformations

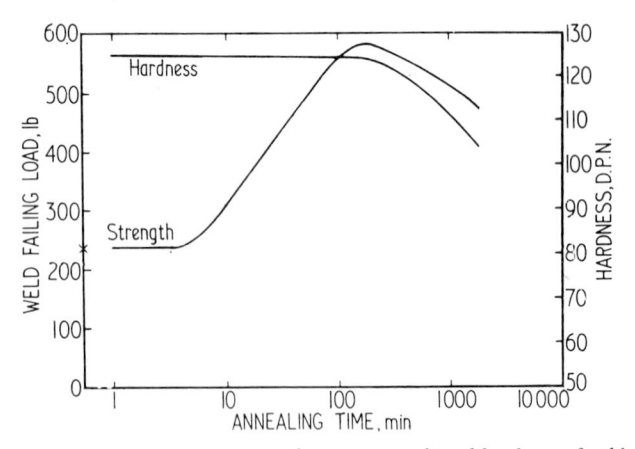

5—Effect of heating at 200°C. on strength and hardness of cold pressure welds in copper. (Welding deformation 55%)

with an accuracy of ± 10°C. Stress relief was carried out at 50°C. intervals in the range 100°–450°C. under the loads previously stated. Each treatment was repeated three times and the mean value was plotted.

Experimental Results

Heat treatment of copper welds

Cold welds in copper gave an interface consisting of primary discontinuities and areas of bonding (Fig. 1); complete resolution of the discontinuities was not possible with the weld in this condition.

Figure 2 illustrates the relationship between the strength of cold welds in copper and the extent of true metallic contact at the weld interface expressed as a linear percentage. This agrees with the results of other workers.[3]

A series of welds made in copper with a constant 55% reduction in thickness were heat treated *in vacuo* at 400°, 600°, 800° and 900°C. The changes in strength so produced are presented in Fig. 3. To avoid confusion, the individual points in this and in many other figures have not in general been plotted. The averaging of three determinations for every point has meant a very good fit for all the curves.* The

---

* A copy of the tabulated data has been deposited with the Institute of Welding.

points for the curve in Fig. 3 for the 400°C. heat treatment have been plotted so as to show the normal order of scatter. The general trend of the changes in strength is the same for each temperature. The overall picture is of an increase to a maximum value followed by a fall in strength and a rise to a final equilibrium value. The results of isothermally heat treating welds made with varying deformations (Fig. 4) indicate that the final equilibrium values of weld strength were very similar although the as-welded strengths varied markedly.

Annealing a cold worked metal causes softening by recovery or recrystallisation processes. Hardness measurements, made as close to the line of the interface as possible, showed an appreciable incubation period before the onset of softening at 200°C. (Fig. 5) and 400°C. (Fig. 6), but during this period there was a marked increase in weld strength. An examination of the primary discontinuities at high magnifications showed no changes; nor did their micro-hardness undergo any change.

The onset of thermal softening was coupled with an immediate decrease in weld strength. Metallographic examination showed that not only was there recrystallisation of the weld metal but significant changes were also taking place within the primary discontinuities. Initially the latter consisted of heavily cold worked metal with cracks visible where welding had failed to take place. As the weld metal softened, so the discontinuities recrystallised to give an extremely fine-grained structure containing many black and roughly spherical areas (Fig. 7). The nature of these areas is not known with certainty; they may be either oxide particles, porosity, or a mixture of both, as will be discussed in detail later, but for convenience they are referred to as pores. Oxide particles would sooner or later give rise to pores since the oxygen would dissolve into copper during heat treatment.

The minimum weld strength was coincident with complete recrystallisation, and thereafter the strength tended to increase owing to the elimination of porosity at the interface. This appeared to be most rapid in the recrystallised discontinuities and it was noticeable that at elevated temperatures, where secondary recrystallisation was marked, the grain size in the primary discontinuities remained extremely fine until

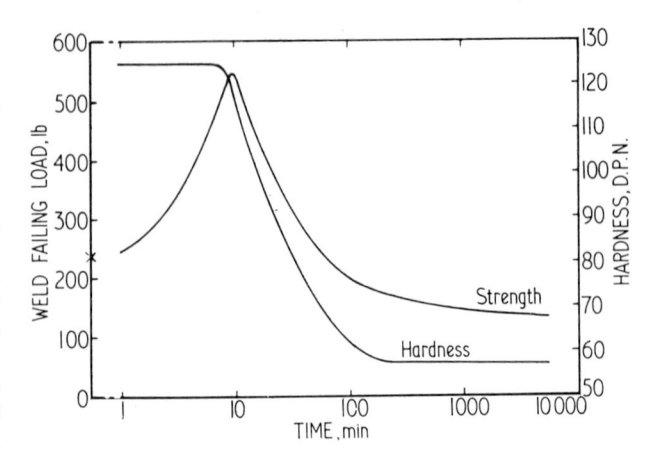

6—Effect of heating at 400°C. on strength and hardness of cold pressure welds in copper. (Welding deformation 55%)

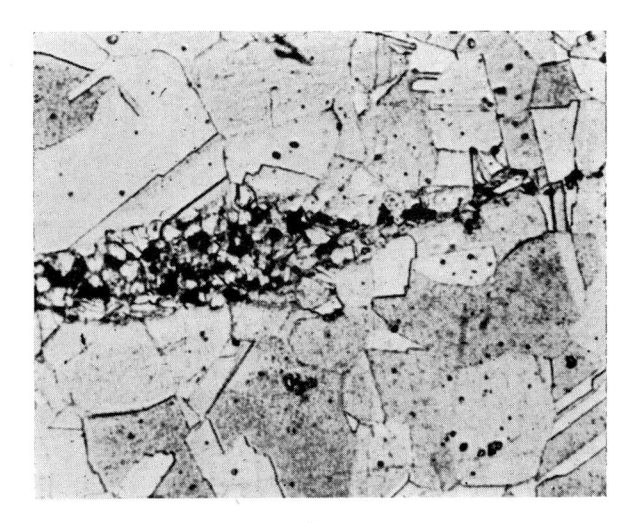

7—*Cold pressure weld in copper after heating for 70 min at 600°C., showing a recrystallised primary discontinuity. (Welding deformation 55%)* ×1050

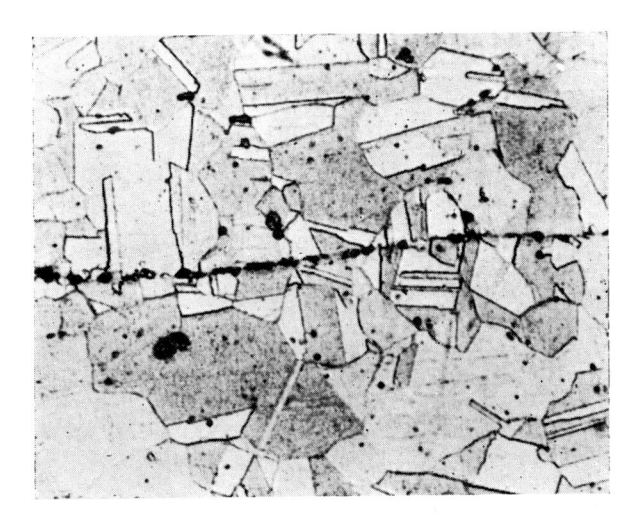

8—*Cold pressure weld in copper after heating at 60°C. for 300 min, showing a line of discrete pores along the original interface. (Welding deformation 55%)* ×550

all the pores had been eliminated. After grain growth had occurred, the only trace of the primary discontinuities was a line of pores (Fig. 8).

Measurements of this interfacial porosity, made by measuring the linear porosity present in a given length of weld interface, showed it to decrease with long heat-treatment times and increasing temperature (Fig. 9). Initially the pores tended to be elongated along the line of the interface but heat treatment resulted in their spheroidisation, and in some cases the long pores appeared to break up into a number of much smaller, more nearly spherical pores. A survey of pore dimensions and number showed a decrease in the number of pores present and a gradual increase in the size of those remaining (Fig. 10). Lengthy annealing completely eliminated the porosity (Fig. 11) and its disappearance corresponded to the final equilibrium strength attained by the welds.

Welds made with a deformation of 55% and heat treated at 600°, 800° and 900°C., showed that after the disappearance of the primary discontinuities the strength of the weld could be directly related to the porosity at the interface (Fig. 12). Such a relationship applies only when welds of constant deformation are considered.

Heat treatment of aluminium welds

Aluminium has much the same cold pressure welding characteristics as copper, giving a weld interface marked with a line of lens-shaped discontinuities with areas of bonding between. The discontinuities, remnants of the scratch-brushed layer, consist of severely cold-worked metal with entrapped oxide particles.

A series of welds in commercial purity aluminium, made with a 50% welding deformation, were heat treated at temperatures in the range 200°–600°C. The changes in strength so produced, shown in Fig. 13, bear some comparison with the changes brought about by heating copper welds (Fig. 3); the incubation period, the increase in weld strength to a maximum value followed by a decrease in strength, are all similar

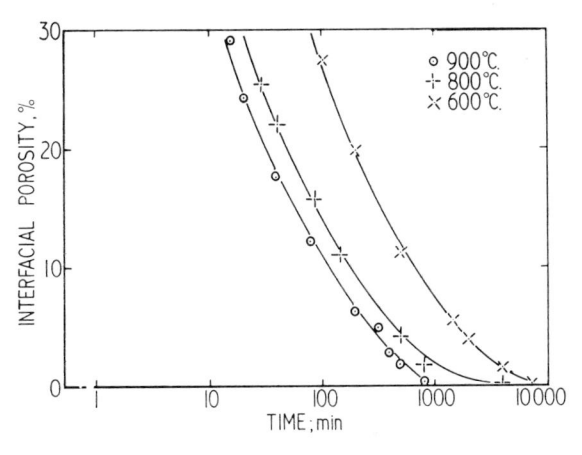

9—*Effect of heat treatment on removal of porosity in copper pressure welds made at room temperature. (Welding deformation 55%)*

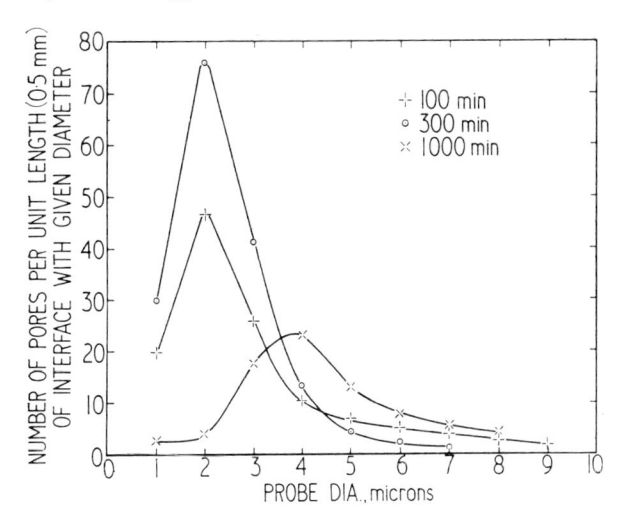

10—*Effect of heating at 600°C. on size and distribution of interfacial porosity in cold pressure welds in copper. (Welding deformation 55%)*

effects. There was, however, no final increase in strength, which may be attributed to the different behaviour of the oxide. Alumina would not be soluble in aluminium, unlike cuprous oxide in copper, and therefore no pores would form to coalesce and diffuse away.

Investigations on the heat treatment of welds in aluminium at 300°C. showed a well-marked period in which no softening was observed, but during which time the strength of the weld increased (Fig. 14), as with copper at 200°C. No micro-structural changes were detected in either the weld metal or the discontinuities to account for such an increase. The onset of softening was accompanied by a decrease in weld strength, considerable softening occurring before recrystallised grains were seen, unlike copper in which the presence of recrystallised grains was noted almost as soon as softening commenced. The attainment of the final equilibrium weld strength coincided with complete recrystallisation. Oxide within the discontinuities underwent no change and the persistent line of discontinuities prevented growth across the interface, even after prolonged heating (Fig. 15).

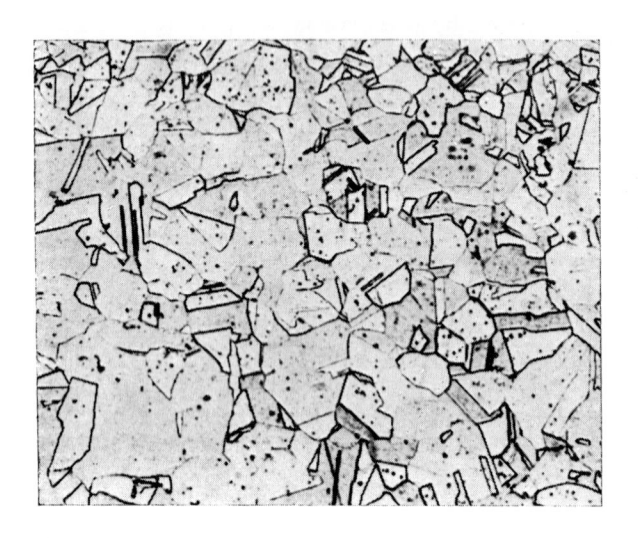

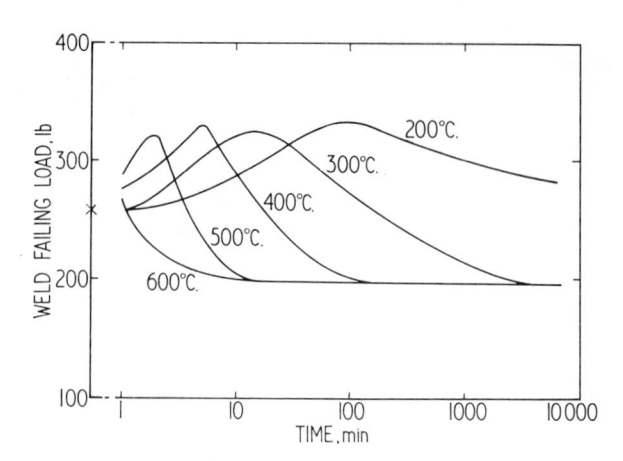

13—*Changes in strength of cold pressure welds in commercial purity aluminium brought about by heating for various times and at various temperatures in range 200°–600°C. (Welding deformation 50%)*

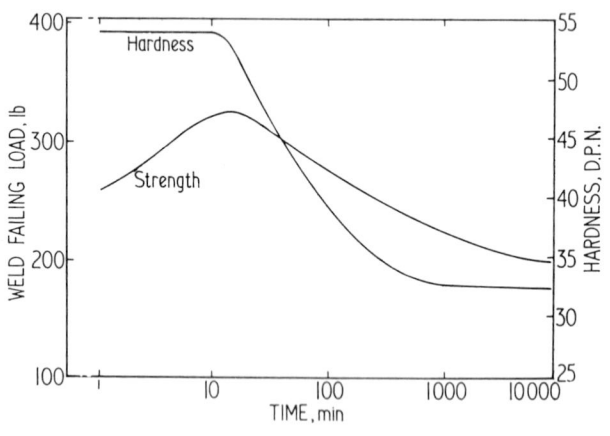

14—*Effect of heating at 300°C. on strength and hardness of cold pressure welds in aluminium. (Welding deformation 50%)*

11—*Cold pressure weld in copper after heating at 600°C. for 3000 min, showing complete removal of interface. (Welding deformation 55%)* ×300

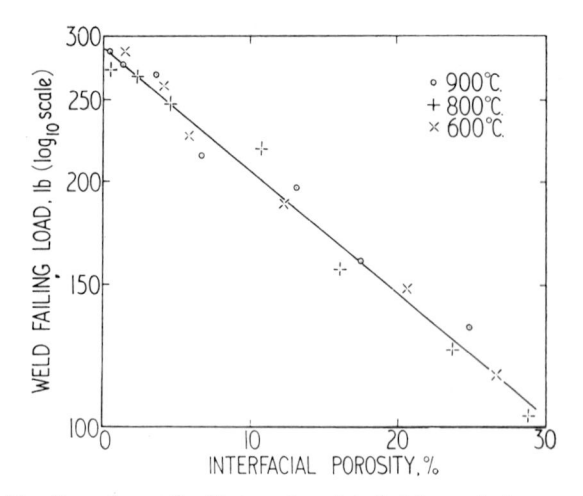

12—*Dependence of weld strength on interfacial porosity in copper welds heated at 600°, 700° and 900°C. for varying times. (Welding deformation 55%)*

The rate of decrease in weld strength is related to the welding deformation, since recrystallisation rates are a function of the degree of cold work (Fig. 16). The strength in the as-welded condition is a function solely of the area of true metallic contact, no contribution being made by the primary discontinuities. The initial increase in strength also appears to be inversely proportional to the deformation; this suggests an improvement in the poorly welded metal/metal contact areas, which must exist in greater numbers at the lower deformations. Since no visible changes occurred in the discontinuities, a series of welds were made without scratch-brushing in an attempt to find the reason for the initial increase in strength. Strips of pickled commercial-purity aluminium were baked for 6 hr at 500°C., cooled to room temperature in a dessicator, and welded. The threshold deformation for welding was increased and the strength of welds made at a given deformation was lower than with conventional surface preparation, which is in agreement with the results of other workers.[2] Welds with this surface preparation were made with 60% deformation and were heat treated at 400°C. Figure 17

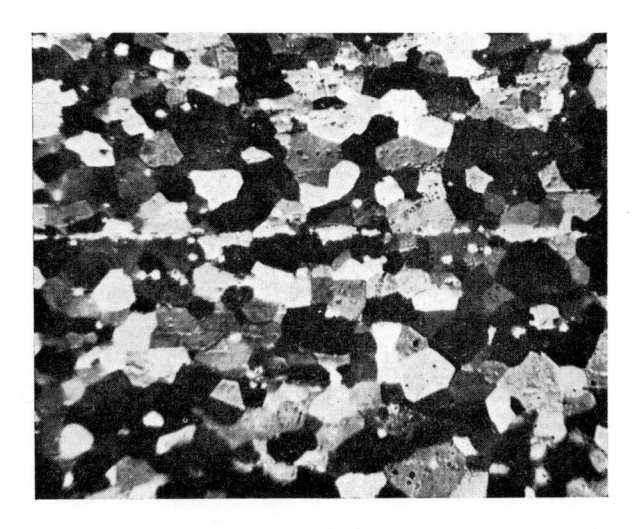

15—*Cold pressure weld in aluminium after heating for 33 hr at 400°C., showing comparative absence of trans-interfacial grain growth. (Welding deformation 50%)*
×300, *viewed under polarized light*

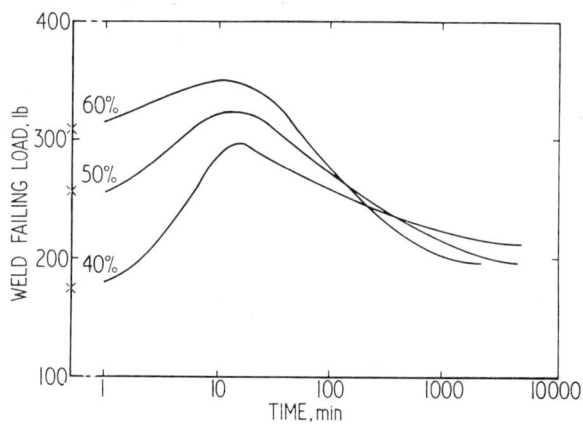

16—*Effect of welding deformation on strength of cold pressure welds in aluminium after heating at 300°C.*

compares the changes in the strength of welds on sheet that was merely baked, with changes in weld strength after conventional surface preparation with similar deformations (60% reduction). The general trend of the change in strength is the same in both cases. The difference in strength is due to the greater length of the metal/metal contact areas in the case of the scratch-brushed surface.

Figure 18 illustrates the changes produced by heating welds made with different types of aluminium: super- and commercial-purity aluminium and strip made from sintered aluminium powder (S.A.P.). The welds were made with a deformation of 50% and annealed at 400°C., but because of the different strip thicknesses the results are not strictly comparable. The differences in purity between the commercial- and super-purity grades gave marked differences in recrystallisation rates and strengths. The changes in S.A.P. were due to the unusual thermal stability of this material. There is no difference in strength between the fully worked and the fully heat-treated material, as the main resistance to deformation is provided by the fine dispersion of alumina. The strengthening must be brought about by the same mechanism that produces the initial strengthening in all the other welds investigated.

Heat treatment of ferrous welds

Figure 19 shows the effect on welds in mild steel of a low-temperature heat treatment performed under some fraction of the welding load. Heating under an applied load at temperatures below 200°C. caused virtually no change in strength. Above this temperature there was a marked increase in strength, proportional to the applied load. After complete removal of the welding load, heating gave a continuously increasing strength at all temperatures. These results were quite unexpected. However, steel of the composition used (0·08% C) is extremely prone to strain ageing and therefore some of the experiments were repeated using a decarburised steel (Fig. 20). Although the strengths were considerably lower, the overall form of the curves were similar. Only above 400°C. did the welds heated under load exceed the strength of those heated without an applied load.

Metallographic examination of the welds provided no information as to the possible reasons for such behaviour. The simple hypothesis that the process of

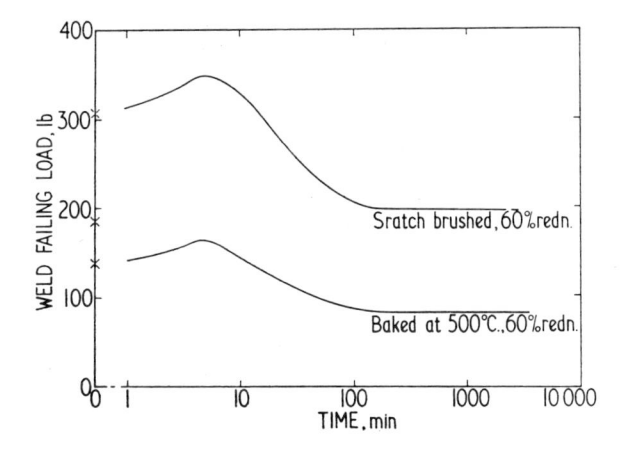

17—*Effect of surface preparation on strength of welds in commercial-purity aluminium heat treated at 400°C.*

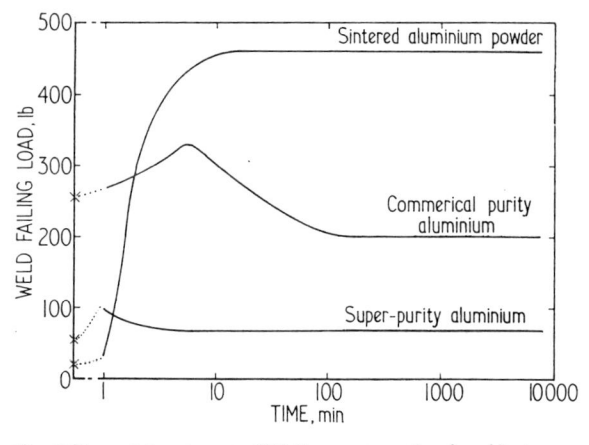

18—*Effect of heating at 400°C. on strength of welds in commercial- and super-purity aluminium, and S.A.P. strip. (Welding deformation 50%)*

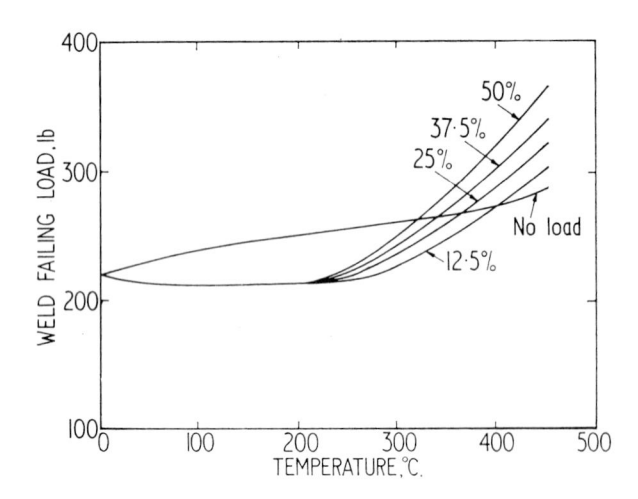

19—*Effect of low-temperature stress relief performed on cold welds in low carbon steel under various fractions of welding load. (Welding deformation 75%. Specimens merely raised to temperatures indicated and allowed to cool)*

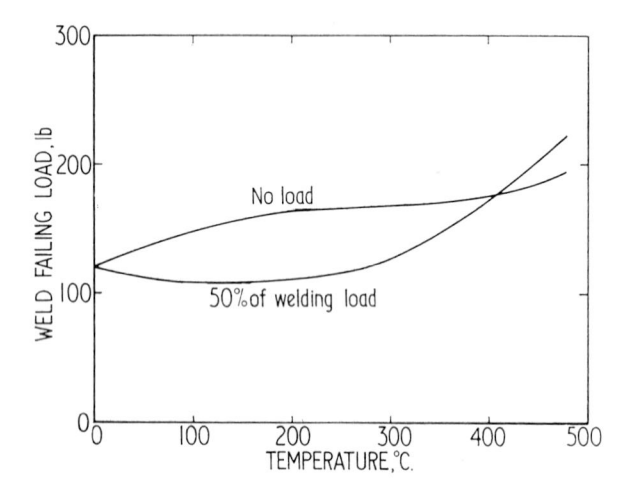

20—*Effect of low-temperature stress relief performed on cold welds in decarburised steel under 50% of welding load and after removal of welding bond. (Welding deformation 75%)*

elastic recovery causes bonds to break as the welding load is progressively removed is not borne out by the behaviour of welds heated after complete removal of the load.

Discussion of Results

Cold welds in copper and aluminium consist of areas of metal/metal contact and primary discontinuities (Fig. 1). As the deformation increases, both the area of metal contact and the weld strength increase. In discussing the changes which take place on heat treatment it is necessary to distinguish carefully between the changes taking place in the metal/metal contact areas and those in the discontinuities. In copper, heat treatment produces an initial rapid rise in strength which is soon followed by a decrease. After long periods, this decrease is followed by a second increase. The initial increase is the more pronounced the lower the deformation, although welds made with the higher deformations give the maximum strength. There is no change in material hardness during the first period of increasing strength nor is there any visible change in the microstructure.

In the case of aluminium the initial stages are the same as for copper—a rise in weld strength followed by a fall. But there is no secondary increase, which suggests that this phenomenon is due to some change that is peculiar to copper.

It seems that in both metals, and with deformations greater than about 60%, the metal/metal areas are well bonded and undergo little change on heating until the normal recrystallisation process occurs. With lower deformations, the metal/metal contact areas are at first poorly bonded, and it is probable that, as Erdmann-Jesnitzer and others[12] have shown on rocksalt, bonds are formed only between favourably situated lattice points. The bonded areas are increased in size by thermally activated surface diffusion along small gaps or via vacant sites. Figure 21 (after Sauerwald[13]) shows the probable situation at the interface after welding. Subsequent heating and surface diffusion may cause a metal atom to reach a region within a sufficiently small distance of both surfaces, which Sauerwald called a 'trap', where it is exposed to the attraction of atoms on both sides. From experiments made by Kuczynski[14] it can be assumed that further lattice units are built up between trapped atoms and the two surfaces into wedge-shaped areas of material. Erdmann-Jesnitzer was able to show, by means of X-ray work on rocksalt, that there were distorted areas between the two surfaces which gave rise to asterism in the Laue spots, but no material of completely different orientation.

A careful examination of Fig. 3 shows that, for a constant deformation of 55%, the rate of increase in strength of welds in copper increases with tempera-

21—*Diffusion of surface atoms (after Sauerwald[13])*

22—*Micro-structure of S.A.P., showing 'flow' of oxide particles in sheet material* ×200

ture. This suggests a thermally activated process, and in order to test it the logarithm of the rate in lb/min was plotted against the reciprocal of the absolute temperature. Whilst only four points were available these fell near a straight line, giving an activation energy of 8,200 cal/g. atom. Similar calculations were made for the results on aluminium, shown in Fig. 13. The four points obtained in this case fell precisely on a straight line giving an activation energy of 8,500 cal/g. atom. Although not too much importance must be attached to so few points, it is clear that a single thermally activated process is involved; the order of energy suggests one of surface diffusion. A recent determination of the surface diffusion constants for nickel by Blakely and Mykura[15] gave an activation energy of 14,000 cal/g. atom, which is about one fifth of the activation energy for volume diffusion. Since the activation energy for volume diffusion of copper is about 47,000 cal/g. atom a figure of the order of 10,000 would be expected; this is in good agreement with that found here. However, such an activation energy is not incompatible with the process of desorption of a contaminant, as recently postulated by Nicholas and Milner.[16]

So much for the first stage. The second stage, as Fig. 5 shows, coincides with a decrease in the hardness and the onset of the processes of recovery and recrystallisation. The reduction in the weld strength is, therefore, in part at least, ascribed to thermal softening of the base material.

A third stage, which in this work occurs only in copper and then only at temperatures of 600°C. and above, results in a final increase of strength. This increase occurs because the dissolution of oxide taking place at elevated temperatures results in porosity, which contributes to the reduction of strength in the second stage. The pores tend to diffuse away after long times thus causing an increase in strength.

This third stage is absent in aluminium. After recrystallisation there appeared to be complete bonding across the primary discontinuities which had completely changed in character; and although particles of alumina remained visible, no voids could be observed. The presence of the recrystallised discontinuities contributed in some way to the strength of the weld, as the welds made with a lower deformation and which, therefore, contained more discontinuities, were slightly stronger (Fig. 16). This is no doubt due to the effect of the dispersed oxide phase having a strengthening effect on the weld when subjected to shear. The presence of oxide particles has prevented grain growth across the interface.

The heating of welds made in thermally stable sintered aluminium powder showed, as expected, no change after the initial increase in strength. But the increase in strength at 400°C. with 50% deformation is far greater than with either pure or commercially pure aluminium. The initial welding characteristics of S.A.P. are obviously far inferior, probably owing to the greater oxide content and higher hardness. There would be a higher proportion of poorly bonded regions within the metal/metal contact areas and, therefore, a greater propensity for improvement on heating. According to Bloch[17] the oxide flakes which are present in S.A.P. have a thickness of about 100 Å, and are aligned in the direction of working (Fig. 22).

It would be expected that during cold welding the oxide flakes present in the freshly exposed metal in the metal/metal contact areas would reduce the weldability of these regions. The oxide flakes would not be expected to have such good welding properties as the metal itself. But it is possible for the oxide near the surface, which later comprises much of the material of the discontinuities, to become hydrated during exposure before and even after scratch brushing. In commercial-purity aluminium sheet the oxide is to a large extent dehydrated during scratch brushing. In the case of S.A.P. with its more complex oxide metal structure it is possible that this process is not complete and that residual water molecules are transferred to the metal/metal contact zones during cold deformation, as suggested by Nicholas and Milner.[16] If this is so, it is possible that the subsequent improvement that takes place on heating is similar to that taking place during the sintering of aluminium powder itself. It is suggested by Bloch that the original oxide is hydrated and that at elevated temperatures the water separates according to the equation:

$$Al_xO_y(OH)_z \rightarrow Al_xO_{(y+n)}(OH)_{(z-2n)} + nH_2O$$

The water then reacts with the aluminium, liberating hydrogen:

$$3Al + 3H_2O \rightarrow Al_2O_3 + 3H_2$$

If small aluminium powder compacts are placed in a furnace at 550°C. and above, it is possible to observe superheating in the compact by as much as 100°C., suggesting that the process is exothermic. This reaction also results in a change in the oxide, causing it to fragment. Thus it would seem that by welding and then heating at 400°C. the reaction taking place on the welded surface in the first two minutes is highly exothermic and results in enhanced oxide break-up and improved adhesion. After this initial increase no further change would be expected, as S.A.P. does not anneal owing to the presence of the dispersed oxide phase.

It is not possible to obtain high-strength joints in S.A.P. by cold welding since the material is embrittled and cracked by deformations exceeding about 60%. For this reason hot welding or the two-stage process are the only methods that are applicable.

It is possible that the mechanism outlined by Bloch for the sintering of S.A.P. could also play some part in the improvement that occurs during the heat treatment of welds in aluminium. But the effect is not likely to be as great and would be difficult to distinguish from the other factors involved. Nicholas and Milner[16] have in fact recently made use of a very similar reaction to explain the effect of water vapour on the welding of aluminium. The fact that this type of reaction is not applicable to a more noble metal like copper, and that welds in copper when heat treated behave in much the same way as those in aluminium, shows that some more generally applicable hypothesis is required.

The interpretation of the results of the experiments designed to investigate the role of elastic recovery in both mild steel and decarburised steel presents some difficulties. Welds heated after complete removal of the welding load showed a continuous improvement in strength over the range 50°–450°C. However, welds heated under some fraction of the welding load showed no such improvement until a temperature of 225°C. was used. Thereafter, the increase in strength was directly proportional to the applied load and to the temperature. The strength of welds heated under an applied load exceeded the strength of those heated after its removal only when the temperature was higher than 400°C. It appears very unlikely that elastic recovery has the great influence that has been ascribed to it. Even if it has a detrimental effect it would seem that this test is not capable of revealing it.

In the absence of information on the strain-ageing of iron under pressure at low temperatures, and since there is little difference in trend between Figs. 19 and 20, it seems best to ignore any role that strain-ageing might have. It must be remembered that the material has been cold deformed 75% and is in a relatively brittle condition compared with the other materials investigated. It would be expected that, after the load has been removed, the effect of heat treatment would be that of stress relief—a process which is known to take place on a micro-scale at only slightly elevated temperatures. Under load, however, this process is inhibited, and the weld is effectively maintained in a state of equilibrium until a minimum temperature of 200°C. is reached. Although no further overall reduction in thickness occurs under load, further deformation can be envisaged, taking place in a very localised area at the interface. Under these conditions both temperature and load become effective in increasing the weld strength.

Conclusions

(1) Cold pressure welds in aluminium and copper made with up to 60% reduction were increased in strength by a heat treatment which had no detectable effect on micro-structure or on the mechanical properties (*i.e.*, hardness) of the material comprising the weld. Such an increase is considered to be due to local atomic rearrangement at the interface.

(2) The onset of recovery or recrystallisation produced a decrease in strength.

(3) The behaviour of welds after recrystallisation is complete is governed by the characteristics of the metal oxide. Where oxide dissolution was possible, as in copper, prolonged heat treatment improved the weld strength owing to the elimination of pores arising from areas of high oxide content at the interface. When oxide dissolution was not possible, as in aluminium, the final equilibrium strength was attained once recrystallisation was complete.

(4) The effect of long-term heat treatment on aluminium welds is governed by the welding deformation. Welds made with low deformations were improved whilst those made with large deformations were weakened.

(5) The heat treatment of cold welds in both 0·08% C and decarburised steel gave a steady rise in strength with increasing temperature. Welds in such material, heat treated in the range 225°–450°C. under various fractions of the welding load, gave an improved weld strength above 225°C. The degree of improvement was a function of the applied load and temperature and was in no way connected with the prevention of elastic recovery.

Acknowledgments

Thanks are due to the British Welding Research Association for grants towards the cost of this work and to High Duty Alloys Limited for the supply of S.A.P. sheet.

REFERENCES

1. M. G. NICHOLAS and D. R. MILNER: *Brit. Welding J.*, 1961, vol. 8, pp. 375–383.
2. L. R. VAIDYANATH and D. R. MILNER: *Brit. Welding J.*, 1960, vol. 7, pp. 1–6.
3. L. R. VAIDYANATH, M. G. NICHOLAS and D. R. MILNER: *Brit. Welding J.*, 1959, vol. 6, pp. 13–28.
4. R. F. TYLECOTE: *Trans. Inst. Weld.*, 1948, vol. 11, p. 95s.
5. R. F. TYLECOTE: *Brit. Welding J.*, 1954, vol. 1, pp. 117–135.
6. K. J. B. MCEWAN and D. R. MILNER: *Brit. Welding J.*, 1962, vol. 8, pp. 406–420.
7. W. H. BRUCKNER and J. H. SAYLES: *Welding J.*, 1956, vol. 35. pp. 501S.
8. F. ERDMANN-JESNITZER and W. WICHMANN: *Zeit. Metallk.*, 1955, vol. 46, pp. 854–859.
9. D. PUGH: Ph.D. Thesis, University of Birmingham, 1954.
10. C. F. WILFORD and R. F. TYLECOTE: *Brit. Welding J.*, 1960, vol. 7, pp. 708–712.
11. R. F. TYLECOTE, D. HOWD and J. E. FURMIDGE: *Brit. Welding J.*, 1958, vol. 5, pp. 21–38.
12. F. ERDMANN-JESNITZER and F. GUNTHER: *Zeit. Metallk.*, 1955, vol. 46, pp. 801–809.
13. F. SAUERWALD and E. THILO: Actuelle Probleme der Physikalische Chemie, Berlin, 1953.
14. G. C. KUCZYNSKI: *J. Appl. Phys.*, 1950, vol. 21, pp. 632–635.
15. J. M. BLAKELY and H. MYKURA: *Acta Met.*, 1961, vol. 9, pp. 23–31.
16. M. G. NICHOLAS and D. R. MILNER: *Brit. Welding J.*, 1962, vol. 9, pp. 469–475.
17. E. A. BLOCH: *Met. Rev.*, 1961, vol. 6, pp. 193–237.

Diffusion Welding of Molybdenum*

Tatsuya HASHIMOTO and Kinji TANUMA

It has been well known that the ductile joint of molybdenum is difficult to be made by fusion welding because of its embrittleness due to recrystallization and grain growth.

This investigation shows that molybdenum can be joined without recrystallization by means of diffusion welding techniques using an intermediate such as iron, nickel, copper and silver. It is also found out that the remelting temperature of the joint made at 1000—1100°C for 15 minutes is quite higher than the melting one of the above mentioned intermediates.

1. Introduction

Diffusion welding method can be considered to be one of techniques making joint below the melting point of parent metal. The joining technique is similar to brazing or hot pressure welding. Especially when qualifying deformation associated with welding or welding temperature, it would be required to establish a suitable diffusion welding technique of the material to be welded. At present, however, it is difficult to clear all the mechanism of diffusion welding in all cases.

The purpose of this investigation is to make ductile joint of pure molybdenum without recrystallization and extreme deformation by autogenous or diffusion welding using an intermediate material and to establish fundamental technique for diffusion welding.

2. Materials and Experimental Procedure

Pure molybdenum sheets produced through sintering or arc-casting method were used in this investigation. The sheets are 1.0 mm thick. Any trace of the dirt, grease and oxide should not be left on the specimen surface in order not to impair the weldability for diffusion welding. Therefore, the surface to be welded was polished by emery paper up to #0/4 and most of them were further cleansed by benzine, and some were treated with the electrolytic polishing (87% sulfuric acid, 13% water, at 20°C, and the current density of 40A/dm^2). Electroplating was adopted to stick intermediate material on the surface of one of the sheets to be welded. Iron, nickel, copper, silver and cadmium were tested as the intermediate. The technical date of the electroplating and bath conditions are given in **Table 1**. The thickness of the intermediate was not measured, but only calculated from a nomograph prepared in advance with respect to the electroplating condition.

Welding was performed in a high vacuum welding chamber. **Fig. 1** illustrates the main parts of the

Table 1. Electroplating conditions.

Material	Plating bath	Anode	Bath condition	Current density (A/dm^2)
Iron	Ferrous sulfate 250g/l Ferrous chloride 30 ″ Ammonium chloride 23 ″	Iron	40°C Stirred	1.1
Nickel	Nickelous sulfate 150g/l Ammonium chloride 15 ″ Boric acid 15 ″	Nickel	Room temperature and Stirred	1.1
Copper	Cupric sulfate 200g/l Sulfuric acid 50 ″	Copper	″	1.0
Silver	Silver cyanide 5g/l Silver potassium cyanide 60 ″ Potassium carbonate 15 ″	Silver	″	0.4
Cadmium	Cadmium cyanide 45g/l Sodium cyanide 90 ″ Sodium hydride 35 ″ Peptone 0.5 ″	Cadmium	″	1.0

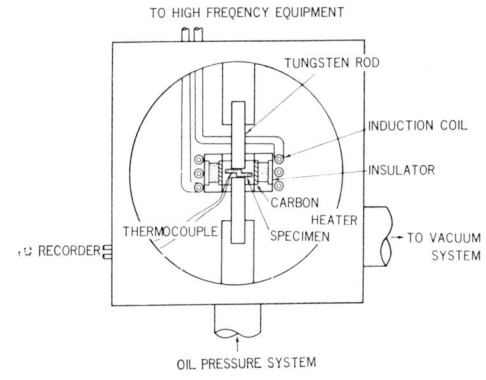

Fig. 1. Schematic diagram of welding equipment.

---

* This paper was published originally in Japanese in Journal of the Japan Welding Society, 37 (1968) 1345.

vacuum chamber. The size of the chamber is about 40×40×40 cm and the vacuum system consists of 10 in. diffusion pump and 5 in. ejector pump backed by a rotary pump having capacity of 1600 l/min. The degree of vacuum attained in the chamber was about 1×10^{-4} torr. The specimen was placed between tungsten rods of 10 mm in diameter. Loading was applied by oil pressure system through the tungsten rods. The load during the diffusion welding was measured continuously with oil pressure gauge. The specimen was heated by radiation from a carbon-filament heated by suitable induction coil with high frequency oscillator of 30 kc. Heating was started after loading in high vacuum in order to avoid contamination of the surfaces to be welded. The temperature of the specimen was measured and controlled by a thermocouple which was percussively welded in the vicinity of the interface of the specimens. The time elape between the surface finishing and heating in high vacuum of $2-3\times10^{-4}$ torr was about five minutes. The welding time was equal to the duration, for which the specimen was heated at the maximum temperature. The welded specimens were used for metallographic observation, material testing, measurements of the distribution of hardness near the interface and remelting temperature of the joint. The shape and dimensions of the specimen used for shear test are shown in **Fig. 2**. Stainless steel sheet of 1.0 mm

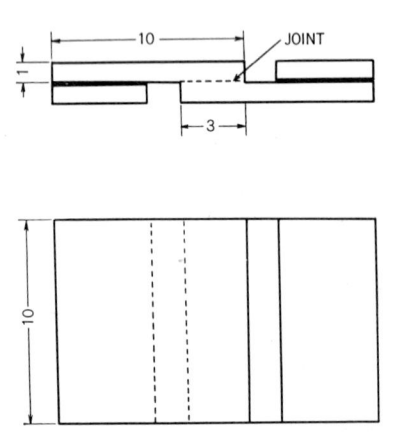

Fig. 2. Design of welding specimen.

thick was adhered by binding agent to the end of the specimen mounted on tensile testing machine, so that shearing forse was applied correctly to the interface. An etching solution developed by Murakami was used for metallographical observation of molybdenum.

3. Experimental Results and Discussions

3.1. Autogenous welding of molybdenum

The autogenous welding of molybdenum was carried out as a preliminary experiment on diffusion welding using an intermediate material. In general, the mechanical properties of the pressure welds are improved with the increase of deformation during the pressure welding[1]. It has been reported[2] that molybdenum can be joined with not less than 50% deformation at 700–1000°C by the pressure welding method. Our experiment was, however, a preliminary one with respect to diffusion welding and in which only small deformation up to 10% was treated. The thickness of the weld was measured by micrometer and the weld deformation was calculated from the following definition

Weld deformation $(\%) = 100(1-t/2T)$

where, T is the thickness of a sheet specimen before welding and t is that of the weld after welding. The influences of the welding temperature on the strength of molybdenum joint, welded as a spot of 4 mm in diameter, are shown in **Fig. 3**. The welding time for

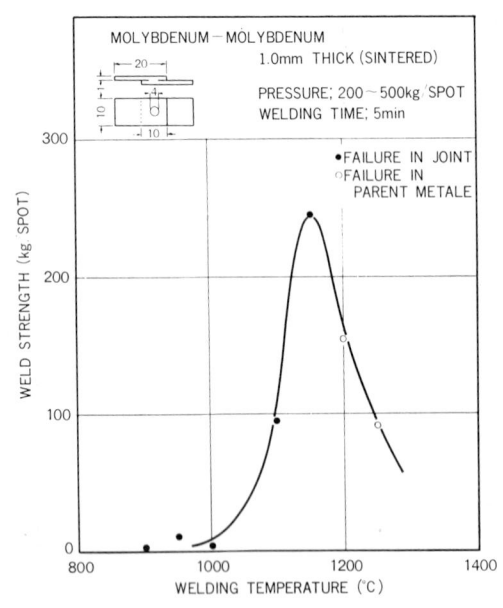

Fig. 3. Correlation between welding temperature and strength of autogenous welds of molybdenum.

each specimen was taken constant as five minutes. The amount of the weld deformation was affected more or less by the welding temperature, but it did not exceed 10%. The weld made at 900°C had not suitable strength at room temperature and the strength was increased sharply by increasing the welding temperature. On the other hand, when the welding temperature was higher than 1200°C, the strength of the weld rather lowered and fracture took place in the parent metal because of embrittleness due to the recrystallization. However, in all the welds made below 1150°C, the

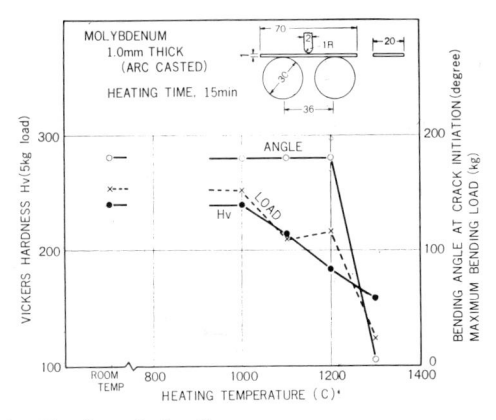

Fig. 4. Results of bending test at room temperature of molybdenum sheets after heating at various temperatures.

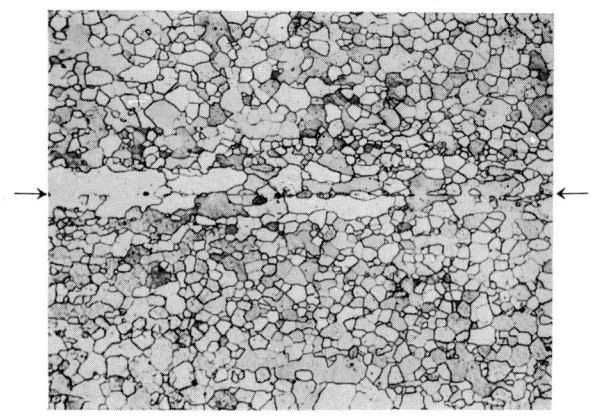

Photo. 1. Cross-section of autogenous weld of molybdenum with 5% deformation made at 1250°C for 5 minutes. (×100) 1/1

fracture occurred along the interface.

Fig. 4 shows the results of the bending test and the hardness measured at room temperature of the molybdenum sheet used after heating at various temperatures. The size of specimen used in the bending test is 20 mm wide, 70 mm long and 1.0 mm thick. After surface polishing by emery paper, the specimens were heated at 1000–1300°C for 15 minutes in the chamber filled with pure argon at 1 atm. As-received specimen was also tested in this experiment. The mechanical properties at room temperature was not affected by heat treatment for 15 minutes at 1000°C and below, but the hardness and bending load of the sheets was decreased by the heat treatment for 15 minutes above 1000°C compared with the as-received sheet. It is found from the above results and micro-structure inspection that recrystallized structure does not generate in the molybdenum used at 1000°C, but a bit of it at 1100–1200°C and completely recrystallized structure exist at 1300°C when the heating lasts 15 minutes in every case. However, the recrystallization temperature of the surface layer of the molybdenum sheet used (about 50 microns in thickness) was higher than that of the interior part, as shown in Photo. 2 or Fig. 8. This phenomenon was probably caused by difference of draft between the surface layers and the interior parts during metal production. Therefore, the specimens heated for 15 minutes at 1100 and 1200°C might be bent completely without crack because of the presence of the not recrystallized surface layer of the specimen. This experimental result shows that the temperature, at which the strength of the joint rise steeply, is nearly equal to its recrystallization one, similarly in the case of pressure welding of various metals[3].

Photo. 1 shows the micro-structure on the cross-section of the molybdenum welded autogenously at 1250°C for 5 minutes with 5% deformation. The recrystallization took place, and the grains on the interface coarsened remarkably, and the grain growth occurred across the interface indicated by arrows in the photograph. It can be said, therefore, that the metallugically sound weld of molybdenum can be obtained at higher welding temperature than 1250°C by autogenous welding. It was reported[4] that such grain growth across the interface in Photo. 1 was accomplished by gas pressure bonding under 4.5 ton/in² at 1260—1440°C for three hours. It is found that the welding temperature required for the trans-interfacial grain growth in molybdenum weld is nearly half of the milting point of molybdenum in Kelvin scale and that this temperature is far lower than the corresponding one of carbon steel[5] and copper[6]. Therefore, from this point of view, molybdenum may have good weldability compared with the other metals, except embrittleness due to its recrystallization. This investigation shows that the molybdenum joint without recrystallization can not be obtained by autogenous welding when the deformation is not more than 10%. Therefore, the molybdenum joint without recrystallization with not more than 10% deformation is expected to be obtained only by diffusion welding using proper intermediate material.

3.2 Diffusion welding of molybdenum

In diffusion welding, the cleanness and flatness of surfaces to be welding are essential factor affecting the weldability. Therefore, the effects of the surface treatment and the thickness of the intermediate on the mechanical strength of the joint are examined, and the results are given in **Table 2**. The surfaces of specimens polished electrolytically or by emery paper of #0 or #0/4 were welded each other without intermediate under the

Table 2. Effect of surface treatment on strength of autogenous weld of molybdenum joint.

Surface treatment	Weld strength (kg/joint)	Fracture type
0 grade emery paper polishing	168	C
	180	B
0/4 grade emery paper polishing	182	C
	192	B
	180	B
Electrolytic polishing	176	C
	216	A
	190	A

Welding pressure ;   15 kg/mm^2
Welding temperature ;   1100°C
Welding time ;   15 min

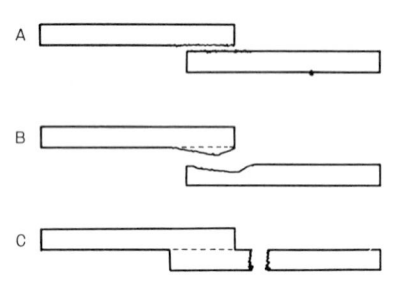

Fig. 5. Typical aspects of joint fracture.

condition given in Table 2. The fracture of the joint can be divided into three types as showen in **Fig. 5.** Type A, where fracture occurred along the interface, was often observed in unsound joint. The strength of joint was not much sensiteve to the surface treatment adopted in this experiment. Consequently, polishing by emery paper of #0/4 was taken as only way of surface treatment throughout this investigation. The effect of the thickness of copper intermediate on the strength of joint formed for 15 minutes at 1100°C under 15 kg/mm^2 is shown in **Fig. 6.** The strength of the joint decreased with the increasing thickness of the intermediate within a range from 2.5 to 30 microns and the type of fracture varied from B to A according to the thickness increase of the intermediate. Therefore, the thickness of the intermediate was chosen as 2.5 microns. It should be noted that copper is quite insoluble in molybdenum. In case that the intermediate is soluble in molybdenum, the effect of the thickness of it on the joint strength may be slightly different from the result in Fig. 6, where the intermediate is quite insoluble. The result in Fig. 6 seems to correspond to the case that the intermediate still remains along the interface after welding because of insufficient alloying during the diffusion welding.

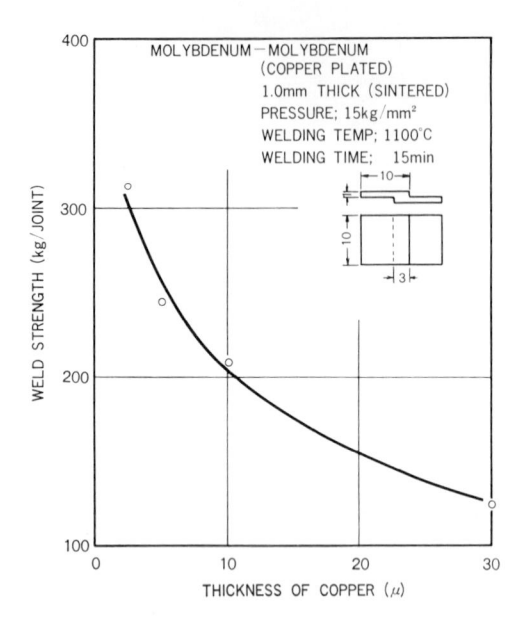

Fig. 6. Effects of thickness of copper intermediate on weld strength of molybdenum joints made at 1100°C for 15 minutes.

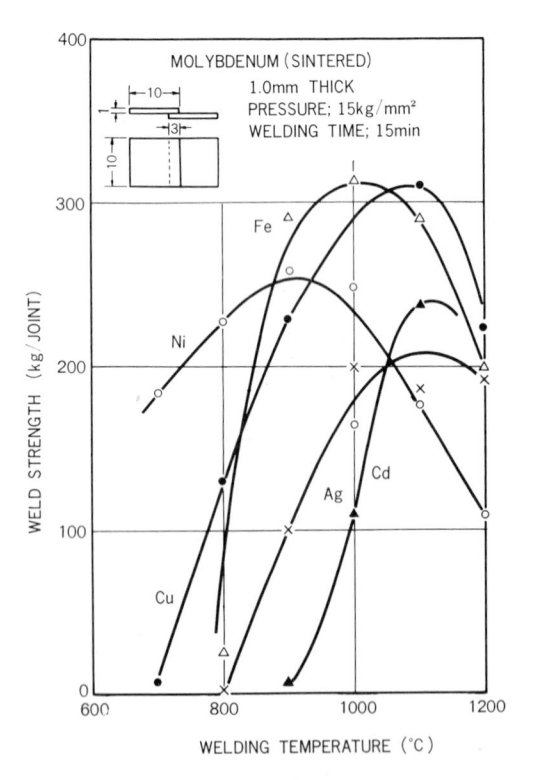

Fig. 7. Correlation between welding temperature and weld strength of molybdenum joints using various intermediates of 2.5 microns thick.

Fig. 7 shows the effects of the variety of intermediate and welding temperature on the strength of the joint with intermediate of 2.5 microns in thickness. It is clear from Fig. 7 that the diffusion weld with an intermediate is accomplished at about 700°C lower than the temperature required for the autogenous weld, and the temperature obtained the maximum strength of weld lowers by about 100—200°C than that of autogenous welding. In case of the joint formed at 1200°C, fracture occurred in the parent metal such as the type C in Fig. 5, and fracture in the joint whose strength was more than 200 kg occurred in the parent metal near the interface such as the type B in Fig. 5. When iron or copper is used as intermediate material, the strongest joint is obtained at 1000—1100°C without recrystallization. While, at lower welding temperature than 800°C, nickel was more suitable than iron or copper as intermediate, but the strength of the joint made above 900°C decreased with increasing welding temperature. Silver or cadmium resulted in giving not so strong joint, compared with other intermediate materials. Moreover, cadmium was the intermediate material that required the highest temperature among those used in this experiment. It may be closely connected with the melting temperature of the intermediate to require comparatively higher temperature in welding using silver or cadmium intermediate. The melting point of cadmium is 321°C and is very low compared with practical temperature of diffusion welding. Thus, in diffusion welding under the condition of high vacuum and pressure, cadmium should evaporete or flow out of the interface during the heating process. Therefore, it is probable that diffusion welding through the use of cadmium intermediate corresponds to autogenous one without intermediate. It is clear from the results in Fig. 7 that diffusion welding using silver intermediate shows similar phenomena to that using cadmium, although the melting point of silver is higher than that of cadmium. This experiment also shows that sound joint of molybdenum can be formed even below recrystallization temperature by diffusion welding through the use of proper intermediates.

Photos. 2 (a), (b) and (c) show the microstructures on cross-sections of molybdenum joints made with nickel intermediate for 15 minutes at 1000, 1100 and 1200°C, respectively. In the joint made at 1000°C a little nickel intermediate still remained along the interface and no recrystallization occurred. On the other hand, in the joint made at 1100 and 1200°C, nickel scarcely remained. In the case of 1100°C, recrystallization did not occur limitedly in the layer of about 100 microns as shown in the central figure, in the case of 1200°C it existed throughout the interface. The results of X-ray microprobe analysis are shown in **Fig. 8**, which is taken in the vertical direction to the interface of the specimen given in Photo. 2 (b). This figure shows that the thickness of nickel at the interface is only about 5 microns

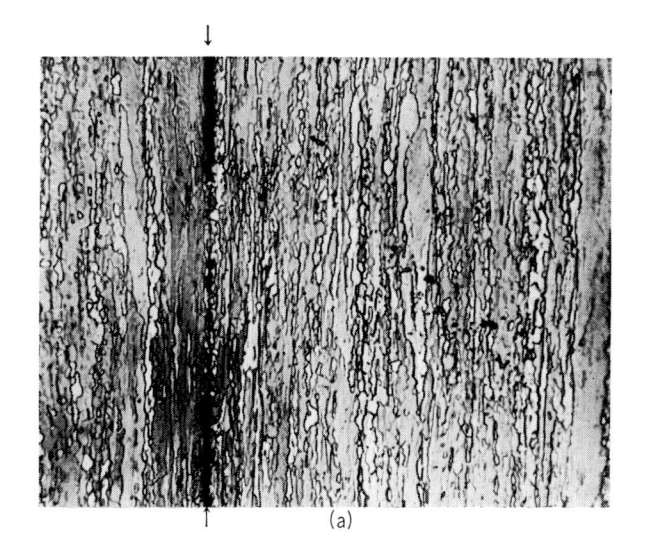

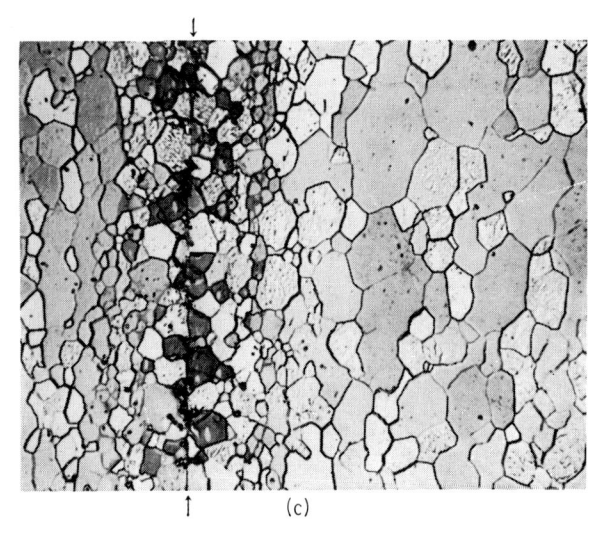

Photo. 2. Cross-sections of molybdenum joints with nickel intermediate at (a) 1000°C, (b) 1100°C, (c) 1200°C, respectively. (×400) 1/1

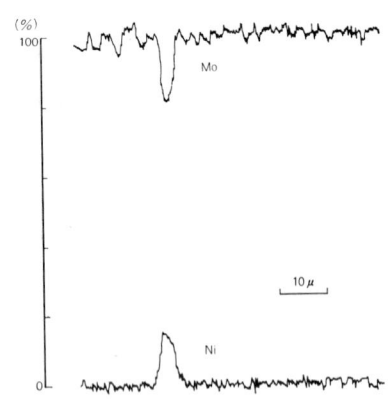

Fig. 8. X-ray microanalysis of molybdenum joint with nickel intermediate made at 1100°C for 15 minutes.

due to its insufficient diffusion during the welding process. Therefore, the existence of not recrystallized layer near the interface shown in Photo. 2 (b) seems not to be caused by formation of an alloy resulted from the diffusion of nickel into molybdenum, but to be attributable of some peculiar properties of the molybdenum sheet used in this experiment. **Fig. 9** shows

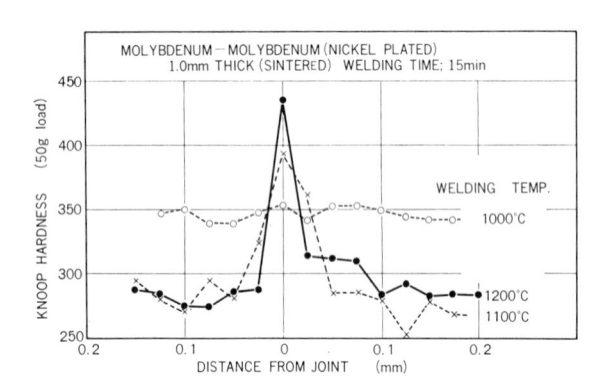

Fig. 9. Hardness distributions of molybdenum joints with nickel intermediate made at various temperatures.

hardness distributions in the joints made with nickel intermediate for 15 minutes at various temperatures. The hardness in the vicinity of the interface was higher than that of the parent metal and reised with increasing temperature of diffusion welding. The reason why the hardness at the position 25 microns distant from the center of the joint was higher than that of the parent metal was probably the difference of the recrystallization due to the heterogeneity of the specimen used. It should be noted that molybdenum constitutes an intermetallic compound with nickel. A special example of brittleness of such compounds formed for 4 hours at 1100°C is illustrated in **Photo. 3**. The alloyed layer in the joint was extremely hard and had many micro-cracks in itself. It is probable that the strength of joint made

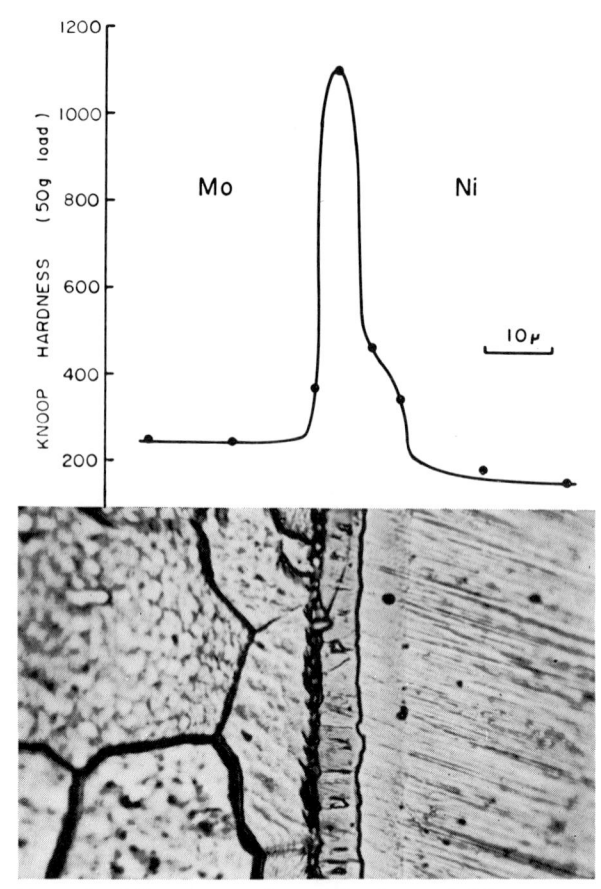

Photo. 3. Microphotograph and hardness of joint between molybdenum and nickel after holding at 1100°C for 4 hours.

with nickel intermediate above 900°C decreases because of formation of brittle intermetallic compound. Moreover, nickel does not dissolve into molybdenum by any more than 0.5% in weight at 1200°C. It was, therefore, impossible in this experiment to make the solid solution between molybdenum and nickel during welding at the recrystallization temperature of molybdenum or lower.

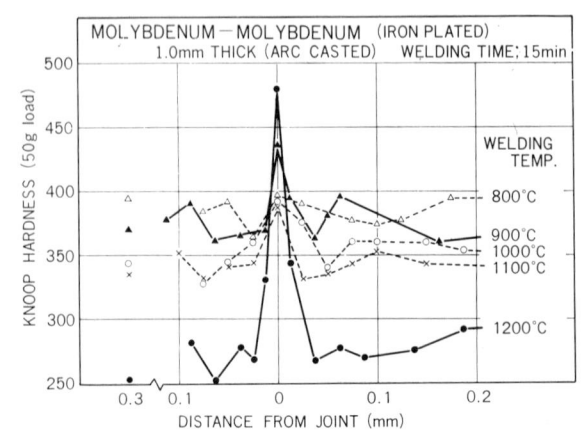

Fig. 10. Hardness distributions of molybdenum joints with iron intermediate made at various temperatures.

It is found that nickel intermediate is available for diffusion welding of molybdenum only at relatively low temperature where any intermetallic compound does not grow up. While, the microstructure of the joint made with iron intermediate was similar to that with nickel one. Therefore, only the hardness distributions of the joint made with iron intermediate are shown in **Fig. 10**. **Photos. 4** (a) and (b) show the microstructures of joints made at 1000 and 1200°C through the use of silver intermediate, respectively. Silver appears in these photographs as a fairly black constituent along the interfaces. The thickness of the intermediate was not constant along the joint. These phenomena were probably caused by welding at higher temperature than the melting point of the intermediate. Furthermore, even at the welding temperature as high as 1200°C, diffusion of silver into molybdenum was not observed, in similar to the case of nickel. The hardness distributions of these joints are shown in **Fig. 11,** the center of the joint formed at 900°C is rather soft comprared with

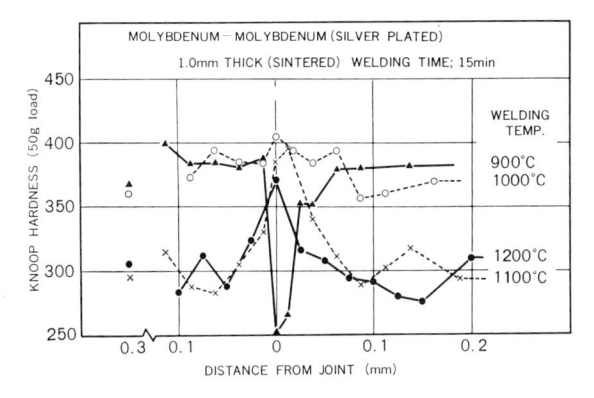

Fig. 11. Hardness distributions of molybdenum joints with silver intermediate made at various temperatures.

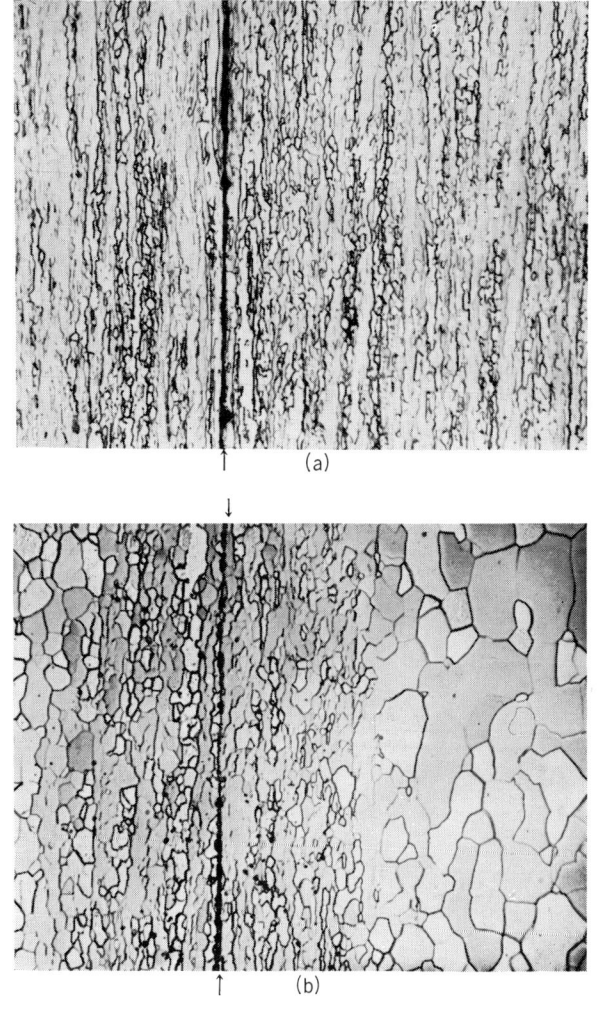

Photo. 4. Cross-sections of molybdenum joints with silver intermediate made at (a) 1000°C, (b) 1200°C, respectively. (×400) 1/1

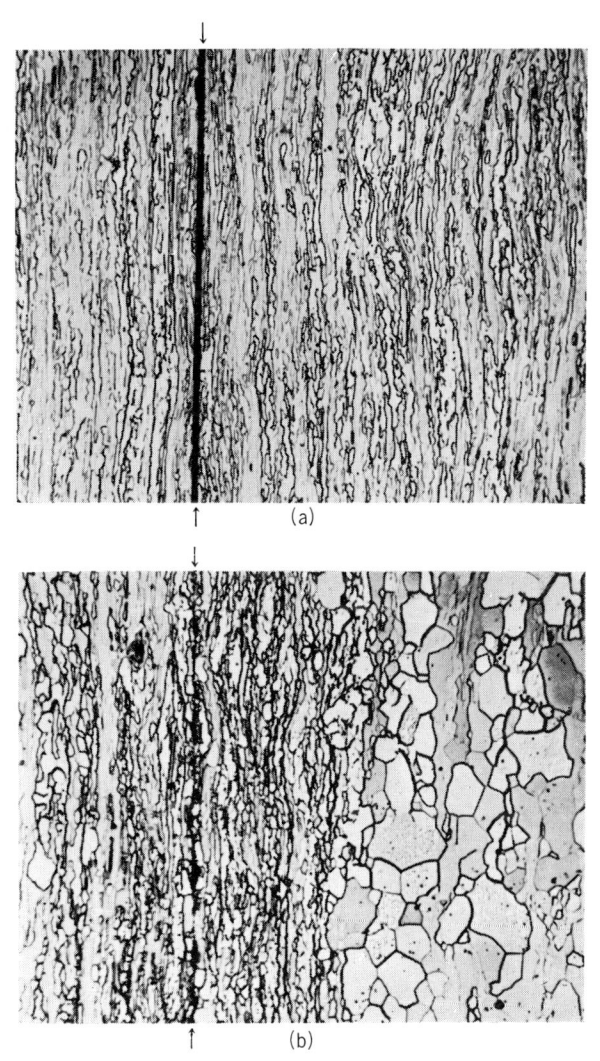

Photo. 5. Cross-sections of molybdenum joints with copper intermediate made at (a) 1000°C, (b) 1100°C, respectively. (×400) 1/1

the parent metal, but that above 1000°C becomes harder as the result of alloying similar to that through the use of nickel or iron intermediate. It has been reported[7] that copper is an effective filler metal for brazing of molybdenum although it does not form an alloy phase with molybdenum. **Photos. 5** (a) and (b) show the microstructures of the joints made at 1000 and 1100°C with copper intermediate, respectively. The joint made at 1000°C was obtained together with an uniform layer of copper intermediate and that made at 1100°C was bonded with discontinuous intermediate layer. It is clear from the results mentioned above that molybdenum can be joined without recrystallization and grain growth by means of diffusion welding technique using proper intermediate. Molybdenum is often used in the environment of high temperature because it has high melting point and excellent strength at elevated temperature. Therefore, molybdenum joint made by diffusion welding must be available for high temperature service. In this experiment the joint was heated up to 2000—2500°C and their remelting temperature was measured. The dimensions of the specimen used for the measurement of remelting temperature is shown in Fig. 2. The procedure of the remelting experiment is also shown in **Fig. 12**. The specimen was placed in

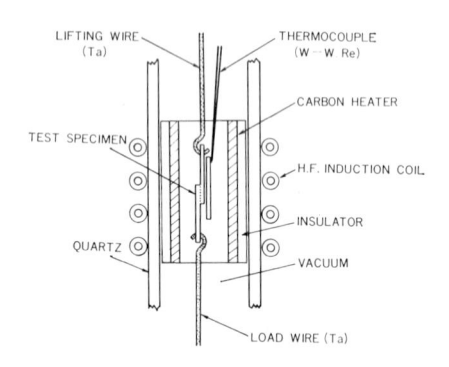

Fig. 12. Procedure of remelting test.

carbon-heater by means of a supporting wire made of tantalum and loaded by 15 g. The specimen was heated by radiation from a high frequency induction heated carbon-filament, the outer face of which was enveloped with a quartz tube. The temperature was measured by a tungsten tungsten-rhenium thermocouple of 0.5 mm in diameter which was spot-welded to molybdenum sheet near the test specimen. The heating was conducted at a rate of 50°C/sec and under the condition of 1×10^{-2} torr and the maximum temperature was hold for 10 sec. **Table 3** shows the results of the remelting test of the joint. All the joint, welded at 1100°C for 15 minutes with intermediate of 2.5 microns in thickness

Table 3. Remelting test of molybdenum joints with various intermediate materials.

Welding condition.*	Intermediate Material	Thickness (μ)	Remelting temperature† (°C)
1100°C 15 min	Cd	2.5	>2000
	Ag	″	>2400
	Cu	″	>2500
	Ni	″	>2000
	Fe	″	>2000
1000°C 15 min	Cu	5	>2000
	Cu	30	1100

* Welding pressure; 15 kg/mm²
† Rate of reheat; 50°C/sec
　Tested under shearing load of 15 g

such as cadmium, silver, copper, nickel and iron, is not broken despite of reheating above 2000°C. The joints made at 1000°C for 15 minutes with copper intermediate of 5 and 30 microns in thickness were tested, and the former did not fail by reheating at 2000°C and the later failed at about 1100°C, which was nearly equal to the melting temperature of copper. It is found from the above results that the joint with copper, which does not alloy with molybdenum, stand for about 2000°C as high as those with the other intermediates, probably owing to the local formation of autogenous weld along the interface as shown in Photo. 5 (b), although these joints may be forced during reheating process for measurment of remelt temperature of joint. It seems that these joints, which is not fracture by reheating of this experiment, has the same remelting temperature as the melting one of molybdenum. **Photo. 6** shows the microphotograph of the joint formed with silver intermediate after the remelting test at 2400°C. Although all the grains became coarse, the interface coincided

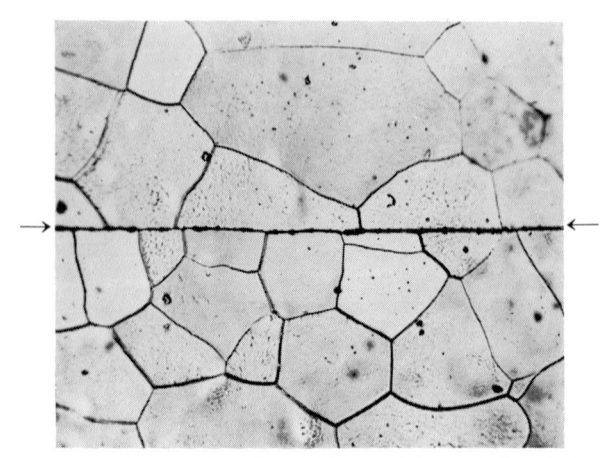

Photo. 6. Cross-section of molybdenum joint with silver intermediate after remelting test at 2400°C. (×400) 1/1

with grain boundary. These phenomena were observed also in the case of the joint with the other intermediates. On the contrary, the grain boundaries of autogenous weld of molybdenum moved even at 1250°C across the interface as shown in Photo. 1. The migration of grain boundaries across the weld interface was probably affected by the amount of weld deformation and the presence of intermediate. The above mentioned results show that molybdenum can be joined without recrystallization and considerable deformation through the use of intermediate such as iron, nickel, and copper, and that the joint has remelting temperature of higher than 2000°C. It may be required in practical use to choose a suitable intermediate fitting well to service conditions. For example, the joint made at 800—900°C with nickel or at 900—1000°C with iron or copper intermediate can be considered suitable for the use at room temperature. While, for the use at high temperature for long duration, the joint formed with nickel intermediate may not be suitable because of the formation of intermetallic compound, and that with copper intermediate seems rather suitable. In use at above its recrystallization temperature at which substantial embrittleness occurs, the suitable joints can be obtained with any intermediate used in this experiment, but also can be obtained even without an intermediate because molybdenum has fairly good weldability of autogenous welding.

In this investigation only several kinds of intermediates are dealt with respecting the diffusion welding of molybdenum. It can be expected further that chromium, titanium and niobium are also available as intermediate for the diffusion welding of molybdenum.

4. Conclusions

The followings are concluded.

1. The autogenous weld of molybdenum without intermediate are attained at higher temperature than its recrystallization temperature with 5—10% deformation, but the joint made at higher temperature than 1200°C are brittle due to recrystallization. Grain growth across the interface is observed to occur at 1250°C for 5 minutes and migration of grain boundaries take place easily at relatively lower temperature compared with those for various metals.

2. It is found that the formation of molybdenum joint without recrystallization and large deformation is accomplished with intermediate such as iron, nickel, copper and silver prepared by electroplating. The remelting temperature of the joint is much higher than the melting one of the intermediate. Although nickel forms an intermetallic compound with molybdenum, nickel intermediate make stronger joint at low temperature than the others. Copper is also found to be applicable as an intermediate, but it does not alloy with molybdenum.

References

1) T. Hashimoto and K. Tanuma: Trans. NRIM, **11** (1969) 116.
2) A. R. Moss: J. Inst. Metals, **82** (1953—54) 374.
3) J. M. Parks: Welding J., **32** (1953) 209s.
4) R. F. Tylecote: Weld. & Met. Fab., **36** (1968) 67.
5) T. Hashimoto and K. Tanuma: Trans. NRIM, **10** (1968) 253.
6) T. Hashimoto and K. Tanuma: Umpublished.
7) J. H. Johnston, H. Udin and J. Wulff: Welding J., **33** (1954) 449s.

DIFFUSION BONDED COLUMBIUM PANELS

FOR THE SHUTTLE HEAT SHIELD

L. J. Korb, C. S. Beuyukian, and J. Rowe

ABSTRACT

Several different processes have been considered for joining metallic radiative heat shield panels for shuttle orbiter, including brazing, inert gas welding, electron beam welding, and diffusion bonding. Each method of joining has its distinct advantages and limitations. Diffusion bonding, however, has some rather unique advantages which make it quite attractive in certain types of designs. In this process, thick sections can be joined to thin sections without shrinkage or distortion. High strength joints are attained which are capable of 2500°F service. Diffusion bonded columbium parts readily accept current oxidative protective coatings without difficulties. Ultrasonic inspection is sufficiently developed to provide acceptance criteria for the majority of the joined areas. This paper presents a review of the efforts at North American Rockwell in the design of columbium shuttle orbiter heat shield panels and in the development of a satisfactory method of diffusion bonding these panels.

1. INTRODUCTION

The Space Shuttle is the key to the economical development of space by future manned systems. The Space Shuttle, as presently visualized, consists of two vehicles: a booster to provide the thrust for launching the system and an orbiter for carrying men and logistic supplies to and from earth orbit. One of the critical design problems facing the orbiter is the development of a satisfactory, reusable light-weight heat shield to protect the vehicle during entry. This paper discusses one of several approaches to this problem under study at the Space Division of North American Rockwell (NR), namely the design and fabrication of diffusion bonded columbium heat shield panels.

While the requirements for the orbiter are still being defined, some general guidelines were assumed for this study. First, the orbiter must be capable of 100 flights, with minimum refurbishment and minimum turnaround time. Second, the orbiter would enter as a spacecraft, and, along the latter portion of its trajectory, would be capable of conventional aircraft maneuvering and landing. Third, the orbiter may be boosted in a "piggy-back" configuration with the booster vehicle.

Heat shield studies which have been undertaken at the NR Space Division include designs using hot metallic structures, radiative metallic panels, ablatives, and external rigidized ceramic insulation. In these studies, typical vehicles were defined and trajectories were selected which were considered "optimum" for a particular design approach. Heat shield panels and substructure were then designed, fabrication processes were developed (as necessary), and full scale panels were manufactured. Environmental evaluation of these panels is now being conducted. In this manner, the spectrum of engineering and manufacturing problems was defined, areas requiring further development were pointed out and some comparison of cost, complexity, and reusability were obtained. Because of the limited funding and tight schedules, it was not possible with any particular system to achieve a truly optimum design; however, sufficient knowledge is being acquired to assess cost and schedule risks of any specific type of approach.

In the metallic area, heat shields were made from conventional superalloys (Haynes 188), dispersion hardened superalloys (TD-NiCr), and refractory metals (Cb 752). The development of the joining parameters for brazing, diffusion bonding, fusion welding and electron beam welding was undertaken in each alloy system. In the case of columbium, full scale panels were designed and fabricated by both electron beam welding and diffusion bonding techniques.* Both methods appear satisfactory and each have unique advantages and disadvantages which constrain the design.

Recent phase B Shuttle studies at NR have been concentrating on the development of rigidized external insulation (REI), considering a system based upon mullite. These studies will provide a direct comparison to the metallic approaches mentioned above.

2. DISCUSSION

Vehicle and Environment

The vehicle chosen for the columbium heat shield development was an early delta winged version of the shuttle orbiter (July 1970 baseline at NR) designated as the SSV 134C and shown in Figure 1. The orbiter would ride "piggy-back" on a booster with its nose 9 inches behind the booster's nose as shown in Figure 2. The total configuration weight was approximately 3,500,000 pounds. The booster was 262 feet in length and the orbiter was 177 feet long. In size, it could be roughly compared to mounting a Boeing 707 on a Boeing 747. The orbiter launch weight was 760,000 pounds and the total wetted surface area was 21,355 square feet.

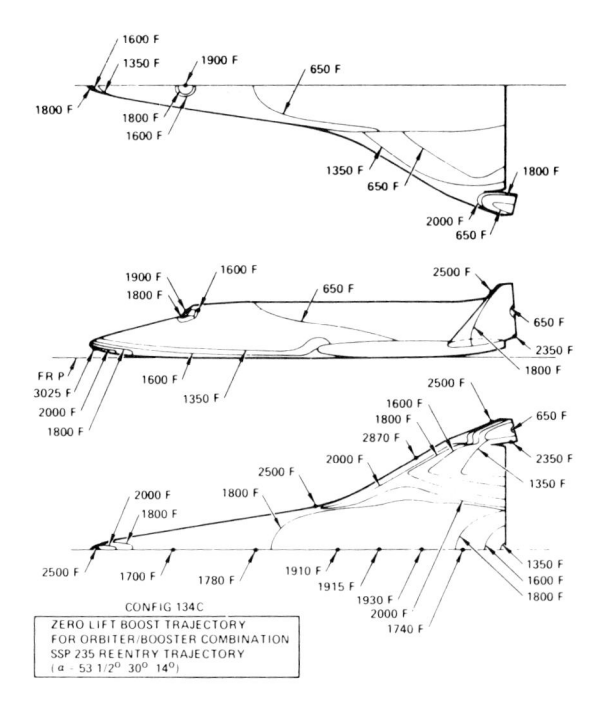

FIGURE 1
NR ORBITER 134C MAXIMUM SURFACE - ISOTHERM MAP

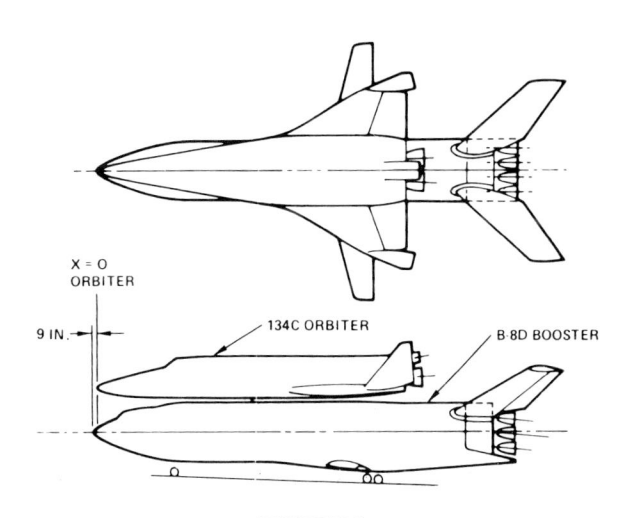

FIGURE 2
BASELINE ORBITER/BOOSTER POSITION

---

*NR is also supporting the Convair Aerospace Division of General Dynamics in a more detailed study involving the design, fabrication, and testing of a coated columbium heat shield for the Space huttle under Langley Research Center contract NAS1-9793.

An area of the vehicle, typical of the requirements for columbium service, was selected for detailed design. It was located approximately 4 feet aft of the orbiter nose along the centerline on the underside of the vehicle. The vehicle cross-sectional diameter at the panel location was 64 inches. For manufacturing purposes, we considered that this area could be represented by a flat panel.

Design Criteria

The heat shield design criteria were defined by specifying a limiting set of operational environments, the heat shield and substructure temperature limitations, and the reuse and inspection requirements.

A design temperature limit of 2400 F was specified for the coated columbium heat shield and 650 F for the titanium load-carrying substructure. The following design criteria were used:

Maximum Maneuvering Load During Subsonic Flight

(1) Maximum load during subsonic flight = 2.5g (Federal Aviation Regulation Part 25 for Transportation Aircraft)

Boost-Phase Design Load Conditions and Temperatures:

(1) Maximum negative pressure (acting outward)   (2 psi limit)

(2) Maximum positive pressure (acting inward)   (3 psi limit)

(3) Temperature at maximum load   70 F

(4) Maximum panel deflection   0.25 in.

Flutter Criteria:

Panel flutter criteria shall be as defined in References 1 and 2.

Creep Criteria:

Materials shall not exhibit cumulative creep strain leading to rupture, creep buckling of compression members, or cumulative permanent deformation in excess of 1% during the vehicle mission life. Mission life is based on 100 flights times the appropriate factor of safety. The value of 1% was arbitrarily chosen based on material properties and acceptable deflections that result.

Factors of Safety:

The following factors of safety were used to determine design loads from the limit design conditions above:

(1) Yield   1.0

(2) Thermal gradient only   1.25

(3) Combined load and thermal gradient
    Loads   1.5
    Thermal gradient   1.0

(4) Flutter - aerodynamic pressure   1.5

(5) Mission Life
    Creep   1.0
    Fatigue   1.5

Panel Design

The major design considerations were that the heat shield panels (1) have minimum weight and volume, (2) are reusable, (3) permit external removal from the vehicle, (4) permit coating inspection, (5) accommodate thermally induced stresses, deflections, and rotations, and (6) sustain imposed limit loads without permanent set and ultimate loads without failure. Generally, the boost and entry environments design the outer heat shield panels. The boost environment sizes the heat shield and its supports and the entry environment causes the more severe thermal stresses. The outer heat shield panels are sized so that they sustain normal air loads, do not permit permanent distortion, and are of sufficient stiffness to resist panel flutter. The standoffs are sized to transmit panel normal pressure and inertia loads to the structure, provide slip joints to accommodate thermal expansion, and minimize the number of thermal shorts. Initially the panels and their components are sized to withstand the boost pressure loads which occur near maximum dynamic pressure and at low material temperatures. The panel size and gages are then checked for combined mechanical and thermal stresses which occur during the entry phase of the trajectory. In all cases the panel size and the heat shield components are limited to combined stresses resulting from air loads, inertia loads, and thermal loads which are below the selected material allowables for crippling, compressive yield, tensile ultimate, and creep. In sizing, all gages are increased to compensate for subsequent coating diffusion. Once the panel size has been established on a strength-stability-creep basis, the heat shield is examined to determine that no combinations of load (including dynamic load) and temperature produce deflections or rotations that would result in permanent set at limit loads or create hot spots.

Out-of-plane deflections occur during the boost and entry phases due to differential pressure and thermal gradient loadings respectively. In addition, there are deflections that result from permanent creep deformations. A panel deflection of 0.25 inch for an 18-inch span was assumed to be the maximum allowed. The value was based on the results of preliminary estimates of the effect of a perturbed aerodynamic surface on convective heating.

Specific Design Details

The panel size was limited primarily by the coating facilities available to a 36-inch maximum dimension. From the point of view of accommodating expansion and minimizing structural supports, an 18 x 36 inch panel was chosen. Various panel cross-sections were considered. NR eventually chose an I-stiffened design as optimum from a combined structural and manufacturing approach. This particular panel concept provides an open section ideally suited for coating and coating inspection. The total design effort included not only the design of the panel but also the design of all support structure. The total weight of the heat shield, including the panel, supports, and insulation was 3.8 pounds per square foot. Only a cursory description of the supporting structure considered pertinent to this discussion will follow.

In essence normal air pressure and inertial loads imposed on the panel are transmitted to the spacecraft primary load carrying structure by channel supports located on the long sides of the panels. The panels are basically free floating on the channel supports, thus isolating thermal strains and movements. The panels are connected to the channel supports with two pins located at the center of the long-dimension - the pin on one side is inserted in a slotted hole to permit thermal expansion growth. Externally applied "crushing" loads are transmitted to the channel supports by panel "beam" elements and are subsequently reacted by the primary structure. Internally applied "burst loads," such as could occur during boost, are transmitted by the panel to pi-straps which in turn are bolted to channel supports. The pi-straps form an external "picture frame" around the panels. Channel supports are composed of columbium and TD-NiCr. The latter is used below 2000°F. This material permits reduced heat shield height due to its high thermal resistivity.

Initial optimization of the panel cross section is shown in the Figure 3 along with the modification to permit manufacturing by diffusion bonding. Changing the section from the "optimized" to the "manufactured" section increased the panel weight by 29% while increasing its moment of inertia by 55 percent. The resulting panel design weighed 2.33 pounds per square foot.

The panel design incorporates a continuous metal border around the underside of the panel to give the panel bending and torsional rigidity during handling so that the coating will not receive localized straining prior to and during assembly operations.

Alloy Cb-752 was used for sizing the panel based upon a combination of its high temperature creep properties at 2400°F and its low density. Mechanical properties at room temperatures were as follows:

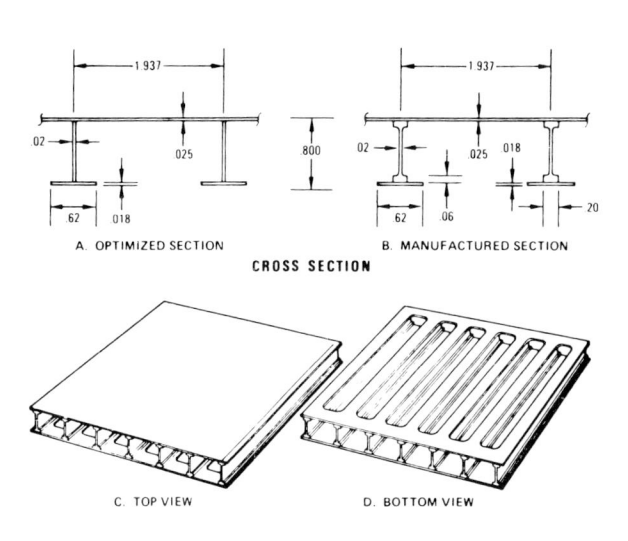

$$F_{tu} = 64 \text{ ksi}$$
$$F_{ty}, F_{cy} = 50 \text{ ksi}$$
$$E \ \& \ F_c = 15.5 \times 10^6 \text{ psi}$$

FIGURE 3
PANEL DESIGN

High temperature data are shown in figures 4 and 5. Initially it was believed that the creep strength of columbium would be the limiting factor in such a design, however, the structural analysis showed that the out-of-plance deflection which occurs during the early portion of the reentry cycle was the critical design condition. At this time, because of the rapid external heating, the panels were subjected to high temperature gradients between the inner and outer surfaces which caused maximum deflections. Panel deflection was outward, i.e., away from the support structure.

Since excessive deflections could trip the boundary layer resulting in turbulent rather than laminar flow and excessive heating, NR reexamined the deflection criterion of 0.25 inch which had to be assumed. Boundary layer thickness calculations were made at several points along the trajectory including the early portion of reentry. These calculations verified that the .250 inch maximum deflection criterion was very near the limit which could be tolerated.

On this basis, it was incumbent on the manufacturing process to hold panels to the maximum possible flatness as fabricated. Out of flatness from all sources, including as-fabricated. tolerances, thermal strains and cyclic creep strain must be accommodated within this deflection limit.

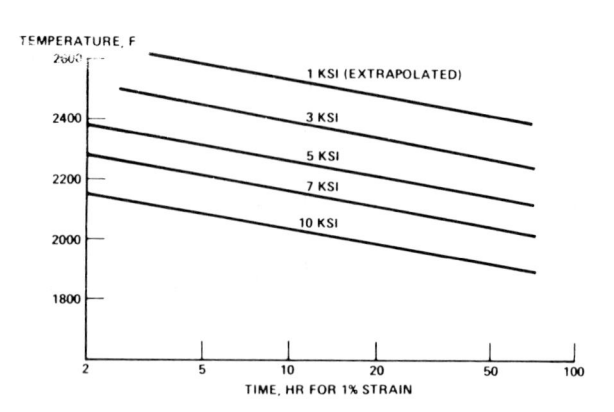

FIGURE 5
COLUMBIUM Cb-752 PRELIMINARY DESIGN PROPERTIES

Panels were analyzed for burst and crushing loads, buckling, flange stability, thermal gradients and thermal stresses, cumulative creep strain, flutter and maximum cumulative panel deflections. Calculation took into account the loss in cross section due to coating diffusion during 100 entry cycles. Thermal analyses were extended to all supporting structure.

3. DIFFUSION BONDING

The joining process used for the heat shield panels should result in as little distortion and built-in stresses as possible, for not only is panel flatness of great importance to the design, but it is also necessary to minimize locked-in thermal stresses - which may invalidate the thermal structural analysis. Potential joining processes included electron beam welding, fusion welding, diffusion bonding, and brazing. Electron beam welding, with its low distorting characteristics, would be preferred to fusion welding in this application. (Electron beam welding, because it is accomplished under a vacuum, also runs little risk of embrittlement of the columbium during welding, whereas the control of shielding for fusion welding is critical.)

Compared to electron beam welding, diffusion bonding offers the following advantages:

(1) Faying surfaces can be joined over the full area of the surface.

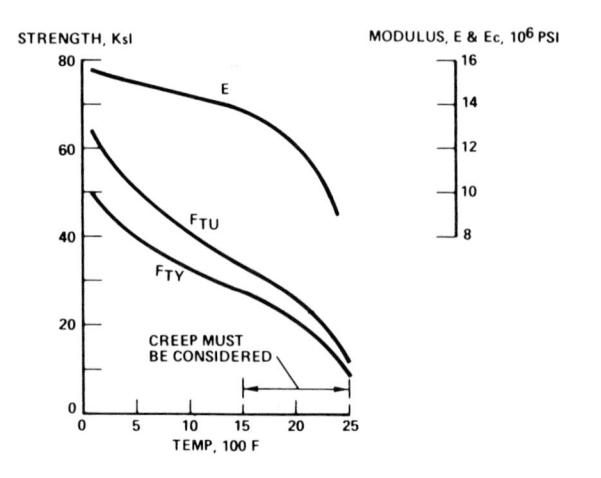

FIGURE 4
PRELIMINARY DESIGN PROPERTIES Vs TEMPERATURE Cb-752

(2) The parts are stress free. A solid state reaction occurs in which no shrinkage takes place. Close tolerances can be attained. (Electron beam welded parts involve fusion and solidification and highly localized heating resulting in some localized distortion and built-in stresses).

(3) Parts which vary considerably in thickness can be joined. (In electron beam welding of columbium it is difficult to join members with consistent quality welds when ratios exceed approximately 3:1, particularly in thicker sections.)

(4) The process can be carried out on conventional press equipment (although platens must be modified for high temperatures). Parts, therefore are not particularly size limited.

(5) All joints can be made simultaneously. Joints where electron beam access is impossible can be made. Often more than one panel can be made at one press operation (depending on design configuration).

Electron beam welding, on the other hand, is ideal for butt-type joints, uses less expensive tooling, is more easily repaired, and in some cases, results in lighter designs because of smaller joint areas.

While brazing is also a consideration in the joining of columbium, it offers the following disadvantages when compared to diffusion bonding.

(1) As a general rule, brazing is performed well above the maximum service temperature of columbium, such as 2800-3000 F, whereas diffusion bonding can be carried out below 2000 F.

(2) A high temperature vacuum furnace is almost certainly needed whereas diffusion bonding can be done in low cost retorts.

(3) Control of brazing alloy flow and its coating or oxidative protective capabilities remain undefined.

Diffusion bonding of columbium is not a new process. It was used with limited success in the early sixties by the Los Angeles Division of North American in the fabrication of columbium honeycomb panels (Reference 3). In this program, diffusion bonding was achieved using interleaving foils of titanium or vanadium. Greater success was obtained with titanium. Diffusion bonding of columbium has also been successfully carried out at the Solar Division of the International Harvester Company since 1965, using both titanium and vanadium foil for attaining the bond. (Reference 4). Solar has actually developed two different diffusion bonding processes. One is called Yield Strength Diffusion Bonding (YSDB) and was developed in 1965. The other is called Continuous Seam Diffusion Bonding (CSDB) and was developed in 1970. YSDB utilizes refractory block electrodes approximately 3/8" x 1-1/4." A continuous joint is effected by overlapping bonds and is always accomplished in an inert atmosphere. CSDB is similar to a seam weld, but does not result in overlapping spots as the electric current is pulsed in such a manner that a constant temperature is maintained at the point being joined and just immediately ahead of that point. This process also utilizes an inert atmosphere when joining refractories, but titanium may be bonded in air.

NR Diffusion Bonding Process

The approach to diffusion bonding taken by North American was similar in many respects to that used in the diffusion bonding of aluminum cold plates for the Apollo. In this process the parts were heated in retorts under vacuum or containing an inert gas to avoid oxidation during diffusion bonding. Pressures were applied by heated platens, and bonding was accomplished at approximately 1050-1100 F. Faying surface pressures of 200 psi were used.

At NR we considered the retort approach to be the best production approach to diffusion bonding of columbium. While the processes employed by Solar have the advantages of lower initial tooling cost and relative ease in making repairs, some difficulties are encountered with the occurrence of thermal distortions and in joining gages of less than 0.020 inch. The Solar processes are also limited by the need for electrode access.

Using the NR process, several joints can be made simultaneously and, in some cases, more than one panel at a time. Panels can be made essentially free of distortion, and no face sheet gage limitations have been experienced. On the other hand, repair methods for bonds made by the NR process have not been developed. Futhermore, production runs are essential to make the NR process economical.

The process details which had to be worked out for diffusion bonding columbium included:

(1) The bonding parameters, i.e., interleaf material, temperature, time, pressure, and vacuum.

(2) The development of high temperature platens, capable of up to 2500 F service and pressures of 3000 psi (at 2500 F).

(3) The control of heating uniformity.

(4) The selection of heating electrodes for extended service life.

(5) the selection of proper tooling materials.

(6) The design of reusable internal tooling.

In an attempt to establish a suitable diffusion bond process, foils of pure columbium, titanium, and vanadium were used to bond small lap-joint specimens over a wide range of temperatures, pressures, and times. A paramount consideration was to keep the maximum temperature of the bonded part below 2000 F for several reasons:

(1) to permit the use of low-cost, stainless steel retorts,

(2) to minimize tooling problems,

(3) to minimize differential coefficient of expansion problems, and

(4) to give longer life to heaters and platens.

Successful, high quality bonds were best achieved by using titanium foil, temperature up to the 1950-2000 F range and faying surface pressures up to 1500 psi. Temperatures and pressures were held for 5-7 hours to permit adequate titanium diffusion. Retorts were built from AISI 321 stainless steel sheet which were sealed by resistance seam welding and/or fusion welding. Sheet type retorts such as these could be built for under $100 each. Retorts were checked for leak tightness of 5×10^{-6} cc/sec of helium prior to use. During bonding a vacuum of 5 microns or less was maintained. Under these conditions parts emerge from the retort bright and shiny with no visible or detectable surface oxidation. It was also determined that pressures are not critical to the bond itself; however, since diffusion bonding is a solid state reaction, physical contact of the columbium part with the one-mil thick titanium foil was needed to insure a bond. Figures 6 and 7 illustrate diffusion bonds in columbium using titanium and vanadium foils, respectively.

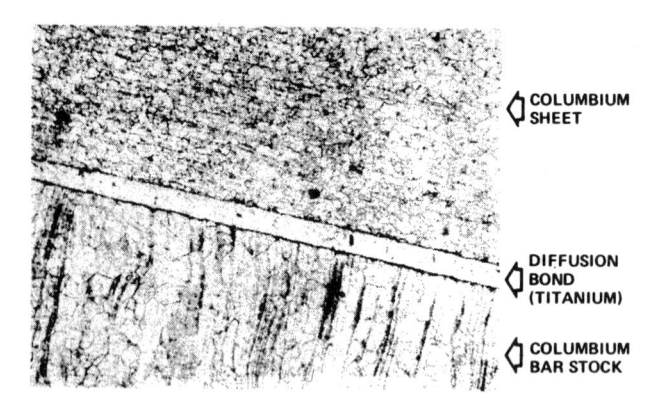

FIGURE 6
COLUMBIUM Cb-752 DIFFUSION BONDED WITH PURE TITANIUM FOIL

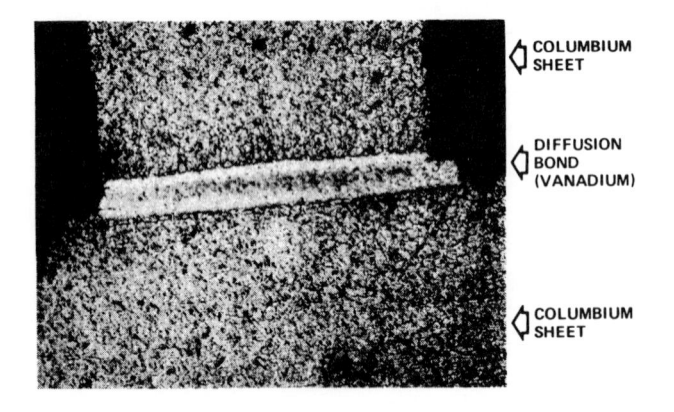

FIGURE 7
DIFFUSION BONDED COLUMBIUM WITH A VANADIUM FOIL INTERLEAF 150X

Originally Masrock* platens were used but they could not withstand the combination of pressures and temperature. During the heating cycle the platens can be expected to reach temperatures, in some cases, several hundred degrees above the part temperatures. Silicon carbide platens, approximately 4 inches thick proved quite serviceable. Later in the process Mullfrax,** an

---

*Masrock is a registered trademark of the Glasrock Company.

**Mullfrax and Fiberfrax are registered trademarks of the Carborundum Company.

alumina-silica ceramic product from Carborundum Co., was used to insulate the press platens. It also provided an excellent hard backing for the silicon carbide platens.

Heating was accomplished by using cylindrical hollow-cored silicon carbide heating elements (5/8" diameter) inserted through holes penetrating the entire length of the platens. The heaters were insulated from the platen at uniform intervals by ceramic rings. Six zones of temperature control were used. To insure uniformity of part temperatures, two carbon steel plates, 1-½ inches thick, were used as thermal distribution plates between the platens and retort, one on each side of the retort. These were instrumented with 6-8 thermocouples to record temperature uniformity. In addition, two thermocouples were inserted within the retort at opposite sides of the part. Hand controlling of the heating zones kept part temperature uniformity within 50 F. The exposed edges of the retort had to be packed with insulation (Fiberfrax) to minimize localized loss of heat.

Parts were laid up in the retorts and held in place by tooling made from a variety of stainless steel and refractory metal alloys. The titanium foil interleaf was tack welded in place along its entire length to maintain its location during handling and heat-up cycles. Originally aluminum oxide was flame sprayed on the columbium tooling used in the retorts, however, it was later eliminated because of the difficulty in controlling its thickness and because it was abrasive.

Tooling materials were required to hold the parts in contact while allowing for differential expansion and contraction without scoring the part surfaces. When expansion or contraction is restricted, the final part may be warped, or may exhibit puckers or wrinkles. Tooling should also be designed for ease of removal and reusability. The latter requirement is most important since tooling costs are high and must be amortized over several parts.

Based upon our experience, cost projections assuming moderate tool life would run $500-800 per square foot depending on part design and complexity. These figures compare very well with the fabrication costs for brazed precipitation hardened stainless steel used on the Apollo and B-70 programs.

Metallurgy

The columbium diffusion bond using titanium has not been characterized to any great extent. Titanium and columbium appear to diffuse into one another forming a solid solution interface which is roughly twice the original

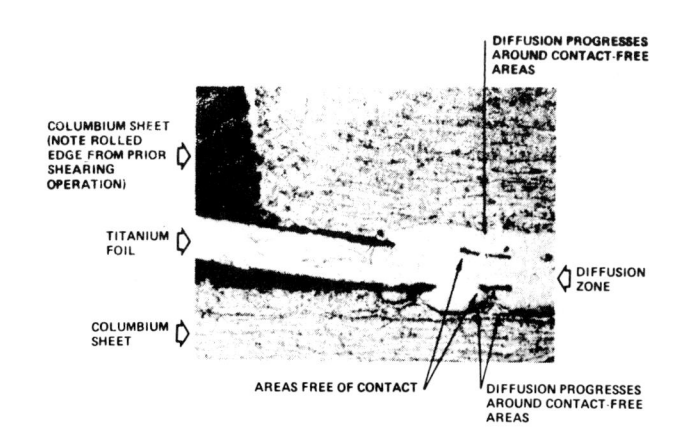

FIGURE 8
IMPROPER EDGE CONDITION OF DIFFUSION BOND

titanium thickness. Diffusion seems to progress around small areas where contact was not made, as shown in a substandard edge detail in Figure 8. Notice in this figure that diffusion bonding did not occur up to the edge of the part. In this case, the edge had been sheared and the "rolled over" edge was not in contact with the foil for about .005 inch. Machining the edges of surfaces to be bonded allows contact to within one to two mils of the part edge.

The strength of the columbium diffusion bond using titanium has never been adequately measured. Typically, the diffusion bonding area is 15 to 20 times the cross-sectional areas of the sheet. Failures in thin gage specimens always occur outside of the bonded area. High temperature creep tests (2500 F) on coated, diffusion bonded specimens have shown no detectable creep in the bond joints nor any preferential oxidation attack. Figure 9 illustrates how a coating such as R512E will penetrate into the small edge void or notch area. In this figure, a specimen was used for coating in which the foil had been offset up to .006 inch from edge to determine the penetration

FIGURE 9
CROSS-SECTION OF A COATED DIFFUSION BONDED SPECIMEN

characteristics of the coating. The coating penetrates the one mil thick void area fully. (Note: Defects which appear in the coating are artifacts resulting from poor polishing techniques and not indicative of the integrity of the coating.)

One unusual feature within the bond is that there appears to be microfissure formation perpendicular to the diffused joint as shown in Figure 10. In this direction microcracks are parallel to any load across the bond and seem totally harmless. In addition, two parallel paths of linear microporosity appear as though they nucleate at the former titanium foil surface. These are suspected to be the result of a Kirkendall effect in the diffusion zones.

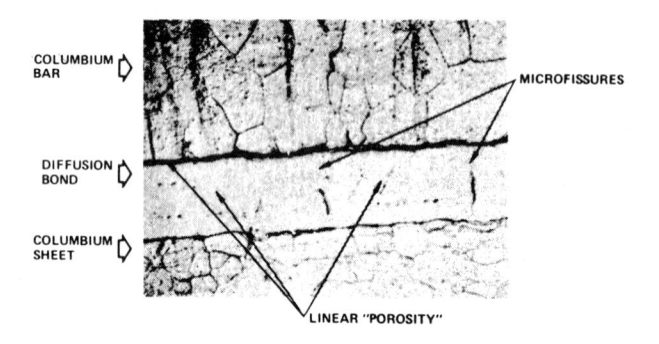

FIGURE 10
DIFFUSION BONDING ANOMALIES

4. PANEL FABRICATION AND QUALITY CONTROL

Initial efforts to make an .020-inch thick sheet panel with standing legs of the same gage as shown in Figure 3A were not successful. It was too difficult to get full contact along a .020-inch wide line. Furthermore, foil placement was difficult to make and hold. Therefore, the standing legs were redesigned to have a miniature I-beam cross section allowing .200 inch of width for the diffusion bond at the beam flanges. This worked extremely well. NR has not established how narrow a width can properly be bonded. However, it is expected to be somewhat less than .200 inch.

Panels 4" x 18" long were diffusion bonded to check the tooling concept and bond integrity. One such panel is shown in Figure 11. The scratches on the panel were caused by efforts to withdraw tooling. Eventually tooling was removed by slitting the ends of the panel. Figure 12 shows the layup of a typical diffusion bonding pack in the retort. Figure 13 shows an ultrasonic C-scan of the bonded joint. Several points are immediately obvious. First, the joints

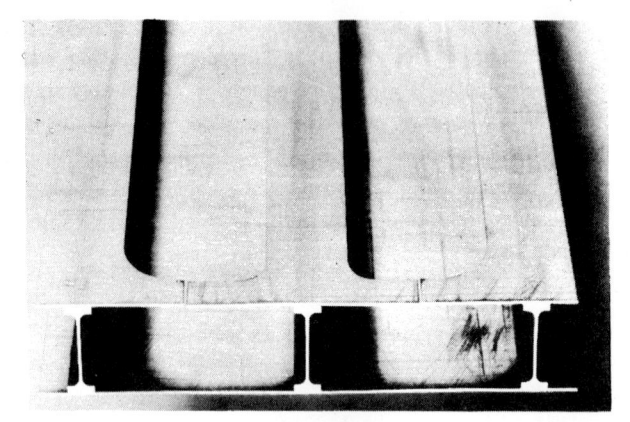

FIGURE 11
END VIEW OF A DIFFUSION BONDED COLUMBIUM PANEL, 4" x 18"

FIGURE 12
LAYUP OF A DIFFUSION BONDING PACK IN A RETORT

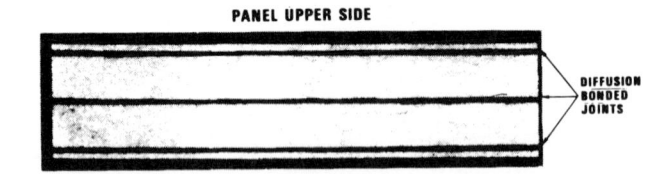

FIGURE 13
ULTRASONIC C-SCAN OF DIFFUSION BONDED COLUMBIUM PANEL

appear sound. Second, the sensitivity of the scan is limited, hense there is no method of determining if an edge void as deep as perhaps .015 to .020 exists. Third, this inspection technique can pick up panel flatness variations as shown by the signals between the bonded joints on the lower picutre in Figure 13 (Panel Lower Side). These indications are

306

puckers up to approximately .040 inch in depth caused by uneven heating control and tooling problems. Heating was more closely controlled in future panels. A completely new tooling design concept was also pursued, the details of which remain proprietary. A 16 x 18 inch panel was diffusion bonded using the new tooling concept. The panel was free of wrinkles and distortion. Figures 14 and 15 show ultrasonic scans of this panel.

Figures 16 through 18 show our final panel configuration after diffusion bonding. The part is 18 inches by 36 inches. Flatness has been held within .015 inch throughout the part. Total panel thickness tolerance has been held to .005 inch. An ultrasonic scan of this panel shows the panel to be free of puckered areas and to contain sound bonds.

FIGURE 14
ULTRASONIC C-SCAN RECORDING OF THE RIB-TO-FACE BONDS OF THE SOLID FACING PANEL SIDE

FIGURE 15
ULTRASONIC C-SCAN RECORDING OF THE RIB-TO-FACE BONDS OF THE SLOTTED FACING PANEL SIDE

FIGURE 16
CROSS-SECTION OF AN 18" x 36" DIFFUSION BONDED COLUMBIUM PANEL

FIGURE 17
EXTERNAL SURFACE OF DIFFUSION BONDED PANEL

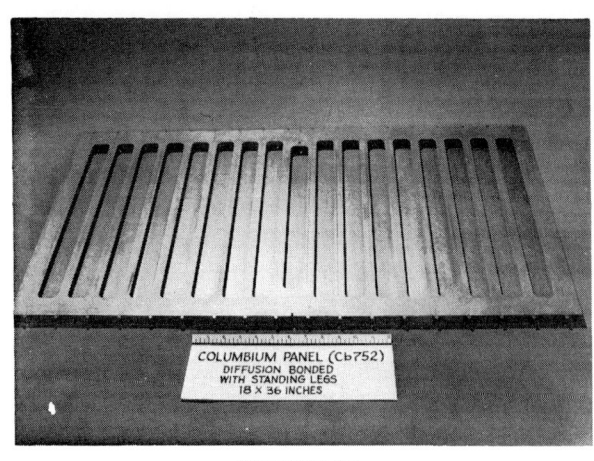

FIGURE 18
INTERNAL SURFACE OF DIFFUSION BONDED PANEL

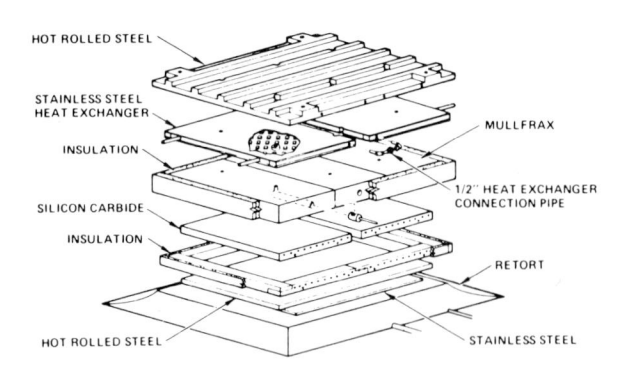

FIGURE 19
HIGH TEMPERATURE PLATEN ASSEMBLY

Figure 19 illustrates a typical high temperature press assembly used at NR for diffusion bonding.

5. SUMMARY

Diffusion bonding of columbium alloys, using a titanium foil interleaf can be accomplished at competitive costs and on a production basis. The bond is stress free and can result in flat parts. Parts which have been bonded develop the full strength of the sheet metal gages anticipated for the shuttle. The bond does not appear to degrade under creep stress levels expected on shuttle heat shields up 2500 F. Coating of the diffusion bond appears to present no problems. The diffusion bonding process offers the ability to join sections of widely different thicknesses and to seal faying surfaces. Before the diffusion bonding process could be used reliably in spacecraft, however, additional engineering effort will be required to characterize the bond more fully and to develop methods for repair.

Acknowledgements. The authors are indebted to J. M. Dieter, C. B. Blumer, R. Heisman, D. D. Helman, W. S. Harmon and M. J. Mitchell - whose contributions to this study have been considerable.

References

(1) NASA Space Vehicle Design Criteria, Panel Flutter Volume III, Structures NASA SP-8004.

(2) Lemley, Clark E., Design Criteria for the Prediction and Prevention of Panel Flutter, Vol. I McDonnell Douglas Corporation, AFFDL-TR-67-140, August 1968.

(3) J. P. King et al, Final Report on Diffusion Bonded Honeycomb Sandwich Panels, Technical Report No. AFML-TR-64-329, 1964.

(4) Technical discussions between J. Dieter, NR and Dr. A G. Metcalfe, Solar on 8/19/71.

Mr. Korb is Supervisor of Advanced Materials and Processes at the Space Division of North American Rockwell. Mr. Rowe and Mr. Beuyukian are members of the Technical Staff at NR.

ADVANCED DIFFUSION-WELDING PROCESSES

W. A. Owczarski and D. S. Duvall
Pratt & Whitney Aircraft Div.
United Technologies Corp.
East Hartford, Conn.

Diffusion welding is an art/science that has been practiced since early Egyptian times. Today its importance remains as its applications grow in electronics, aerospace, and nuclear fabrication. The development of advanced high-performance materials in these industries and the need to use them in complex fabricated structures has stimulated the use of diffusion welding to complement other forms of joining. The demands for low manufacturing cost, high reliability, and precision have produced an increased process-development activity in diffusion welding, with recent emphasis on utilizing liquid or semiliquid interface states to aid in achieving joint quality. An understanding of the basic diffusion-welding-process mechanisms is also evolving to aid in the expansion of its application.

Diffusion welding: what is it? This simple question no longer has a simple answer. Diffusion welding is one of the solid state joining processes, which include friction welding, explosive welding, cold welding and others, as shown in Fig. 1. All basically form joints between metals without melting, through the application of either heat or pressure (sometimes both). The means of applying this energy takes many forms: thermal, high-speed mechanical motion, intense isostatic pressure, and even acoustic energy. Occasionally, fusion of the participating members can occur during the nominally solid state joining processes to aid in the coalescence process. This fact is becoming better recognized as a way of sometimes enhancing the practical applications of these processes.

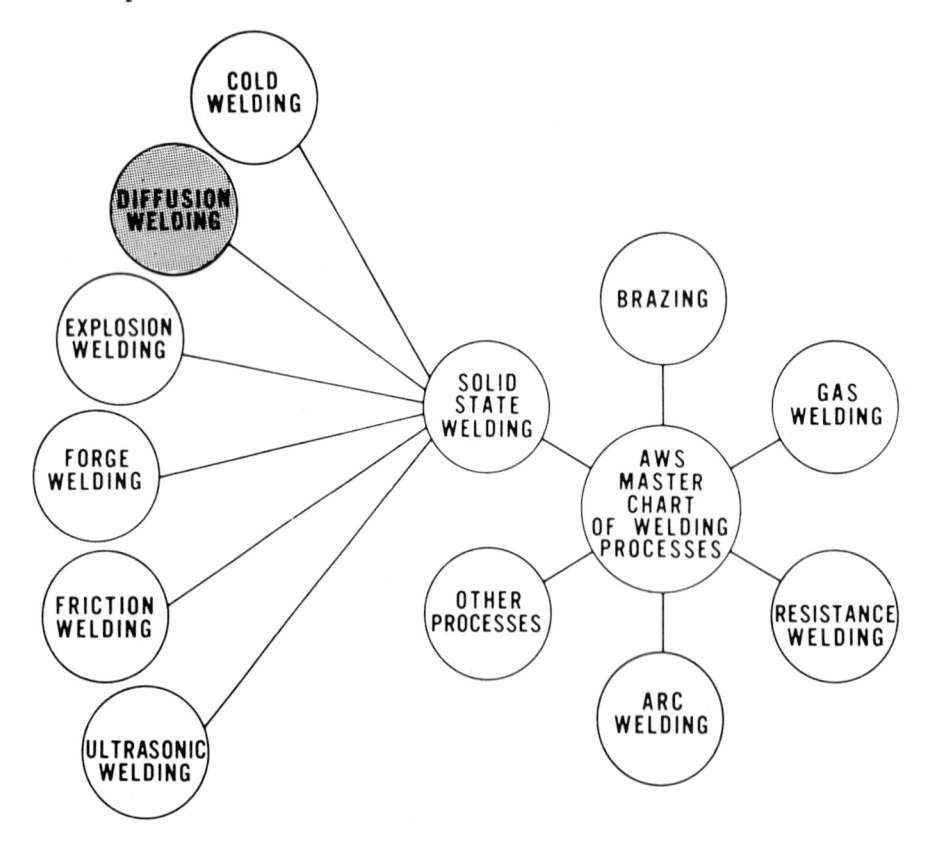

Fig. 1. The American Welding Society chart of welding processes (1), showing the relationship of diffusion welding to other solid state welding processes

Diffusion welding can be generally distinguished from other forms of solid state joining processes in that it produces coalescence of closely contacting joint surfaces by applying pressure at temperatures sufficiently elevated to ensure rapid atom mobility (usually 0.5 to 0.9 Tm). The process does not require relative motion between the parts, and does not involve macroscopic deformation (although some localized, controlled plastic strain is involved), nor does it involve melting of the components (although localized interface fusion may be used).

The diffusion-welding process, in its earliest forms, can be traced back to ancient Egyptian metalsmiths who used heat, pressure and surface coatings in elaborate and sophisticated combinations to form and join precious-metal ornamental pieces (2). Examples of jewelry cases dated to circa 1500 B.C. that display this early application of diffusion welding are shown in Fig. 2. Precious metals, because of their malleability and relative freedom from adherent surface films, were easy to form and join for the early artisans. Where increased complexity and ornate detail were desired, beads of dissimilar alloys were frequently added, often "plated" with diffusion aids to permit bonds to occur without extensive hammering or distortion.

Today, diffusion welding is used not to join simple precious-metal ornaments, but rather to produce useful structures from complex alloy materials where cost, performance requirements, and design require unique fabrication characteristics. The materials so joined are often difficult or impossible

to fabricate by other methods. Material with special
properties achieved by precipitation, texture or
other metallurgical characteristics that may be
destroyed or altered through conventional fabrication
methods are often involved. Applications today,
which will be shown later, are being made in the
aerospace, nuclear and electronics industries.

This chapter will first consider solid state
diffusion welding, and then discuss recent derivatives
of the process that involve liquid phases. Each
topic will be introduced by discussing the fundamental
characteristics of the method and then will be
followed by a description of how it is applied in

Fig. 2. Jewelry cases, dated to circa 1500 B.C.,
that were fabricated by an early form of
solid state welding

practical use. This chapter is not intended to be
a comprehensive and detailed reference document
about diffusion welding. It is rather hoped to be
sufficiently descriptive to provide a broad understanding of its characteristics and use.

Solid State Diffusion Welding

Clean, smooth, contacting metallic surfaces
will tend to unite or coalesce spontaneously to
reduce surface energy at the contacting or faying
surfaces. In fact, copper surfaces have been made
to coalesce even at room temperatures under low
pressure when surface films are removed (3). However,
most engineering materials are not clean and smooth
enough to coalesce spontaneously. This is clearly
beneficial; otherwise all engineering mechanisms
would be cursed with enormous seizure and galling
problems. Real surfaces are irregular, frequently
oxidized and dirty, as diagrammatically illustrated
in Fig. 3. The contaminants and surface roughness
cause practical barriers to the easy formation of

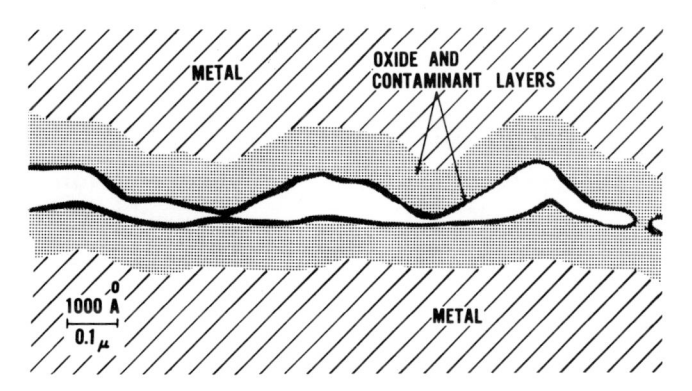

Fig. 3. Diagrammatic view of contacting metal
surfaces, showing local asperities and
layers of oxide (light gray) and contaminant
(dark gray), which affect diffusion welding

metallic bonds. To minimize these inherent difficulties, surfaces to be diffusion welded are prepared to be as flat and as clean as possible. Then the application of pressure and thermal energy, usually done in a protective environment to prevent further oxidation and/or contamination, allows the deformation of surface irregularities and the activation of diffusional processes to accelerate bond formation.

The sequence in which these events occur has been previously described (4) and is shown in Fig. 4. As the pieces are initially contacted, micro-asperities come into contact and partially deform, thus increasing metal-to-metal contact area. Holding at temperature and under pressure causes continued growth of contact asperities through creep, while the influence of temperature alone causes surface boundaries to begin to migrate into more energetically stable arrays. As the effects of temperature and pressure continue, the contact area approaches 100%

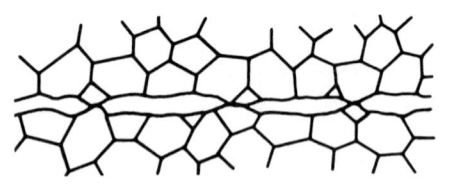

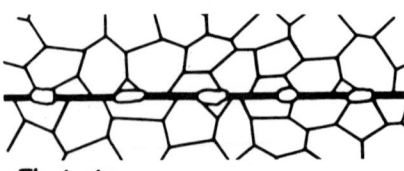

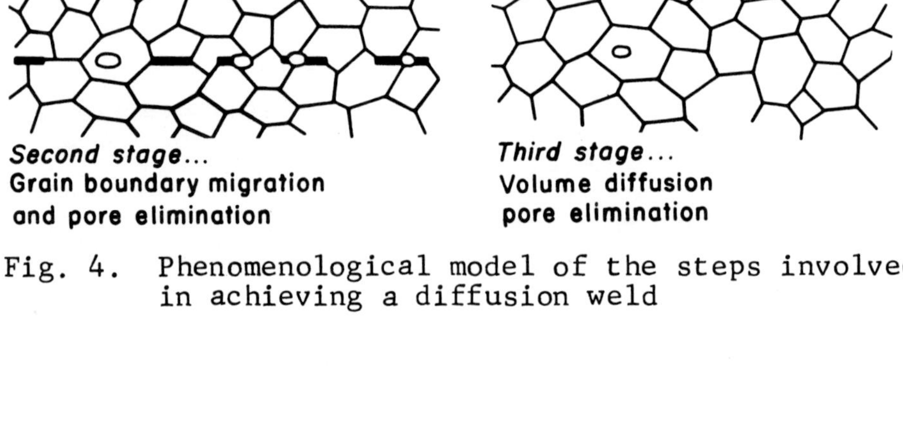

Initial asperity contact

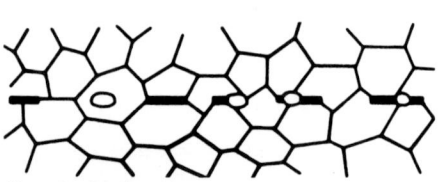

First stage...
Deformation forming interfacial boundary

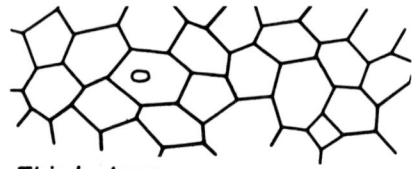

Second stage...
Grain boundary migration and pore elimination

Third stage...
Volume diffusion pore elimination

Fig. 4. Phenomenological model of the steps involved in achieving a diffusion weld

and only small voids are left behind at the original faying-surface location. Further, as interfacial or grain boundaries leave the original faying surface, the joint becomes relatively indistinguishable except for residual voids, which finally disappear through continued diffusion processes.

Metallurgical verification of this hypothesized model is shown in Fig. 5. Micrographs and fractures of diffusion welds made in titanium and iron are presented that show the welds at varied stages of completion. The titanium welds illustrate the changes in appearance that occur in metallographic cross

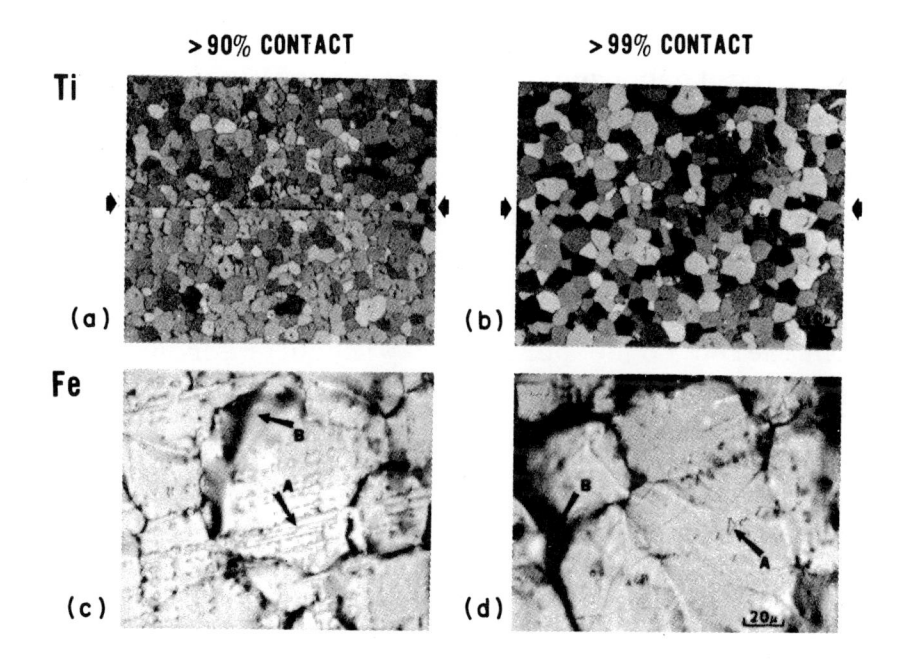

Fig. 5. Diffusion welds with varying amounts of surface contact and bonding: (a), (b) cross sections of titanium welds (arrows mark weld interface); (c), (d) iron samples fractured through weld (arrows "A" indicate voids; arrows "B" indicate migrated interfacial boundaries)

sections as the welding process nears completion. The iron welds were fractured to provide a third-dimensional view of voids and boundary mobility in welds at a similar stage of completion. (Iron was utilized because brittle/nondistorted interface fractures were obtainable at liquid-nitrogen temperatures.) Seen in Fig. 5 are extensive planar boundaries and about 10% interfacial voids, while the welds are in early phases of formation.* With more extensive time at the joining temperature, weld formation is virtually complete, although a few residual voids (arrows "A" in Fig. 5) denote the prior interface location. At locations marked "B" in Fig. 5, "out of focus" boundaries can be observed that have moved into nonplanar, energetically favored locations consistent with the model of Fig. 4.

A point should be mentioned here regarding the relative roles of diffusion and deformation in diffusion welding. Diffusion welding is usually considered to be performed without macroscopic deformation. However, it is clear that deformation has a significant role at least on a local basis. In fact, the overall processes of diffusion and deformation undoubtedly interact, with deformation enhancing the local diffusion processes as it causes grain boundary adjustment. Much more needs to be

---

*A cross-hatched pattern of voids is clearly evident on the interface of the fractured iron weld (Fig. 5). This results from directional polishing scratches formed on the faying surfaces during preweld surface preparation. The specimen halves were then aligned with their polishing scratches approximately perpendicular for subsequent welding.

learned about this interrelation quantitatively to provide a firmer basis for future process development and control.

The degree of weld completion clearly affects the subsequent properties of a diffusion weld. The effects can be complex and inconsistent, requiring a careful evaluation of a broad spectrum of weld properties before ultimate engineering utilization of a welded structure can be made. An example of the effect of diffusion-weld quality on properties can be seen in titanium welds (5), as shown in Fig. 6. Here, partially completed welds with relatively low total contact area (~80%) failed, in a tensile test, at the weld interface, as expected. However, the fracture strength was surprisingly high (~98 ksi) even though the ductility was low (~3% reduction of area). This

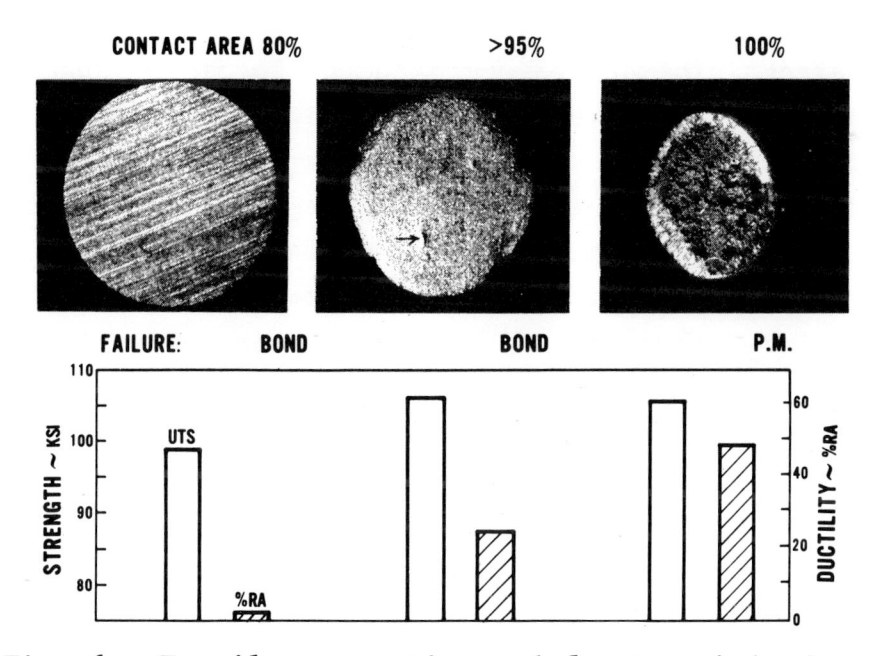

Fig. 6. Tensile properties and fracture behavior of titanium welds of varying quality. Arrow in middle photograph denotes defect on fractured weld interface.

strength is very nearly that of the parent-metal strength (106 ksi) that is reached in a properly made diffusion weld. In the intermediate case, where a weld with ~95% contact area was tested, failure was through the weld but at an ultimate tensile strength (106 ksi) equivalent to that of the parent metal. The ductility was approximately 20%. In the fully completed weld, with ~100% contact area, fracture took place away from the weld at normal parent-metal strength and ductility (48%).

The major conclusion here is that fracture strength in this case is not an effective indicator of weld quality, whereas ductility is. By noting the fracture surfaces, the degree of weld quality is very evident, with multiple defects present in the intermediate case and evidence of original grinding markings even visible on the poor weld. The nature, frequency and distribution of interfacial defects ultimately govern the performance capability of a diffusion weld. Thorough consideration of critical service requirements is mandatory prior to use of diffusion welds for engineering structures.

Most of the examples and discussion thus far have focused on materials having simple behavior. In more complex metals, other factors enter into the diffusion-welding picture in addition to voids, grain-boundary mobility, etc. One example is precipitation-hardened nickel-base superalloys (6). These alloys are designed for stability at elevated temperature, which includes creep resistance and oxidation-corrosion resistance. Because of these intrinsic characteristics, special problems arise in diffusion welding. In Fig. 7(a), a superalloy diffusion weld is shown that is completely welded, i.e., has 100% contact area. However, the continuous, horizontal, interfacial grain boundary possesses undesirable property characteristics, such as low ductility, that would impair the utility of this type of weld in elevated-temperature service.

The boundary itself remains planar because it is pinned by the early formation, during welding, of refractory intermetallic compounds such as Ti(C, N) (6). They prevent the boundary from migrating to a more stable condition and, further, provide brittle-fracture initiation sites. Such compounds form spontaneously at welding temperatures

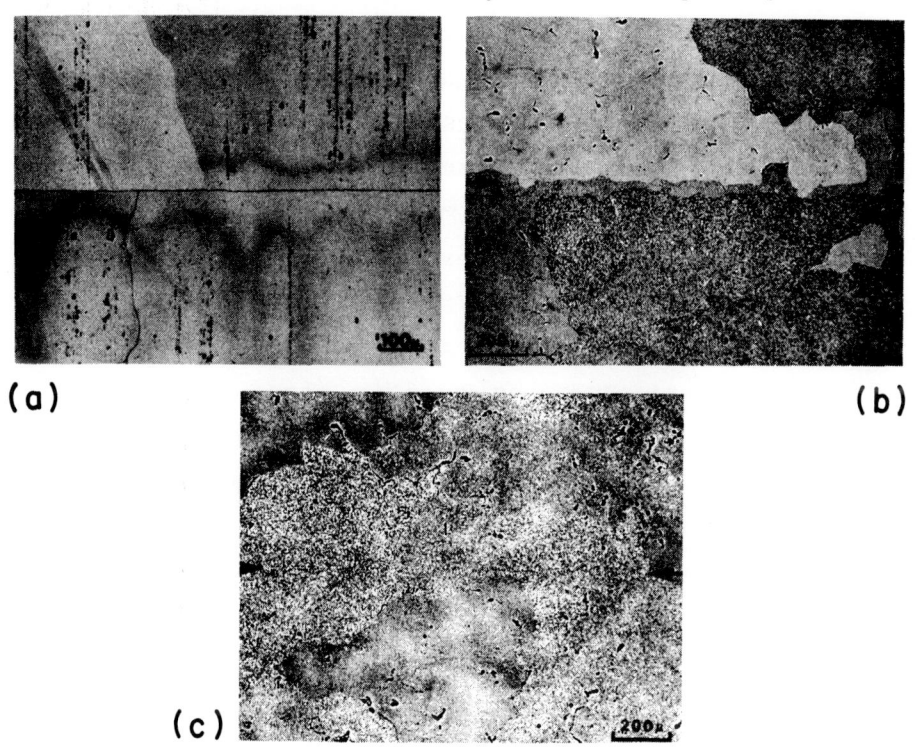

Fig. 7. Diffusion welds in the nickel superalloy Udimet 700, welded with (a) no intermediary layer, (b) 0.0001-in. Ni-Co electroplate and <1% deformation, and (c) 0.0001-in. Ni-Co electroplate and ~7% deformation

and appear to be caused by the segregation of carbon and/or nitrogen at the initial faying surface, where it combines with the titanium present in the alloy to form the stable precipitates shown. When such compounds tend to form, means must be found to effectively prevent their formation.

Figure 7(b) illustrates the beneficial effect of introducing a thin (e.g., 0.0001-in.) plating of Ni-30 Co on the faying surfaces of this alloy prior to diffusion welding (7). This prevents the formation of interface particles during the subsequent diffusion weld. However, the overall weld quality is not yet completely satisfactory because of the relatively immobile interface boundary that remains. In this case, in order to achieve adequate boundary migration it was also necessary to increase local interfacial deformation during the welding process. The weld in Fig. 7(c) was made with similar weld parameters, including the surface plate; but pressure was added to produce 7% increase in interface cross-sectional area during welding, as opposed to 1% for the weld in Fig. 7(b). The result was a more favorable weld microstructure with improved weld stress-rupture properties. In both cases, chemical homogeneity of the weld interface region has been effectively achieved by diffusion during the joining process so that the plated Ni-Co diluent has no residual harmful effect on properties.

When dissimilar metal joints are made, other problems must be considered. If two dissimilar metals are in contact, undesirable diffusional effects may occur. Illustrated in Fig. 8 are two possible consequences of such events. Diffusion welds made between metals with widely differing diffusion coefficients are subject to Kirkendall porosity. In the weld between nickel and copper shown in Fig. 8(a), in which there were two original interfaces, the porosity formed in the copper, as can be predicted from the greater diffusivity of copper in nickel than of nickel in copper. It is possible to alter these effects by increased pressure or decreased temperature and time (8, 9) as long as the conditions chosen will still achieve a weld of satisfactory quality.

The second concern with dissimilar metals is the formation of any stable intermediate phases that form between the metals involved. Intermediate

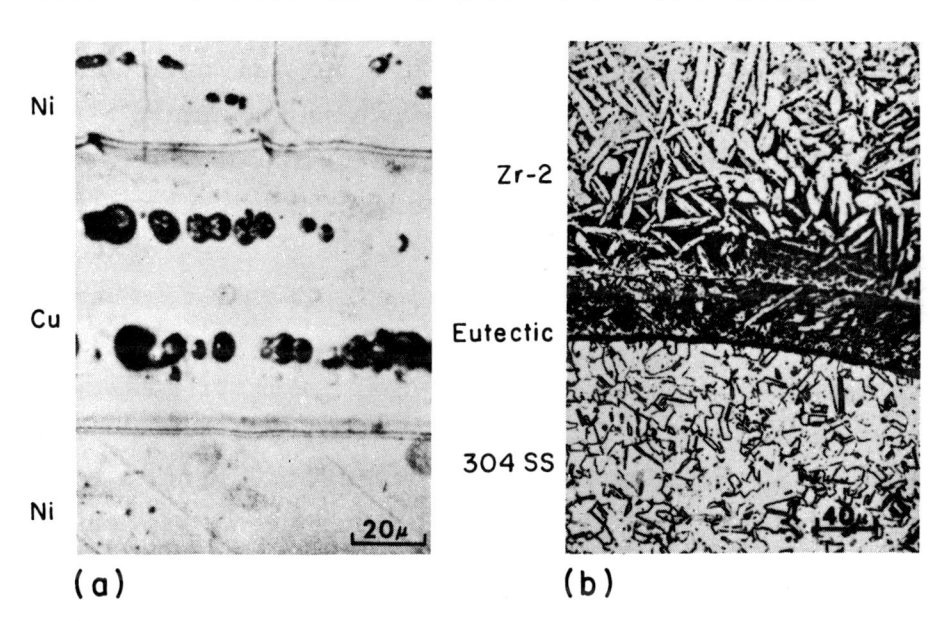

Fig. 8. Diffusion welds between dissimilar metals: (a) weld of copper to nickel showing Kirkendall porosity (8); and (b) eutectic compound formed in weld of Zircaloy to stainless steel (10)

Source: *New Trends in Materials Processing*, ASM, 1976

phase formation can take two forms. The first, as shown in Fig. 8(b), is the occurrence of low-melting liquid-phase fields. If the metals in contact are subject to this phenomenon, selection of a diffusion-welding temperature above the minimum liquidus for the system will result in formation of interface liquid. While this is not necessarily harmful, it is potentially so for several reasons. First, liquid reactions in a bimetal system will continue as long as both metals are available as sources of alloying. This can lead to excess liquid and uncontrolled consumption of the pieces being welded. Secondly, liquid phases when cooled frequently form complex phase structures characterized by brittleness. It should be pointed out that this liquid reaction can be beneficial, as in the example shown, in forming joints between certain metals if done properly. This subject will be discussed again later.

The second type of phase reaction occurs in bimetallic combinations that exhibit intermetallic compounds. Here layers can form corresponding to all phase fields in the appropriate phase diagram. Phases or layers in simple binary systems are reasonably predictable but those of higher order systems are not. Considerable study may be needed when working in a particular multimetal system to know what the consequences of this phenomenon are. It should be pointed out that various parameters of the welding process, such as pressure, temperature, and the use of intermediary layers, can substantially alter the characteristics of phases formed (11).

Applications

Diffusion welding is normally used to fabricate parts when high-strength welds are required and when one or both of the following conditions exist: (a) the part shapes are intricate and would be costly or impossible to manufacture by conventional means, or (b) the materials used possess unique properties that interfere with, or are difficult to maintain during, conventional fabrication processing. To illustrate both features and describe the practical aspects of diffusion welding, two examples will be further discussed.

The first example, shown in Fig. 9, is that of a large aircraft gas-turbine fan disk. In this

Fig. 9. Hollow-rim gas-turbine fan disk prior to diffusion welding. Two 180° spacer-ring segments are welded into the outer rim of the disk to complete the disk structure.

application, a weight and performance improvement is achieved by designing a titanium-alloy fan disk with a hollow-box geometry at the rim. A weight savings of 100 lb is obtained compared to a disk of conventional solid-rim design. Creating the hollow rim of the disk requires initial removal of the inner material followed by subsequent replacement of the outer hoop by welding. Diffusion welding is employed for this joining operation.

In this case, the outer hoop is a split ring that matches the slot from which the inner chamber is machined. Requirements for such an assembly include exceptional dimensional control in that many surfaces are not accessible to machining after welding. Also required are exceptional reliability, reproducibility and high properties in all weld regions around the large faying-surface area ($\sim$200 sq in.). The diffusion-welding process is conducted in a large vacuum hot press, shown in Fig. 10, wherein the environment, temperature, pressure, and total dimensional control in the weld region are all controlled and monitored throughout the 4-hr diffusion-weld process cycle. Vacuum levels are maintained at pressures of 10^{-4} torr or lower. In Fig. 11, the finish machined disk is shown ready for engine installation.

A second example illustrates a part fabricated by gas-pressure bonding, a form of diffusion welding. The major difference in gas-pressure bonding is that the pressure applied to the part is transmitted using a heated inert gas. It is applied in such a manner that all surfaces of the part are isostatically enveloped in the pressure medium, thus requiring no external tooling or shaping devices. A part prior to gas-pressure bonding can be seen in Fig. 12. Here several nuclear-material segments are assembled

Fig. 10. Hollow-rim turbine-engine fan disk (arrow) assembled in vacuum hot press for diffusion welding

Fig. 11. Diffusion welded and machined fan disk ready for installation in an aircraft gas turbine

Fig. 12. Nuclear-fuel-element components prior to fabrication by gas-pressure bonding (12)

with appropriate cover and separator plates and then inserted in an evacuated metallic envelope. The entire assembly is placed in a furnace that is contained in a cooled-wall pressure vessel filled with argon or helium gas (Fig. 13). Once assembled and closed, the vessel is pressurized (typically to 5 to 15 ksi) and the furnace is heated. For this type of component, the furnace, inert gas and parts are usually heated to temperatures of 2000 to 2200 F, held for 2 to 6 hr, and then cooled and depressurized.

Figure 14 shows the same assembly as Fig. 12 after gas-pressure bonding. Note several features of this assembly. All the surfaces become bonded in all directions as a result of the isostatic pressure. It is possible to bond hollow shapes, but proper methods of assembly and/or leachable inserts are required to prevent collapse of the hollow cavities from the isostatic pressure.

Summary

In perspective, this process has had limited but important usage. It has unique capabilities to

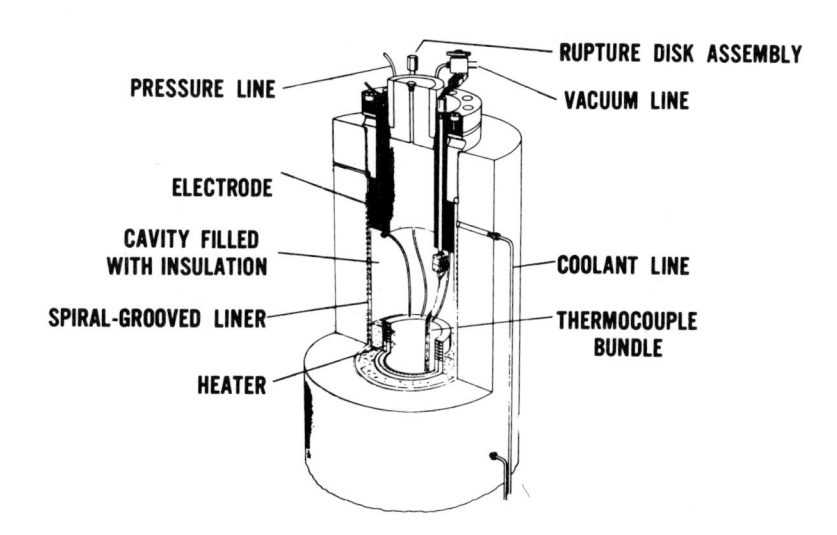

Fig. 13. Typical autoclave employed for the gas-pressure bonding process (12)

Fig. 14. Gas-pressure-bonded nuclear fuel element (12)

produce outstanding joint properties and characteristics, to join complex shapes and large surface areas and to produce stress-free, distortion-free welds. It can weld materials that may otherwise be unweldable, in dissimilar or similar combinations, and can preserve unique metallurgical structures such as dispersion-strengthened or aligned microstructures.

It is not, however, a panacea destined to replace all other processes. While simple in concept, it can be complex in practice, with tooling, preweld preparation and process-parameter control all being potentially difficult. All these problems can require precise controls and often expensive equipment. Solid state diffusion welding is, therefore, mainly used today for low-volume, high-performance parts and materials where quality requirements or part complexity provide sufficient incentive for the barriers that must be overcome.

Liquid-Phase Diffusion Welding: A New Tool

The search to simplify the tooling and inspection requirements typical of solid state diffusion welding has led to the recent development, introduction and application of several variant diffusion-welding techniques utilizing interlayers that partially and/or temporarily melt during the joining process. The major aim is to reduce or eliminate the need for significant application of pressure during welding. This is accomplished by having a thin liquid layer promote the necessary wetting contact so that subsequent diffusional processes can promote sound bonds that permanently unite the faying surfaces. The work in this area has promoted a surge of new terminology, and phrases such as liquid-phase diffusion welding, diffusion brazing, eutectic bonding, activated diffusion bonding and transient liquid-phase diffusion welding have all been used to describe related phenomena. Trade names such as Rohr

Bonding(TM), Nor-Ti-Bond and TLP(R) diffusion bonding have also become well known.

The results achieved have been very encouraging and have led to several applications, in particular in the aerospace industry. Figure 15 shows examples of liquid-phase diffusion welding in a nickel-base superalloy (by TLP bonding) and in a titanium alloy (by Rohr bonding). Characteristic of these welds is a joint region that does not contain evidence of liquid in its microstructure following the welding process. In appearance and performance, the results achieved often compare favorably with conventional solid state diffusion welds (compare Fig. 7c and 15a, for example).

Generally, two metallurgical approaches are employed with liquid-phase diffusion welding. These are shown graphically in Fig. 16. With most processes, a thin layer of a material that melts at a temperature

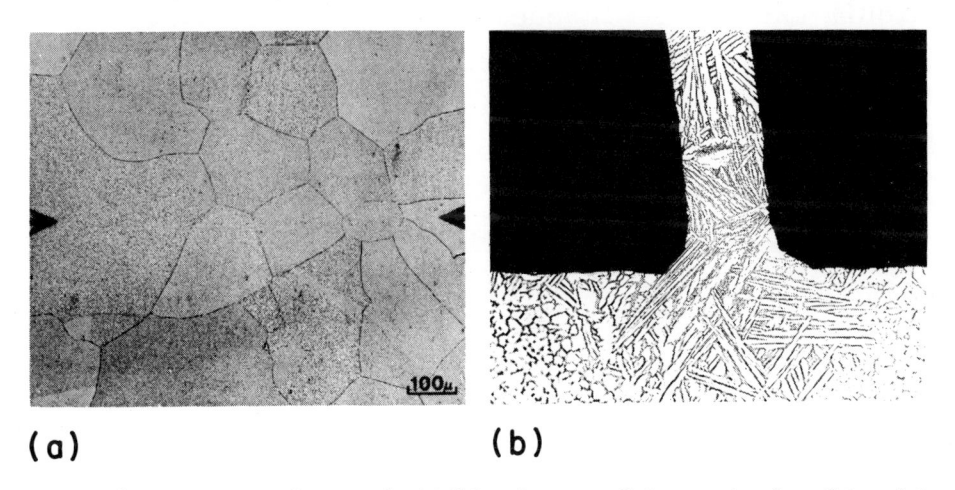

Fig. 15. Examples of diffusion welds made by liquid-phase processes: (a) TLP bond in the nickel superalloy Udimet 700, and (b) titanium joined by the Rohr-bond process (courtesy of Rohr Industries)

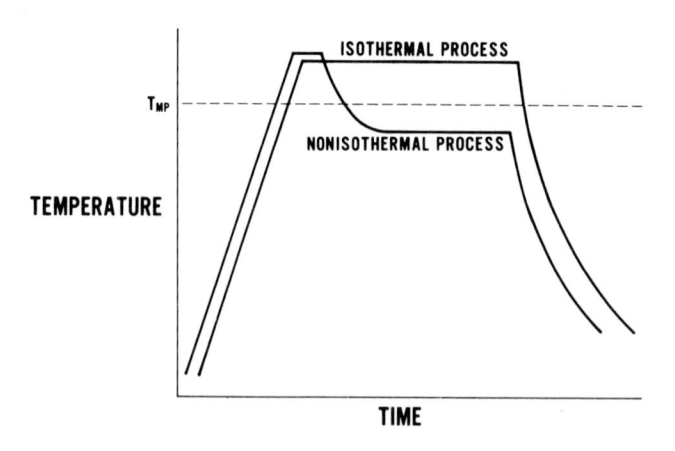

Fig. 16. Thermal cycles illustrating the two basic types of liquid-phase processes in relation to the melting point (T_{mp}) of the interfacial material

below that of the parent metals to be joined, but above that for rapid diffusion, is placed between the faying surfaces. When heated above the interlayer melting temperature (T_{mp}), the joint region is filled and wetted with a liquid or partially liquid material. In one form of liquid-phase diffusion welding, the parts are cooled to just below T_{mp} to permit the liquid zone to resolidify athermally. The parts are usually held at this temperature ($T < T_{mp}$), where homogenization and microstructural stabilization can occur. In the other form of liquid-phase diffusion welding, as discussed in Ref 13, the parts are held for the entire process at a temperature above T_{mp} where isothermal or diffusional solidification is promoted.

The features of isothermal solidification are displayed in Fig. 17. Across the bottom of this figure is a series of pseudo-binary-phase fields that trace the compositional changes in the liquid-phase diffusion-welding system under question. It shows that the initial interlayer composition, C_i, exists at a low melting position in the phase field for the system. When the assembly to be joined is heated to the welding or bonding temperature, T_B, the interlayer melts and coexists initially with solid parent metal. Early in the process, interdiffusion between the liquid layer and the parent metal dilutes the liquid composition to that of composition C_1, the equilibrium composition for liquid to coexist with solid solution C_s (Fig. 17b). With further time, interdiffusion continues, and eventually the liquid shrinks in size (Fig. 17c) and disappears (Fig. 17d) as the temperature-depressant

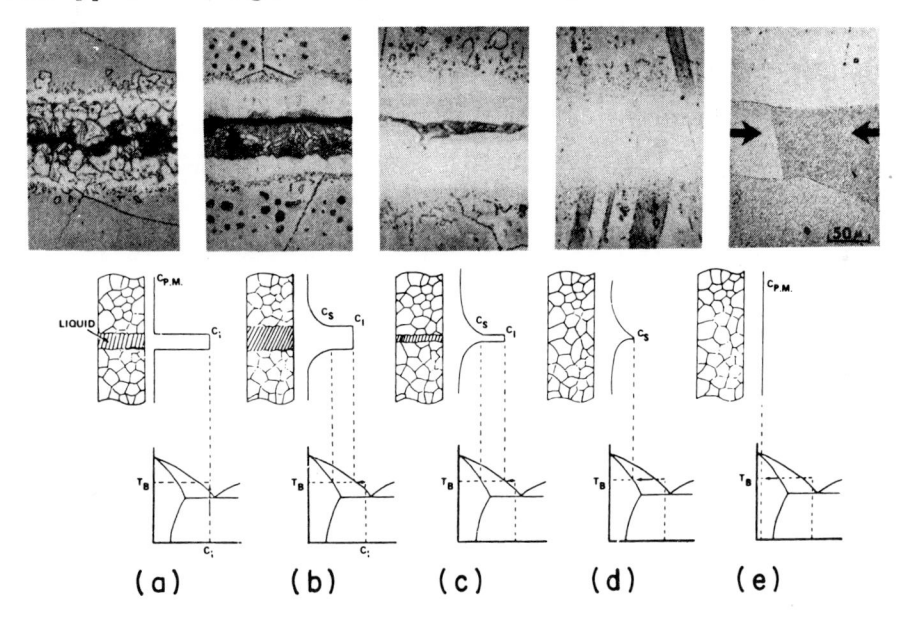

(a) (b) (c) (d) (e)

Fig. 17. (a through e) Sequential chemistry and phase changes at the interface region during isothermal solidification of a liquid-phase diffusion weld. "Stop action" micrographs of welds interrupted at the indicated stages of completion are shown at top.

elements in the liquid are diffused away and exhausted. Further diffusion completely eliminates the diffusion gradient so that a fully uniform composition is achieved at the interface (Fig. 17e).

The micrographs at the top of Fig. 17 depict a "stop-motion" view of these metallurgical events, showing the initial melted zone, gradual narrowing of the liquid region, complete isothermal solidification and, finally, complete homogenization. These micrographs are for the alloy Udimet 700 (Ni-15 Cr-15 Co-5 Mo-4.5 Ti-3.5 Al-0.1 C) being joined using a Ni-Cr-Mo-B interlayer. A unique feature of isothermal solidification is that it prevents the athermal formation of intermediate phases that are typical of phase systems exhibiting substantial melting-point depressants and reactive elements. In a practical sense, such compounds are difficult to subsequently disperse or diffusionally homogenize and, consequently, contribute to reduced diffusion-weld properties.

The liquid-phase diffusion-welding approach has been explored or is being utilized for joining nickel, cobalt, titanium, iron and zirconium alloy systems. Work to date has shown that selection of proper interlayer composition is crucial. Appropriate phase relations must be found that have the right combination of melting-point depression, solid solubilities, stability and diffusional mobility for this concept to be successfully applied to engineering materials and structures.

Application

One illustrative example of liquid-phase diffusion welding is shown in Fig. 18. In this case, individual aircraft gas-turbine vanes (see bottom of Fig. 18a) are joined by the process into clusters of three (see top of Fig. 18a). The vanes are assembled

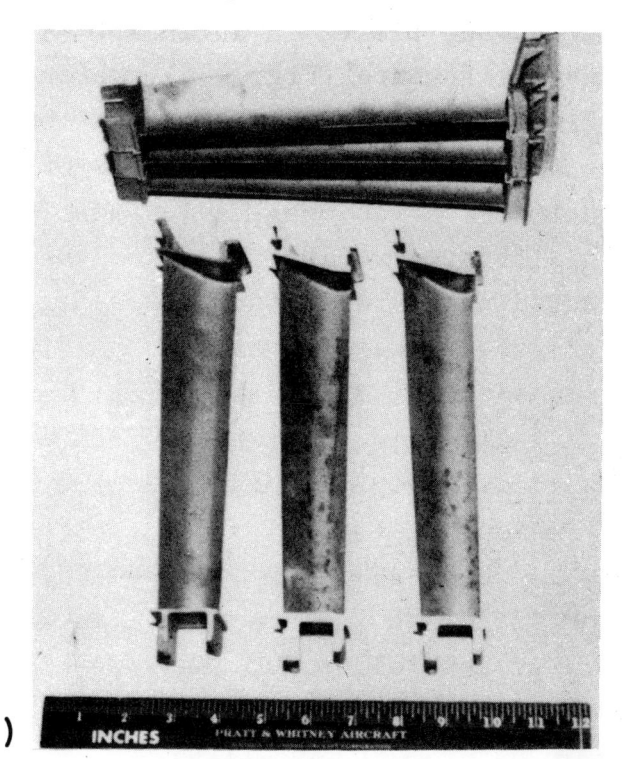

Fig. 18. Gas-turbine vanes joined by liquid-phase diffusion welding: (a) individual vanes and vane cluster welded together at shroud ends, and (b) cluster of vanes assembled in a ceramic fixture for vacuum-furnace diffusion welding

by placing preforms of the interlayer foil material at the two faying interfaces and then aligning and stacking the vanes into a cluster. The parts are readily aligned and placed in an inexpensive, castable ceramic fixture (Fig. 18b), which helps preserve alignment and applies a modest compressive load to the faying interfaces. Batches of parts are then placed in a conventional vacuum heat treating furnace for diffusion welding. Figure 19 shows several hundred parts ready for processing in a typical furnace. Simplicity and the ability to do large batch quantities are tributes to the successful achievement of the major objective of liquid-phase diffusion welding, i.e., simplification of diffusion welding to an inexpensive, repetitive production operation. In the particular application shown, more than 80,000 assemblies have been made to date, with a 99.7%-successful production yield. Thus, in this form the diffusion-welding process has achieved a high-volume production capability.

Summary

Liquid-phase diffusion welding has the advantage of obtaining the quality of a solid state weld while achieving considerable simplicity of processing and tooling mechanics. It lends itself to volume or batch processing, and is performed in simplified equipment. The presence of a temporary liquid allows relaxation of the critical preweld faying-surface preparation often needed with solid state processes. It also enhances the detectability of defects and, thereby, the inspection of parts. It has economic advantages over conventional solid state diffusion welding and even rivals brazing in cost.

It is not, however, necessarily inexpensive. It still requires substantial process monitoring and

Fig. 19. Typical production vacuum furnace utilized for liquid-phase diffusion welding large batches of turbine vane clusters

control. Its application to date has generally been limited to relatively exotic or high-performance materials and structures. Applying it to other materials systems and to situations where costs are most critical requires considerable additional development.

Diffusion Welding: Perspective

This specialized welding process has found considerable acceptance in manufacturing aerospace, nuclear and electronics components. Its roots are in ancient metal-forming practice but its contemporary application is sophisticated and highly technological. It can now perform specialized quality-fabrication tasks that are impossible by any other means. It is a process area that will find broader use with time but requires more development to reach this expanded status. It will not likely ever replace common welding or fastening methods in normal fabrication areas.

References

1 "The AWS Master Chart of Welding Processes", American Welding Society, Miami, Florida

2 E. C. Rollason, Brit Weld J, 6, No. 1, 1 (1959)

3 M. G. Nicholas, Trans AIME, 227, 250 (1963)

4 W. H. King and W. A. Owczarski, Weld J, 47, 444s (1968)

5 W. A. Owczarski et al, Weld J, 48, 377s-383s (1969)

6 D. S. Duvall et al, Weld J, 51, 41s-49s (1972)

7 W. A. Owczarski et al, U. S. Patent No. 3,530,568, Sept 1970

8 R. S. Barnes and D. J. Mazey, Acta Met, 6, 1 (1958)

9 K. J. B. McEwan and D. R. Milner, Brit Weld J, 9, 406 (1962)

10 W. A. Owczarski, Weld J, 41, 78s-83s (1962)

11 J. M. Gerken and W. A. Owczarski, "A Review of Diffusion Welding", Bulletin No. 109, Welding Research Council, New York, 1965

12 E. S. Hodge, Metals Eng Quart, 1, No. 4, 3 (1961)

13 D. S. Duvall et al, Weld J, 53, 203 (1974)

ROLL DIFFUSION BONDING OF BORON ALUMINUM COMPOSITES

G. S. Doble and I. J. Toth

ABSTRACT

A low cost process has been developed for the primary manufacture of boron-6061 aluminum monotapes and multi-ply panels by roll diffusion bonding in air. Composites produced by preheating in either air or argon and subsequently roll diffusion bonded in air had properties equivalent to higher cost vacuum press-bonded material. Average tensile strength of roll bonded monotapes was 207 ksi and panels made from these monotapes had longitudinal strengths of 201 ksi and transverse strength of 19 ksi. Transverse strength was increased to 46-70 ksi by introducing 18-20 volume percent of interleaved titanium foil during primary or secondary roll bonding. Cost projections predict processing costs of $40, $19 and $13/lb for roll diffusion bonded boron-aluminum monotape at annual production rates of 2,000, 20,000 and 200,000 pounds.

G. S. Doble is a principal engineer with Materials Technology of TRW Inc., Cleveland, Ohio 44117.

I. J. Toth is Manager, Materials Development with Materials Technology of TRW Inc., Cleveland, Ohio 44117.

INTRODUCTION

Filamentary reinforced metal matrix composites are of interest for propulsion and aircraft structural applications because of high specific stiffness, strength, and design versatility. High primary fabrication costs are currently limiting the more widespread employment of these materials. The present paper summarizes a laboratory development program on roll diffusion bonding of boron-aluminum with the objective of lowering primary fabrication costs.

The method presently used to produce filamentary reinforced metal matrix composites is press diffusion bonding. Typically, press dwell times are extensive and a vacuum environment is employed. Consolidation and bonding take place by the mechanism of time-dependent plastic flow, that is, by creep. In roll diffusion bonding, consolidation is flow-stress controlled because of the high unit stress generated in the roll contact arc. The consolidation process is therefore very rapid. The primary fabrication costs associated with roll bonding are low because of reduced cycle time, low capital cost structure, and elimination of the vacuum atmosphere.

The purpose of the study was to demonstrate the potential low cost and high properties associated with roll bonding. The program selected 50 volume percent reinforced 5.6 mil boron-6061 aluminum as a demonstration system. The first section of the program evaluated the influence of roll bonding parameters and procedures upon the properties of direct rolled monotapes and multi-ply panels, referred to as primary fabrication. Later portions of the program determined the mechanical properties of panels made from these monotapes by secondary press bonding in air. All work was performed using conventional heating and rolling mill equipment. A final program task was the projection of the rolling process costs in comparison with press bonding.

EXPERIMENTAL PROCEDURE

The experimental procedure employed for roll diffusion bonding contained some features in common with the press diffusion process. The boron filaments were collimated into mats by drum winding with a fugitive binder of polystyrene. After winding, this mat is quite handleable and the fiber spacing is very uniform. The filament mat was placed between pieces of 6061 aluminum foil which had been lightly etched prior to assembly. Either monotapes, consisting of foil filament sandwiches, or panels which consisted of alternate layers of foil and filament mat, were assembled. When monotapes were to be rolled, a number of monotapes could be rolled in a single package by inserting reusable stainless separators in between layers. The assembled composite package was placed in between 1/4-inch thick stainless steel cover plates which were joined at the corners to provide a clamping load for maintenance of filament alignment. The initial monotape size was 4 inches wide by 16 inches long which was later increased to 6 inch wide by 2 feet long. The maximum size produced to date has been 6 inch by 4 feet.

Rolling variables included temperature, roll pressure, rolling speed, and heating atmosphere. The rolling temperature and roll pressure were found to be the most important variables. The assembled package was heated for one hour in either an air or a flowing argon atmosphere using a 6-inch diameter tube inside a standard furnace. In the case of an air atmosphere, the package was wrapped with a protective foil. The heating furnace was adjacent to the mill so that transfer time was minimal and because the cover plates were sufficiently thick, no chilling of the composite occurred either during transfer or during rolling. All rolling was performed in air using a two-high rolling mill with 16-inch diameter by 30-inch wide rolls. Rolling mill speeds of 10-120 inch per minute were used.

The roll consolidation pressure on the composite was adjusted by the size of the roll opening. The roll opening was set to various sizes less than the thickness of the fully dense composite and cover plates. The resulting elastic deflection of the rolls provided the force for consolidation, the smaller the roll opening the greater the elastic roll deflection and the greater the rolling force. The rolling pressure was kept low enough to prevent significant elongation of the stainless steel cover plates. Rolling pressure was not measured directly but is reported in terms of the elastic roll deflection. The initial roll bonding experiments were performed using monotapes with step-wise shims to prove four reductions within a single run.

After rolling the package was air cooled and the composite evaluated by metallography, radiography and tensile testing. Metallography was used to determine the degree of consolidation while radiography indicated the absence of filament breakage. Tensile testing of monotapes or panels used 1/4-inch wide x 4-inch long specimens with tabs adhesively bonded at the grip ends.

RESULTS AND DISCUSSION

Monotape Rolling

Monotapes were roll diffusion bonded using temperatures of 950°F-1150°F. The optimum rolling temperature is a compromise between a temperature low enough to prevent filament-matrix reaction while being high enough to allow ease of plastic flow in the matrix. The rolling pressure was also varied but the maximum pressure was limited by the pressure at which the stainless steel cover plates deformed. Representative results showing the relationship between temperature and rolling pressure for full consolidation are listed in Table 1 and corresponding microstructures shown in Figure 1. Results are as one would expect, increasing either temperature or pressure produced complete consolidation. Too much pressure leads to filament breakage because of the elongation of the composite. A temperature of at least 1050°F was required for consistent consolidation while 1150°F produced consolidation at very low pressures. A typical microstructure of a monotape illustrating full consolidation and a clean bond line is shown in Figure 2. The monotapes displayed excellent surface finish and radiography indicated that good filament alignment had been maintained.

TABLE 1

CONSOLIDATION OF BORON-ALUMINUM MONOTAPES

ELASTIC ROLL DEFLECTION (THOUSANDTHS)	5	10	15	20	25	30	35	40
ROLLING TEMPERATURE °F								
950		-	-	-	-			
1050			-	-	-	0	0	x
1100			-	-	0	0	0	
1150	-	0	0	0				

- not consolidated
0 consolidated
x filament breakage

Tensile tests were performed on all consolidated and some partially-consolidated monotapes and the results are illustrated in Figure 3. The precipitous drop at the 1050°F temperature was due to filament breakage. Two observations brought out by this figure deserve emphasis; the optimum 1050°F rolling temperature and the strength loss above 1100°F. The optimum 1050°F temperature for rolling is much higher than temperatures used for press diffusion bonding. For example, the temperature producing the highest strength during a one-hour press cycle time, shown in Figure 4, is 875°F [1]. The strength at higher temperatures is reduced because of matrix-filament interaction. Higher temperature may be tolerated in roll diffusion bonding, in spite of a one-hour heating time, because the consolidation time is very short. The low strength of fully-consolidated monotapes rolled at 1100°F and 1150°F is most likely due to the presence of a liquid phase and hence high reaction rate during rolling. The solidus and liquidus temperature of 6061 aluminum alloy are 1080°F and 1200°F, respectively, so that a certain amount of liquid is formed at 1100-1150°F. Scanning electron micrographs of monotape tensile fracture surfaces are compared in Figure 5. A reaction zone is apparent in the matrix which is much larger at 1150°F than at 1050°F. The surface of the filament is also irregular with

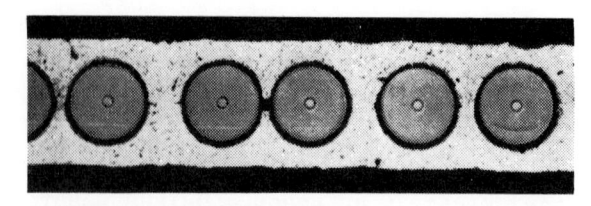

(a) 0.025" Elastic Roll Deflection

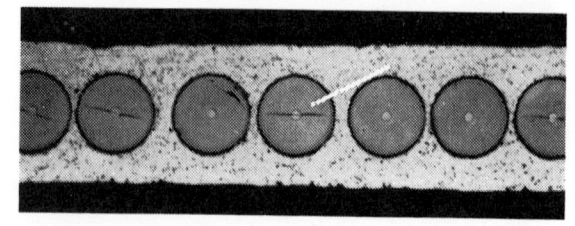

(b) 0.035" Elastic Roll Deflection

(c) 0.040" Elastic Roll Deflection

FIGURE 1   MICROSTRUCTURE OF BORON-ALUMINUM MONOTAPES AFTER ROLLING AT 1050°F AND VARIOUS ELASTIC ROLL DEFLECTIONS   75X

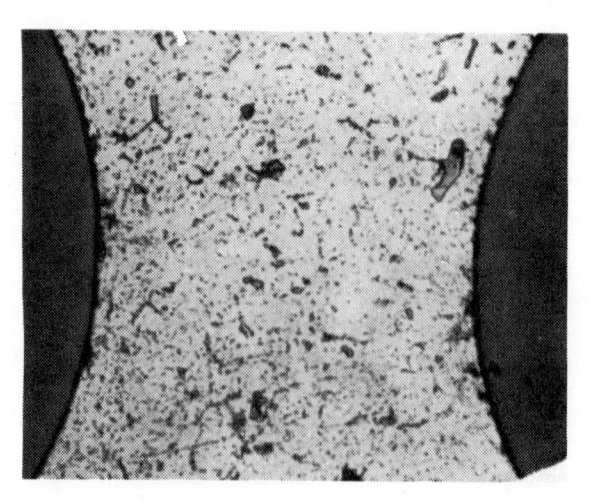

FIGURE 2   MICROSTRUCTURE OF ROLL DIFFUSION BONDED BORON-6061 ALUMINUM MONOTAPE   750X

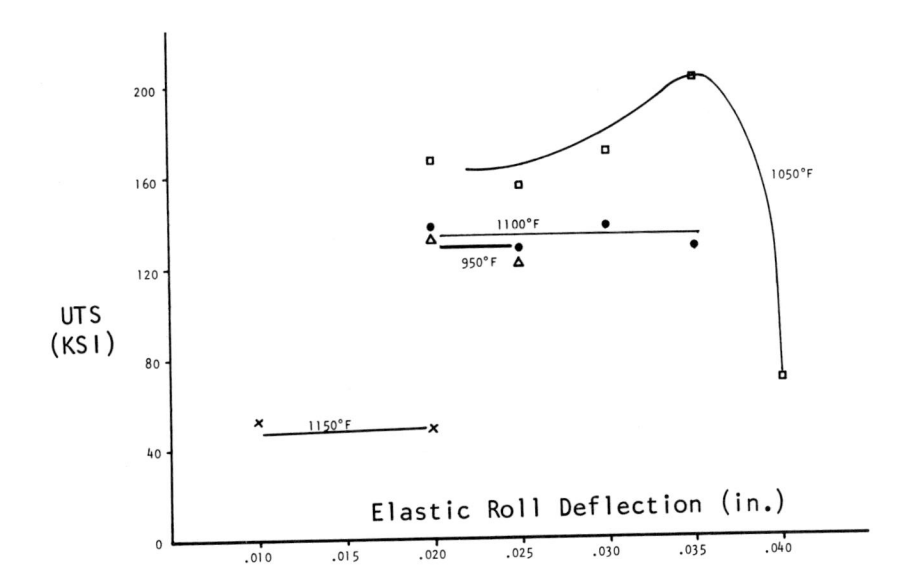

FIGURE 3   ULTIMATE TENSILE STRENGTH OF BORON-ALUMINUM MONOTAPES AT VARIOUS ROLLING TEMPERATURES AND PRESSURES

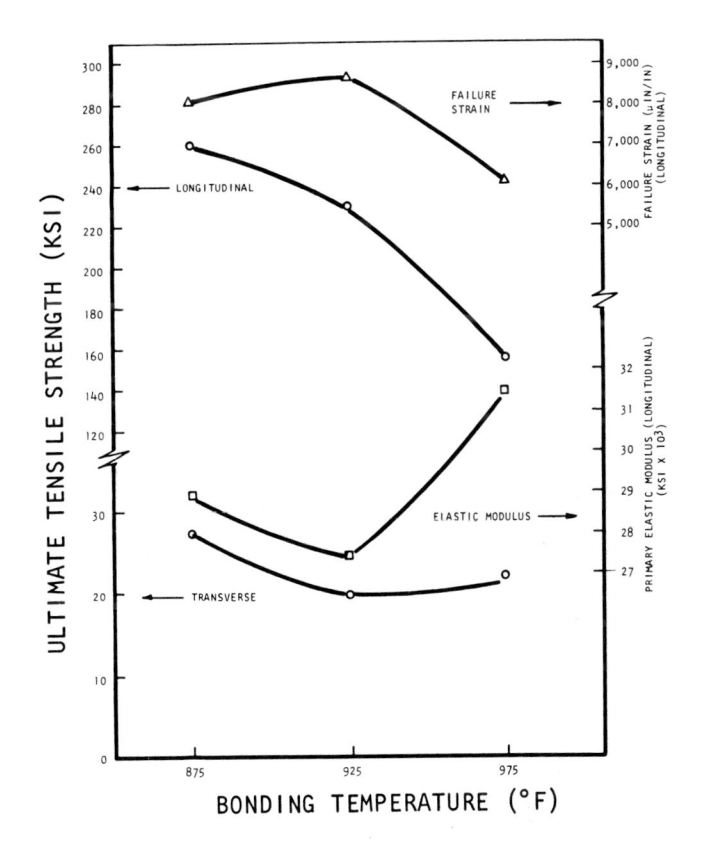

FIGURE 4   THE EFFECT OF BONDING TEMPERATURE ON TENSILE BEHAVIOR OF 60 V/O B-6061AL (1)

(a)
Roll Diffusion Bonded at 1050°F

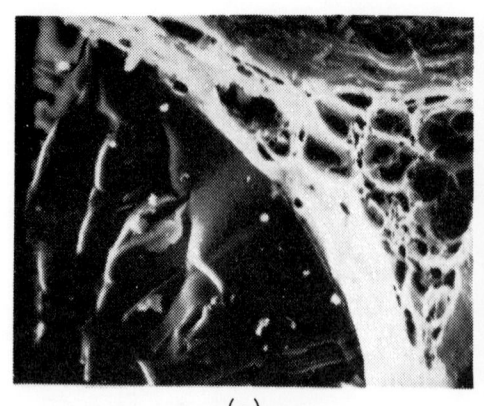

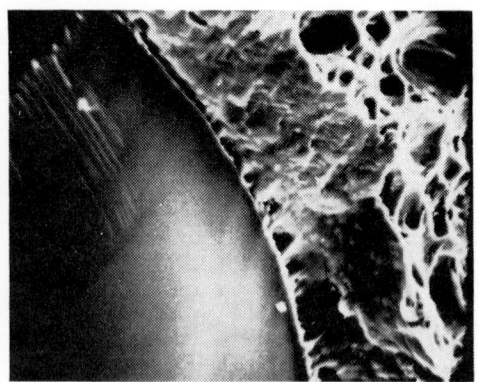

(b)
Roll Diffusion Bonded at 1150°F

FIGURE 5 TENSILE FRACTURE SURFACES OF BORON-ALUMINUM MONOTAPES ROLL BONDED AT 1050°F AND 1150°F   1200X

matrix voids present at the interface after 1150°F rolling, suggesting some form of chemical attack.

After selecting 1050°F as the rolling temperature, a number of runs were made using various sizes and numbers of monotapes. A summary of monotape tensile strength of every run which passed visual inspection is presented in Table 2. The average ultimate tensile strength of the roll bonded monotapes is 207 ksi with a standard deviation of 17.5 ksi.

The tensile strengths of panels which were made from roll-bonded monotapes are summarized in Table 3. These panels were made by secondary press bonding of monotapes in air. The average tensile strength of all tests is 201 ksi longitudinal and 19 ksi transverse. This value of longitudinal tensile strength is in good agreement with the 207 ksi value obtained for the longitudinal tensile strength of monotapes. The variation in tensile strength with position in the panel was also determined. There was no significant difference in strength between specimens taken from the lead or tail end of roll-bonded monotapes.

Atmosphere

A vacuum atmosphere is commonly used when consolidating boron-aluminum composites. In the present work all rolling was carried out in air, although an argon atmosphere was employed in most runs

TABLE 2

TENSILE-STRENGTH OF BORON-ALUMINUM MONOTAPES HEATED IN ARGON PRIOR TO ROLL BONDING IN AIR

AVERAGE OF 20 RUNS
207,000 psi

STANDARD DEVIATION
17,500 psi

MONOTAPE SIZES
4"x4" to 6"x24"

ROLLING PARAMETERS
1050°F
0.035" Elastic
  Roll Deflection

TABLE 3

TENSILE STRENGTH OF BORON-ALUMINUM PANELS PRESS BONDED FROM ROLL BONDED MONOTAPES

AVERAGE OF 6 RUNS
201,000 psi longitudinal
18,700 psi transverse

STANDARD DEVIATION
18,400 psi longitudinal
4,000 psi transverse

for heating. Some runs were also made after heating in air for comparison. There are two potential problems that are presented by the presence of air; first, oxidation of the filament with subsequent loss of strength, and, second, oxidation of the matrix producing poor matrix-matrix or matrix-filament bonding. Under the heating conditions used, one hour at 1050°F, the filament strength did not deteriorate. This is illustrated by the high tensile strength found in monotapes which were heated in air prior to rolling. Typical tensile test results, summarized in Table 4, show an average ultimate tensile strength of 208 ksi for monotapes which had been heated in air and roll bonded. This is similar to the results shown for monotapes heated in argon which were listed in Table 2. The high monotape

TABLE 4

TENSILE STRENGTH OF BORON-ALUMINUM HEATED
IN AIR PRIOR TO ROLL BONDING IN AIR

AVERAGE OF 3 MONOTAPE RUNS

207 psi

AVERAGE OF 3 PANELS PRESS BONDED
FROM ROLL BONDED MONOTAPES

231 psi longitudinal
20.6 psi transverse

tensile strength values could only be reached if the filament strength were high. The presence of air during heating did not prevent matrix-matrix or matrix-filament bonding. Metallographic examination indicated excellent matrix-matrix bonding and scanning electron microscopy of tensile fracture surfaces revealed no filament matrix debonding or filament pullout. An electron microprobe trace showed no significant oxygen or carbon concentration either at the bond line or at the matrix-filament interface.

The strengths of panels made from monotapes heated in air are also shown in Table 4. Both longitudinal and transverse strength levels were generally high.

Rolling Speed

Using a rolling temperature of 1050°F, the rolling speed was varied between 10 and 120 inches/minute to determine if there were any appreciable differences in consolidation dynamics within this range of strain rate. The degree of consolidation was unchanged as a function of rolling speed.

Multilayer Panels

Rolling conditions for the direct rolling of multilayer panels were also investigated in order to assess the capability of the rolling process to produce multilayer basic mill shapes. Rolling is particularly attractive in producing long lengths of such panels having uniform properties. The rolling parameters are different from those used for rolling monotapes because the pressure for consolidation changes with the number of plies. All panels were prepared using unidirectional reinforcement.

The longitudinal and transverse tensile strengths of panels direct rolled at 1050°F are listed in Table 5. The significance of these data is, one, that generally high strength levels are

TABLE 5

TENSILE STRENGTH OF DIRECT ROLLED BORON-ALUMINUM PANELS

(1050°F ROLLING TEMPERATURE)

NUMBER OF PLIES	ELASTIC ROLL DEFLECTION (IN.)	LONGITUDINAL UTS (KSI)	TRANSVERSE UTS (KSI)
2	0.025	180	17
	0.030	164	19
	0.035	227	15
3	0.025	243	19
	0.030	175	19
	0.035	228	19
5	0.025	226	23
	0.030	190	17
	0.035	188	17

produced, and, two, the highest strength levels occur at different roll deflections or rolling pressures, depending upon the number of plies. A summary of the roll setting for optimum tensile strength as a function of number of plies is shown in Figure 6. The larger

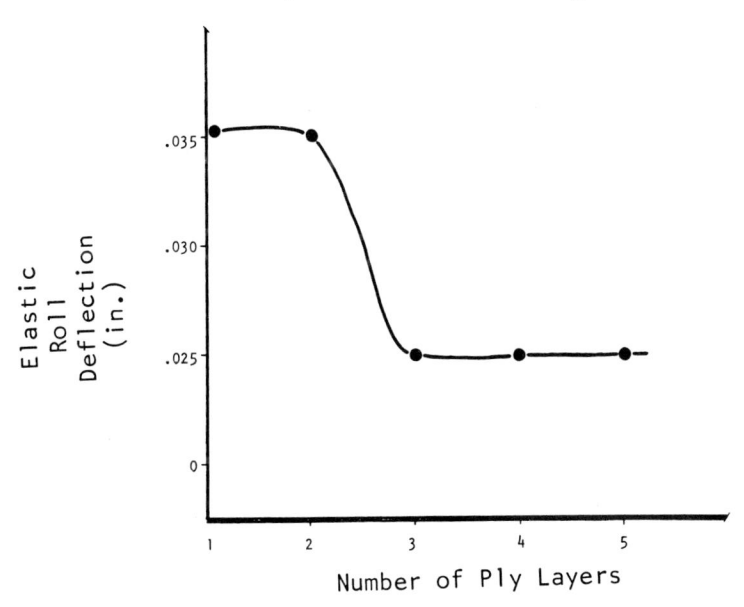

FIGURE 6   ELASTIC ROLL DEFLECTION VERSUS NUMBER OF PLY LAYERS FOR OPTIMUM STRENGTH OF DIRECT ROLLED PANELS

2 Ply

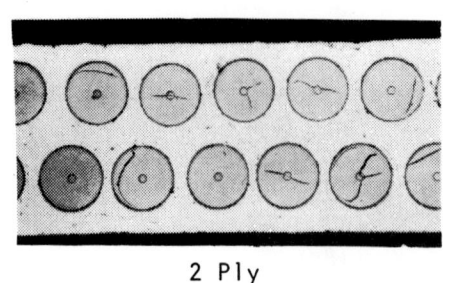

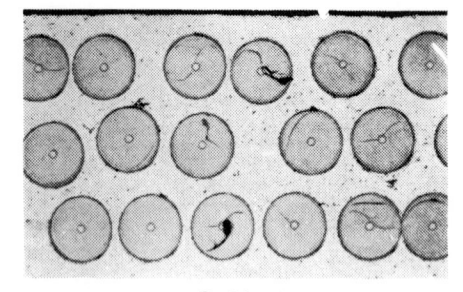

3 Ply

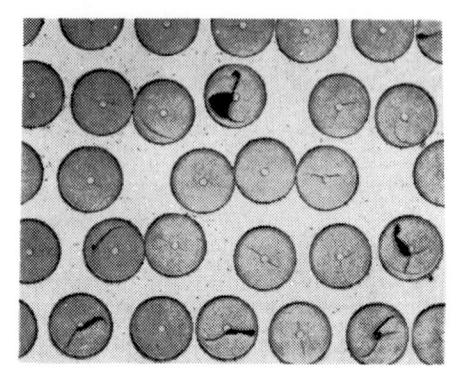

5 Ply

FIGURE 7   DIRECT ROLLED BORON-ALUMINUM PANELS   60X

number of plies requires less rolling pressure. The filament distribution as a function of numbers of plies is illustrated in Figure 7. The 2-ply panels have an excellent distribution, while some filament movement is apparent in the 3-ply panels. Considerable filament movement was observed in the 5-ply panels. Even with this filament movement, however, the strength of the 5-ply panels was high.

Cross-Ply Panels

The feasibility of primary roll bonding of cross-plied panels was also examined during the program. A 4-ply, ±15° panel was direct rolled at 1050°F and a 0.025 inch reduction. Consolidation and bonding were complete. A potential problem in direct roll bonding with cross plies is filament breakage where overlap occurs. Observation of a section of the panel with the matrix etched away indicated that filament breakage is not present. Tensile properties of the direct-rolled and press-bonded panels were comparable.

Hybrid Panels

Certain applications of metal matrix composites benefit from the presence of a third alloy component. Interleaving or cladding of stainless steel or titanium may be employed to increase shear strength, impact resistance, erosion resistance or transverse strength. In the case of boron-aluminum, the addition of 18-26 v/o interleaved titanium provides a 2 or 3-fold increase in transverse strength. Such a strength increase would eliminate the need for cross plying in many applications. The incorporation of layers of titanium may be performed at very low incremental cost, either during direct rolling of panels or during secondary bonding.

Direct-rolled, 5-ply panels with four layers of titanium-6Al-4V alloy, amounting to 18 volume percent, were rolled using the same parameters as had been used for boron-aluminum 5-ply panels. The panels produced were fully bonded and contained an excellent filament distribution. Tensile properties of these panels are listed in Table 6, the longitudinal strength being 200 ksi and the transverse strength being 48-64 ksi. A typical microstructure is shown in Figure 8.

TABLE 6

TENSILE STRENGTH OF BORON-ALUMINUM PANELS
CONTAINING INTERLEAVED TI(64-) FOIL

PANEL	VOL % TI(6-4)	LONGITUDINAL UTS (KSI)	TRANSVERSE UTS (KSI)
DIRECT ROLLED 5 PLY	18	199	48
SECONDARY PRESS BONDED 5 PLY	25	201	64

Cost Analysis

A preliminary cost analysis was performed to compare the roll-diffusion bonding process with press diffusion bonding. To facilitate comparison, many of the assumptions used in a recent National Materials Advisory Board Composite Committee report (2) were used. Because the final composite cost depends so much upon the raw-material costs, which in turn depend upon the volume of filament being manufactured, the analysis was performed on the basis of processing cost. The analysis assumed yearly manufacturing volumes of 2,000, 20,000 and 200,000 lbs/year.

The cost comparison between step press bonding and continuous roll bonding in air is shown in Figure 9. The significant features of the projection are the relatively constant differential between the two processes. This differential arises primarily because of the rapid consolidation in rolling compared to press bonding. Depending upon assumptions the upper curve might rise or fall but the differential should remain constant.

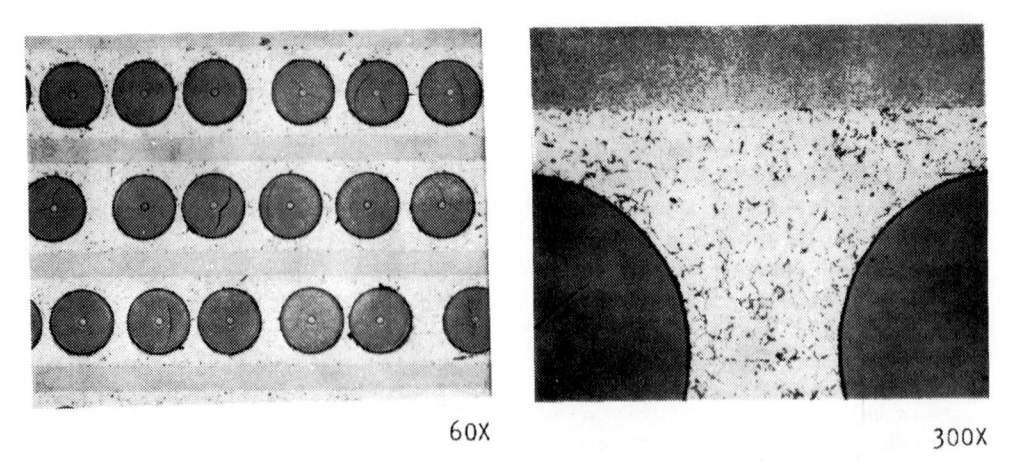

FIGURE 8   BORON-ALUMINUM PANEL OF ROLL BONDED MONOTAPES AND TITANIUM INTERLEAVING PRODUCED BY SECONDARY PRESS BONDING

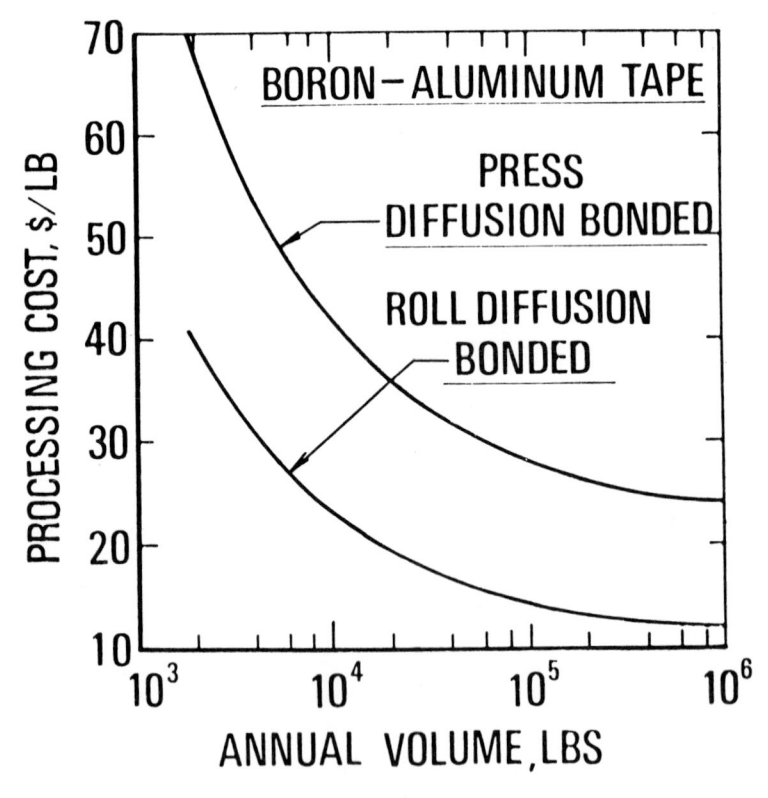

FIGURE 9   PROCESSING COST PROJECTIONS FOR 50 VOLUME PERCENT BORON-ALUMINUM MONOTAPE BONDED IN AIR BY STEP-PRESS OR CONTINUOUS ROLL BONDING. (RAW MATERIAL COST NOT INCLUDED)

SUMMARY

The program has resulted in the development of a process for primary roll-diffusion bonding boron-aluminum composites in air at low cost. Using the boron-6061 aluminum system, it has been demonstrated that quality composites can be roll bonded in air at very short times. The procedure has been used for sizes up to 6 inches wide and 4 feet long and there does not appear to be any fundamental limitation on increased length or width. The rolling temperature providing optimum tensile strength for direct-rolled monotapes or panels was 1050°F which is just below the solidus temperature of the 6061 aluminum alloy. Above this temperature low tensile strength was produced, apparently the result of chemical attack of the boron fiber by the liquid phase. Roll speeds of 10-120 inches per minute were used without affecting the consolidation dynamics of monotapes. Both argon and air were used as atmospheres for heating prior to rolling. High strength in both longitudinal and transverse directions was produced after heating in either atmosphere. A total of 20 runs provided an average ultimate tensile strength of 207 ksi for monotapes and tensile strengths obtained from panels made from these monotapes by press bonding in air averaged 201 ksi longitudinal and 19 ksi transverse. Direct roll bonding was also demonstrated for cross-ply panels. No filament breakage was obtained and properties were equivalent to panels produced by roll bonding monotapes and secondary press bonding.

Panels containing interleaved titanium foil of 18-26 volume percent were produced by both primary and secondary roll bonding. Resultant tensile strengths were about 200 ksi longitudinal and 48-64 ksi transverse. The cost of roll diffusion bonding was projected to be significantly lower than press diffusion bonding.

ACKNOWLEDGEMENTS

This work was sponsored by the Air Force Materials Laboratory under Contract F33615-74-C-5076. Appreciation is expressed to E. Joseph of AFML and P. Melnyk of TRW for their valuable discussions.

REFERENCES

1. Scheirer, S.T. and Toth, I.J., "The Mechanical Behavior of Metal-Matrix Composites," AFML-TR-73-178, May 1973.

2. National Materials Advisory Board, "Metal-Matrix Composites: Status and Prospects," NMAB 313, December 1974.

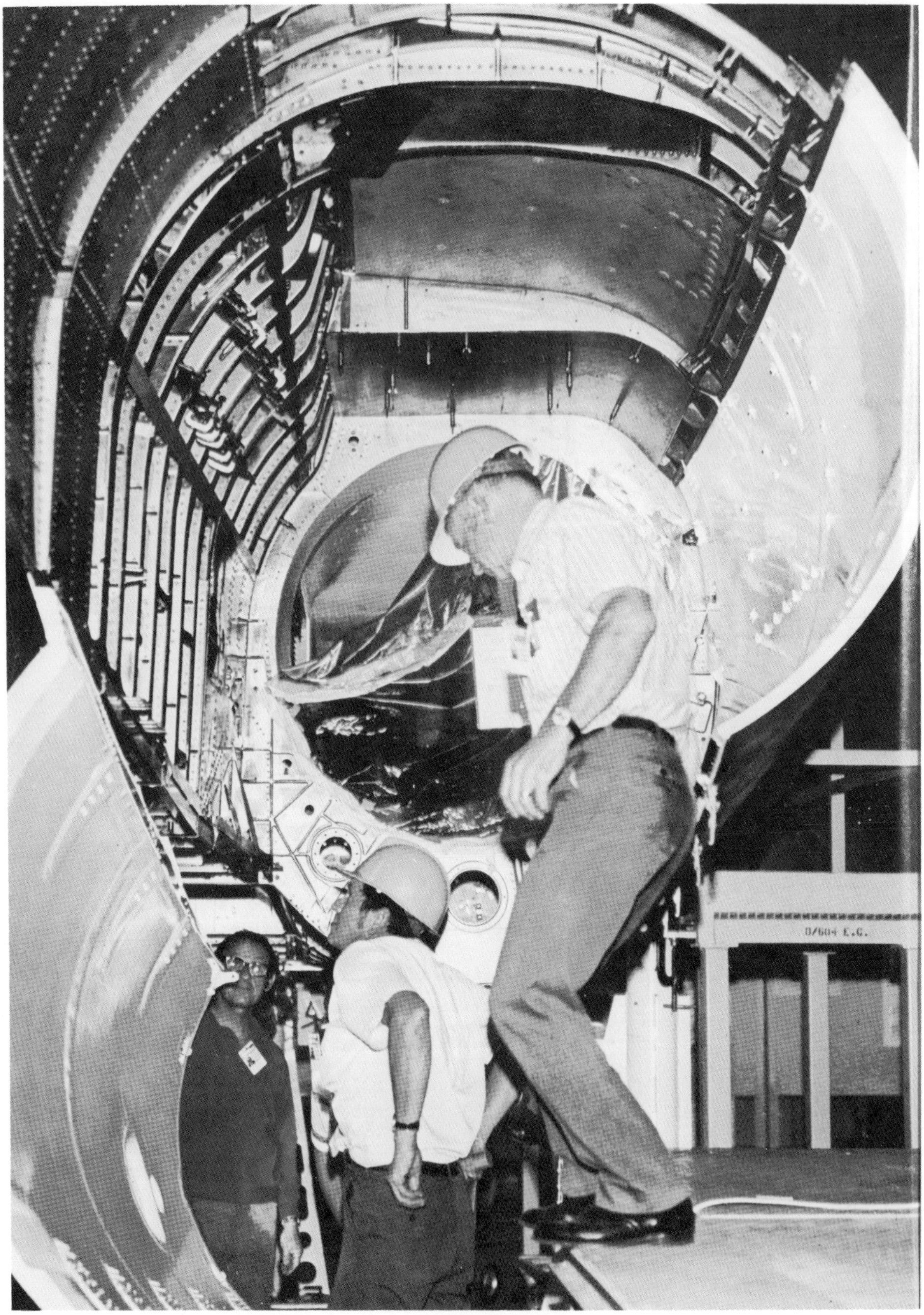

Fabricating Titanium Parts With SPF/DB Process

By Edward D. Weisert and George W. Stacher

THE SPF/DB PROCESS, a single-heat-cycle combination of superplastic forming (SPF) and diffusion bonding (DB), is likely to revolutionize the fabrication of Ti-6Al-4V sheet structures for aircraft applications. Advantages over conventional processing include:
1. A substantial reduction in fabrication cost.
2. More efficient, lighter weight structures.
3. A wider array of structural design possibilities in titanium.

The process is successful because Ti-6Al-4V exhibits superplasticity and is readily diffusion bonded.

The use of the alloy's superplastic behavior to form components from sheet has been reported in *Metal Progress*[1]. Briefly, superplastic behavior is the absence of localized thinning when a material undergoes extensive tensile strain. Elongations of several hundred per cent are typical; some even exceed 1000%.

In superplastic forming, sheet is heated to the SPF temperature in a sealed die. An applied gas pressure then forces the sheet to conform to the shape of the die cavity. Argon is used in titanium SPF to prevent oxidation of the reactive metal. Springback and residual stresses are not problems with SPF parts.

Low Pressures: The optimum temperature for superplastically forming Ti-6Al-4V is 1700 F (925 C). Fortunately, this is the same temperature used to assemble large 6-4 parts, comparable to forgings, by diffusion bonding[2].

In this process, mating surfaces are brought into intimate contact at elevated temperature. Atomic diffusion across the interface produces the bond.

The combination of SPF and DB would thus appear to be a natural one. SPF, however, is a sheet technology requiring the application of relatively low gas pressures of a few hundred pounds per square inch[1,3]. Fabricating large parts by DB, on the other hand, requires press pressures six to ten times greater.

Before a combined process could be actively pursued, we had to determine whether Ti-6Al-4V could be diffusion bonded within SPF's lower pressure regime.

An analysis of the creep accommodation required to obtain intimate surface contact and diffusion bonding of 6-4 sheet was conducted. Data were translated into the theoretical pressure-time curve shown as a solid line in the graph (Fig. 1)[4].

Specimens were then fabricated using various bonding times and pressures. The resultant diffusion bonds were ultrasonically inspected, metallographically examined, and lap shear tested. The experimental data (also plotted in Fig. 1) verified the analytical prediction, indicating that complete bonding could be obtained within reasonable times at low pressures. Lap shear strengths, for example, averaged

The B-1 bomber's lower engine access door will be the largest part attempted by SPF/DB. The 1.5 in. (38 mm) deep expanded sandwich structure will measure 105 in. (2665 mm) long and 55 in. (1395 mm) wide with a 30 in. (760 mm) radius, compound curvature.

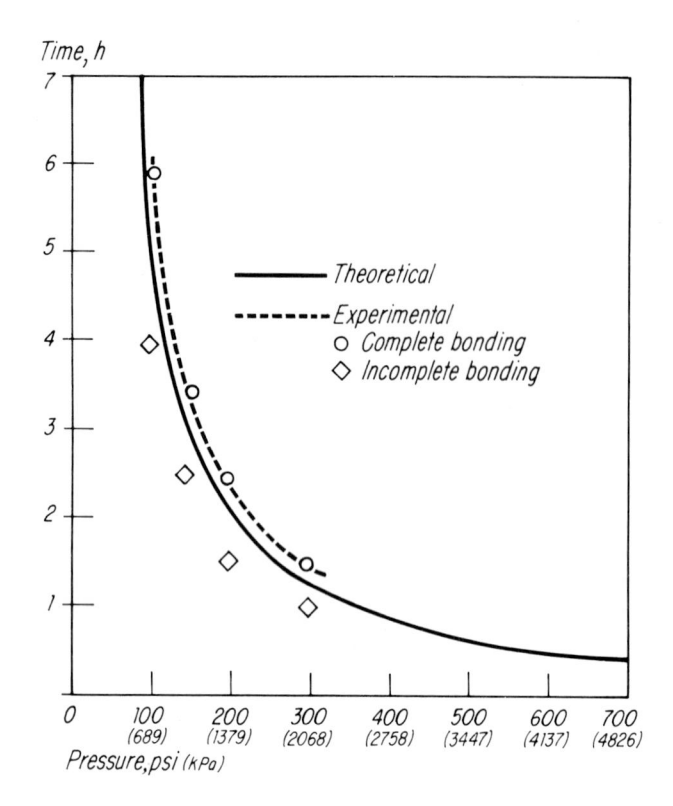

84 000 psi (580 MPa). Typical 6-4 shear strength is 78 000 psi (540 MPa).

SPF/DB Structures Fall Into Three Categories

Independent research at Rockwell has now established three generic types of SPF/DB structures (Fig. 2). In the first type, a superplastically forming sheet encounters titanium details, preplaced in the tooling, and is diffusion bonded to them[5]. It's possible to bond doublers, pads, and other reinforcing or functional members to the formed part. The procedure can also be reversed with forming occurring after bonding because both are done during a single process (heating) cycle.

A prototype similar to the design shown schematically in

Fig. 1 — Success of SPF/DB process hinged on whether Ti-6Al-4V could be diffusion bonded at the relatively low pressures used in superplastic forming. Work proceeded once experimental data bore out a theoretical prediction that quality bonds could be produced.

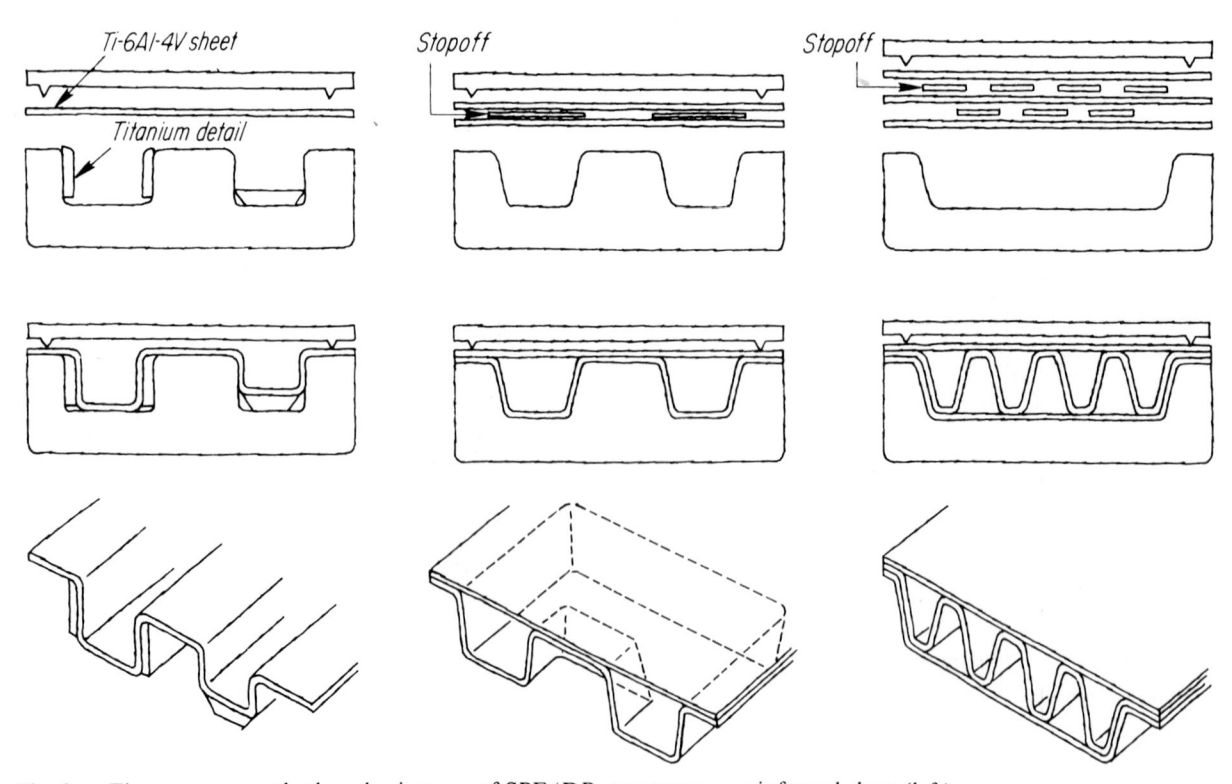

Fig. 2 — There are currently three basic types of SPF/DB structures — reinforced sheet (left), integrally stiffened (center), and sandwich (right).

Fig. 2, bottom left, has been fabricated under a current Air Force Materials Laboratory (AFML) contract[6].

The second type of SPF/DB structure — integrally stiffened — is made by simultaneous processing of two 6-4 sheets (Fig. 2, center). A stopoff compound applied to one sheet prevents bonding in discrete areas. The stopoff pattern corresponds to tooling cavities. When the pack reaches 1700 F (925 C), pressure is applied and those areas not coated with stopoff are bonded. After bonding, gas pressure is introduced which superplastically forms the unbonded areas[5].

Demonstration parts with "hat section" stiffening, similar to that in Fig. 2, bottom center, have been fabricated. Properties of specimens removed from these parts are summarized in Table I. Note that the lap shear strength values match the experimental average of 84 000 psi (580 MPa).

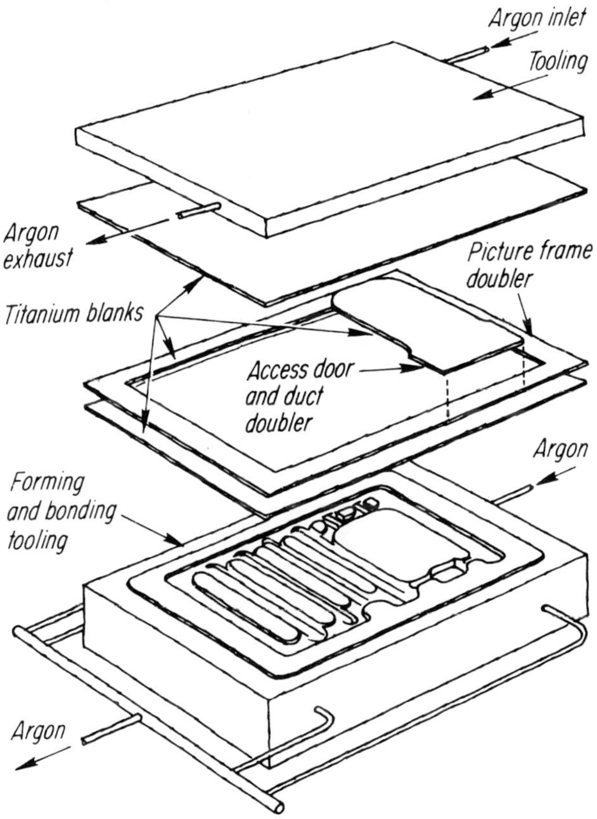

Sandwich Structure: By using selective bonding and three sheets, the SPF/DB process yields the most exciting form of hardware — expanded sandwich (Fig. 2, right)[7].

A variety of demonstration SPF/DB sandwich structures have been fabricated. Although the work is still in an early stage, two important advantages of SPF/DB sandwich can be cited:

1. The external configuration of the fabricated part is obviously determined by the tool cavity and may be a design variable. On the other hand, the core configuration is determined by the stopoff pattern; it may be of infinite variety and can be modified without tooling change.

2. The process inherently provides an edge closure of infinite design. This avoids what is frequently a significant cost factor in applying conventional honeycomb sandwich.

SPF/DB B-1 Parts Will Save Weight and Dollars

The potential for the SPF/DB process is being confirmed by fabricating and evaluating full-scale components designed for possible use in the B-1 bomber. This is another

Fig. 3 — First SPF/DB demonstration part for B-1 was an integrally stiffened version of the APU door (see Fig. 4, top). Blowup illustrates door tooling and sheet blank configuration before forming and bonding. Doublers are used to double sheet thickness in selected areas for enhanced fatigue resistance and other reasons.

Table I — Average Properties of an Integrally Stiffened, SPF/DB Part

Tensile strength, F_{ty} 10^3 psi (MPa)	126.9 (875)
Compressive strength, F_{cy} 10^3 psi (MPa)	138 (950)
Shear strength, F_{su} 10^3 psi (MPa)	
Single lap	83.8 (575)
Double lap	83.9 (580)
Static Peel* 1 lb/in. (kN/m)	1372.5 (245)
Dynamic Peel*	
Maximum Load, lb/in. (kN/m)	Cycles to Failure, 10^3
590 (105)	9
443 (78)	49.3
295 (52)	136
263 (46)	182

* No failures occurred in bonded areas.

Source: *Metal Progress*, Mar 1977

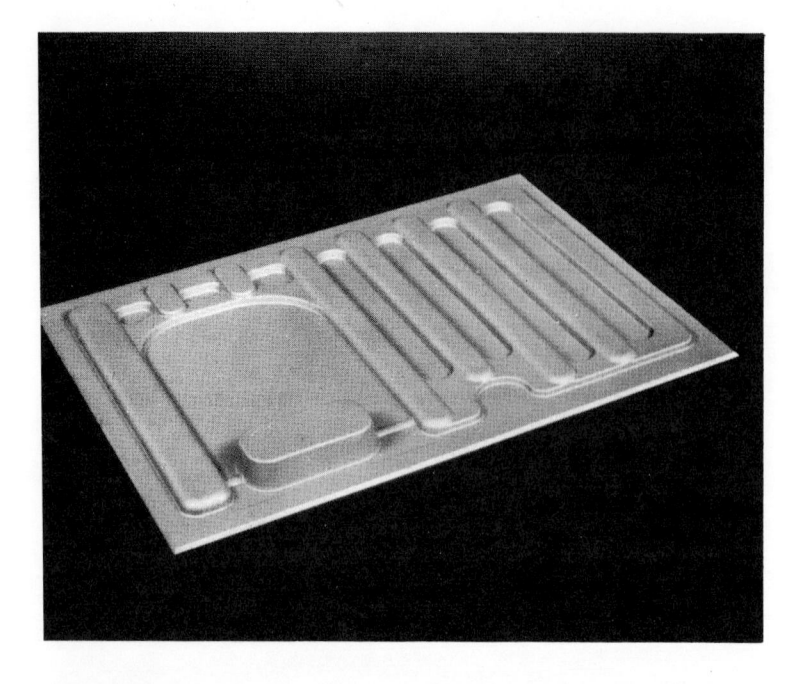

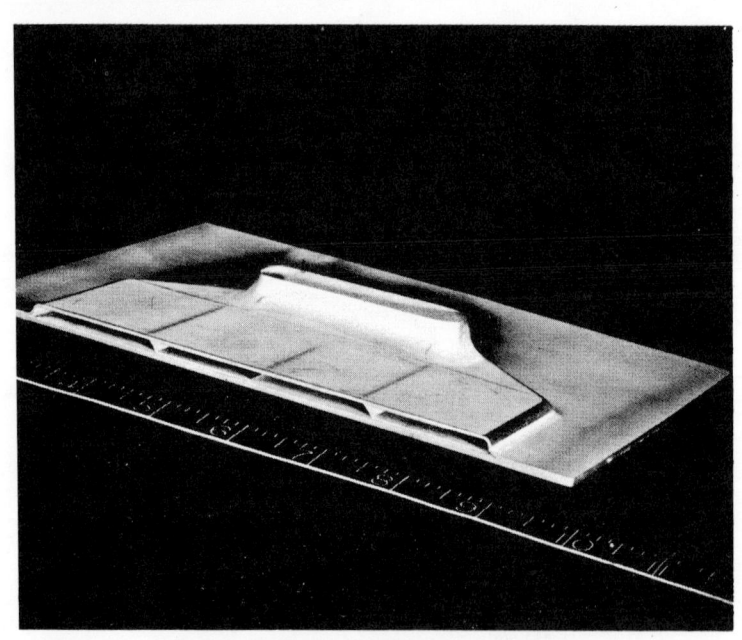

Fig. 4 — The B-1's APU door is shown at top.
An SPF/DB windshield hot air blast nozzle (bottom)
will be incorporated on the fourth B-1.
This complex part is typical of those
that lend themselves to sandwich construction.

part of the AFML contract mentioned above.

The first demonstration part was an integrally stiffened version of the auxiliary power unit (APU) door in each nacelle. The SPF/DB version (Fig. 4, top) measures approximately 22 by 28 in. (560 by 710 mm) with ¾ in. (20 mm) deep hat sections. Two of these doors would replace a single T-section-stiffened door machined from plate.

Results of cost trade-off studies indicate that SPF/DB doors would represent a 50% cost saving and a 31% weight saving over the machined version. The dollar gain derives from a need for less metal and a reduced fabrication cost. The weight advantage stems from the SPF/DB part's more efficient load carrying configuration.

The SPF/DB door is more complex than it first appears. The periphery, for example, has an 0.109 in. (2.8 mm) thick "picture frame" doubler bonded in place, a design feature added to increase resistance to acoustic fatigue loading. The flat area on the door accommodates a smaller access door. It is doubled with an 0.060 in. (1.5 mm) thick sheet to provide an 0.120 in. (3.0 mm) thick frame for the smaller door after machining. The large bulge, truncated to accommodate a duct, is also doubled to 0.120 in. (3 mm) and subsequently formed. The setup for SPF/DB door fabrication is shown in Fig. 3.

The second SPF/DB part is a nacelle beam frame which is currently being fabricated as 12 hot-sized and/or machined details assembled with 81 fasteners. Results of preliminary studies indicate that the essentially monolithic SPF/DB design would weigh 39% less and is expected to cost half as much.

Large Component: The APU door and nacelle frame represent two of the basic types of SPF/DB structures. A third demonstration part representing expanded sandwich structure was recently added to the AFML contract.

The part, a B-1 lower engine access door (see photo, p. 32), measures approximately 105 in. (2665 mm) long and 55 in. (1395 mm) wide with a nominal 30 in. (760 mm) radius, compound curvature. This will be the largest part attempted by SPF/DB.

Windshield Hot Air Blast Nozzle Already Qualified

Test data on SPF/DB coupons are encouraging. There are, however, certain structural tests, particularly acoustic fatigue tests, which must be successfully completed before extensive B-1 applications can be realized.

It's interesting to note that four parts on the fourth B-1 will be of SPF and riveted construction. When SPF/DB is qualified for these components, they will be fabricated on the same tooling used for the SPF versions.

The APU door is one of these parts. Although the SPF/riveted version costs substantially less than the machined component, its weight reduction contribution isn't nearly as great as that of the SPF/DB door.

Aircraft 4 will have one SPF/DB component specifically qualified for its function. This is the windshield hot air blast nozzle, a natural candidate for SPF/DB sandwich construction. The wide, shallow nozzle with internal guide vanes and transitions to inlet tubing posed an extraordinary fabrication challenge that was met with difficulty by three totally different approaches for aircrafts 1 through 3.

The SPF/DB approach to the nozzle is shown in the laboratory demonstration in Fig. 4, bottom. A second part has been successfully tested under simulated operating conditions.

Blue Sky: Advanced design concepts emerging from several studies combine the three basic SPF/DB types into a wide variety of unique structures[8,9].

The wing/fuselage concept shown in Fig. 5, top, combines reinforced sheet with sandwich. The monolithic skin/frame/stringer construction in Fig. 5, bottom, combines integrally stiffened sheet with bonded-in reinforcements.

The goal of these studies is to slash assembly cost by replacing the traditional "bits and pieces" approach to airframe construction with large, monolithic components. Compared with existing components of similar function, the SPF/DB structure in Fig. 5, bottom, for example, could be produced using 75% less tools. Total number of parts needed would be reduced by 87% and the total number of fasteners by 89%. Its cost would be cut by 52% and its weight by 40%.

With incentives like these it's easy to see why SPF/DB is being brought from the laboratory to the production line as rapidly as possible, and why we're predicting a revolution in future aircraft construction.

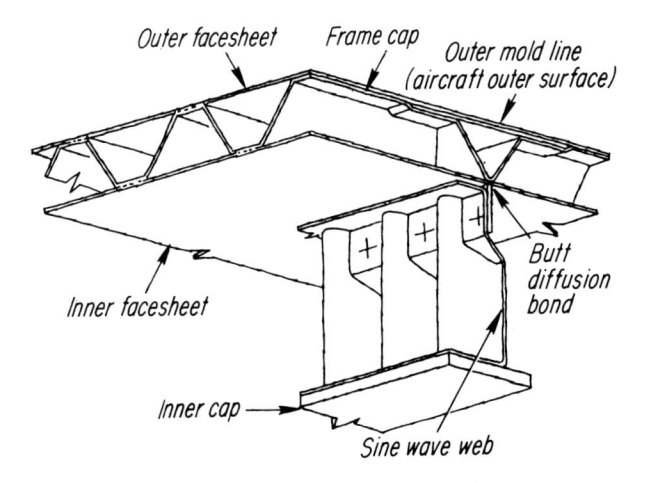

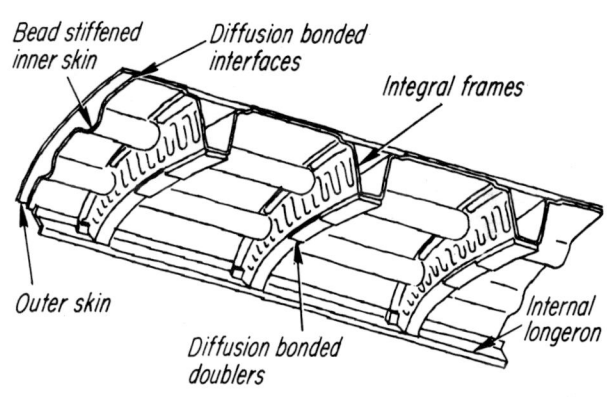

Fig. 5 — Potential SPF/DB applications include wing/fuselage (top) and skin/frame/stringer (bottom) structures. Adoption of these concepts would result in substantial reductions in weight and cost and in numbers of parts, fasteners, and tools needed.

References

1. "Superplastic Forming of Ti-6Al-4V Beam Frames," by C.H. Hamilton and G.W. Stacher, *Metal Progress*, Vol. 109, No. 3, March 1976, p. 34-37.
2. "Fabrication and Evaluation of Diffusion Bonded Laminated Sections," by W.D. Padian and E.C. Supan, AFML-TR-71-131, August 1971.
3. "Controlled Environment Superplastic Forming of Metals," by C.H. Hamilton et al, U.S. Patent 3 934 411, 27 January 1976.
4. "Process Parameter Prediction Methods for Superplastic Forming," by C.H. Hamilton and S. Tsang, NA72-684-1, March 1974.
5. "Method for Superplastic Forming of Metals With Concurrent Diffusion Bonding," by C.H. Hamilton and L.A. Ascani, U.S. Patent 3 920 175, 18 November 1975.
6. "Manufacturing Methods for Superplastic Forming/Diffusion Bonding Process," by E.D. Weisert and G.W. Stacher, AFML Contract F33615-75-C-5058, IR798-5, April 1975-October 1976 (S. Inouye, AFML Monitor; V. Russo, AFML Focal Point).
7. "Method for Making Metallic Sandwich Structures," by C.H. Hamilton and L.A. Ascani, U.S. Patent 3 927 817, 23 October 1975.
8. "Evaluation of Low Cost Titanium Structure for Advanced Aircraft," by J. Pulley, NASA Contract NAS1-14206.
9. "Preliminary Design of Low Cost Advanced Titanium Structure," by J. Pulley, AFFDL Contract F33615-76-C-3066.

Mechanical Properties of Diffusion-Bonded Beryllium Ingot Sheet

CLINTON R. HEIPLE

A technique for producing thick beryllium plate from ingot source material by diffusion-bonding together thin sheets is described. The bonds produced are at least as strong as the matrix, and are produced under conditions where no significant grain growth occurs. The laminated plate has better mechanical properties than the sheet from which it was made, and properties substantially superior to sheet of similar thickness rolled directly from the ingot. Room temperature tensile elongations of approximately 18 pct were observed in the laminated plate. Directly rolled plate of similar thickness normally exhibits 2 to 4 pct elongation, and the sheet from which the laminate was made failed after about 12 pct elongation. The increased ductility of the laminate is a consequence of the compressive creep deformation accompanying bonding and most likely results from a dislocation substructure, produced by the deformation, which differs from the substructure developed by annealing.

THE brittleness of beryllium restricts its use in structural applications. The main cause of the low ductility of polycrystalline beryllium at room temperature is thought to be the lack of available slip systems along the c axis direction. Deformation results in the buildup of stress concentrations at the grain boundaries and results in the formation of bend planes, due to basal slip,[1-3] which leads to cleavage and brittle fractures at small strains, often only 1 to 3 pct. The ductility is generally increased by reducing the grain size.[3-5] In addition, increased ductility in the plane of the sheet is observed with increased basal texture. The increased basal texture, however, simultaneously reduces formability.[5,6]

Grain size reduction is primarily achieved in cast and warm-rolled beryllium by increasing the total reduction in thickness from the original ingot. Beryllium sheets with a thickness of 0.025 in. (0.64 mm) and less, and a grain size* of about 16 μm have been produced.

*The grain size is the average grain diameter from lineal analysis using the procedure given in ASTM:E112-63.

This fine-grained material has mechanical properties substantially superior to those exhibited by material produced from identical ingots rolled to 0.250 in. (0.64 cm) thickness with a typical grain size of about 200 μm. In other metals, notably titanium, it is possible to diffusion-bond thin sheets together into thick sections which retain the fine grain size and the desirable mechanical properties of the thin sheets from which they were made.[7] The present work was undertaken to determine if the same process could produce thick beryllium plate with improved mechanical properties.

For the laminate to retain the properties of the laminae, the bonding temperature must be low enough so that significant grain growth does not occur during bonding. However, the temperature must be high enough so that bonds form in a reasonable time interval. Diffusion bonding is normally impractical below about one-half the absolute melting point (502°C for

CLINTON R. HEIPLE is a Research Metallurgist, Dow Chemical U.S.A., Rocky Flats Division, Box 888, Golden, Colo., 80401.
Manuscript submitted July 19, 1971.

beryllium), and significant grain growth occurs in as-rolled beryllium above about 750°C within 15 min.

EXPERIMENTAL TECHNIQUE

All bonding was conducted in a hot press vacuum chamber at a pressure of about 10^{-5} torr. Specimens were inserted into the cold furnace, and heated to the desired temperature. Heating time to the bonding temperature was about 1 hr 45 min. Pressure was applied for various times and then removed. The samples were then furnace cooled.

Coupons 1 by 1 by $\frac{1}{4}$ in. (2.54 by 2.54 by 0.64 cm) were used to determine if diffusion bonding without grain growth is feasible. The surfaces to be bonded were ground flat and etched (54 parts water, 44 parts nitric acid, 2 parts hydrofluoric acid) prior to insertion into the hot press. Two coupons were used for each bonding test with one exception. In the latter case, six coupons were bonded together in order to make a block thick enough so that tensile specimens with cylindrical gage length and threaded ends with a 0.25 in. (0.64 cm) gage length could be machined from it with their tensile axes normal to the bond planes.

Ten squares 1 by 1 in. (2.54 by 2.54 cm) cut from 0.025 in. (0.64 mm) thick sheet in the as-rolled condition were used to make each laminate for subsequent tensile testing to determine the mechanical properties of laminated plate. The surfaces of the squares were not ground but were etched in the same way as the bonding-test coupons prior to insertion in the hot press. The bonding conditions were 45 min at 700°C under a constant load of 20,000 lb. (9070 kg). Approximately a 45 pct compression occurred during bonding, resulting in a laminate thickness of 0.14 in. (0.36 cm). The pressure during bonding therefore varied from an initial 20,000 psi (14.1 kgf/mm^2) to about 11,000 psi (7.7 kgf/mm^2). These conditions produced successful bonds, as described below. Sheet-type, pin-loaded tensile specimens with a 0.40 in. (1.02 cm) gage length were machined from the laminates so that their tensile axes would be parallel to the bonding planes. All tensile tests were conducted at room temperature at a strain rate of approximately 0.005/min.

Table I. Results of Bonding Tests

Temperature, °C	Time, min	Results
650	60	Partial bond
700	10	No bond
700	30 to 60	Good bond
750	5	Marginal bond, some recrystallization
750	10	Good bond, substantial recrystallization

Load was 20,000 lb. (9070 kg) on orginally 1 sq in. (6.45 cm^2) samples. See text for bond quality criteria.

Table II. Tensile Strengths from Tests of Bonded Block Normal to the Bond Planes

UTS, ksi	(kgf/mm^2)	Remarks
25 ± 2	(17.6)	Unetched
13 ± 2	(9.1)	Broke in threads
21 ± 2	(14.8)	Broke in threads
25 ± 2	(17.6)	
9 ± 2	(6.3)	
30 ± 2	(21.2)	
28 ± 2	(19.7)	

All tensile samples were etched ~0.005 in. (0.13 mm) except as noted.

Fig. 1—A bond (arrow) in a tensile bar after pulling to failure to measure bond strength. Bonding conditions: 1 hr at 690°C under an initial load of 20,000 lb. (9070 kg), reduced during bonding to 8000 lb. (3600 kg) to keep deformation within the range of possible ram travel. Polarized light.

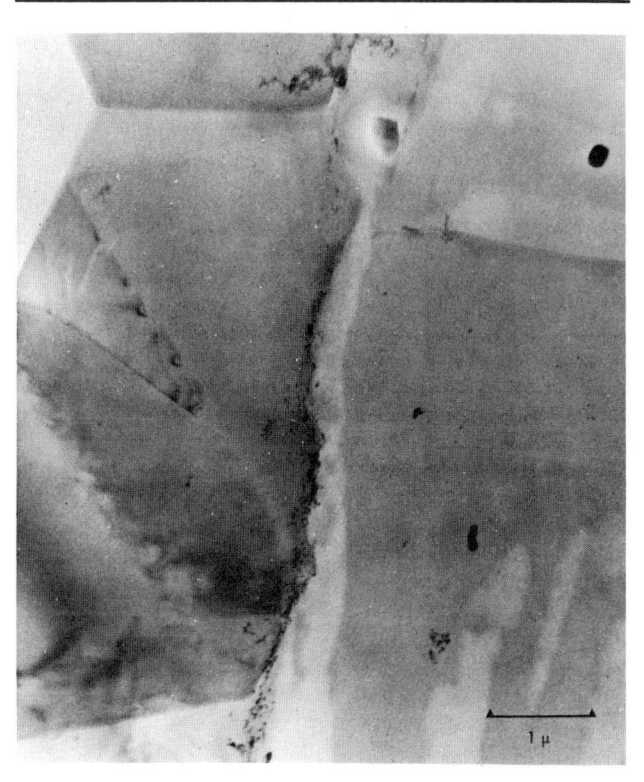

Fig. 2—Electron transmission micrograph of the diffusion bond region.

BONDING TEST RESULTS

The results of the bonding tests are given in Table I. It was not feasible to tensile test all the bonds. Therefore, the quality of those which were not tensile tested was determined by metallographic comparison with the appearance of bonds whose quality was verified by tensile tests. Fig. 1 is a photomicrograph of a successful bond in the block which was tensile tested. The tensile strengths from several tests normal to the bond planes in the bonded block are given in Table II. None of the fractures occurred on the bond planes, indicating that the bonds are at least as strong as the matrix. No elongation was observed in any of the samples, but because gage marks were used to measure plastic deformation, elongations as large as 0.5 pct could have occurred. The low ultimate tensile strength and low elongation observed in the thickness direction in these tests has been reported previously[8] and is to be expected from the basal texture of rolled beryllium sheet.

Because of the marginal bonds produced at 650°C, lower temperature bonding was not attempted. The bonding conditions of 45 min at 700°C under a load of 20,000 lb. (9070 kg) were selected for use in making laminated sheet as a result of the consistently good bonds produced without grain growth under these conditions in the bonding tests. It is not known whether these are necessarily the optimum bonding parameters.

The structure of the bond produced was also investigated by transmission electron microscopy. A typical bond and surrounding material are shown in Fig. 2, while Fig. 3 is a high-magnification picture of the bond region. The bond is clearly a Be-Be bond, and not an oxide-oxide bond. The fine dispersion of particles along the bond is presumably the remains of the oxide layer which was on the beryllium sheet prior to bonding.

TENSILE RESULTS AND DISCUSSION

The properties of the 0.025 in. (0.64 mm) thick sheet from which the laminae were cut were measured for

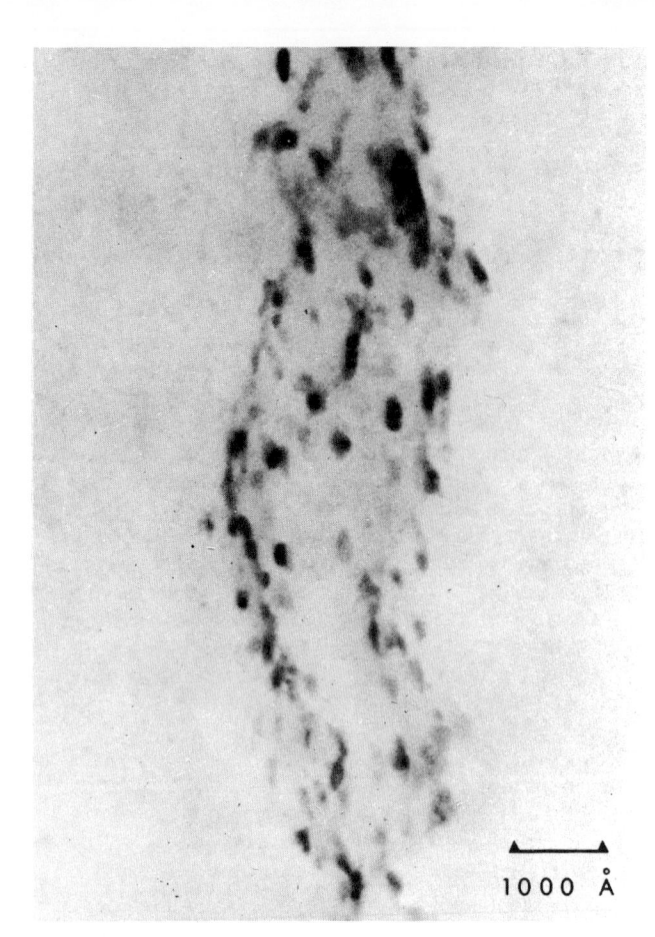

Fig. 3—Electron transmission micrograph of the diffusion bond.

Table IV. Properties of Beryllium of Various Types and with Various Histories

Material	UTS, ksi	(kgf/mm^2)	0.2 Pct Yield, ksi	(kgf/mm^2)	Elongation, Pct
Laminates (this work)	62 ± 3	(44)	32 ± 4	(22)	18 ± 1
0.2 in. (0.51 cm) thick ingot sheet (a)	36.1	(25)	28.5	(20)	1.5
0.1 in. (0.25 cm) thick ingot sheet (b)	46.2	(32)	25.6	(18)	4.5
0.025 in. (0.64 mm) thick ingot sheet (c)	51	(36)	33	(23)	7
QMV Hot-pressed sintered block (d)	47.6	(33)	39.4	(28)	1.4
S200 Cross-rolled sintered block (e)	78	(55)	55.6	(39)	16.1
QMV-200 mesh hot-worked block (f)	65	(46)	–	–	41

NOTES:
(a) Can-rolled ingot sheet annealed 16 to 20 hr at 760°C., etched. Average of 97 tests, longitudinal direction.[9]
(b) Can-rolled, then bare rolled, ingot sheet annealed 16 to 20 hr at 760°C., etched. Average of 129 tests, longitudinal direction.[9]
(c) Can-rolled, then bare-rolled, ingot sheet annealed 1 hr at 650°C., Average of 4 tests, longitudinal direction.[10]
(d) Ref. 11.
(e) Longitudinal direction, material 0.077 in. (0.2 cm) thick.
(f) QMV-200 mesh sintered block. Extruded at 1066°C., then rolled normal to the extrusion direction at 1038°C to a rolling reduction ratio of 10:1 (Ref. 6).

Table III. Average Tensile Properties of Laminates, Annealed Sheet and Deformed Sheet

Material	UTS, ksi	(kgf/mm^2)	0.2 Pct Yield, ksi	(kgf/mm^2)	Elongation, Pct
Laminates	62 ± 3	(44)	32 ± 4	(22)	18 ± 1
Annealed Sheet*	63 ± 2	(44)	44 ± 1	(31)	11.5 ± 1
Annealed Sheet†	69 ± 3	(49)	47 ± 4	(33)	11 ± 1
Deformed Sheet‡	57 ± 4	(40)	28 ± 1	(20)	16 ± 3

NOTES:
*Gage length 1.0 in. (2.54 cm).
†Gage length 0.4 in. (1.02 cm).
‡See text.

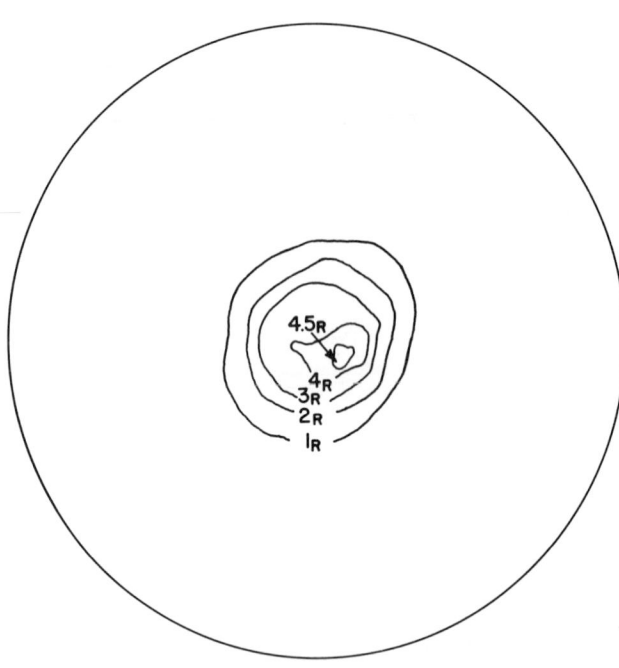

Fig. 4—Basal plane (0002) pole figure for the laminated sheet.

comparison with the properties of the bonded laminate. Prior to testing, the sheet was heat-treated to simulate the thermal history of a laminate during the bonding process so that any differences between the properties of the laminate and the starting material could not be attributed to different annealing histories. This material, after heat treatment, will be subsequently referred to as the annealed sheet. The annealed sheet was tensile tested using two different sample sizes having different gage lengths in order to check for any significant sample size effect on measured elongation. No significant specimen size effect was observed. The results of the tensile tests on the annealed sheet and the bonded laminate are given in Table III. All samples except the small annealed-sheet samples were etched about 0.005 in. (0.13 mm) on all surfaces to remove machining damage before testing. Only machined surfaces were etched on the small sheet samples. Some of the starting sheet was also subjected to the thermomechanical conditions of the bonding cycle and the resulting sheet tensile tested. Because the sheet was only about 0.016 in. (0.41 mm) thick after deformation, it was difficult to etch machining damage from the tensile bars. Consequently, many samples fractured in the grips and there was excessive scatter in the results for the other samples. (The same problems were encountered, to a lesser extent, with the small tensile bars from the annealed sheet.) The tensile results for those samples which did not fail in the grips are given

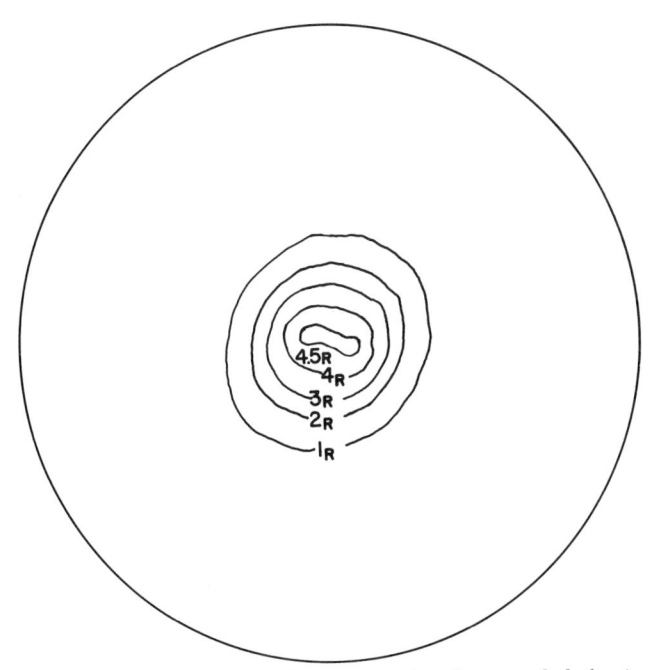

Fig. 5—Basal plane (0002) pole figure for the annealed sheet.

Fig. 6—Micrograph of laminated sheet. Grain size ~16 μm. Polarized light. Arrows indicate bonds.

in Table III and labeled deformed sheet. Tensile results for beryllium from other sources are given in Table IV for comparison with the properties reported here.

The yield strengths for all the samples reported in Table III were measured entirely from the Instron load-time recording chart, except for the large annealed-sheet samples where an extensometer was used. There was some deformation around the grips, nevertheless the agreement between the yield strengths obtained from the large and small annealed sheet specimens indicates that the yield strengths derived from the load-time chart are reasonably accurate.

Elongations were measured from gage marks, with an estimated precision of ±0.5 pct. The tensile properties of the laminate were isotropic in the plane of the laminae, as would be expected since they were stacked randomly with respect to their rolling direction. Deformation of the laminate was three-dimensional. Thinning of 3 to 5 pct was observed perpendicular to the bond plane. The elongations measured in the laminated sheet are, to the author's knowledge, the largest ever seen in directly rolled ingot sheet at room temperature. However, elongations as large as 27 pct have been seen in pack rolled ingot source beryllium with a silver layer between each beryllium sheet.[5] This material had a grain size of 8 μm and a very high basal texture. The texture could not be measured accurately due to the approximately 1.5 vol pct Ag incorporated into the pack. Powder source material worked so as to achieve maximum basal plane concentrations, in the plane of the sheet, of 20 times random has failed after up to 41 pct elongation.[6] The deformation of both these materials is essentially two-dimensional, with no thinning perpendicular to the plane of the sheet.

The dramatic and unexpected increase in elongation of the laminate compared with the annealed sheet prompted a number of tests to determine the origin of the increase. An X-ray texture analysis was performed by Dr. Marilyn S. Werkema at this laboratory. The (0002) pole figure for the laminate is given in Fig. 4, and the (0002) pole figure for the 0.025 in. (0.64 mm) thick annealed sheet is given in Fig. 5. The intensities relative to random were determined by comparison with a random sample made from isostatically pressed powder. For the annealed sheet, several pieces were taped together to make a thicker sample so that accurate intensity comparisons with the laminate could be made. The pole figure presented is for eight such pieces, however, no differences with increasing numbers of pieces were observed for five or more pieces. Fewer than five pieces produced erroneous texture results due to excessive X-ray transmission through the sample.

The pole figures confirm the isotropic nature of the material in the plane of the sheet [(10$\bar{1}$0) pole figures, not shown, also had circular intensity contours centered at the normal direction.] It is also apparent that the deformation accompanying bonding did not significantly alter the degree of basal texture of the material.

The microstructures of the laminate and annealed sheet are reproduced in Figs. 6 and 7. Deformation during bonding does not seem to have altered the microstructure significantly. There is no metallographic evidence that an appreciable amount of recrystallization occurred during bonding. In addition, it is known that the recrystallization texture of ingot source beryllium differs from the rolling texture.[12] The lack of a texture change after bonding indicates that there was little or no recrystallization during bonding. However, transmission electron microscopy reveals a qualitative difference between the materials. The laminate has fewer dislocation tangles, fewer dislocations not associated with subboundaries, and a larger subgrain size. Transmission micrographs of the laminate and annealed sheet are given in Figs. 8 and 9. The black spots are

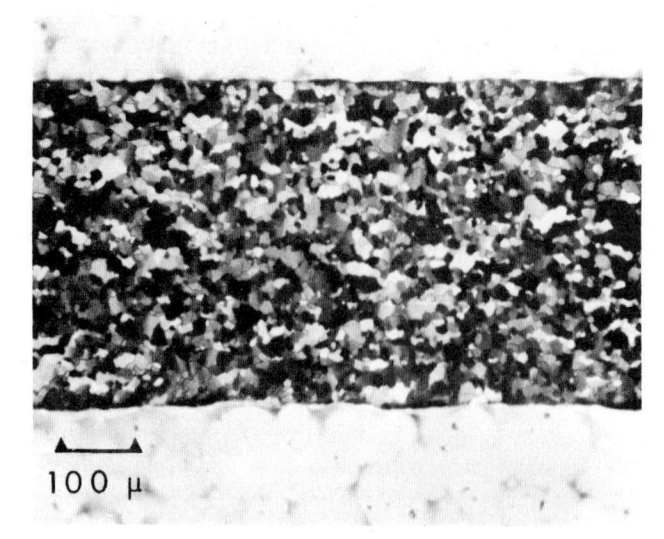

Fig. 7—Micrograph of annealed sheet. Grain size ~16μm. Polarized light.

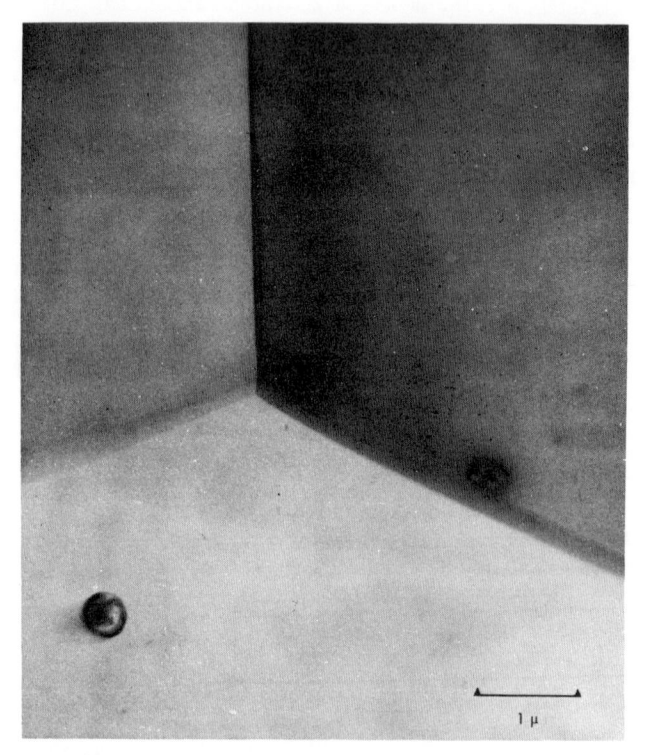

Fig. 8—Electron transmission micrograph of the laminated sheet.

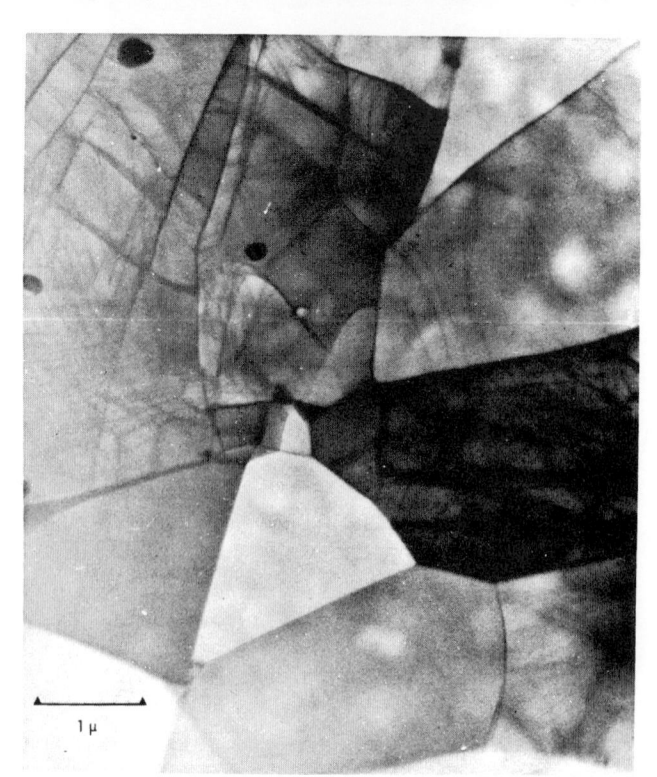

Fig. 9—Electron transmission micrograph of the annealed sheet.

Table V. Notched Charpy Impact Properties of Laminate and Other Beryllium Samples

	Ft-lb/in.²	(joules/cm²)
Laminated sheet	2.1	(0.44)
Laminated sheet	2.4	(0.50)
Average of transverse and longitudinal for .090 in. (0.23 cm) thick, bare rolled, notched, Rocky Flats ingot sheet.[10]	1.9	(0.40)
Brush SR-200-D. Average notched transverse.[10]	7.5	(1.58)
Same as above, but longitudinal.[10]	4.0	(0.84)
Brush S-200-D, Type I.[10]	2.0	(0.42)

precipitates. No dislocation substructure is shown in Fig. 8, nevertheless the compressive creep did not completely eliminate it in the laminate, although the substructure was reduced substantially. Changes in substructure generally occur during creep, and such changes are known to alter room temperature mechanical properties.[13,14] A study of the effect of substructure on the bend ductility of powder-source beryllium and beryllium alloys has recently been reported in which it was found that an increase in subgrain size and reduction in the number of dislocation tangles within the subgrains improved the bend ductility.[15] It is therefore reasonable that the substructure changes observed may be responsible for the increased ductility of the laminate.

Tensile results from the deformed sheet, given in Table III, also point to the deformation accompanying bonding, rather than the existence of the bonds themselves, as being responsible for the increased ductility of the laminates. Evidence that the bond interfaces do not retard crack propagation was obtained from notched Charpy impact tests. The specimens were of standard design and were machined so that fracture would be approximately normal to the laminae. Results of the impact tests are given in Table V along with impact values for beryllium from other sources. The impact strength of the laminate is not significantly higher than normal ingot source beryllium. Additionally, if the bond interfaces retarded crack propagation, one would expect to see cracks extending across one or more laminae in tensile specimens pulled to failure. No such cracks were seen.

The preceding observations all confirm that the ductility increase in the laminate is due to the creep-type deformation accompanying bonding. The deformation produces a dislocation substructure with fewer tangles and subboundaries. It is reasonable that dislocations can move further in such an open substructure before forming pileups and nucleating cleavage. This hypothesis is consistent with the reduced yield strengths observed in the laminates and deformed sheet compared with the annealed sheet. It is also possible that a portion of the increased ductility results from the healing of rolling defects during the compressive deformation, although no evidence for the existence of such defects was noted metallographically.

CONCLUSIONS

The results of this work demonstrate that thick beryllium plate can be fabricated from thin sheet by diffusion bonding and it can have properties equal or superior to those of the sheet from which it was formed. Tensile elongations of 18 pct have been achieved in 0.2 in. (0.51 cm) thick plate compared with 2 to 4 pct usually measured in plate of similar thickness rolled directly from the ingot. Furthermore, the basal texture of the laminate is low enough so that three-dimensional deformation is observed. The deformation accompanying bonding appears to improve the ductility of the beryllium by altering the dislocation structure.

ACKNOWLEDGMENTS

This work was performed under Contract AT(29-1)-1106 for the Albuquerque Operations Office, U.S. Atomic Energy Commission, whose financial support is gratefully acknowledged. The author is also indebted to Arvel W. Brewer for the electron microscopy and Richard M. Simmons for the hot-press work.

REFERENCES

1. A. N. Stroh: *Advan. Phys.,* 1957, vol. 6, p. 418.
2. A. J. Martin and G. C. Ellis: *Conference on the Metallurgy of Beryllium,* p. 3, Inst. of Metals, London, October 1961.
3. J. E. J. Bunce and R. E. Evans: *Conference on the Metallurgy of Beryllium,* p. 246, Inst. of Metals, London, October 1961.
4. C. I. Bort and A. Moore: *Conference on the Metallurgy of Beryllium,* p. 237, Inst. of Metals, London, October 1961.
5. W. Taylor: *The Influence of Grain Refinement and Titanium Alloying Additions on the Mechanical Properties of Beryllium Ingot Sheet,* Kawecki Berylco Industries, Reading, Pa., (Prepared under Contract N 0019-69-0233 for the Dept. of the Navy, Air Systems Command), 1970.
6. J. Greenspan: *Trans. TMS-AIME,* 1959, vol. 215, p. 153.
7. W. Padian, R. Levin, and E. Supan: Report AD 834-301, North American Rockwell Corp., Los Angeles Division, Los Angeles, Calif., May 1969. (There a number of reports from the same group dealing with Ti diffusion bonding.)
8. R. W. Fenn, Jr., D. D. Crooks, W. C. Kinder, and B. M. Lempriere: Tech. Rep. AFML-TR-67-212, Lockheed Missiles and Space Co., Sunnyvale, Calif., October 1967.
9. D. V. Miley and R. P. Brugger: RFP-1525, The Dow Chemical Company, Rocky Flats Division, Golden, Colorado, April 1971.
10. D. R. Floyd: The Dow Chemical Company, Rocky Flats Division, Golden, Colorado, private communication, November 1969.
11. W. J. Salmen and L. P. Gobble: *ASTM Proc.,* 1962, vol. 62, p. 653.
12. M. S. Werkema: *J. Appl. Cryst.,* 1970, vol. 3, p. 265.
13. F. Garofalo: *Fundamentals of Creep Rupture in Metals,* MacMillan, New York, 1965.
14. J. J. Jonas, C. M. Sellars, and W. J. McG. Tegart: *Met. Rev.,* 1969, vol. 14, no. 130, pp. 1-24.
15. F. W. Cooke, V. V. Damiano, G. J. London, H. Conrad, and B. R. Banerjee: *J. Mater.,* 1971, vol. 6, p. 403.

INDEX

A

Activated diffusion bonding. *See* Liquid-phase diffusion welding

Advantages
of friction welding 37
of high-frequency induction resistance welding 205-208
of high-frequency melt welding 243-244, 247
of liquid-phase diffusion welding 334-336

Aerospace industry
liquid-phase diffusion welding 329, 332-334

Aging
effect on macrostructure and hardness in 250 maraging steel butt-welds 87-88

Aircraft parts
liquid-phase diffusion welding 334

Aluminum
cold pressure welding 280-281
heat treatment of cold welds 283-288
measurement of bond strength in roll bonding 264-266
microscopic examination of weld interfaces after roll bonding . 266-269

Aluminum to copper. *See* Copper to aluminum

Aluminum alloy E91E to copper
frictional welding 50-51, 53-54

Aluminum alloy E91E to mild steel
frictional welding 50-51, 53-54

Aluminum alloys, specific types
A1E91E
frictional welding 52
2024-T4A1
friction welding 28-29

Aluminum bronze
friction characteristics 20-21

Aluminum to iron
weld interface in friction welding ... 30-32

Aluminum to nickel. *See* Nickel to aluminum

Aluminum to titanium. *See* Titanium to aluminum

Angular joints
in friction welding 45

Applications
of diffusion bonding 258
of diffusion welding 323
of explosive welding 133, 141-143
of friction welding 37
of liquid-phase diffusion welding 328

A.T.I. Friction Weld Monitor 66-75
as glass scale linear encoder 67-68
as motor load device 67
features 70
for declutch and brake 69
for figuring total distance 69
for forging 68
for heating 68-69
for monitor operation 68-69
for preheating................... 68-69
printout 70-71

Atmosphere
used in roll diffusion bonding of boron/aluminum alloy 6061 composite 344-346

Austenitic stainless steel to steel
bend tests of composite plate 148-149
explosive cladding 147-148
fatigue of composite plate 148-149
tensile shear tests of composite plate 148-150
tensile strength of composite plate 149-150
thermal fatigue tests of composite plate 149

Autogenous welding
of molybdenum 290-297

Automobile parts
development of welder for production of Chevrolet wheel rims 240-242

Auxiliary power unit door
use of SPF/DB process 356

Axial displacement
effect on tensile strength in 250 maraging steel butt-weld tubes 82-83

Axial pressure
in inertia welding 42

B

B-1 bomber
use of SPF/DB process 355-356
use of titanium diffusion bonding 259-262

Bars
inertia welding 41

Bearing inserts
explosion welding 163

Bend tests
of austenitic stainless steel/steel composite plate 148-149
of copper (70,30)/steel composite plate 148-149

Beryllium
bonding tests 359
Charpy impact properties 362
comparison of elongation of laminate with annealed sheet 361
elongation.......................361
microstructure of laminate and annealed sheet 361-362
plate from diffusion bonding sheet 358-363
properties 360
tensile properties of laminate, annealed sheet and deformed sheet ... 360-363
tensile results for deformed sheet 362
yield strength 360-361

Bond
in friction welding 29-32

Bond strength
measurement of in roll bonding of aluminum, copper, tin, lead and zinc 264-266
relation of butt-welded steels and values calculated from weld areas 278
theoretical determination of maximum bond strength attainable 269-271

Bonding temperature
effect on tensile behavior in boron/aluminum alloy 6061 composite 341, 343

Bonding tests
on beryllium bonds from diffusion bonding 359

Boron/aluminum alloy 6061 composites
atmosphere used 344-346
comparison of roll bonding with press diffusion bonding.......... 349-350
cost analysis 349-350
cross-ply panels 348
effect of bonding temperature on tensile behavior 341, 343
elastic roll deflection vs number of ply layers 347
hybrid panels 348-349
microstructure 340-342
monotape rolling 340-344
multilayer panels 346-348
relationship between temperature and rolling pressure 340-341
roll diffusion bonding 339-351
rolling speed 346
tensile strength 341, 343-346

Brazing
compared to diffusion bonding 303

Butt joining
explosion welding 164-165

Butt weld joints
high-frequency melt welding 243-244, 248

365

INDEX

C

Cast iron
  friction welding .39
Casting
  compared with inertia welding 43
Charpy impact properties
  of beryllium 362
Chevrolet Motor Division
  development of welder for production
    of Chevrolet wheel rims 240-242
Cladding, large-plate. *See* Explosive
  cladding
Cluster gears
  cost of inertia welding 48
Cobalt alloys
  joining by liquid-phase diffusion
    welding 332
Coefficient of friction
  as function of surface cleanliness 15
  effect of normal stress 19
  effect of velocity 19
Cold pressure welding (*see also* Cold
  roll bonding; Diffusion bonding;
  Diffusion welding; Gas-pressure
  bonding; Liquid-phase diffusion
  welding; Press diffusion welding;
  Pressure welding; Roll bonding;
  Roll diffusion bonding; SPF/DB
  process)
  background history 279-280
  heat treatment of welds of copper,
    aluminum and low-carbon steel
    strip 282-288
  of copper, aluminum and low-carbon
    steel strip 280-281
Cold roll bonding (*see also* Cold pressure
  welding; Diffusion bonding; Diffusion
  welding; Gas-pressure bonding; Liquid-
  phase diffusion welding; Press diffusion
  welding; Pressure welding; Roll
  bonding; Roll diffusion bonding; SPF/
  DB process)
  application 278
Columbium
  foils in diffusion bonding of
    columbium alloy 304
Columbium alloy Cb 752, diffusion
  bonding .302-308
  compared to brazing 303
  compared to electron beam
    welding 302-303
  diffusion bonded panels for heat
    shield of space shuttle 298-308
  metallurgy of columbium bonded with
    titanium 304-306
  North American Rockwell (NR)
    diffusion bonding process . . . 303-305
  panel design 300-302
  panel fabrication and quality
    control 306-307
  strength of columbium bonded with
    titanium 305-306
  use of columbium and vanadium foils .304

Continuous edge melt weld
  advantages 247
  high-frequency melt welding 246-247
  materials to be welded 247
Conventional friction welding . . .37, 40-41
  comparison with flash welding 40
  control of process variables 40
  examples 40-41
  of exhaust valves 41
  principles of operation 40
Copper
  as intermediate in diffusion welding
    of molybdenum 289-297
  cold pressure welding 280-281
  heat treatment of cold
    welds 282-284, 286-288
  measurement of bond strength in
    roll bonding 264-266
  microscopic examination of weld inter-
    faces after roll bonding 266-269
Copper to aluminum
  explosive cladding 136-141
Copper to aluminum alloy E91E. *See*
  Aluminum alloy E91E to copper
Copper to steel
  explosive welding 112
  frictional welding 50-51
Copper (70,30) to steel
  bend tests of composite plate 148-149
  explosive cladding 147-148
  fatigue of composite plate 148-149
  tensile shear tests of composite
    plate .148-150
  tensile strength of composite plate 149-150
  thermal fatigue tests of composite plate 149
Copper-nickel (80-20) alloy to steel
  explosive welding 135
Cost
  inertia welding vs flash welding 48
  of friction welding 37
  of inertia welding vs other methods
    of fabrication 48
  of roll diffusion bonding of
    boron/aluminum alloy 6061
    composite 349-350
Cross-ply panels
  of boron/aluminum alloy 6061
    composite 348
Cryo Anchor heat pipe
  for Trans-Alaska pipeline 92-95
  prevention of pole jacking and pile
    settling 92-93
  use of friction welding 92-95
Cylindrical surfaces
  explosive welding 121-126

D

Deformation
  in diffusion welding 316-317
Degree of weld completion
  effect on properties of diffusion weld . .317
Diffusion bonding (*see also* Cold pressure
  welding; Cold roll bonding; Diffusion
  welding; Gas-pressure bonding; Liquid-
  phase diffusion welding; Press diffusion
  welding; Pressure welding; Roll
  bonding; Roll diffusion bonding; SPF/
  DB process)
  applications 258
  equipment .257-258
  in B-1 bomber 259-262
  in space shuttle 259-262
  intermediary materials 257
  materials joined 256
  of beryllium plate from diffusion
    bonding sheets 358-363
  process 255-257
  process of titanium diffusion
    bonding 261-262
Diffusion brazing. *See* Liquid-phase
  diffusion welding
Diffusion welding (*see also* Cold pressure
  welding; Cold roll bonding; Diffusion
  bonding; Gas-pressure bonding; Liquid-
  phase diffusion welding; Press diffusion
  welding; Pressure welding; Roll
  bonding; Roll diffusion bonding; SPF/
  DB process)
  application of pressure and thermal
    energy 314-315
  applications 323
  diffusion and deformation 316-317
  ductility in 318
  effect of degree of weld completion
    on properties of diffusion weld . . .317
  for aircraft gas-turbine fan disk . . .323-325
  fracture strength 318
  gas-pressure bonding 324-327
  metallurgy .315-316
  of dissimilar metals 320-322
  of molybdenum 290-297
  of nickel superalloys 318-320
  preparation 313-314
Disadvantages
  of friction welding 37
Dissimilar metals
  diffusion welding 320-322
  friction welding 49-54
Drill-to-shank welding
  cost of inertia welding 48
Ductility
  in diffusion welding 318
DYNACLAD™. *See* Explosive cladding
DYNAWELD™. *See* Explosive welding

E

Efficiency
  of high-frequency induction resistance
    welding 207-208
Elastic roll deflection
  vs number of ply layers in roll
    diffusion bonding of
    boron/aluminum alloy 6061
    composite 347
Electron beam welding
  compared to diffusion bonding . . .302-303

INDEX

Elongation
  comparison of beryllium laminate
    with annealed sheet 361
  of beryllium 361
Encircling inductor
  in high-frequency induction welding
    of pipes 202-204
Equilibrium volumetric displacement
  effect of heating pressure in 250
    maraging steel butt-weld tubes 86
  effect of rotational velocity in 250
    maraging steel butt-weld tubes 86
Equipment
  for high-frequency induction
    resistance welding 208-219
  in diffusion bonding 257-258
Eutectic bonding. *See* Liquid-phase
    diffusion welding
Exhaust valves
  conventional friction welding 41
Explosive cladding
  applications 160, 162-163
  austenitic stainless steel to steel 147
  brass (70,30) to steel 147
  of copper to aluminum 136-141
  of large plates 120-121
  process 132, 136-141
  strength of cladded plates 145-149
Explosive welding
  applications 133, 141-143, 160-165
  critical angle for jet formation 157
  critical flow transition velocity 158
  critical impact pressure 157
  effect of density on wave formation . . . 117
  for butt joining 163
  history 101-102, 129-132
  inclined plate technique 103, 106-107
  kinetic energy of the flyer plate and
    heat dissipation characteristics of
    collision region 158-159
  lap welding 126-127
  low detonation-velocity welding
    parameters 119
  mechanism 102-107
  metallurgy of welds 110-111, 127
  metals that have been explosion welded 133
  no explosive energy source 164-165
  of bearing inserts in diesel engines 163
  of copper to steel 112
  of copper-nickel (80-20) alloy to steel . 135
  of cylindrical surfaces 121-126
  of fiber-reinforced material 126
  of pipe joining 124-126
  of stainless steel to steel 107-108
  of stainless steel (Type 304L) to
    carbon steel 136
  of steel to steel 111
  of titanium to steel 113-114
  of tube plugging 122-124, 163
  of tube-to-tube plate 121-122, 163
  optimum conditions 157
  parallel plate technique 107-108
  process 102-107, 132-136, 152
  wave formation 114
  weldability ratings of metals and
    alloys 159-161
  welding parameters 118-119, 153-157
  welding parameters using Trimonite I . 119

F

Fan disk, gas-turbine
  diffusion welding 323-325
Fatigue
  of austenitic stainless steel/steel
    composite plate 148-149
  of copper (70,30)/steel composite
    plate 148-149
Fiber-reinforced material
  explosive welding 126
Flash welding 236
  compared with conventional friction
    welding 40
  compared with friction welding 38
  compared with friction welding for
    heat pipes 94
  compared with high-frequency
    bar-butt welding 238-239
  vs inertia welding 48
Flexibility
  of high-frequency induction resistance
    welding 206
Flow of weld upset
  design for in friction welding 45-46
Flow stress
  vs temperature in friction welding 27
Flow-transition velocity
  in explosive welding 158
Flywheel energy
  effect of in intertia welding 42-43
Flywheel friction welding 43-44
  compared to gas metal-arc welding 44
Forging
  vs inertia welding 43, 48
Forging pressure
  effect on microstructure of 250
    maraging steel welds 79, 81
  effect on tensile strength of 250
    maraging steel welds 82, 84
Fracture strength
  in diffusion welding 318
Frequency
  of high-frequency induction resistance
    welding 207-208
Friction
  contact conditions for flat surface
    loaded against rough surface 17
  effect of speed and normal loading
    on friction 15
  friction characteristics of Type 304
    stainless steel, AISI 4340 and
    aluminum bronze 20-21
  modern friction theory 12, 14-21
Friction welding
  advantages 37
  angular joints 45
  applications 37
  classification of process 5
  comparison of properties of flash and
    friction welded joints 38
  control of weld quality 46
  conventional friction welding . . . 37, 40-41
  design for flow of weld upset 45-46
  design for heat balance 45
  design for machine size 46
  disadvantages 37-38
  economy in operation and material 37
  effect of varying welding pressure on
    interface temperature 27
  effect of welding speed and
    pressure 25-26
  flow stress vs temperature 27
  flywheel friction welding 37, 43-44
  frictional behavior of metals 49-50
  history 3-4
  inertia welding 37, 41-43
  machine size range 9
  materials 8-9
  joint design 39
  of aluminum alloy E91E to
    copper 50-51, 53-54
  of aluminum alloy E91E to itself 52
  of aluminum alloy E91E to mild
    steel 50-51, 53-54
  of carbon and alloy steel 39
  of cast iron 39
  of copper to steel 50-51
  of dissimilar metals 49-54
  of heat pipes for Trans-Alaska
    pipeline 92-95
  of nonferrous metals and alloys 39-40
  of small missile system components 76-79
  of stainless steels 39
  of 250 maraging steel rocket-motor cases 76
  of 2024-T4 aluminum 28-29
  process 4-7, 13
  process capabilities 37-39
  radial temperature distribution in
    friction welded low carbon steel . . . 23
  sections welded 38-39
  surface conditions 44
  temperature generation and
    distribution 22-25
  time history of friction welding process 24
  tubular welds 44-45
  use of A.T.I. Friction Weld Monitor 66-75
  weld interface of nickel to titanium,
    aluminum to iron, nickel to iron,
    titanium to aluminum, nickel to
    aluminum 30-32
  weld strength 38
  welding parameters 7-8
**Friction welding A-type orbital
    machine** 61-63
**Friction welding B-type orbital
    machine** 61, 63

G

Gas metal-arc welding
  compared with friction welding for
    heat pipes 94

367

INDEX

Gas-pressure bonding (*see also* Cold pressure welding; Cold roll bonding; Diffusion bonding; Diffusion welding; Liquid-phase diffusion welding; Press diffusion welding; Pressure welding; Roll bonding; Roll diffusion bonding; SPF/DB process) 324-327

Gas tungsten-arc welding
   compared with friction welding for heat pipes 94

H

Hardness
   effect of aging in 250 maraging steel butt-weld tubes 87-88
   of molybdenum joints 294-295

Hardness tests
   of HSLA steel Tee welds by high-frequency electric resistance welding 181, 184-188, 194-196

Heat balance
   design for in friction welding 45

Heat dissipation
   of collision region in explosive welding 158-159

Heat pipes
   compared with gas tungsten-arc, gas metal-arc and flash welding 94
   friction welding of joints 92-95

Heat shield
   creep criteria 300
   design criteria 300
   diffusion bonded columbium alloy panels for space shuttle 298-308
   flutter criteria 300
   panel design 300-302
   panel fabrication and quality control 306-307
   safety factors 300

Heat treatment
   of copper, aluminum and steel welds 282-288

Heating pressure
   effect on equilibrium volumetric displacement of 250 maraging steel welds 86
   effect on tensile strength of 250 maraging steel welds 82-83

High-frequency bar-butt welding
   compared with flash welding 238
   development of welder for production of Chevrolet wheel rims 240-242
   of steel strips and stainless steel 239
   process 236-238
   quality of welds 239

High-frequency contact welding
   compared to high-frequency induction resistance welding 234-235
   for pipes and piping 233-234

High-frequency electric resistance welding
   of I-beams 172, 174-175
   of Tee welds using HSLA steels . . 175-197
   process 172-173

tests of HSLA steel welds 178-197
type of sections that can be welded 173

High-frequency induction resistance welding
   advantages 205-208
   basic oscillator 210-211
   compared with high-frequency contact welding 233-234
   effect of material and product on weld speed 220-223
   effect of mill performance on weld speed 222-227
   effect of pipe diameter 235
   effect of scarfing mandrels 235
   effect of welder performance on weld speed 227
   efficiency and frequency 207-208
   equipment used 208-219
   factors affecting weld speed 219-227
   flexibility and maintenance 206
   for pipes and piping 234-235
   impeder and its effect 219, 234-235
   increased weld speed 205
   inductor 216-219
   line voltage regulators 213-214
   metallurgical studies 227-230
   metals to be welded 206
   of nonferrous pipes 235
   of thin-walled piping 206
   on pipeline right-of-way 231-232
   output transformer 215-216
   power control 213
   product quality 206-207
   protective features 211-213
   surface condition 205-206
   transmission line 214-215

High-frequency induction welding
   encircling inductor 202-204
   nonencircling inductor 200-202
   of pipes and piping 199-204

High-frequency melt welding
   advantages 243-244
   for butt weld joints 243-244, 248
   for continuous edge melt welding 246-247, 251
   for laminations 244-246, 249-250
   for lap weld joints 243, 248
   for T-joints 244, 248
   process 243
   simultaneous welding 244

High-strength low-alloy steels. *See* HSLA steels

History
   of explosive welding . . . 101-102, 129-132
   of friction 3-4

HSLA steels
   high-frequency electric resistance welding of Tee welds 175-197
   light load hardness testing of welds 181, 184-188, 194-196
   metallographic examination of weld and heat affected zone 178, 180, 184-188, 194-196
   microhardness testing of welds . . . 180-181, 184-188, 194-196

peel test of welds 178-179
Tee tensile impact tests 181-183, 189-194, 196-197
tests of welds 178-197

Hybrid panels
   of boron/aluminum alloy 6061 composite 348-349

I

I-beams
   high-frequency electric resistance welding 172, 174-175

Impact pressure
   in explosive welding 157

Impeder
   effect on pipe welding 235
   in high-frequency induction resistance welding 219, 234-235

Inclined plate technique
   in explosive welding 103, 106-107

Inductor (*see also* Encircling inductor; Nonencircling inductor)
   for high-frequency induction resistance welding 216-219

Inertia welding 37, 41-43
   axial pressure 42
   comparison to casting or shrink-fit assembly 43
   cost of cluster gear 48
   cost of drill-to-shank welding 48
   cost vs flash welding 48
   cost vs other methods of fabrication 48
   cost of shaft-and-pinion: one-piece forging vs weldment 48
   effect of flywheel energy 42-43
   examples 43
   of 1-in.-diam bars 41
   of powder metallurgy parts 55-58
   of precombustion chamber 46
   of sintered AISI 1020 and AISI 1040 steels 55-58
   of spring-retainer assembly 47
   peripheral velocity of workpiece 42
   principle of operation 41
   process variables 41-43
   use of A.T.I. Friction Weld Monitor 66-75
   vs forging 43, 48
   vs machining 46
   vs pressure gas welding 46
   welding machine 41

Interface temperature
   effect of varying welding pressure in friction welding 27

Intermediary materials
   in diffusion bonding 257

Iron
   as intermediate in diffusion welding of molybdenum 289-297

Iron alloys
   joining by liquid-phase diffusion welding 332

Iron to aluminum. *See* Aluminum to iron

Iron to nickel. *See* Nickel to iron

368

INDEX

Isothermal solidification
   in liquid-phase diffusion welding 330-332

J

Jet formation
   critical angle in explosive welding 157
Joint design
   in friction welding 44-46

K

Kinetic energy
   of flyer plate in explosive welding 158-159

L

Laminations
   high-frequency melt
      welding 244-246, 249-250
Lap weld joints
   high-frequency melt welding 243, 248
Lap welding
   explosive welding 126-127
Lead
   measurement of bond strength in roll
      bonding 264-266
   microscopic examination of weld
      interfaces after roll bonding . 266-269
Line voltage regulators
   for high-frequency induction
      resistance welding 213-214
Liquid-phase diffusion welding (*see also* Cold pressure welding; Cold roll bonding; Diffusion bonding; Diffusion welding; Gas-pressure bonding; Press diffusion welding; Pressure welding; Roll bonding; Roll diffusion bonding; SPF/DB process)
   advantages 334-336
   applications in aerospace
      industry 329, 332-334
   for aircraft gas-turbine vanes 332-334
   for joining nickel, cobalt, titanium,
      iron and zirconium alloys 332
   isothermal solidification 330-332
   metallurgy 329-332
   process 328
Lower engine access door
   use of SPF/DB process 356-357

M

Machine size
   design for in friction welding 46
   range in friction welding 9
Machining
   vs inertia welding 46
Macrostructure
   effect of aging in 250 maraging
      steel butt welds 87-88
Maintenance
   of high-frequency induction resistance
      welding 206

Maraging steel 250
   effect of aging on macrostructure and
      hardness in butt-welds 87-88
   effect of axial displacement during
      heating on tensile strength of butt-
      welded tubes 82-83
   effect of forging pressure on micro-
      structure of welds 79, 81
   effect of forging pressure on tensile
      strength of butt-welded tubes . 82, 84
   effect of heating pressure on
      equilibrium volumetric displacement
      for butt-welded tubes............86
   effect of heating pressure on tensile
      strength of butt-welded tubes .. 82-83
   effect of rotational velocity on
      change in equilibrium volume
      displacement rate with axial
      heating pressure for butt-welds 86
   friction butt welds 82-89
   friction welding after maraging .. 79, 81-82
   friction welding for small rocket-motor
      cases 76-90
   microstructural effects in butt-welds 87-89
   microstructure of welds 79-80
   relationship between strength and
      weld-cycle variables for butt-
      welded tubes 84-85
   strength of butt-welded tubes 85
Material properties
   in friction welding 26-29
Materials
   in friction welding 8-9
Metallographic tests
   of HSLA steel Tee welds by high-
      frequency electric resistance
      welding .178, 180, 184-188, 194-196
Metallurgical studies
   in high-frequency induction resistance
      welding 227-230
Metallurgy
   of diffusion welding 315-316
   of explosive welds 110-114, 127
   of liquid-phase diffusion welding 329-332
Metals
   explosion weldability ratings 159-161
   frictional behavior 49-50
   that have been explosion welded......133
   used in high-frequency induction
      resistance welding 206
Metals, ferrous
   effect on weld speed in high-
      frequency induction resistance
      welding 220-222
Metals, nonferrous
   effect on weld speed in high-
      frequency induction resistance
      welding 235
   friction welding 39-40
   high-frequency induction resistance
      welding 235
Microhardness tests
   of HSLA steel Tee welds by high-
      frequency electric resistance
      welding . 180-181, 184-188, 194-196

Microstructure
   effect of forging pressure on
      microstructure of 250 maraging
      steel welds 79, 81
   effects on 250 maraging steel butt-
      welds 87-89
   of beryllium laminate and annealed
      sheet 361-362
   of boron/aluminum alloy 6061
      composite 340-342
   of molybdenum joints 293, 295-296
   of 250 maraging steel welds 79-80
Mill performance
   effect on weld speed in high-frequency
      induction resistance welding 222-226
Missile components
   friction welding 76-79
Molybdenum
   autogenous welding 290-297
   correlation between welding
      temperature and weld strength
      of joints 292-293
   diffusion welding 289-297
   effect of surface treatment on
      strength of autogenous weld 292
   hardness of joints 294-295
   microstructure of joints 293, 295-296
   remelting of joints 296-297
   using intermediates of iron, nickel,
      copper and silver 289-297
Monitor/controller
   development of A.T.I. Friction Weld
      Monitor 66-75
Monotape rolling
   of roll diffusion bonding in boron/
      aluminum alloy 6061
      composite 340-344
Multilayer panels
   used in roll diffusion bonding of
      boron/aluminum alloy 6061
      composite 346-348

N

Nacelle beam frame
   use of SPF/DB process 356
Nickel
   as intermediate in diffusion welding
      of molybdenum 289-297
Nickel alloys
   joining by liquid-phase diffusion
      welding 332
Nickel superalloys
   diffusion welding 318-320
Nickel to aluminum
   weld interface in friction welding ... 30-32
Nickel to iron
   weld interface in friction welding ... 30-32
Nickel to titanium
   weld interface in friction welding ... 30-32
Nonencircling inductor
   in high-frequency induction welding
      of pipes 200-202
Normal loading
   effect on friction 15

INDEX

North American Rockwell
   development of titanium diffusion bonding for B-1 bomber and space shuttle 259-262
   diffusion bonding of columbium alloy panels for heat shield 298-308
Nor-Ti-Bond. *See* Liquid-phase diffusion welding

O

Orbital welding
   as mathematical solution for friction welding 61
Oscillator
   for high-frequency induction resistance welding 210-211
Output transformer
   for high-frequency induction resistance welding 215-216

P

Parallel plate technique
   in explosive welding 107-108
Peel test
   of HSLA steel Tee welds by high-frequency electric resistance welding 178-179
Peripheral velocity
   of workpiece in inertia welding 42
Pile settling 92-93
Pipes and piping
   effect of pipe diameter on high-frequency induction resistance welding 235
   high-frequency induction welding 199-204
   high-frequency induction resistance welding 205, 234-235
   high-frequency induction resistance welding on pipeline right-of-way 231-232
   joining by explosive welding 124-126
Pipes and piping, thin-walled
   use of high-frequency induction resistance welding 206
Pole jacking 92-93
Post weld heating
   in high-frequency induction resistance welding 228-230
Powder metallurgy
   inertia welding 55-59
Power control
   of oscillator for high-frequency induction resistance welding 213
Precombustion chamber
   inertia welding 46
Press diffusion welding (*see also* Cold pressure welding; Cold roll bonding; Diffusion bonding; Diffusion welding; Gas-pressure bonding; Liquid-phase diffusion welding; Pressure welding; Roll bonding; Roll diffusion bonding; SPF/DB process)

compared with roll diffusion bonding 349-350
Pressure
   in diffusion welding 314-315
Pressure gas welding
   vs inertia welding 46
Pressure welding (*see also* Cold pressure welding; Cold roll bonding; Diffusion bonding; Diffusion welding; Gas-pressure bonding; Liquid-phase diffusion welding; Press diffusion welding; Roll bonding; Roll diffusion bonding; SPF/DB process)
   application of cold bonding 278
   by rolling, ... 263-278
   correlation of initiation and development of bonding with interfacial characteristics 275-276
   effect of roll diameter 274
   effect of roll speed 273
   effect of surface contamination ... 274-275
   effect of temperature 275
   fundamentals of pressure welding relevant to roll bonding 278
   initiation and development of bonding 271-275
   measurement of bond strength for aluminum, copper, tin, lead and zinc 264-266
   mechanism 276-278
   microscopic examination of weld interfaces 266-269
   relation between bond strength and butt-welded steels and values calculated from weld areas 278
   rolling geometry 271-272
   theoretical determination of maximum bond strength attainable 269-271
Properties
   of beryllium 360

Q

Quality
   of high-frequency induction resistance welding 206-207
Quality control
   of diffusion bonded columbium alloy panels for heat shield of space shuttle 306-307

R

Remelting
   of molybdenum joints 296-297
Rocket-motor cases
   friction welding of 250 maraging steel ..76
Rockwell International. *See* North American Rockwell
Rohr Bonding™. *See* Liquid-phase diffusion welding
Roll bonding (*see also* Cold pressure welding; Cold roll bonding; Diffusion bonding; Diffusion welding; Gas-pressure bonding; Liquid-phase diffusion welding; Press diffusion welding; Pressure welding; Roll diffusion bonding; SPF/DB process)
   application of cold roll bonding 278
   correlation of initiation and development of bonding with interfacial characteristics 275-276
   effect of roll diameter 274
   effect of roll speed 273
   effect of surface contamination 274
   fundamentals of pressure welding relevant to roll bonding 278
   initiation and development of bonding 271-275
   measurement of bond strength of aluminum, copper, tin, lead and zinc 264-266
   mechanism 276-278
   microscopic examination of weld interfaces of aluminum, copper, tin, lead and zinc 266-269
   relation between bond strength of butt-welded steels and values calculated from weld areas 278
   rolling geometry 271-273
   theoretical determination of maximum bond strength attainable 269-271
Roll diameter
   effect of in pressure welding 274
Roll diffusion bonding (*see also* Cold pressure welding; Cold roll bonding; Diffusion bonding; Diffusion welding; Gas-pressure bonding; Liquid-phase diffusion welding; Press diffusion welding; Pressure welding; Roll bonding; SPF/DB process)
   atmosphere used 344-346
   compared to press diffusion welding 349-350
   cost analysis 349-350
   cross-ply panels 348
   hybrid panels 348-349
   monotape rolling 340-344
   multilayer panels 346-348
   of boron/aluminum alloy 6061 composites 339-351
   rolling speed 346
Rolling
   in pressure welding 263-278
Rolling geometry
   effect of in pressure welding 271-273
Rolling pressure
   relationship with temperature in boron/aluminum alloy 6061 composite 340-341
Rolling speed
   effect of in pressure welding 273
   used in roll diffusion bonding of boron/aluminum alloy 6061 composite 346

INDEX

S

Safety devices
  for high-frequency induction
    resistance welding 211-213
Scarfing mandrels
  in high-frequency induction
    resistance welding 235
Shaft-and-pinion
  cost of inertia welding 48
Shrink-fit assembly
  compared with inertia welding 43
Silver
  as intermediate in diffusion welding
    of molybdenum 289-297
Simultaneous welding
  high-frequency melt welding 244
Solid state diffusion welding. *See*
  Diffusion welding
Space shuttle
  diffusion bonded columbium alloy
    panels for heat shield 298-308
  use of titanium diffusion bonding 259-262
Speed
  effect on friction 15
SPF/DB process (*see also* Cold
  pressure welding; Cold roll
  bonding; Diffusion bonding;
  Diffusion welding; Gas-pressure
  bonding; Liquid-phase diffusion
  welding; Press diffusion welding;
  Pressure welding; Roll bonding;
  Roll diffusion bonding)
  advantages 353
  aircraft applications 353, 355-357
  cost 355-356
  for auxiliary power unit door 356
  for B-1 bomber 355-356
  for lower engine access door 356-357
  for nacelle beam frame 356
  for windshield hot air blast nozzle 356-357
  integrally stiffened 354-355
  of Ti-6A1-4V sheet 353-357
  reinforced sheet 354-355
  sandwich 354-355
Spring-retainer assembly
  inertia welding 47
Stainless steel to steel
  explosive welding 107-108
Stainless steel Type 304
  friction characteristics 20-21
Stainless steel Type 304L to carbon steel
  explosive welding 136
Stainless steels
  friction welding 39
  high-frequency bar-butt welding 239
Steel, AISI 4340
  friction characteristics 20-21
Steel powders, specific types
  sintered AISI 1020
    fracture toughness 58
    inertia welding 55-58
    pore area fraction 56-57
    porosity character 57
    Vickers hardness 56-57
  sintered AISI 1040
    fracture toughness 58
    inertia welding 55-58
Steel strips
  high-frequency bar-butt welding 239
Steel to aluminum alloy E91E. *See*
  Aluminum alloy E91E to mild steel
Steel to austenitic stainless steel. *See*
  Austenitic stainless steel to steel
Steel to copper. *See* Copper to steel
Steel to copper (70,30). *See* Copper
  (70,30) to steel
Steel to copper-nickel (80-20) alloy.
  See Copper-nickel (80-20) alloy
  to steel
Steel to stainless steel. *See* Stainless
  steel to steel
Steel to stainless steel Type 304L.
  See Stainless steel Type 304L
  to carbon steel
Steel to steel
  explosive welding 111
Steel to titanium. *See* Titanium to steel
Steels
  cold pressure welding 280-281
  friction welding 39
  heat treatment of cold welds 285-288
  radial temperature distribution in
    friction welded low carbon steel ... 23
  relation with butt-welded steels and
    values calculated from weld
    areas 278
Strength
  of bond in roll bonding 264-266
  of 250 maraging steel butt-weld tubes .. 85
  relationship with weld-cycle
    variables in 250 maraging steel
    butt-weld tubes 84-85
Stress, normal
  function of coefficient of friction 19
**Superplastic forming/diffusion
bonding process.** *See* SPF/DB
  process
Surface cleanliness
  function of coefficient of friction 15
Surface conditions
  advantages of high-frequency
    induction resistance welding 205-206
  in friction welding 44
Surface contamination
  effect of in pressure welding 274
Surface treatment
  effect on strength of autogenous
    molybdenum weld 292

T

T-joints
  high-frequency melt welding ... 244, 248
Tee tensile impact tests
  of HSLA steel Tee welds by high-
    frequency electric resistance
    welding . 181-183, 189-194, 196-197
Tee welds
  of HSLA steels by high-frequency
    electric resistance welding .. 175-197
Temperature
  relationship with rolling pressure in
    boron/aluminum alloy 6061
    composite 340-341
  vs flow stress in friction welding 27
Temperature generation and distribution
  in friction welding 22-25
  radial temperature distribution in low
    carbon steel 23
Tensile behavior
  effect of bonding temperature in
    boron/aluminum alloy 6061
    composite 341, 343
Tensile fractures
  in boron/aluminum alloy 6061
    composite 341, 344
Tensile properties
  of beryllium laminate, annealed
    sheet and deformed sheet ... 360-363
Tensile shear tests
  of austenitic stainless steel/steel
    composite plate 148-149
  of copper (70,30)/steel
    composite plate 148-149
Tensile strength
  effect of axial displacement in 250
    maraging steel butt-weld tubes 82-83
  effect of forging pressure in 250
    maraging steel butt-weld tubes 82, 84
  effect of heating pressure in 250
    maraging steel butt-weld tubes 82-83
  of austenitic stainless steel/steel
    composite plate 149-150
  of beryllium bonds 359
  of boron/aluminum alloy 6061
    composite 341, 343-346
  of copper (70,30)/steel composite
    plate 149-150
Thermal energy
  in diffusion welding 314-315
Thermal fatigue tests
  of austenitic stainless steel/steel
    composite plate 149
  of copper (70,30)/steel 149
Time history
  of friction welding process 24
Tin
  measurement of bond strength in roll
    bonding 264-266
  microscopic examination of weld inter-
    faces after roll bonding 266-269
Titanium
  diffusion bonding 261-262
  foils in diffusion bonding of
    columbium alloy 304-306
  in B-1 bomber 259-262
  in space shuttle 259-262
  metallurgy of columbium bond with
    titanium 304-306

371

INDEX

Titanium *(continued)*
   strength of columbium bond with
      titanium 305-306
   testing after bonding 262
Titanium alloys
   joining by liquid-phase diffusion
      welding 332
Titanium to aluminum
   weld interface in friction welding ... 30-32
Titanium to nickel. *See* Nickel
   to titanium
Titanium to steel
   explosive welding 113-114, 135
TLP^R diffusion bonding. *See* Liquid-phase
   diffusion welding
Transient liquid-phase diffusion
   welding. *See* Liquid-phase
   diffusion welding
Transmission line
   for high-frequency induction resistance
      welding 214-215
Trimonite I
   in explosive welding 119
Tube plugging
   explosive welding 122-124, 163
Tube-to-tubeplate
   explosive welding . 121-122, 123-124, 163
Tubes and tubing. *See* Pipes and
   piping
Tubular welds
   in friction welding 44-45

V

Vanadium
   foils in diffusion bonding of columbium
      alloy 304

Vanes, gas turbine
   liquid-phase diffusion welding ... 332-334
Velocity vs coefficient of friction 19

W

Wave formation
   effect of density in explosive welding . 117
   in explosive welds 110-111, 114
Weld interfaces
   microscopic examination of roll
      bonding of aluminum, copper, tin,
      lead and zinc 266-269
Weld quality
   design for in friction welding 46
   in high-frequency bar-butt welding 239
Weld roll stands
   in high-frequency induction resistance
      welding 226
Weld speed
   advantage of high-frequency induction
      resistance welding 205
   effect of material and product in
      high-frequency induction resistance
      welding 220-223
   effect of mill speed in high-frequency
      induction resistance welding 222-227
   effect of welder performance in high-
      frequency induction resistance
      welding 227
Weld strength
   correlation with weld temperature of
      molybdenum joints 292-293
   of friction welding 38
Welder performance
   effect on weld speed in high-frequency
      induction resistance welding 227

Welding machine
   in inertia welding 41
Welding parameters
   in explosive welding .. 118-119, 153-157
   low detonation-velocity welding
      parameters in explosive welding .. 119
   using Trimonite I in explosive welding 119
Welding pressure
   effect on friction welding 25-26
   effect on interface temperature in
      friction welding 27
Welding speed
   effect on friction welding 25-26
Welding temperature
   correlation with weld strength of
      molybdenum joints 292-293
Wheel rims
   development of welder for production
      of Chevrolet wheel rims 240-242
Windshield hot air blast nozzle
   use of SPF/DB process 356-357

Y

Yield strength
   of beryllium 360-361

Z

Zinc
   measurement of bond strength in roll
      bonding 264-266
   microscopic examination of weld inter-
      faces after roll bonding 266-269
Zirconium alloys
   joining by liquid-phase diffusion
      welding 332